Modern Power System Analysis

Step into the captivating world of power systems with *Modern Power System Analysis, Third Edition* by acclaimed author Turan Gönen and revised and updated by Chee-Wooi Ten and Yun-he Hou. This illuminating book offers a comprehensive examination of power system analysis. Whether you're a curious non-specialist, a voracious reader seeking knowledge, or a librarian or bookseller searching for a valuable resource, Gönen's masterpiece is sure to keep you engaged. This book is an excellent source to begin your journey.

An in-depth understanding of the concepts and techniques involved in power system analysis is provided in this comprehensive guide. The book covers a wide range of topics, including fundamental modeling of power transmission network, power flow analysis, and fault analysis. Dr. Gönen elucidates the mathematical foundations and computational methods necessary for analyzing and optimizing power systems. Readers will gain insights into advanced topics such as power system harmonics, transient stability, and power system protection. Furthermore, the text explores emerging areas like renewable energy integration, smart grid technologies, and the application of artificial intelligence in power system analysis. Dr. Gönen's meticulous approach combines theoretical explanations, practical examples, and real-world case studies to provide readers with a comprehensive and up-to-date resource. With its focus on modern techniques and advancements, this book is an invaluable reference for engineers, researchers, and students venturing into the exciting realm of power system analysis. This book also includes a new chapter on power system restoration, which reviews methodologies corresponding to different utilities and practices.

A cutting-edge compilation of the latest developments in power system analysis is presented in *Modern Power System Analysis, Third Edition*. While the challenges and issues have evolved, the book emphasizes the enduring importance of classical methods as the foundation for understanding. It integrates today's advancements, addresses contemporary issues, and provides readers with a comprehensive grasp of the most current techniques and approaches for analyzing, optimizing, and managing complex power systems. With practical examples, real-world case studies, and a strong focus on emerging areas like renewable energy integration and smart grids, this invaluable resource empowers engineers, researchers, and students to navigate the dynamic landscape of modern power system analysis confidently.

Modern Power System Analysis
Third Edition

Originally authored by Turan Gönen
Revised and updated by Chee-Wooi Ten and Yunhe Hou

CRC Press
Taylor & Francis Group
Boca Raton London New York

CRC Press is an imprint of the
Taylor & Francis Group, an **informa** business

Third edition published 2024
by CRC Press
2385 NW Executive Center Drive, Suite 320, Boca Raton FL 33431

and by CRC Press
4 Park Square, Milton Park, Abingdon, Oxon, OX14 4RN

CRC Press is an imprint of Taylor & Francis Group, LLC

© 2024 Chee-Wooi Ten, and Yunhe Hou

First edition published by Wiley 1988
Second edition published by CRC Press 2013

ISBN: 978-0-367-65506-8 (hbk)
ISBN: 978-0-367-65507-5 (pbk)
ISBN: 978-1-003-12976-9 (ebk)
ISBN: 978-1-032-71570-4 (eBook+)

DOI: 10.1201/9781003129769

Typeset in Nimbus font
by KnowledgeWorks Global Ltd.

Publisher's note: This book has been prepared from camera-ready copy provided by the authors.

Dedication

This is dedicated to our parents who have given us life, allowing us to realize these wonderful things in electrical engineering.

Contents

Preface

The smart grid vision was promoted in 2009, and the values have been sustained. As a result, new technical books have been published ever since. Many authors focus on the renewable generation and electronics aspects of innovation. These books represent the modeling of renewable energy and advanced sensors, e.g., phasor measurement units (PMU) and enriching the curriculum, which would make new courses possible for instructors to adopt the texts. This can be very exciting for graduate courses where master's and doctoral students would be able to follow the details of the book and reproduce the contents through tutorials given in the books.

Transmission circuits have been researched for operational planning, and methodologies have been well-established for decades. There were ambitious visions to establish a direct current (DC) macro grid to connect major US interconnections as a superhighway to improve overall grid resilience with the flexibility to reconfigure the grids when a major threat has occurred to save the grids from major cascading failure. Electricity market participation will change from wholesale to a mix of retailers from prosumers (consumers who produce power to sell). Supporting infrastructure on enhancing the control is deployed with advanced sensing networks, which has introduced opportunities for hackers to intrude electronically into the control system. These are the pressing challenges new generations of power engineers will face. Inevitably, the power area students will explore the system fundamentals that often begin with transmission modeling and analysis.

Fundamentally, this book provides the modeling of power transmission networks, which has been one of the core subjects for power engineers to inculcate in the junior year of college after the first two circuit theory courses in the electrical engineering curriculum. The network connectivity for the transmission network is beyond two connected points, but is represented in a three-phase format. The circuitry contents on problem-solving skills are for undergraduate students to advance their mastery in tackling different complexities of network topology. The notational consistency in electrical engineering with many other textbooks would be important for student learning. Most importantly, the examples associated with each subsection are often important for students to follow.

The title of this book, *Modern Power System Analysis*, is intended to focus on transmission circuitry for network modeling with emerging subjects such as control center applications. From the system perspective, it is the modeling that will be used for power flow analysis. This is a well-known subject for power engineering that emphasizes the non-linearity of the solutions using iterative methods. This material allows us to nexus a new chapter introduced in this book, i.e., power system restoration. We anticipate that the "modern" should provide the restoration manner in the senior elective courses or even introductory graduate courses. We moved the power system protection into the electronic copy because this is a more localized subject for substation instrumentation. The two electrical short circuit chapters would provide an adequate introduction to the undergraduate and graduate students using the symmetrical components.

We have extensively added new references for some chapters where an update is necessary. Some proprietary old figures are removed from the text. We retain useful examples associated with each chapter where they would help students learn. Typographical errors are meticulously revised with help from students and CRC Press staff.

We sincerely hope this book will rejuvenate some of the important subjects fundamentally, which will inspire young generations of students aspiring to power engineering.

Authors,
Chee-Wooi Ten and Yunhe Hou

Authors

Turan Gönen, a distinguished figure in the field of electrical engineering, held the esteemed positions of professor of electrical engineering and director of the Electrical Power Educational Institute at California State University, Sacramento. Gönen's educational background includes BS and MS degrees in electrical engineering from Istanbul Technical College, followed by a PhD in electrical engineering from Iowa State University. His academic pursuits extended further as he achieved an MS in industrial engineering and a PhD with a co-major in industrial engineering from Iowa State University, alongside an MBA from the University of Oklahoma. Dr. Gönen first authored this book in 1988. His contributions to academia and research were substantial, marked by a portfolio of over 100 technical papers. His expertise was recognized as he held the distinction of IEEE Senior Member. He passed away in 2014.

Chee-Wooi Ten (senior member, IEEE) is a professor of electrical engineering at Michigan Tech. He focuses on investigating rare events using system risk models and data science, with a keen interest in untangling complex interactions between power grids and transportation systems to advance decarbonization. His efforts center on crafting cyber-aware security strategies for bulk power systems, validated through steady-state and dynamic analyses, emphasizing the need to anticipate disruptions in the context of distributed generation. Dr. Ten earned degrees from Iowa State University and a PhD from University College Dublin. He joined Michigan Tech in 2010 after working as a power application engineer at Siemens. His research covers critical cyber infrastructure modeling and SCADA automation applications in power grids, and he's edited for IEEE and Elsevier journals, authoring the textbook *Electric Power: Distribution Emergency Operation*.

Yunhe Hou (senior member, IEEE) received BE and PhD degrees in electrical engineering from the Huazhong University of Science and Technology, Wuhan, China in 1999 and 2005, respectively. He was a postdoctoral research fellow with Tsinghua University, Beijing, China from 2005-2007, and a postdoctoral researcher with Iowa State University and the University College Dublin from 2008-2009. He was also a visiting scientist with the Laboratory for Information and Decision Systems, Massachusetts Institute of Technology in 2010. He has been a guest professor with the Huazhong University of Science and Technology since 2017 and an academic adviser with China Electric Power Research Institute since 2019. In 2009, he became part of the faculty of The University of Hong Kong, where he is currently an associate professor in the Department of Electrical and Electronic Engineering. Dr. Hou was an associate editor for *IEEE Transactions on Smart Grid* from 2016-2021, and is currently an associate editor for *IEEE Transactions on Power Systems and Journal of Modern Power Systems and Clean Energy*.

1 General Considerations of Bulk Power Systems

He is not only dull himself; he is the cause of dullness in others.

Samuel Johnson

Some cause happiness wherever they go; others, whenever they go.

Oscar Wilde

1.1 INTRODUCTION

Energy generation prior to the 20th century relied on the combustion of fuel at the point of use. In 1882, Thomas Edison established the first electric power station, Pearl Street Electric Station in New York City, which led to the development of the electric utility industry and the spread of generating stations throughout the country [1]. There are several reasons for the rapid increase in demand for electrical energy. One reason is that electrical energy is a highly convenient form of energy that can be transported via wire to the point of consumption and converted into various other forms of energy such as mechanical work, heat, radiant energy, light, or others. However, electrical energy cannot be effectively stored, which has contributed to its increasing use.

To meet the peak energy demand, generating facilities are designed accordingly. In the late 1950s and early 1960s, peak demand was during winter when there were longer nights and increased lighting and heating needs. However, heavily promoting off-peak use, such as summer use of air conditioning, was economically viable. This promotion was so effective that summer is now the peak time. The rate structure has also contributed to the increasing use of electrical energy, where rate reductions are offered to attract industrial consumers. With the predicted energy needs and available fuels for the next century, energy is expected to be increasingly converted into electricity in the near future.

The structure of the electric power or energy[1] system is vast and intricate. However, it can be divided into five main stages, components, or subsystems, as shown in Figure 1.1. The power grid has become more intelligent with advanced sensing technologies such as phasor measurement units (PMU) and substation automation [2–4]. The first major component is the energy source or fuel that undergoes conversion. The energy source may be coal, gas, or oil burned in a furnace to heat water and produce steam in a boiler. Alternatively, it may be fissionable material, which heats water in a nuclear reactor to produce steam, water in a dam, or oil or gas burned in a combustion turbine.

The second major component of the electric power system is the energy converter or generation system, which converts the energy from the energy source into electrical energy. Typically, this is achieved by converting the fuel's energy into heat energy, which is then used to create steam to drive a steam turbine, which subsequently drives a generator to produce electrical energy. However, there are other potential energy conversion methods as well. The third main component is the transmission system, which transmits the bulk electrical energy from the generation system to the primary load

---

[1]The term *energy* is being increasingly used in the electric power industry to replace conventional term *power*. Here, they are used interchangeably.

DOI: 10.1201/9781003129769-1

```
┌──────────┐   ┌──────────┐   ┌──────────┐   ┌──────────┐   ┌──────────┐
│  Energy  │   │  Energy  │   │          │   │          │   │   Load   │
│  source  │ → │ converter│ → │Transmission│ → │Distribution│ → │ (energy  │
│  (fuel)  │   │(generation│   │  system  │   │  system  │   │  sink)   │
│          │   │  system) │   │          │   │          │   │          │
└──────────┘   └──────────┘   └──────────┘   └──────────┘   └──────────┘
```

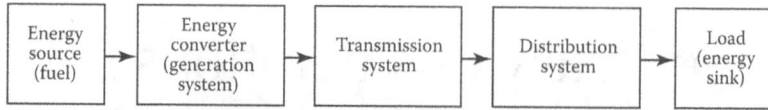

Figure 1.1 Structure of an electrical power system.

centers via high-voltage lines. The fourth component is the distribution system, which distributes this energy to consumers via lower-voltage networks. Finally, the fifth component is the load or energy sink, which uses the energy by transforming it into a useful or desired form, such as lights, motors, heaters, or other devices, alone or in combination.

It's important to note that energy is never used up; instead, it is successively transformed into various forms. Thus, in keeping with the principle of conservation of energy, all energy use ultimately results in unrecoverable waste heat, with the final heat sink for the Earth being radiation to space. The utilization of energy depends on two factors: the availability of resources and the technological ability to convert those resources into useful heat and work. While energy resources have always been generally available, technological devices capable of converting energy to useful work have been a recent historical development [5].

There are currently four different groups of energy conversion methods available. The first group comprises conventional and technologically feasible conversion methods, which generate more than 99.

The fuel supply system industry is an enormous industry. Energy sources are not readily available to consumers and must be discovered, processed or mined, and transported to demand centers. A typical fuel supply system includes the exploration, extraction, processing, and transportation stages, as depicted in Figure 1.2. Each of these stages causes significant environmental harm. The fuel supply industry creates pollution problems similar to those of other industrial processes, except for radioactive waste, land damage, and oil spills. The disposal of these wastes creates serious problems that must be considered in conjunction with biological safety. For some fuels, some of the stages shown in Figure 1.2 may not exist. For example, in the case of hydropower, exploration is still relevant, while combustion and its by-products are not. In determining the availability of a fuel for producing electrical energy, all by-products, and their handling must be thoroughly considered and evaluated.

Most forms of energy utilization ultimately lead to the production of waste heat, which places a greater thermal burden on the biosphere. However, this is not the case for an invariant energy source like hydropower, as it cycles through the terrestrial water cycle. Unfortunately, hydropower cannot meet a significant portion of energy needs. Solar energy could also be an invariant energy resource if either the solar cells or the converters were located on the earth's surface. All other forms of energy utilization, except for hydropower and solar conversion, are noninvariant, and hence lead to the release of waste heat into the environment. This includes geothermal energy, which originates from the earth itself.

The various stages illustrated in Figure 1.2 contribute to the overall cost of fuel to varying degrees, depending on the type of fuel being used. Unless nuclear fuel is used to generate electricity at a plant near a load center, the transportation of the energy source, in one form or another, represents a significant portion of the total cost of electricity. For electricity generation, fossil fuels such as coal, gas, and oil must be transported from the source to the generating plant, or the electrical energy must be transported by wire to the load centers when electricity is generated at the source of fuel, such as in the case of mine-mouth plants (coal), wellhead plants (gas), hydropower (water), or plants near refineries (residual fuel oil). The choice of energy transport is usually based on economics and its environmental impact.

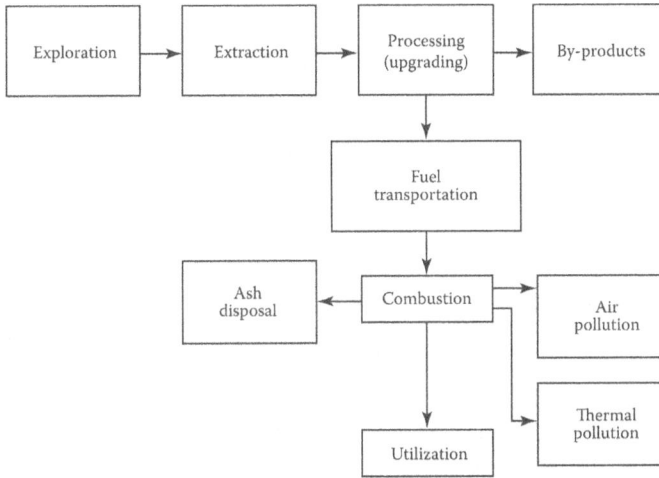

Figure 1.2 General fuel supply system.

Energy conversion refers to the process of changing energy from one form to another. However, currently, all energy conversions are more or less inefficient, and there are losses at the power plant, in transmission, and at the point of application of the power. In the case of fuels consumed in end-use, the loss occurs at the point of use. The waste heat produced at electric power plants enters the biosphere, and nearly all of the electrical energy that is carried to various points of use is eventually converted to heat.

Increasing the technical efficiency of energy conversion devices is crucial because it could make the same useful products available with less fuel. Although the average heat rate of electric power plants in the United States has increased significantly over time, it is not likely to increase further due to government regulations and environmental concerns.

Many electric utilities are currently unable to generate enough cash earnings to cover their common-stock dividends, and they finance the payouts through depreciation, borrowings, or the sale of more stock. Construction costs for power plant projects have grown substantially, and problems such as delays due to construction issues, environmental regulations, inflationary equipment and labor costs, recent recession problems, or the industry's inability to raise money have further increased projected costs. Figures 1.3–1.5 show the total growth in annual electric power system capital, the growth in electric utility plants in service, and the actual and projected investment growth in electric utility plants by investor-owned electric utilities in the past, respectively.

According to a report from an Institute of Electrical and Electronics Engineers committee [6], the current challenges faced by the electric utility industry have necessitated changes in the utility planning process. The industry has been forced to alter its planning objectives, planning methods, and even organizational structures to adapt to the constantly changing planning environment. Due to the complexity of the problem, long-range power system planning has become more of an art than a science.

In particular, the rising costs of electric power generation and the environmental impact of power generation and distribution have led utilities to implement new planning strategies. One approach is to adopt integrated resource planning, which considers all available energy resources and technologies, as well as energy conservation and demand-side management, to identify the most cost-effective means of meeting demand. Another approach is to engage in least-cost planning, which seeks to meet demand at the lowest cost to the utility and customers while still meeting reliability and environmental requirements.

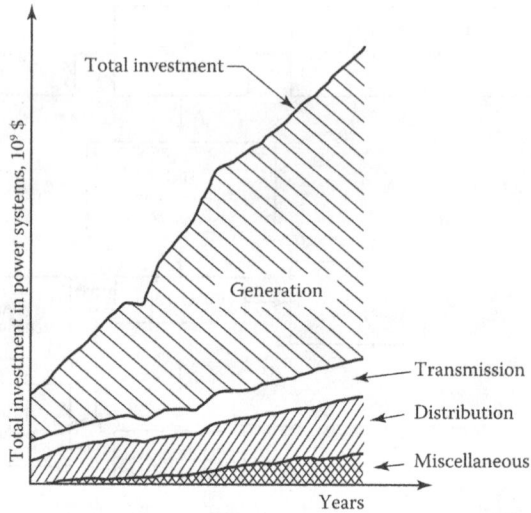

Figure 1.3 Growth in total annual electric power system expenditures in the past.

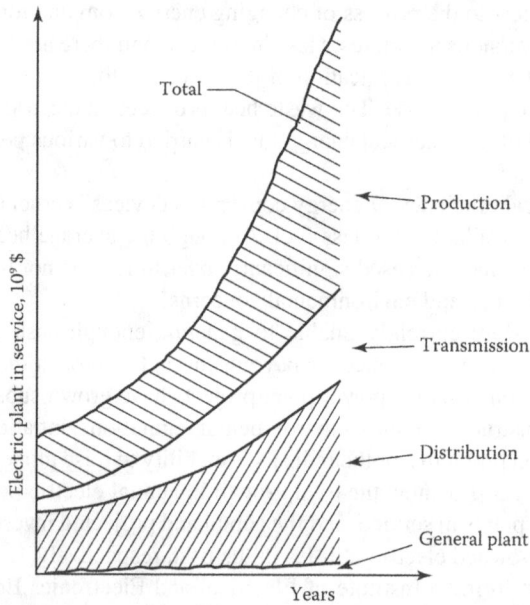

Figure 1.4 Growth in electric utility plants in service in the past.

However, these planning strategies face challenges due to the uncertainty of future demand growth, technological innovation, and regulatory changes. The complexities of these challenges have made long-range power system planning a more difficult task for electric utilities, requiring them to be flexible and adaptable in their planning processes. Overall, the electric utility industry must continue to evolve and adapt to meet the changing needs of society while balancing economic, environmental, and social considerations.

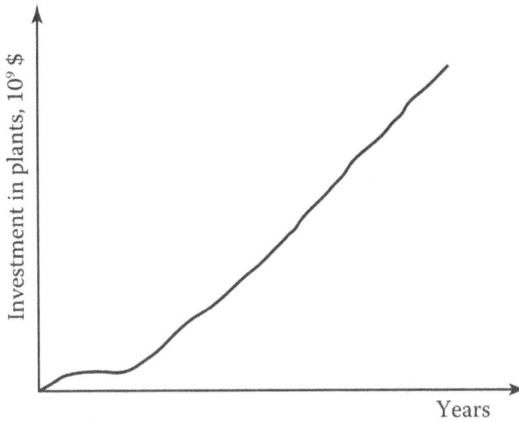

Figure 1.5   Investment growth in electric utility plants by investor-owned electric utilities in the past.

1.2   RECENT DEVELOPMENT

Planning for transmission growth over optimal time horizons has been established for decades using a computational approach [7]. To maximize grid reliability, analytical methods have been developed with the objective of incorporating distributed generation planning into the optimization frameworks [8]. Multistage system expansion in planning provides the net present values for a utility company [9]. Planning criteria include transitioning to low-carbon alternatives with renewable energy and demand response [10], well-coordinated periods of transmission and generation expansion [11], new synchronized phasor measurement applications [12], and cybersecurity measures for the grid [13]. System resiliency has become more important due to rare events such as extreme catastrophes [14–16] and intentional cyberattacks [15]. Sensing technologies have improved with advanced devices and data-driven methodologies [17–19]. New models and applications have been developed to improve resource allocation against cyberattacks [20–22], and investment in strengthening system resilience is a crucial strategy to anticipate these threats, with attempts made to federate testbeds [23, 24]. With the era of big data and ubiquitous advanced sensors, placement methodologies have been developed, including renewable penetration studies, as well as the potential disruptive threat and potential cascading challenges [25–30]. New features of grid attribution in grid expansion have been associated with a deep understanding of total generation and loads, where high levels of uncertainty may not allow for accurate financial estimates for grid resources in the long haul [31]. Statistical analysis of the mean-max expectation of the aggregated values over the past can help enhance resilient grid planning against abnormal events and high-impact, low-probability events (HILP) [32–35]. Grid resilience has been extensively researched [36–38].

1.3   SYSTEM PLANNING

According to the Institute of Electrical and Electronics Engineers [39], the primary responsibility of the electric utility industry is to provide an adequate and reliable supply of electric energy at a reasonable cost. This is crucial for maintaining a healthy economy, ensuring the well-being of citizens, and national security. Therefore, power system planning is critical to meeting the growing demand for electrical energy. In the future, electric utilities will need fast and cost-effective planning tools to evaluate the consequences of different alternatives and their impact on the rest of the system, ensuring economic, reliable, and safe energy delivery to consumers [40].

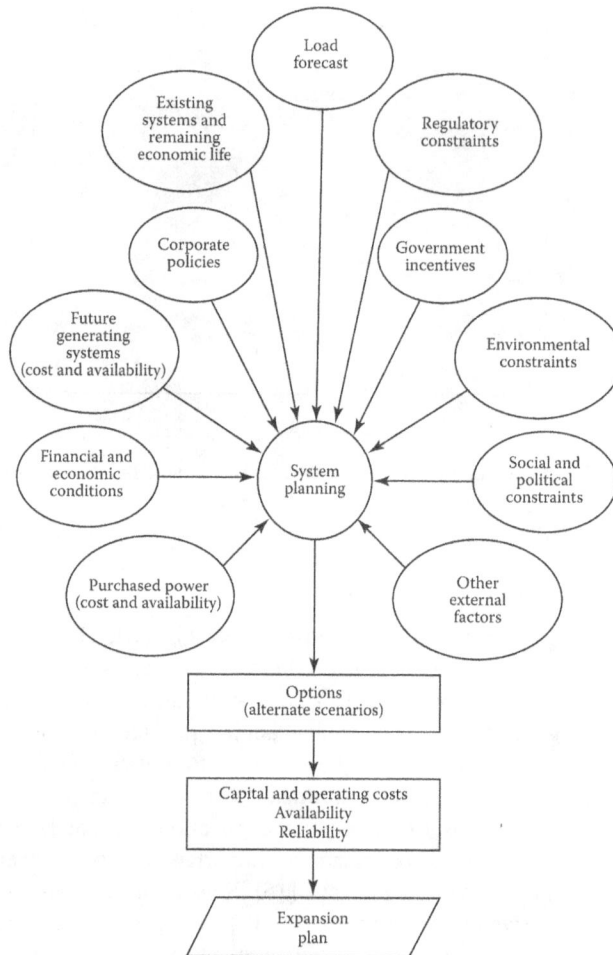

Figure 1.6 Factors affecting a typical system planning process.

Figure 1.6 illustrates the factors affecting system planning and the steps involved in the process. Additionally, Figure 1.7 shows the organizational chart of the power system planning function in a modern public utility company.

Nowadays, every electric utility company engages in short-term and long-term planning. Short-range planning refers to the analytical process of evaluating alternative courses of action against desired objectives and selecting a recommended course of action for the immediate future, which requires immediate commitments.

Long-range planning refers to an analytical process that evaluates alternative courses of action against desired objectives and selects a recommended course of action extending beyond the period requiring immediate commitments. With the fast-paced changes in technology, fuel availability, and environmental constraints, the need for long-range system planning has become more critical than ever. It allows planners to explore different alternatives in supplying electrical energy and provides them with a guide for making short-term decisions and actions. Long-range planning must be both quantitative and qualitative and may cover up to 15 to 20 years into the future. However, some power plant additions may have planning horizons extending up to 30 years.

The aim of power system planning is to minimize the cost while ensuring a sufficient electrical energy supply. System planning activities are generally divided into three categories: (1) synthesis,

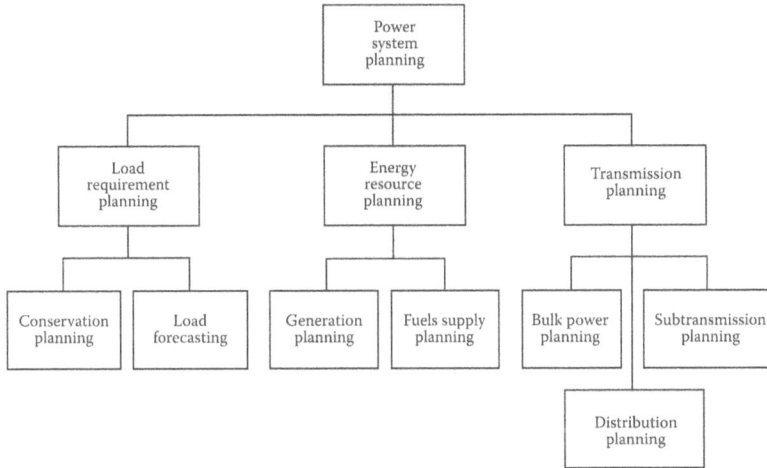

Figure 1.7 Modern organizational chart of a power system planning function.

which involves creating initial plans for studying the system; (2) analysis, which entails the technical evaluation of system operations, such as reserve requirements, load flow, and stability, under simulated conditions; and (3) optimization, which involves economic evaluations of various alternatives to determine the most cost-effective option. Figure 1.8 illustrates that system planning may encompass the following activities:

1. Load forecasting
2. Generation planning
3. Transmission planning
4. Subtransmission planning
5. Distribution planning
6. Operations planning
7. Fuel supply planning
8. Environmental planning
9. Financial planning
10. Research and development (R&D) planning

Load forecasting is a critical component of the system planning process as it predicts load increases and system reactions to these increases. Typically, long-range load forecasting has a horizon of 15 to 20 years, while short-range forecasting ranges from 1 to 5 years. The primary function of load forecasting is to analyze historical load data and develop models of peak demand for capacity planning, energy for capacity planning, and energy for production costing. Advanced techniques are available that allow for the analysis of load components separately and the determination of their effects on factors like weather, economic conditions, and per capita income.

Sophisticated load forecasting methods can result in load projections with probability distributions. However, the degree of uncertainty that planners perceive in their load forecasts has increased in the past decade. The risk of planning based on too low load projections is poor system reliability, while the risk of planning based on too high load projections is an uneconomical operation.

Generation planning helps identify the technology, size, and timing of the next generating plants to be added to the power system. It aims to ensure that adequate generation capacity is available to meet future electricity demands and that the cost of power generation remains economical.

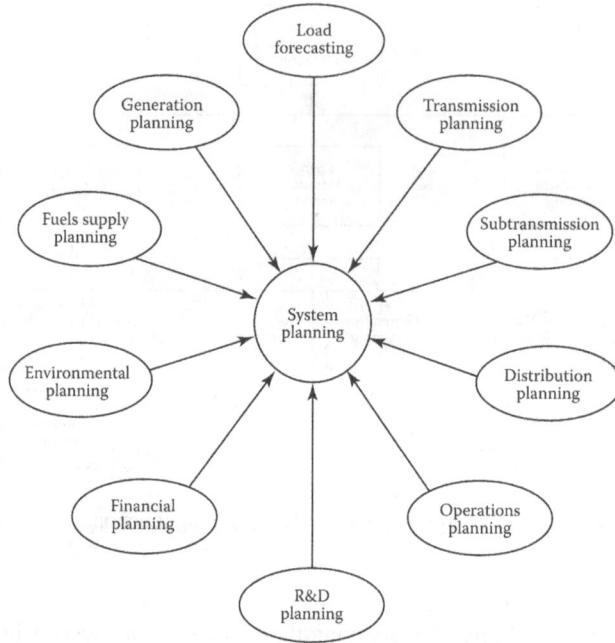

Figure 1.8 Power system planning activities.

The planning of generation includes (1) capacity planning, (2) production costing, and (3) calculating investment and operating and maintenance costs. Capacity planning involves combining load and generation models, considering scheduled maintenance, and determining the capacity required to meet the system reliability criterion. Production costing involves determining the costs of generating energy requirements of the system, taking into account maintenance and forced outages. Finally, investment and operating and maintenance costs are calculated, and their present value is predetermined.

The objective is to achieve optimum generation planning that combines the above functions into one, developing economically optimal generation expansion patterns year by year over the planning horizon. System planners have a variety of synthesis tools available, including (1) target plant mix, (2) expert judgment, and (3) computer-based mathematical programming models. Simulation models are employed to predict power system reliability, capital and production costs, and power plant operation for each year of the planning horizon based on a given generation expansion plan.

Transmission planning is closely linked with generation planning and aims to create yearly plans for the transmission system by considering existing systems, future load and generation scenarios, right-of-way constraints, construction costs, line capabilities, and reliability criteria. Transmission lines serve two primary purposes: (1) to transmit electrical energy from generators to loads within a single utility and (2) to establish pathways for electrical energy to flow between utilities, also known as "tie lines," enabling utility companies to collaborate and achieve benefits that would be difficult to achieve individually. Interconnections, or the installation of transmission circuits across utility boundaries, impact the generation and transmission planning of each utility involved [41–44].

When power systems are interconnected through transmission lines, it is essential that they operate at the same frequency, and the alternating current pulse should be coordinated. This coordination of generator speeds is necessary for the generators to operate "in parallel" or in synchronism, "which is considered a stable" condition.

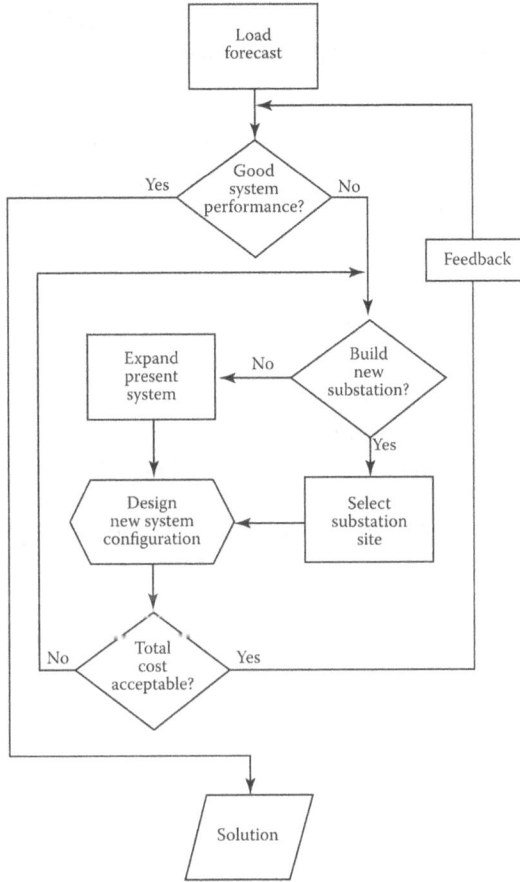

Figure 1.9 Block diagram of a typical distribution system planning process. (From Gönen, T., *Electrical Power Distribution System Engineering*, 2nd ed., CRC Press, Boca Raton, FL, 2008.)

Changes in loading at a generator can cause fluctuations in frequency. However, when a generator is interconnected with other generators, they can help absorb the effect of the changed loading, resulting in negligible frequency changes and maintaining the system's stability. Therefore, the installation of an interconnection significantly affects generation planning, including the amount of generation capacity required, reserve generation capacity, and the type of generation capacity required for operation.

Interconnections can also impact generation planning by allowing the installation of jointly owned apparatus between neighboring utilities and the planning of generating units with greater capacity than feasible for a single utility without interconnections. Interconnection planning also affects transmission planning by requiring bulk power deliveries to or from interconnection substations, which are bulk power substations, and often the addition of circuits on a given utility's network [45].

According to Gonen [46], subtransmission planning involves planning activities for bulk stations, subtransmission lines from stations to distribution substations, and the high-voltage portion of distribution substations. Distribution planning, on the other hand, considers substation siting, sizing, number of feeders, voltage levels, and the type and size of the service area while also coordinating with overall subtransmission and transmission planning efforts to ensure reliable and cost-effective system design. Computer programs such as load flow, radial or loop network load, short-circuit, voltage drop, voltage regulation, load forecasting, regulator setting, and capacitor planning are commonly used by distribution system planners. A functional block diagram of the distribution system planning process is shown in Figure 1.9 [47].

The customer level is where planning begins, with load characteristics dictating the type of distribution system required. Secondary lines connecting to distribution transformers are defined, and their loads are combined to determine primary distribution system demands. Substation assignments are based on these loads, with the size and location of substations and the route and capacity of associated subtransmission lines determined accordingly.

Fuel supply and operations planning aim to provide generation planning with necessary information for power system operation modeling, estimating fuel prices and availability, gathering long-term fuel contract information, and providing information on heat rates, plant capacity factor restrictions, maintenance, and energy sales and purchases [48].

Environmental planning considers regulatory responsibilities and available fuel types to determine plant type, location, and design requirements, limiting available alternatives in developing system expansion plans. Financial planning uses corporate models to develop financial reports and cash flow information on system expansion, including tax and regulatory constraints. R&D planning provides information on alternative energy sources and future technological developments in generation [48].

2 Basic Concepts and Power System Modeling

2.1 INTRODUCTION

In this chapter, a brief review of fundamental concepts associated with steady-state alternating current (AC) circuits, especially three-phase circuits, is presented. It is hoped that this review provides a sufficient common base, in terms of notation and references, for readers to follow subsequent chapters. This chapter also elaborates on how a circuit system can be modeled from the electric circuit standpoint, and why the non-linearity of power systems has been characterized from the instrumentation standpoint in power system communication. An equivalent circuit with a connection to a one-line diagram is also discussed in this chapter. By doing so, this provides a high-level abstraction of system modeling, together with the introduction of the per unit system for normalizing calculations between different parts of a power system.

2.2 COMPLEX POWER IN BALANCED TRANSMISSION LINES

A polyphase system of power transmission network is treated as balanced system where a simplified version of per phase analysis has been widely employed in many energy management system (EMS). Such software system implemented in programming platform where it would allow efficient algorithms that rapidly determine the power flow solutions[1]. Such tradition has been established by power engineers where the other two phases of the transmission circuit can be determined quickly based on the conditions and foundation already established.

Figure 2.1a shows a per-phase representation (or a one-line diagram) of a short three-phase balanced transmission line connecting buses i and j [49]. Here, the term *bus*[2] defines a specific nodal point of a transmission network. Assume that the bus voltages $\mathbf{V}_i$ and $\mathbf{V}_j$ are given in phase values (i.e., line-to-neutral values) and that the line impedance is $\mathbf{Z} = R + jX$ per phase. Since the transmission line is a short one, the line current $\mathbf{I}$ can be assumed to be approximately the same at any point in the line. However, because of the line losses, the complex powers $\mathbf{S}_{ij}$ and $\mathbf{S}_{ji}$ are not the same. Thus, the complex power per phase* that is being transmitted from bus i to bus j can be expressed as

$$\mathbf{S}_{ij} = P_{ij} + jQ_{ij} = \mathbf{V}_i\mathbf{I}^* \qquad (2.1)$$

Similarly, the complex power per phase that is being transmitted from bus j to bus i can be expressed as

$$\mathbf{S}_{ji} = P_{ji} + jQ_{ji} = \mathbf{V}_j(\mathbf{-I})^* \qquad (2.2)$$

Notice both voltages and currents can be represented in polar form, i.e., magnitude and angle. The magnitudes of both voltages and currents can be demined from the sensors in substation. However, not all angles can be determined. Hence, the product of both quantities has resulted in non-linearity. Further, the current can be interepreated as follow:

$$\mathbf{I} = \frac{\mathbf{V}_i - \mathbf{V}_j}{\mathbf{Z}} \qquad (2.3)$$

---

[1] The iterative power flow analysis is introduced in Chapter 9.

[2] Bus defined here is the node of a power system network. This can also be interpreted as the juncture representing it as a power substation.

DOI: 10.1201/9781003129769-2

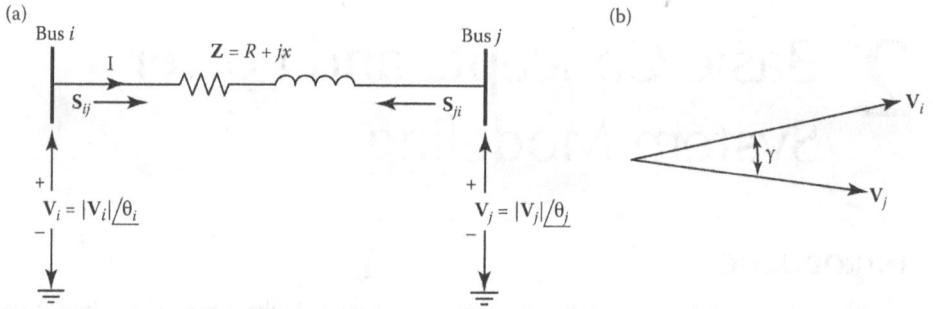

Figure 2.1 Per-phase representation of short transmission line.

By substituting Equation 2.3 into Equations 2.1 and 2.2, the following is expanded.

$$\mathbf{S}_{ij} = \mathbf{V}_i \left(\frac{\mathbf{V}_i^* - \mathbf{V}_j^*}{Z} \right)$$
$$= \frac{|\mathbf{V}_i|^2 - |\mathbf{V}_i||\mathbf{V}_j|\angle\theta_i - \theta_j}{R - jX} \tag{2.4}$$

* For an excellent treatment of the subject, see Elgerd [50,51].

$$\mathbf{S}_{ji} = \mathbf{V}_j \left(\frac{\mathbf{V}_j^* - \mathbf{V}_i^*}{Z} \right)$$
$$= \frac{|\mathbf{V}_j|^2 - |\mathbf{V}_j||\mathbf{V}_i|\angle\theta_j - \theta_i}{R - jX} \tag{2.5}$$

However, as shown in Figure 2.1b, if the power angle (i.e., the phase angle between the two bus voltages) is defined as

$$\gamma = \theta_i - \theta_j \tag{2.6}$$

then the real and the reactive power per-phase values can be expressed, respectively, as

$$P_{ij} = \frac{1}{R^2 + X^2} \left(R|\mathbf{V}_i|^2 - R|\mathbf{V}_i||\mathbf{V}_j|\cos\gamma + X|\mathbf{V}_i||\mathbf{V}_j|\sin\gamma \right) \tag{2.7}$$

and

$$Q_{ij} = \frac{1}{R^2 + X^2} \left(X|\mathbf{V}_i|^2 - X|\mathbf{V}_i||\mathbf{V}_j|\cos\gamma + R|\mathbf{V}_i||\mathbf{V}_j|\sin\gamma \right) \tag{2.8}$$

Similarly,

$$P_{ji} = \frac{1}{R^2 + X^2} \left(R|\mathbf{V}_j|^2 - R|\mathbf{V}_i||\mathbf{V}_j|\cos\gamma + X|\mathbf{V}_i||\mathbf{V}_j|\sin\gamma \right) \tag{2.9}$$

and

$$Q_{ji} = \frac{1}{R^2 + X^2} \left(X|\mathbf{V}_j|^2 - X|\mathbf{V}_i||\mathbf{V}_j|\cos\gamma + R|\mathbf{V}_i||\mathbf{V}_j|\sin\gamma \right) \tag{2.10}$$

The three-phase real and reactive power can directly be found from Equations 2.7 through 2.10 if the phase values are replaced by the line values.

In general, the reactance of a transmission line is much greater than its resistance. Therefore, the line impedance value can be approximated as

$$\mathbf{Z} = jX \tag{2.11}$$

by setting $R = 0$. Therefore, Equations 2.7 through 2.10 can be expressed as

$$P_{ij} = \left(\frac{|\mathbf{V}_i||\mathbf{V}_j|}{X} \right) \sin\gamma \tag{2.12}$$

$$Q_{ij} = \frac{1}{X}(|\mathbf{V}_i|^2 - |\mathbf{V}_i||\mathbf{V}_j|\cos\gamma) \tag{2.13}$$

and

$$P_{ji} = \left(\frac{|\mathbf{V}_i||\mathbf{V}_j|}{X} \right) \sin\gamma = -P_{ij} \tag{2.14}$$

$$Q_j = \frac{1}{X}(|\mathbf{V}_j|^2 - |\mathbf{V}_i||\mathbf{V}_j|\cos\gamma) \tag{2.15}$$

Example 2.1: Assume that the impedance of a transmission line connecting buses 1 and 2 is 100 $\angle 60° \Omega$, and that the bus voltages are 73,034.8 $\angle 30°$ and 66,395.3$\angle 20°$ V per phase, respectively. Determine the following:

a. Complex power per phase that is being transmitted from bus 1 to bus 2
b. Active power per phase that is being transmitted
c. Reactive power per phase that is being transmitted

Solution:

a.

$$\mathbf{S}_{12} = \mathbf{V}_1 \left(\frac{\mathbf{V}_1^* - \mathbf{V}_2^*}{\mathbf{Z}^*} \right)$$

$$= (73,034.8\angle 30°) \left(\frac{(73,034.8\angle - 30° - 66,395.3\angle - 20°)}{100\angle - 60°} \right)$$

$$= 10,104,280.7667\angle 3.56°$$

$$= 10,085,280.6 + j627,236.51 \text{ VA}$$

b. Therefore,

$$P_{12} = 10,085,280.6 \text{ W}$$

c.

$$Q_{12} = 627,236.5 \text{ VArs}$$

2.3 ONE-LINE DIAGRAM

Figure 2.2a shows a one-line diagram of electrical power systems, which is commonly used to represent the connectivity of control variables and elements in the power control center. This visual representation is effective for operators in the control center, where circuit breaker statuses can be evaluated based on online topology processors. Such software establishments in supervisory control and data acquisition (SCADA) are essential for intuition, allowing operators to quickly assimilate relevant events from alarm processing applications and make decisions based on application recommendations.

The one-line diagram is also referred to as the single-line diagram. In usual situations, the need for the three-phase equivalent impedance diagram, as shown in Figure 2.2b, is almost non-existent. This is because a balanced three-phase system can always be represented by an equivalent impedance diagram per phase, as shown in Figure 2.2c. The per-phase equivalent impedance can be simplified by neglecting the neutral line and representing system components by standard symbols rather than by their equivalent circuits [52]. Table 2.1 provides some of the symbols used in one-line diagrams. Additional standard symbols can be found in Institute of Electrical and Electronics Engineers Standard 315-1971 [53, 54]. Peripheral apparatus such as instrument transformers, protective relays, and lightning arrestors may also be shown on the one-line diagram as needed.

The details shown on a one-line diagram depend on its purpose. For example, one-line diagrams used in load flow studies do not show circuit breakers or relays, contrary to those used in stability studies. Furthermore, those used in unsymmetrical fault studies may even show the positive-, negative-, and zero-sequence networks separately.

Note that buses shown in Figure 2.2a have been identified by their bus numbers. Also, the neutral of generator 1 has been "solidly grounded," while the neutral of generator 2 has been "grounded through impedance" using a resistor or inductance coil. Usually, the neutrals of transformers used in transmission lines are solidly grounded. Proper generator grounding is facilitated by burying a ground electrode system made of grids of buried horizontal wires. Sometimes, a metal plate is buried instead of a mesh grid (especially in European applications) [55].

Transmission lines with overhead ground wires have a ground connection at each supporting structure to which the ground wire is connected. In some circumstances, a "counterpoise" is buried under a transmission line to decrease the ground resistance. The equivalent circuit of the transmission line shown in Figure 2.2c has been represented by a nominal π [3]. The line impedance has been lumped in terms of the resistance and the series reactance of a single conductor for the length of the line. The line-to-neutral capacitance (or shunt capacitive reactance) for the length of the line has been calculated, and half of this value has been put at each end of the line. The transformers have been represented by their equivalent reactances, neglecting their magnetizing currents and consequently their shunt admittances. The resistance values of the transformers and generators are also neglected because their inductive reactance values are much greater than their resistance values.

The ground resistor is not shown in Figure 2.2c because no current flows in the neutral under balanced conditions. The impedance diagram shown in Figure 2.2c is also known as the positive-sequence network or diagram. This is because the phase order of the balanced voltages at any point in the system is the same as the phase order of the generated voltage, and they are positive. The per-phase impedance diagrams may represent a system given in ohms or in per unit.

Peripheral apparatus, such as instrument transformers (i.e., current and voltage transformers), protective relays, and lighting arrestors, may also be shown on the one-line diagram as needed. Therefore, the details displayed on a one-line diagram depend on its purpose. For instance, the one-line diagrams used in load flow studies do not show circuit breakers or relays, unlike those used in stability studies. Additionally, those utilized in unsymmetrical fault studies may even exhibit the positive-, negative-, and zero-sequence networks separately [57].

2.4 PER-UNIT SYSTEM

In power system analysis calculations, it is customary to use per-unit values for impedances, currents, voltages, and powers due to their various advantages over physical values in ohms, amperes, kilovolts, and megavolt-amperes (MVA), megavars, or megawatts. Per-unit values are scaled or normalized values that make it easier to compare quantities. The per-unit system defines the per-unit value of any quantity as the ratio of the quantity to an arbitrarily chosen base (or reference) value with the same dimensions [58]. Therefore, the per-unit value of any quantity can be expressed as:

$$\text{Quantity in per unit} = \frac{\text{Physical quantity}}{\text{Base value of quantity}} \tag{2.16}$$

---

[3]Further information is available in Chapter 3.

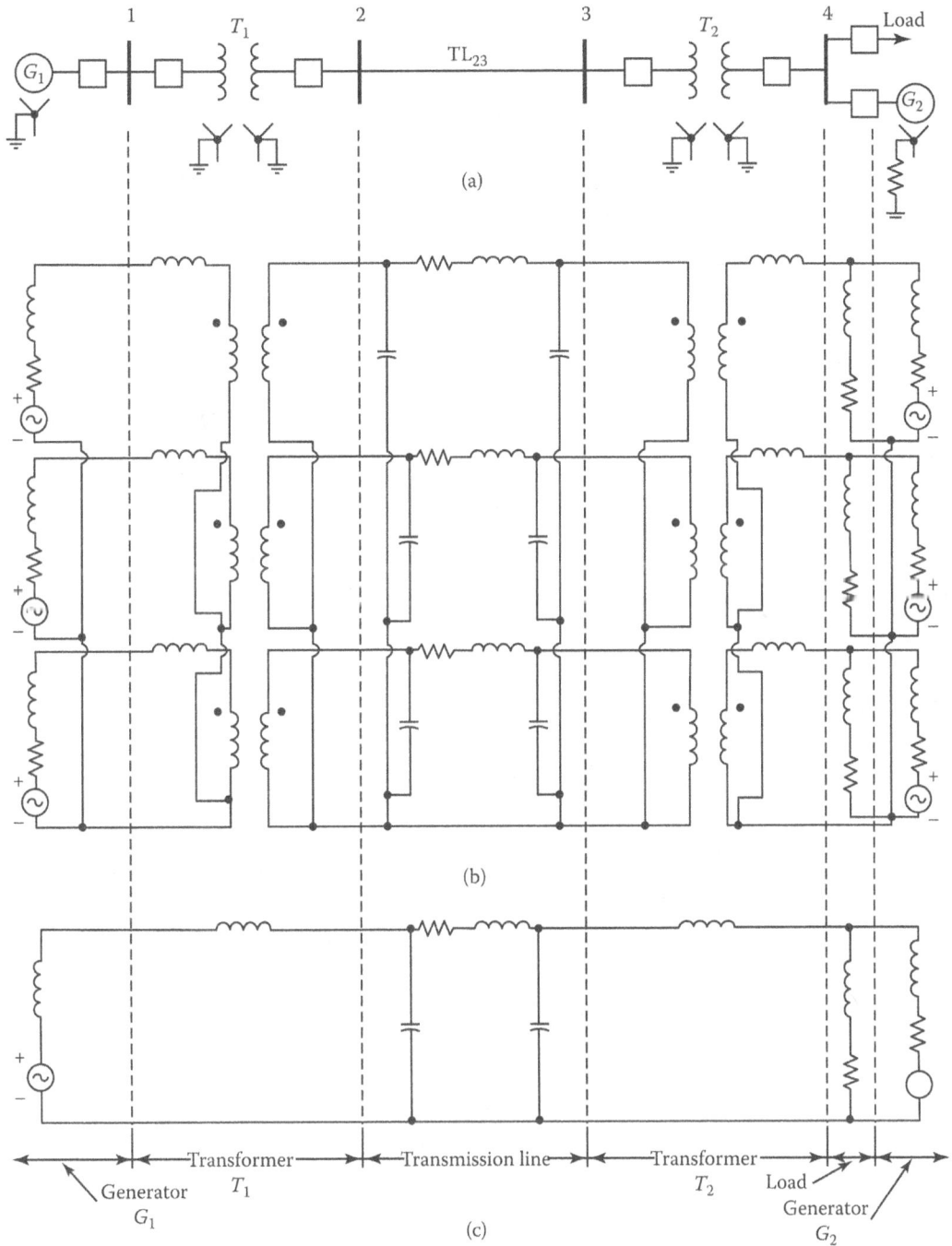

Figure 2.2 Power system representations: (a) one-line diagram; (b) three-phase equivalent impedance diagram; (c) equivalent impedance diagram per phase.

where *physical quantity* refers to the given value in ohms, amperes, volts, etc. In the per-unit system, the *base value* is equivalent to the "unit value," which has a value of 1 or unity. This means that a base current is also referred to as a unit current. As both the physical quantity and base quantity

Table 2.1

Symbols Used in One-Line Diagrams [56]

Symbol	Usage	Symbol	Usage
◯	Rotating machine	▭	Circuit breaker
	Bus		Circuit breaker (air)
	Two-winding transformer		Disconnect
	Three-winding transformer	or	Fuse
△	Delta connection (3φ, three wire)		Fused disconnect
Y	Wye connection (3Φ, neutral ungrounded)		Lightning arrester
Y	Wye connection (3Φ, neutral grounded)		Current transformer (CT)
	Transmission line	or	Potential transformer (VT)
	Static load		Capacitor

have the same dimensions, the resulting per-unit value expressed as a decimal has no dimension and is indicated by a subscript pu. The base quantity is denoted by a subscript B. The per-unit symbol is typically expressed as pu or 0/1. To convert from the per-unit system to the percent system, the per-unit value is multiplied by 100. Therefore,

$$\text{Quantity in percent} = \frac{\text{Physical quantity}}{\text{Base value of quantity}} \times 100 \qquad (2.17)$$

The percent system can be challenging to work with and more prone to errors since it is necessary to remember that the quantities have been multiplied by 100. As a result, the factor of 100 must be continuously added or removed, which may not be immediately apparent. For instance, multiplying 40% reactance by 100% current yields 4000% voltage, which must be corrected to 40% voltage. Therefore, the per-unit system is generally preferred in power system calculations.

The advantages of using the per-unit system in power system calculations are as follows [59,60]:

1. Network analysis is simplified since all impedances of a given equivalent circuit can be added directly regardless of the system voltages.
2. The need for $\sqrt{3}$ multiplications and divisions, which are necessary when representing balanced three-phase systems by per-phase systems, is eliminated. The factors $\sqrt{3}$ and 3, which are associated with delta and wye quantities in a balanced three-phase system, are directly considered by the base quantities.
3. The impedance of an electrical apparatus is often given in percent or per unit by its manufacturer on the basis of its nameplate ratings, such as its rated volt-amperes and voltage.
4. Differences in the operating characteristics of many electrical apparatus can be estimated by comparing their constants expressed in per units.
5. Average machine constants can be obtained easily since the parameters of similar equipment tend to fall within a relatively narrow range and are comparable when expressed as per units according to rated capacity.
6. The use of per unit quantities is more convenient in calculations that involve computers.

2.4.1   SINGLE-PHASE SYSTEM

If any two of the four base quantities (base voltage, base current, base volt-amperes, and base impedance) are specified, the remaining two can be immediately determined. However, the term "arbitrarily" is slightly misleading since the base values are typically chosen to produce results within specified ranges. For example, the base voltage is selected such that the system voltage is usually close to unity. Similarly, the base volt-ampere is usually chosen as the kilovolt-ampere (kVA) or megavolt-ampere (MVA) rating of one of the machines or transformers in the system or a convenient round number such as 1, 10, 100, or 1,000 MVA, depending on the size of the system. Once the base volt-amperes and base voltages are determined, the other base values are fixed. For example, the current base can be calculated as:

$$I_B = \frac{S_B}{V_B} = \frac{VA_B}{V_B} \tag{2.18}$$

where
I_B = current base in amperes
$S_B = VA_B$ = selected volt-ampere base in volt-amperes
V_B = selected voltage base in volts

Note that,
$$S_B = VA_B = P_B = Q_B = V_B I_B \tag{2.19}$$

Similarly, the impedance base* can be determined as

$$Z_B = \frac{V_B}{I_B} \tag{2.20}$$

where

$$Z_B = X_B = R_B \tag{2.21}$$

Similarly,

$$Y_B = B_B = G_B = \frac{I_B}{V_B} \tag{2.22}$$

* Defined as that impedance across which there is a voltage drop that is equal to the base voltage if the current through it is equal to the base current.

Note that by substituting Equation 2.18 into Equation 2.20, the impedance base can be expressed as

$$Z_B = \frac{V_B}{VA_B/V_B} = \frac{V_B^2}{VA_B} \tag{2.23}$$

or

$$Z_B = \frac{(kV_B)^2}{MVA_B} \tag{2.24}$$

where

kV_B = voltage base in kilovolts

MVA_B = volt-ampere base in MVA

The per-unit value of any quantity can be found by the *normalization process*, that is, by dividing the physical quantity by the base quantity of the same dimension. For example, the per-unit impedance can be expressed as

$$Z_{pu} = \frac{Z_{physical}}{Z_B} \tag{2.25}$$

or

$$Z_{pu} = \frac{Z_{physical}}{V_B^2/(kVA_B \times 1,000)} \tag{2.26}$$

or

$$Z_{pu} = \frac{(Z_{physical})(kVA_B)(1,000)}{V_B^2} \tag{2.27}$$

or

$$Z_{pu} = \frac{(Z_{physical})(kVA_B)(1,000)}{V_B^2} \tag{2.28}$$

or

$$Z_{pu} = \frac{(Z_{physical})}{(kV_B)^2(1,000)} \tag{2.29}$$

or

$$Z_{pu} = \frac{(Z_{physical})(MVA_B)}{(kV_B)^2} \tag{2.30}$$

Similarly, the others can be expressed as

$$I_{pu} = \frac{(I_{physical})}{I_B} \tag{2.31}$$

or

$$V_{pu} = \frac{(V_{physical})}{V_B} \tag{2.32}$$

or

$$kV_{pu} = \frac{(kV_{physical})}{kV_B} \tag{2.33}$$

or

$$VA_{pu} = \frac{(VA_{physical})}{VA_B} \tag{2.34}$$

or

$$kVA_{pu} = \frac{(kVA_{physical})}{kVA_B} \tag{2.35}$$

or

$$MVA_{pu} = \frac{(MVA_{physical})}{MVA_B} \tag{2.36}$$

Note that, the base quantity is always a real number, whereas the physical quantity can be a complex number. For example, if the actual impedance quantity is given as $Z\angle\theta$ Ω, it can be expressed in the per-unit system as

$$Z_{pu} = \frac{Z\angle\theta}{Z_B} = Z_{pu}\angle\theta \tag{2.37}$$

that is, it is the magnitude expressed in per-unit terms. Alternatively, if the impedance has been given in rectangular form as

$$\mathbf{Z} = R + jX \tag{2.38}$$

then

$$\mathbf{Z}_{pu} = R_{pu} + jX_{pu} \tag{2.39}$$

where

$$R_{pu} = \frac{R_{physical}}{Z_B} \tag{2.40}$$

and

$$X_{pu} = \frac{X_{physical}}{Z_B} \tag{2.41}$$

Similarly, if the complex power has been given as

$$\mathbf{S} = P + jQ \tag{2.42}$$

then

$$\mathbf{S}_{pu} = P_{pu} + Q_{pu} \tag{2.43}$$

where

$$P_{pu} = \frac{P_{physical}}{S_B} \tag{2.44}$$

and

$$Q_{pu} = \frac{Q_{physical}}{S_B} \tag{2.45}$$

If the actual voltage and current values are given as

$$\mathbf{V} = V\angle\theta_v \tag{2.46}$$

and

$$\mathbf{I} = I\angle\theta_{\mathbf{I}} \tag{2.47}$$

then the complex power can be expressed as

$$\mathbf{S} = \mathbf{VI}^* \tag{2.48}$$

or

$$S\angle\theta = (\mathbf{V}\angle\theta_v)(I\angle-\theta_{\mathbf{I}}) \tag{2.49}$$

Therefore, dividing through by S_B,

$$\frac{S\angle\phi}{S_B} = \frac{(\mathbf{V}\angle\theta_v)(I\angle-\theta_I)}{S_B} \tag{2.50}$$

However,

$$S_B = V_B I_B \tag{2.51}$$

Thus,

$$\frac{S\angle\phi}{S_B} = \frac{(\mathbf{V}\angle\theta_v)(I\angle - \theta_I)}{V_B I_B} \tag{2.52}$$

or

$$S_{\text{pu}}\angle\theta = (V_{\text{pu}}\angle\theta_v)(I_{\text{pu}}\angle - \theta_{\mathbf{I}}) \tag{2.53}$$

or

$$S_{\text{pu}} = V_{\text{pu}} I_{\text{pu}}^* \tag{2.54}$$

2.4.2 CONVERTING FROM PER-UNIT VALUES TO PHYSICAL VALUES

The physical values (or system values) and per-unit values are related by the following relations:

$$\mathbf{I} = \mathbf{I}_{\text{pu}} \times I_B \tag{2.55}$$

$$\mathbf{V} = \mathbf{V}_{\text{pu}} \times V_B \tag{2.56}$$

$$\mathbf{Z} = \mathbf{Z}_{\text{pu}} \times Z_B \tag{2.57}$$

$$R = R_{\text{pu}} \times Z_B \tag{2.58}$$

$$X = X_{\text{pu}} \times Z_B \tag{2.59}$$

$$VA = VA_{\text{pu}} \times VA_B \tag{2.60}$$

$$P = P_{\text{pu}} \times VA_B \tag{2.61}$$

$$Q = Q_{\text{pu}} \times VA_B \tag{2.62}$$

2.4.3 CHANGE OF BASE

Generally, the per-unit impedance of a power apparatus is based on its own volt-ampere and voltage ratings, and thus, its own impedance base. When such an apparatus is utilized in a system with its own bases, it is necessary to reference all given per-unit values to the system base values. Suppose the per-unit impedance of the apparatus is provided based on its nameplate ratings:

$$Z_{\text{pu(given)}} = (Z_{\text{physical}})\frac{\text{MVA}_{B,(\text{given})}}{[kV_{B(\text{given})}]^2} \tag{2.63}$$

and that it is necessary to refer the very same physical impedance to a new set of voltage and volt-ampere bases such that

$$Z_{\text{pu(new)}} = (Z_{\text{physical}})\frac{\text{MVA}_{B,(\text{new})}}{[kV_{B(\text{new})}]^2} \tag{2.64}$$

By dividing Equation 2.63 by Equation 2.64 side by side,

$$Z_{\text{pu(new)}} = (Z_{\text{pu(given)}})\left[\frac{\text{MVA}_{B(\text{new})}}{\text{MVA}_{B(\text{given})}}\right]\left[\frac{kV_{B(\text{given})}}{kV_{B(\text{new})}}\right] \tag{2.65}$$

In certain situations, it is more convenient to use subscripts 1 and 2 instead of the subscripts "given" and "new," respectively. Then, Equation 2.65 can be expressed as

$$Z_{\text{pu(2)}} = Z_{\text{pu(1)}}\left[\frac{\text{MVA}_{B(2)}}{\text{MVA}_{B(1)}}\right]\left[\frac{kV_{B(1)}}{kV_{B(2)}}\right]^2 \tag{2.66}$$

In the event that the kilovolt bases are the same but the MVA bases are different, from Equation 2.65,

$$Z_{pu(new)} = Z_{pu(given)} \frac{MVA_{B(new)}}{MVA_{B(given)}} \tag{2.67}$$

Similarly, if the MVA bases are the same but the kilovolt bases are different, from Equation 2.65,

$$Z_{pu(new)} = Z_{pu(given)} \left[\frac{kV_{B(given)}}{kV_{B(new)}} \right]^2 \tag{2.68}$$

Equations 2.65 through 2.68 must only be used to convert the given per-unit impedance from the base to another but not for referring the physical value of an impedance from one side of the transformer to another [61].

2.4.4   THREE-PHASE SYSTEMS

The three-phase problems involving balanced systems can be solved on a per-phase basis. In that case, the equations that are developed for single-phase systems can be used for three-phase systems as long as per-phase values are used consistently [62]. Therefore,

$$I_B = \frac{S_{B(1\phi)}}{V_{B(L-N)}} \tag{2.69}$$

or

$$I_B = \frac{VA_{B(1\phi)}}{V_{B(L-N)}} \tag{2.70}$$

and

$$Z_B = \frac{V_{B(L-N)}}{I_B} \tag{2.71}$$

or

$$Z_B = \frac{[kV_{B(L-N)}]^2 (1,000)}{kVA_{B(1\phi)}} \tag{2.72}$$

or

$$Z_B = \frac{[kV_{B(L-N)}]^2}{MVA_{B(1\phi)}} \tag{2.73}$$

where the subscripts 1ϕ and L–N denote per phase and line to neutral, respectively. Note that, for a *balanced system*,

$$V_{B(L-N)} = \frac{V_{B(L-L)}}{\sqrt{3}} \tag{2.74}$$

and

$$S_{B(1\phi)} = \frac{S_{B(3\phi)}}{3} \tag{2.75}$$

However, it has been customary in three-phase system analysis to use line-to-line voltage and three-phase volt-amperes as the base values. Therefore,

$$I_B = \frac{S_{B(3\phi)}}{\sqrt{3}V_{B(L-L)}} \tag{2.76}$$

or

$$I_B = \frac{kVA_{B(3\phi)}}{\sqrt{3}kV_{B(L-L)}} \tag{2.77}$$

and

$$Z_B = \frac{V_{B(\text{L-L})}}{\sqrt{3}I_B} \tag{2.78}$$

$$Z_B = \frac{[kV_{B(\text{L-L})}]^2(1{,}000)}{\text{kVA}_{3\phi}} \tag{2.79}$$

or

$$Z_B = \frac{[kV_{B(\text{L-L})}]^2}{\text{MVA}_{3\phi}} \tag{2.80}$$

where the subscripts 3 ϕ and L–L denote per three phase and line, respectively. Furthermore, base admittance can be expressed as

$$Y_B = \frac{1}{Z_B} \tag{2.81}$$

or

$$Y_B = \frac{\text{MVA}_{B(3\phi)}}{[kV_{B(\text{L-L})}]^2} \tag{2.82}$$

where

$$Y_B = B_B = G_B \tag{2.83}$$

The data for transmission lines are usually given in terms of the line resistance R in ohms per mile at a given temperature, the line inductive reactance X_L in ohms per mile at 60 Hz, and the line shunt capacitive reactance X_c in megohms per mile at 60 Hz. Therefore, the line impedance and shunt susceptance in per units for 1 mi of line can be expressed as*

$$\mathbf{Z_{pu}} = (\mathbf{Z}, \Omega/\text{mi})\frac{\text{MVA}_{B(3\phi)}}{[kV_{B(\text{L-L})}]^2}\text{pu} \tag{2.84}$$

where

$$\mathbf{Z} = R + jX_L = Z\angle\theta \ \Omega/\text{mi}$$

and

$$B_{pu} = \frac{[kV_{B(\text{L-L})}]^2 \times 10^{-6}}{[\text{MVA}_{B(3\phi)}][X_c, \text{M}\Omega/\text{mi}]} \tag{2.85}$$

In the event that the admittance for a transmission line is given in microsiemens per mile, the per-unit admittance can be expressed as

$$Y_{pu} = \frac{[kV_{B(\text{L-L})}]^2 \times (Y, \mu\text{S})}{[\text{MVA}_{B(3\phi)}] \times 10^6} \tag{2.86}$$

Similarly, if it is given as reciprocal admittance in megohms per mile, the per-unit admittance can be found as

$$Y_{pu} = \frac{[kV_{B(\text{L-L})}]^2 \times 10^{-6}}{[\text{MVA}_{B(3\phi)}][Z, \text{M}\Omega/\text{mi}]} \tag{2.87}$$

* For further information, see Anderson [63].

Figure 2.3 shows conventional three-phase transformer connections and associated relations between the high-voltage and low-voltage side voltages and currents. The given relations are correct for a three-phase transformer as well as for a three-phase bank of single-phase transformers. Note that, in the figure, n is the turns ratio, that is,

$$n = \frac{N_1}{N_2} = \frac{V_1}{V_2} = \frac{I_2}{I_1} \tag{2.88}$$

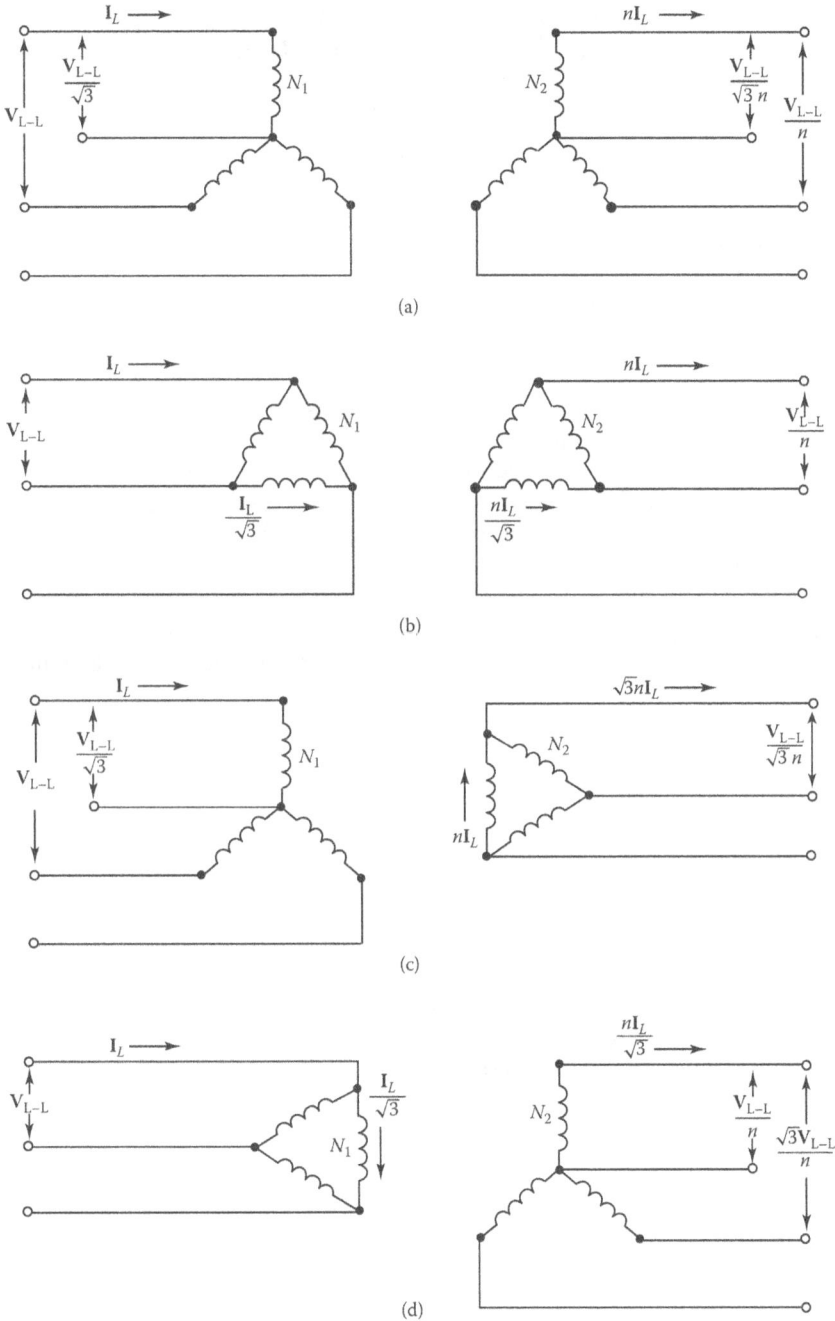

Figure 2.3 Conventional three-phase transformer connections: (a) wye-wye connection; (b) delta-delta connection; (c) wye-delta connection; (d) delta-wye connection.

where the subscripts 1 and 2 are used for the primary and secondary sides. Therefore, an impedance Z_2 in the secondary circuit can be referred to the primary circuit provided that

$$Z_1 = n^2 Z_2 \tag{2.89}$$

Thus, it can be observed from Figure 2.3 that in an ideal transformer, voltages are transformed in the direct ratio of turns, currents in the inverse ratio, and impedances in the direct ratio squared; and power and volt-amperes are, of course, unchanged. Note that, a balanced delta-connected circuit of Z_Δ Ω/phase is equivalent to a balanced wye-connected circuit of Z_Y Ω/phase as long as

$$Z_Y = \frac{1}{3}Z_\Delta \qquad (2.90)$$

The per-unit impedance of a transformer is independent of whether physical impedance values are obtained by referring to the high-voltage side or low-voltage side. This is possible by selecting appropriate bases for each side of the transformer, regardless of whether it is connected in wye-wye, delta-delta, delta-wye, or wye-delta, as long as the same line-to-line voltage ratings are used. Thus, the designated per-unit impedance values of transformers are based on their coil ratings.

Since the ratings of coils remain unaffected by a simple change in connection (e.g., from wye-wye to delta-wye), the per-unit impedance remains the same regardless of the three-phase connection. Although the line-to-line voltage for the transformer may differ, the per-unit impedances calculated in various sections can be combined on one impedance diagram without considering whether the transformers are connected in wye-wye or delta-wye, owing to the method of selecting the base in different sections of the three-phase system.

* This subject has been explained in greater depth in an excellent review by Stevenson [61].

Example 2.2: Given a three-phase transformer with a nameplate rating of 30 MVA, 230Y/69Y kV voltage ratings, 10% leakage reactance, and a wye-wye connection, determine the following using a base of 30 MVA and 230 kV on the high-voltage side:

a. Transformer reactance in per units
b. High-voltage side base impedance
c. Low-voltage side base impedance
d. Transformer reactance referred to the high-voltage side in ohms
e. Transformer reactance referred to the low-voltage side in ohms

Solution:

a. The reactance of the transformer in per units is 10/100, or 0.10 pu. Note that, it is the same whether it is referred to the high-voltage or the low-voltage side.
b. The high-voltage side base impedance is

$$Z_{B(HV)} = \frac{[kV_{B(HV)}]^2}{MVA_{B(3\phi)}}$$
$$= \frac{230^2}{30} = 1763.3333\Omega$$

c. The low-voltage side base impedance is

$$Z_{B(LV)} = \frac{[kV_{B(LV)}]^2}{MVA_{B(3\phi)}}$$
$$= \frac{69^2}{30} = 158.7\Omega$$

d. The reactance referred to the high-voltage side is

$$X_{\Omega(HV)} = X_{pu} \times X_{B(HV)}$$
$$= (0.10)(176.3333) = 176.3333 \ \Omega$$

e. The reactance referred to the low-voltage side is

$$X_{\Omega(LV)} = X_{pu} \times X_{B(LV)}$$
$$= (0.10)(158.7) = 15.87 \ \Omega$$

or, from Equation 2.89,

$$X_{\Omega(LV)} = \frac{X_{\Omega(HV)}}{n^2}$$
$$= \frac{176.3333\Omega}{\left(\frac{230/\sqrt{3}}{69/\sqrt{3}}\right)^2} = \frac{176.3333 \ \Omega}{(3.3333)^2} \cong 15.87 \ \Omega$$

where n is defined as the turns ratio of the windings.

Example 2.3: Suppose a three-phase transformer is rated at 30 MVA with voltage ratings of 230Y kV/69ΔkV and a leakage reactance of 10%, and it is connected in wye-delta. We select a base of 30 MVA and 230 kV on the high-voltage side. The task is to find the turns ratio of windings, the transformer reactance referred to the low-voltage side in ohms, and the transformer reactance referred to the low-voltage side in per units.

a. Determine the turns ratio of the transformer windings.
b. Find the transformer reactance referred to the low-voltage side in ohms.
c. Calculate the transformer reactance referred to the low-voltage side in per units.

Solution:

a. The turns ratio of the windings is

$$n \triangleq \frac{V_{HV(\phi)}}{V_{LV(\phi)}} = \frac{230/\sqrt{3}}{69} = 1.9245$$

b. Since the high-voltage side impedance base is

$$Z_{B(HV)} = \frac{[kV_{B(HV)}]^2}{MVA_{B(3\phi)}}$$
$$= \frac{[230kV]^2}{30} = 1763.3333\Omega$$

and

$$X_{\Omega(HV)} = X_{pu} \times X_{B(HV)}$$
$$= (0.10)(1763.3333\Omega) = 176.3333 \ \Omega$$

Thus, the transformer reactance referred to the delta-connected low-voltage side is

$$X_{\Omega(LV)} = \frac{X_{\Omega(HV)}}{n^2}$$
$$= \frac{176.3333}{(1.9245)^2} = 47.61 \ \Omega = X_\Delta$$

c. From Equation 2.90, the reactance of the equivalent wye connection is

$$Z_Y = \frac{Z_\Delta}{3}$$

$$= \frac{47.61\ \Omega}{3} = 15.87\ \Omega = X'_{\Omega(LV)}$$

where $X'_{\Omega(LV)}$ = reactance per phase at a low voltage of the equivalent wye. Similarly,

$$Z_{B(LV)} = \frac{[kV_{B(LV)}]^2}{MVA_{B(3\phi)}}$$

$$= \frac{69^2}{30} = 158.7\ \Omega$$

Thus,

$$X_{pu} = \frac{X'_{\Omega(LV)}}{Z_{B(LV)}}$$

$$= \frac{15.87\ \Omega}{158.7\ \Omega} = 0.10\text{pu}$$

Alternatively, if the line-to-line voltages are used,

$$X_{(LV)} = \frac{X_{\Omega(HV)}}{n^2}$$

$$= \frac{176.3333\ \Omega}{\left(\frac{230}{69}\right)^2} = 15.87\ \Omega$$

and thus,

$$X_{pu} = \frac{X'_{\Omega((LV)}}{Z_{B(LV)}}$$

$$= \frac{15.87\Omega}{158.7\Omega} = 0.10\text{pu}$$

as before.

Example 2.4: Let us consider the previous example where a three-phase transformer has a name-plate rating of 30 MVA, voltage ratings of 230Y kV/69ΔkV, and a leakage reactance of 10%. Suppose that the transformer connection is changed from delta-wye to wye-wye. Select a base of 30 MVA and 230 kV on the high-voltage side for the delta-wye transformer and solve it by converting to its equivalent wye-wye connection first. Determine the following:

a. Turns ratio of windings
b. Transformer reactance referred to the low-voltage side in ohms
c. Transformer reactance referred to the low-voltage side in per units

Solution:
First converting the delta low-voltage to its corresponding wye low-voltage as

$$\sqrt{3}(69\text{kV}) = 119.5115\text{kV}$$

a. The turns ratio of the windings is

$$n \triangleq \frac{V_{HV(\phi)}}{V_{LV(\phi)}} = \frac{230/\sqrt{3}}{119.5115/\sqrt{3}} = \frac{230}{119.5115} = 1.9245$$

b. Since the high-voltage side impedance base is

$$Z_{B((HV)} = \frac{[kV_{B((HV)}]^2}{MVA_{B(3\phi)}}$$

$$= \frac{[230kV]^2}{30} = 1763.3333\Omega$$

and

$$X_{\Omega(HV)} = X_{pu} \times X_{B(HV)}$$
$$= (0.10)(1763.3333\Omega) = 176.3333\ \Omega$$

Thus, the transformer reactance referred to the delta-connected low-voltage side is

$$X_{\Omega(LV)} = \frac{X_{\Omega(HV}}{n^2}$$
$$= \frac{176.3333}{(1.9245)^2} = 47.61\ \Omega = X_\Delta$$

c. From Equation 2.90, the reactance of the equivalent wye connection is

$$Z_Y = \frac{Z}{3}$$
$$= \frac{47.61\Omega}{3} = 15.87\Omega = X'_{\Omega(LV)}$$

where $X'_{\Omega(LV)}$ = reactance per phase at low-voltage of equivalent wye. Similarly,

$$Z'_{B(LV)} = \frac{[kV_{B(kV)}]^2}{MVA_{B(3\phi)}}$$
$$= \frac{(119.5115)^2}{30} = 476.1\Omega$$

Thus,

$$X_{pu} = \frac{X_{\Omega((LV)}}{Z'_{B((LV)}}$$
$$= \frac{47.61\Omega}{476.1\Omega} = 0.10pu$$

Example 2.5: Resolve Example 2.3 but violate the definition of turns ratio. Use it as the ratio of the line-to-line voltage of the wye-connected primary voltage to the line-to-line voltage of the delta-connected secondary voltage. Since the transformer is rated as 230Y kV/69Δ kV, solve it without converting to its equivalent wye–wye connection first and determine the following:

a. Turns ratio of windings
b. Transformer reactance referred to the low-voltage side in ohms

c. Transformer reactance referred to the low-voltage side in per units

Solution:

a. The turns ratio of the windings is

$$n \triangleq \frac{V_{HV(L-L)}}{V_{LV(L-V)}} = \frac{230}{69} = 3.3333$$

Here, of course, the definition of the turns ratio has been violated.

b. Since the high-voltage side impedance base is

$$Z_{B((HV))} = \frac{[kV_{B((HV))}]^2}{MVA_{B(3\phi)}}$$

$$= \frac{[230kV]^2}{30} = 1763.3333\Omega$$

and

$$X_{\Omega(HV)} = X_{pu} \times X_{B(HV)}$$
$$= (0.10)(1763.3333\Omega) = 176.3333 \ \Omega$$

Thus, the transformer reactance referred to the delta-connected low-voltage side is

$$X_{\Omega(LV)} = \frac{X_{\Omega(HV}}{n^2}$$
$$= \frac{176.3333}{(3.3333)^2} = 15.8703 \ \Omega = X_\Delta$$

c. From Equation 2.90, the reactance of the delta connection is

$$Z_{B((LV))} = \frac{[kV_{B((LV))}]^2}{MVA_{B(3\phi)}}$$
$$= \frac{(69)^2}{30} = 158.7\Omega$$

Thus,

$$X_{pu} = \frac{X_{\Omega((LV))}}{Z_{B(LV)}}$$
$$= \frac{15.8703\Omega}{158.7\Omega} = 0.10pu$$

This method is obviously a shortcut, but one should apply it carefully.

Example 2.6: Figure 2.4 shows a one-line diagram of a three-phase system. Assume that the line length between the two transformers is negligible and the three-phase generator is rated 4160 kVA, 2.4 kV, and 1,000 A and that it supplies a purely inductive load of $I_{pu} = 2.08\angle-90°$ pu. The three-phase transformer T_1 is rated 6000 kVA, 2.4Y–24Y kV, with leakage reactance of 0.04 pu. Transformer T_2 is made up of three single-phase transformers and is rated 4000 kVA, 24Y–12Y kV, with leakage reactance of 0.04 pu. Determine the following for all three circuits, 2.4-, 24-, and 12-kV circuits:

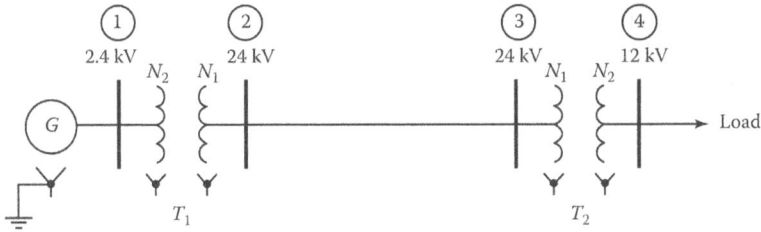

Figure 2.4 One-line diagram for Example 2.4.

a. Base kVA values
b. Base line-to-line kilovolt values
c. Base impedance values
d. Base current values
e. Physical current values (neglect magnetizing currents in transformers and charging currents in lines)
f. Per-unit current values
g. New transformer reactances based on their new bases
h. Per-unit voltage values at buses 1, 2, and 4
i. Per-unit apparent power values at buses 1, 2, and 4
j. Summarize results in a table

Solution:

a. The kVA base for all three circuits is arbitrarily selected as 2080 kVA.
b. The base voltage for the 2.4-kV circuit is arbitrarily selected as 2.5 kV. Since the turns ratios for transformers T_1 and T_2 are

$$\frac{N_1}{N_2} = 1 \quad \text{or} \quad \frac{N_2}{N_1} = 0.10$$

and

$$\frac{N_1'}{N_2'} = 2$$

the base voltages for the 24- and 12-kV circuits are determined to be 25 and 12.5 kV, respectively.

c. The base impedance values can be found as

$$Z_B = \frac{(kV_{B(L-L)})^2(1,000)}{kVA_{B(3\phi)}}$$

$$= \frac{(2.5kV)^2 1,000}{2080kVA} = 3.005\Omega$$

and

$$Z_B = \frac{(25kV)^2 1,000}{2080kVA} = 300.5\Omega$$

and

$$Z_B = \frac{(12.5kV)^2 1,000}{2,080kVA} = 75.1\Omega$$

d. The base current values can be determined as

$$I_B = \frac{kVA_{B(3\phi)}}{\sqrt{3}kV_{B(L-L)}}$$

$$= \frac{2,080kVA}{\sqrt{3}(2.5kV)} = 480 \text{ A}$$

and

$$I_B = \frac{2,080kVA}{\sqrt{3}(25kV)} = 48 \text{ A}$$

and

$$I_B = \frac{2,080kVA}{\sqrt{3}(12.5kV)} = 96 \text{ A}$$

e. The physical current values can be found based on the turns ratios as

$$I = 1,000 \text{ A}$$

$$I = \left(\frac{N_2}{N_1}\right)(1,000 \text{ A}) = 100 \text{ A}$$

$$I = \left(\frac{N_1'}{N_2'}\right)(100 \text{ A}) = 200 \text{ A}$$

f. The per-unit current value is the same, 2.08 pu, for all three circuits.

g. The given transformer reactances can be converted on the basis of their new bases using

$$Z_{pu(new)} = Z_{pu(given)} \left(\frac{kVA_{B(new)}}{kVA_{B(given)}}\right) \left(\frac{kV_{B(given)}}{kVA_{B(new)}}\right)^2$$

Therefore, the new reactances of the two transformers can be found as

$$Z_{pu(T_1)} = j0.04 \left(\frac{2,080kVA}{6,000kVA}\right) \left(\frac{2.4 \text{ kV}}{2.5 \text{ kV}}\right)^2 = j0.0128pu$$

and

$$Z_{pu(T_2)} = j0.04 \left(\frac{2,080kVA}{4,000kVA}\right) \left(\frac{12 \text{ kV}}{12.5 \text{ kV}}\right)^2 = j0.0192pu$$

h. Therefore, the per-unit voltage values at buses 1, 2, and 4 can be calculated as

$$V_1 = \frac{2.4kV\angle 0°}{2.5kV} = 0.96\angle 0°pu$$

$$V_1 = \frac{2.4kV\angle 0°}{2.5kV} = 0.96\angle 0°pu$$

$$V_2 = V_1 - I_{pu}Z_{pu(T_1)}$$
$$= 0.96\angle 0° - (2.08\angle -90°)(0.0128\angle 90°) = 0.9334\angle 0° \text{ pu}$$

$$V_4 = V_2 - I_{pu}Z_{pu(T_2)}$$
$$= 0.9334\angle 0° - (2.08\angle -90°)(0.0192\angle 90°) = 0.8935\angle 0° \text{ pu}$$

Table 2.2
Results of Example 2.4

Quantity	2.4-kV Circuit	24-kV Circuit	12-kV Circuit
$kVA_{B(3\phi)}$	2080 kVA	2080 kVA	2080 kVA
$kV_{B(L-L)}$	2.5 kV	25 kV	12.5 kV
Z_B	3.005 Ω	300.5 Ω	75.1 Ω
I_B	480 A	48 A	96 A
$I_{physical}$	1000 A	100 A	200 A
I_{pu}	2.08 pu	2.08 pu	2.08 pu
V_{pu}	0.96 pu	0.9334 pu	0.8935 pu
S_{pu}	2.00 pu	1.9415 pu	1.8585 pu

i. Thus, the per-unit apparent power values at buses 1, 2, and 4 are

$$S_1 = 2.00 \text{ pu}$$
$$S_2 = V_2 I_{pu} = (0.9334)(2.08) = 1.9415 \text{pu}$$
$$S_4 = V_4 I_{pu} = (0.8935)(2.08) = 1.8585 \text{pu}$$

j. The results are summarized in Table 2.2.

2.5   CONSTANT IMPEDANCE REPRESENTATION OF LOADS

Usually, the power system loads are represented by their real and reactive powers, as shown in Figure 2.5a. However, it is possible to represent the same load in terms of series or parallel combinations of its equivalent constant-load resistance and reactance values, as shown in Figure 2.5b and 2.5c, respectively [63, 64].

In the event that the load is represented by the series connection, the equivalent constant impedance can be expressed as

$$\mathbf{Z}_s = R_s + jX_s \tag{2.91}$$

where

$$R_s = \frac{|\mathbf{V}|^2 \times P}{P^2 + Q^2} \tag{2.92}$$

$$X_s = \frac{|\mathbf{V}|^2 \times Q}{P^2 + Q^2} \tag{2.93}$$

where
  R_s = load resistance in series connection in ohms
  X_s = load reactance in series connection in ohms
  $\mathbf{Z}_s$ = constant-load impedance in ohms
  V = load voltage in volts
  P = real, or average, load power in watts
  Q = reactive load power in VArs

The constant impedance in per units can be expressed as

$$\mathbf{Z}_{pu(s)} = R_{pu(s)} + j\mathbf{X}_{pu(s)} \text{pu} \tag{2.94}$$

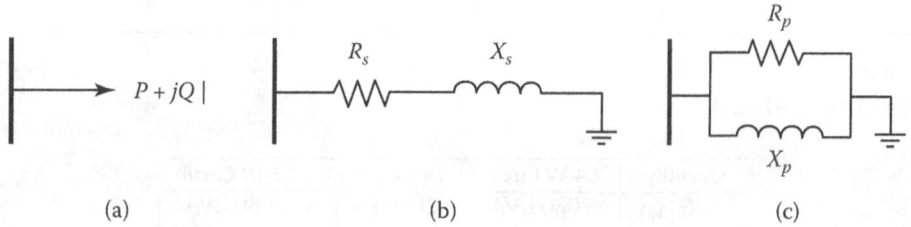

Figure 2.5 Load representations as: (a) real and reactive powers; (b) constant impedance in terms of series combination; (c) constant impedance in terms of parallel combination.

where

$$R_{\text{pu(s)}} = (P_{\text{physical}}) \frac{S_B \times (V_{\text{pu}})^2}{P^2 + Q^2} \text{pu} \tag{2.95}$$

and

$$X_{\text{pu(s)}} = (Q_{\text{physical}}) \frac{S_B \times (V_{\text{pu}})^2}{P^2 + Q^2} \text{pu} \tag{2.96}$$

If the load is represented by the parallel connection, the equivalent constant impedance can be expressed as

$$Z_p = \left(\frac{1}{R_p} + \frac{1}{jX_p} \right)^{-1} = \left(\frac{R_p \times jX_p}{R_p + jX_p} \right) \tag{2.97}$$

where

$$R_p = \frac{V^2}{P}$$

and

$$X_p = \frac{V^2}{Q}$$

where

R_p = load resistance in parallel connection in ohms
X_p = load reactance in parallel connection in ohms
Z_p = constant-load impedance in ohms

The constant impedance in per units can be expressed as

$$Z_{\text{pu(p)}} = j \frac{R_{\text{pu(p)}} \times X_{\text{pu(p)}}}{R_{\text{pu(p)}} + X_{\text{pu(p)}}} \text{pu} \tag{2.98}$$

where

$$\mathbf{R}_{\text{pu(p)}} = \frac{S_B}{P} \left(\frac{V}{V_B} \right)^2 \text{pu} \tag{2.99}$$

or

$$\mathbf{R}_{\text{pu(p)}} = \frac{V_{\text{pu}}^2}{P_{\text{pu}}} \text{pu} \tag{2.100}$$

and

$$\mathbf{X}_{\text{pu(p)}} = \frac{S_B}{Q} \left(\frac{V}{V_B} \right)^2 \text{pu} \tag{2.101}$$

or

$$\mathbf{X}_{\text{pu(p)}} = \frac{V_{\text{pu}}^2}{Q_{\text{pu}}} \text{pu} \tag{2.102}$$

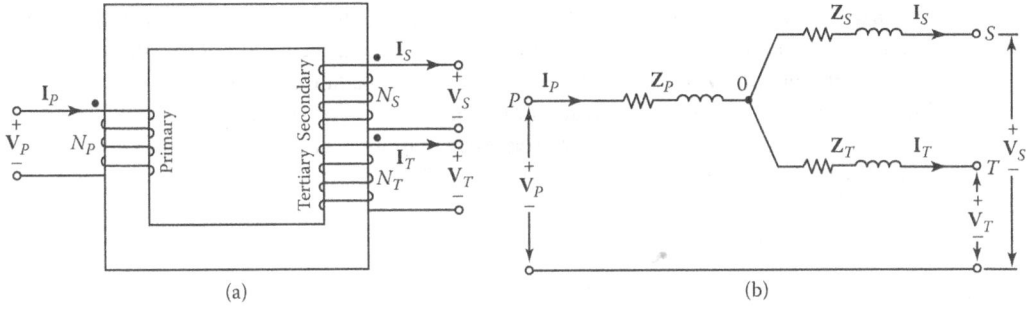

Figure 2.6 Single-phase, three-winding transformer: (a) winding diagram; (b) equivalent circuit.

2.6 THREE-WINDING TRANSFORMERS

In Figure 2.6a, a single-phase three-winding transformer is depicted, which is commonly utilized in bulk power (transmission) substations to lower the transmission voltage to the subtransmission voltage level. Assuming that excitation impedance is disregarded, the three-winding transformer's equivalent circuit can be represented by a wye of impedances, as shown in Figure 2.6b. In this figure, P, S, and T denote the primary, secondary, and tertiary windings, respectively.

It is essential to note that the common point 0 is imaginary and unrelated to the system's neutral. Typically, the tertiary windings of a three-phase and three-winding transformer bank are linked in delta and are employed for (1) offering a path for zero-sequence currents, (2) distributing power within a facility, and (3) utilizing power-factor-correcting capacitors or reactors. By analyzing the short-circuit impedance between pairs of windings with the third winding left open, we can determine the impedance of any of the branches shown in Figure 2.6b. Therefore,

$$Z_{PS} = Z_P + Z_S \tag{2.103a}$$

$$Z_{TS} = Z_T + Z_S \tag{2.103b}$$

$$Z_{PT} = Z_P + Z_T \tag{2.103c}$$

where

$$Z_P = \frac{1}{2}(Z_{PS} + Z_{PT} - Z_{TS}) \tag{2.104a}$$

$$Z_S = \frac{1}{2}(Z_{PS} + Z_{TS} - Z_{PT}) \tag{2.104b}$$

$$Z_T = \frac{1}{2}(Z_{PT} + Z_{TS} - Z_{PS}) \tag{2.104c}$$

where
  Z_{PS} = leakage impedance measured in primary with secondary short-circuited and tertiary open
  Z_{PT} = leakage impedance measured in primary with tertiary short-circuited and secondary open
  Z_{ST} = leakage impedance measured in secondary with tertiary short-circuited and primary open
  Z_P = impedance of primary winding
  Z_S = impedance of secondary winding
  Z_T = impedance of tertiary winding

In the majority of large transformers, the value of Z_S is exceedingly small and may even be negative. Unlike in two-winding transformers, the kVA ratings of the three windings in a three-winding transformer bank are typically not equivalent. As a result, all impedances, as defined earlier, must be expressed on the same kVA basis. In the case of three-phase transformer banks with delta- or wye-connected windings, the positive- and negative-sequence diagrams remain unchanged. However, the corresponding zero-sequence diagrams are illustrated in Figure 5.10.

2.7 AUTOTRANSFORMERS

Figure 2.7a illustrates a typical two-winding transformer. However, the same transformation of voltages, currents, and impedances can also be achieved using the connection depicted in Figure 2.7b when viewed from the terminals. This is why in an autotransformer, only one winding is utilized per phase, with the secondary voltage tapped off the primary winding, as shown in Figure 2.7b. The winding located between the low-voltage terminals is referred to as the *common winding*, while the remaining winding, exclusive to the high-voltage circuit, is called the *series winding*. When combined with the common winding, the series winding forms the *series-common winding* located between the high-voltage terminals [45].

In essence, an autotransformer is essentially a conventional two-winding transformer connected in a specific way. The sole structural difference is that the series winding must be equipped with additional insulation. In a *variable autotransformer*, the tap is adjustable. Autotransformers are becoming more popular in interconnecting two high-voltage transmission lines that operate at varying voltages. An autotransformer has two distinct sets of ratios, namely circuit ratios and winding ratios. In the case of circuit ratios, consider the ideal autotransformer's equivalent circuit (ignoring losses) shown in Figure 2.7b. When viewed from the terminals, the voltage and current ratios can be expressed as:

$$a = \frac{V_1}{V_2} \tag{2.105a}$$

$$a = \frac{N_c + N_s}{N_c} \tag{2.105b}$$

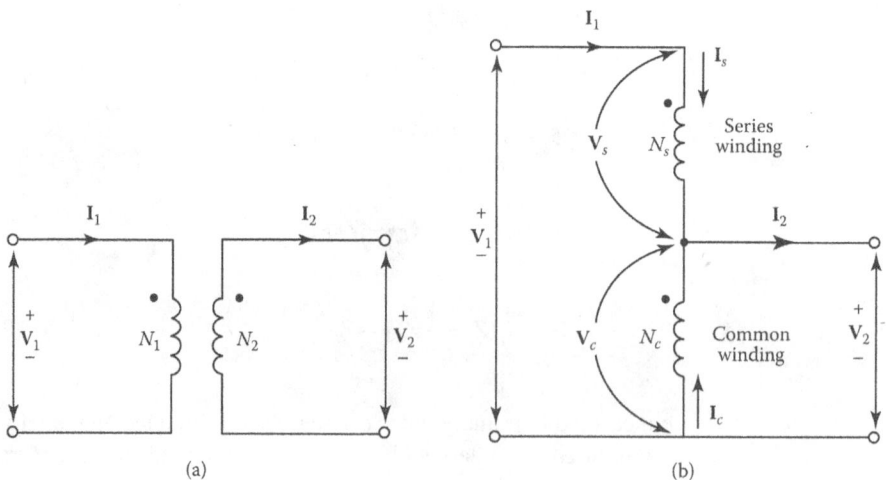

(a)   (b)

Figure 2.7 Schematic diagram of ideal (step-down) transformer connected as: (a) two-winding transformer; (b) autotransformer.

$$= \frac{N_c + N_s}{N_c} \tag{2.105c}$$

and

$$a = \frac{I_2}{I_2} \tag{2.106}$$

Equation 2.105c indicates that the ratio a is always greater than 1. In terms of winding ratios, the voltages and currents of the series and common windings depicted in Figure 2.7b should be taken into account. As a result, the voltage and current ratios can be expressed as:

$$\frac{V_s}{V_c} = \frac{N_s}{N_c} \tag{2.107}$$

and

$$\frac{I_c}{I_s} = \frac{I_2 - I_1}{I_1} \tag{2.108a}$$

$$= \frac{I_2}{I_1} - 1 \tag{2.108b}$$

From Equation 2.105c,

$$\frac{N_s}{N_c} = a - 1 \tag{2.109}$$

Therefore, substituting Equation 2.109 into Equation 2.107 yields

$$\frac{V_s}{V_c} = a - 1 \tag{2.110}$$

Similarly, substituting Equations 2.106 and 2.109 into Equation 2.108b simultaneously yields

$$\frac{I_c}{I_s} = a - 1 \tag{2.111}$$

For an ideal autotransformer, the volt-ampere ratings of circuits and windings can be expressed, respectively, as

$$S_{\text{circuits}} = V_1 I_1 = V_2 I_2 \tag{2.112}$$
$$S_{\text{windings}} = V_s I_s = V_c I_c \tag{2.113}$$

The advantages of autotransformers are lower leakage reactances, lower losses, smaller exciting currents, and less cost than two-winding transformers when the voltage ratio does not vary too greatly from 1 to 1. For example, if the same core and coils are used as a two-winding transformer and as an autotransformer, the ratio of the capacity as an autotransformer to the capacity as a two-winding transformer can be expressed as

$$\frac{\text{Capacity as autotransformer}}{\text{capacity as two-winding transfomer}} = \frac{V_1 I_1}{V_s I_s} = \frac{V_1 I_1}{(V_1 - V_2)I_1} = \frac{a}{a - 1} \tag{2.114}$$

Hence, the greatest advantage is obtained when the difference between the voltages on the two sides is relatively small (for instance, 161 kV/138 kV, 500 kV/700 kV, and 500 kV/345 kV). Consequently, significant reductions in size, weight, and cost can be achieved in comparison to a two-windings per-phase transformer. However, an autotransformer has several drawbacks, such as the absence of electrical isolation between the primary and secondary circuits, and the likelihood of a greater short-circuit current than that of a two-winding transformer.

In general, three-phase autotransformer banks feature wye-connected main windings, with the neutral typically being solidly connected to the earth. Furthermore, it is customary to include a third winding, referred to as the tertiary winding, which is connected in delta.

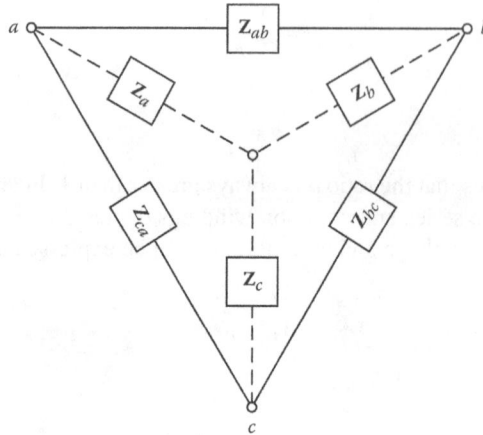

Figure 2.8 Delta-wye or wye-delta transformations.

2.8   DELTA-WYE AND WYE-DELTA TRANSFORMATIONS

The three-terminal circuits encountered so often in networks are the delta and wye* configurations, as shown in Figure 2.8. In some problems, it is necessary to convert delta to wye or vice versa. If the impedances $\mathbf{Z}_{ab}$, $\mathbf{Z}_{bc}$, and $\mathbf{Z}_{ca}$ are connected in delta, the equivalent wye impedances $\mathbf{Z}_a$, $\mathbf{Z}_b$, and $\mathbf{Z}_c$ are

$$\mathbf{Z}_a = \frac{Z_{ab}Z_{ca}}{Z_{ab} + Z_{bc} + Z_{ca}} \tag{2.115}$$

$$\mathbf{Z}_b = \frac{Z_{ab}Z_{bc}}{Z_{ab} + Z_{bc} + Z_{ca}} \tag{2.116}$$

$$\mathbf{Z}_c = \frac{Z_{bc}Z_{ca}}{Z_{ab} + Z_{bc} + Z_{ca}} \tag{2.117}$$

If $\mathbf{Z}_{ab} = \mathbf{Z}_{bc} = \mathbf{Z}_{ca} = \mathbf{Z}$,

$$\mathbf{Z}_a = \mathbf{Z}_b = \mathbf{Z}_c = \frac{\mathbf{Z}}{3} \tag{2.118}$$

* In Europe, it is called the *star configuration*.

On the other hand, if the impedances $\mathbf{Z}_a, \mathbf{Z}_b$, and $\mathbf{Z}_c$ are connected in wye, the equivalent delta impedances $\mathbf{Z}_{ab}, \mathbf{Z}_{bc}$, and $\mathbf{Z}_{ca}$ are

$$\mathbf{Z}_{ab} = \mathbf{Z}_a + \mathbf{Z}_b + \frac{\mathbf{Z}_a\mathbf{Z}_b}{\mathbf{Z}_c} \tag{2.119}$$

$$\mathbf{Z}_{bc} = \mathbf{Z}_b + \mathbf{Z}_c + \frac{\mathbf{Z}_b\mathbf{Z}_c}{\mathbf{Z}_a} \tag{2.120}$$

$$\mathbf{Z}_{ca} = \mathbf{Z}_c + \mathbf{Z}_a + \frac{\mathbf{Z}_c\mathbf{Z}_a}{\mathbf{Z}_b} \tag{2.121}$$

If $\mathbf{Z}_{ab} = \mathbf{Z}_{bc} = \mathbf{Z}_{ca} = \mathbf{Z}$, then

$$\mathbf{Z}_{ab} = \mathbf{Z}_{bc} + \mathbf{Z}_{ca} + 3\mathbf{Z} \tag{2.122}$$

2.9 SHORT-CIRCUIT MVA AND EQUIVALENT IMPEDANCE

When introducing a new circuit to an existing bus in a complex power system, it is often necessary to determine the short-circuit MVA (or kVA) for that bus. This data provides the equivalent impedance of the power system up to that bus. Once the short-circuit MVA has been determined, the short-circuit impedance of the system can be easily calculated. Separate short-circuit MVA values are determined for three-phase faults and single line-to-ground faults.

2.9.1 THREE-PHASE SHORT-CIRCUIT MVA

At a given three-phase bus, three-phase short-circuit MVA can be determined from the following equation:

$$MVA_{sc(3\phi)} = \frac{\sqrt{3}(kV_{L\text{-}L})I_{f(3\phi)}}{1,000} \tag{2.123}$$

where

$$kV_{L\text{-}L} = \text{system line-to-line voltage in kV (rated system voltage)}$$
$$I_{f(3\phi)} = \text{total three-phase fault current in amperes}$$

If three-phase short-circuit MVA is known, the fault current can be determined from

$$I_{f(3\phi)} = \frac{1,000 \, MVA_{sc(3\phi)}}{\sqrt{3}kV_{(L\text{-}L)}} \tag{2.124}$$

then the equivalent impedance can be found as

$$Z_{sc} = \frac{V_{L\text{-}N}}{I_{f(3\phi)}} \, \Omega \tag{2.125a}$$

or

$$Z_{sc} = \frac{1,000 \, kV_{L\text{-}L}}{\sqrt{3}I_{f(3\phi)}} \, \Omega \tag{2.125b}$$

or

$$Z_{sc} = \frac{kV_{L\text{-}L}^2}{MVA_{sc(3\phi)}} \, \Omega \tag{2.125c}$$

Since base impedance is found from

$$Z_B = \frac{kV_{B(3\phi)}^2}{MVA_{B(3\phi)}} \tag{2.126}$$

then the per unit impedance can found from

$$Z_{pu} = \frac{Z_{\Omega(sc)}}{Z_B} \tag{2.127a}$$

or

$$Z_{pu} = \frac{MVA_{3\phi} \times Z_{sc}}{kV_{B(L\text{-}L)}} \tag{2.127b}$$

Thus, the positive-sequence Z to the fault location can be found as

$$Z_{1(\text{pu})} = \frac{\text{MVA}_{B(3\phi)}}{\text{MVA}_{sc}}\text{pu} \tag{2.128}$$

In general,

$$Z_1 = Z_2 \tag{2.129}$$

Also, it is assumed that $Z_1 = X_1$ unless the X/R ratio of the system is known so that the angle can be found.

2.9.2 SINGLE-PHASE-TO-GROUND SHORT-CIRCUIT MVA

At a given single-phase bus, single-phase-to-ground short-circuit MVA can be determined from the following equation:

$$\text{MVA}_{f((\text{SLG})\text{SC}} = \frac{\sqrt{3}(kV_{\text{L-L}})I_{f(\text{SLG})}}{1,000} \tag{2.130}$$

where
  $kV_{\text{L-L}}$ = system line-to-line voltage in kV (rated system voltage)
  $I_{f(\text{SLG})}$ = total single-phase fault current in amperes

If single-Phase short-circuit MVA is known, the fault current can then be determined from

$$I_{f(\text{SLG})} = \frac{1,000\,\text{MVA}_{f(\text{SLG})\text{SC}}}{\sqrt{3}kV_{(\text{L-L})}}\text{A} \tag{2.131}$$

But,

$$I_{f(\text{SLG})} = I_{a0} + I_{a1} + I_{a2} \tag{2.132}$$

or

$$I_{f(\text{SLG})} = \frac{3V_{\text{L-N}}}{Z_0 + Z_1 + Z_2} \tag{2.133}$$

or

$$I_{f(\text{SLG})} = \frac{3V_{\text{L-N}}}{Z_G} \tag{2.134}$$

where

$$Z_G = Z_0 + Z_1 + Z_2 \tag{2.135}$$

From Equations 2.125c and 2.127

$$Z_G = \frac{3kV_{\text{L-L}}^2}{\text{MVA}_{f(\text{SLG})\text{SC}}}\Omega \tag{2.136}$$

and

$$Z_G = \frac{3\text{MVA}_B}{\text{MVA}_{f(\text{SLG})\text{SC}}}\text{pu} \tag{2.137}$$

From

$$Z_0 = Z_G - Z_1 - Z_2 \tag{2.138}$$

or in general,

$$X_0 = X_G - X_1 - X_2 \tag{2.139}$$

since the resistance involved is usually very small with respect to the associated reactance value.

Example 2.7: A short-circuit (fault) study shows that at bus 15 in a 132-kV system, on a 100 MVA base, short-circuit MVA is 710 MVA and the single-line-to-ground short-circuit MVA is 825 MVA. Determine the following:

a. The positive and negative reactances of the system
b. The X_G of the system
c. The zero-sequence reactance of the system

Solution:

a. The positive- and negative-reactances of the system are

$$X_1 = X_2 = \frac{100\text{MVA}}{710\text{MVA}} = 0.1408 \text{ pu}$$

b. The XG of the system is

$$X_G = \frac{300\text{MVA}}{825\text{MVA}} = 0.3636 \text{ pu}$$

c. The zero-sequence of the system is

$$X_0 = 0.3636 - 0.1408 = 0.2228 \text{ pu}$$

Note that, all values above are on a 100-MVA, 132-kV base.

PROBLEMS

1. Assume that the impedance of a line connecting buses 1 and 2 is 50 $\angle 90°\Omega$ and that the bus voltages are 7560 $\angle 10°$ and $7200\angle 0°$ V per phase, respectively. Determine the following:
   a. Real power per phase that is being transmitted from bus 1 to bus 2
   b. Reactive power per phase that is being transmitted from bus 1 to bus 2
   c. Complex power per phase that is being transmitted
2. Solve Problem 1 assuming that the line impedance is $50\angle 26°\Omega$/phase.
3. Verify the following equations:
   a. $V_{pu(L-N)} = V_{pu(L-L)}$

   b. $VA_{pu(1\phi)} = VA_{pu(3\phi)}$

   c. $Z_{pu(Y)} = V_{pu(\Delta)}$
4. Verify the following equations:
   a. Equation 2.24 for a single-phase system
   b. Equation 2.80 for a three-phase system
5. Show that $Z_{B(\Delta)} = 3Z_{B(Y)}$
6. Consider two three-phase transmission lines with different voltage levels that are located side by side in close proximity. Assume that the bases of VA_B, $V_{B(1)}$, and $I_{B(1)}$ and the bases of VA_B, $V_{B(2)}$, and $I_{B(2)}$ are designated for the first and second lines, respectively. If the mutual reactance between the lines is $X_m\Omega$, show that this mutual reactance in per unit can be expressed as

$$X_{pu(m)} = (\text{physical}X_m)\frac{\text{MVA}_B}{(kV_{B(1)})(kV_{B(2)})}$$

7. Consider Example 2.3 and assume that the transformer is connected in delta–wye. Use a 25-MVA base and determine the following:
   a. New line-to-line voltage of low-voltage side
   b. New low-voltage side base impedance
   c. Turns ratio of windings

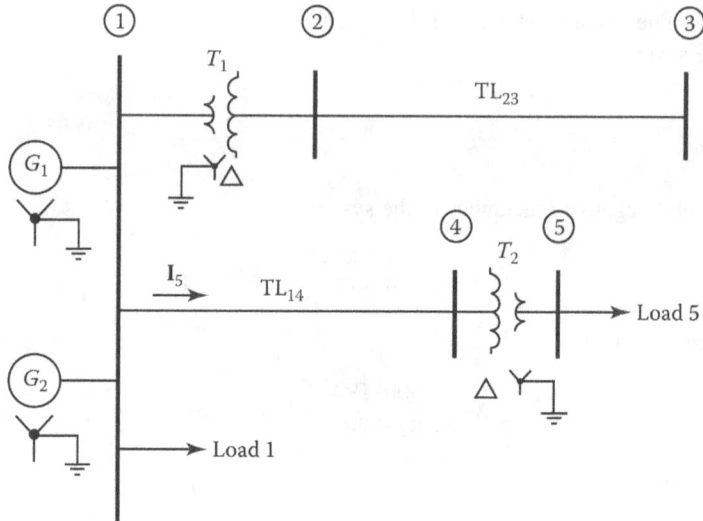

Figure 2.9 One-line diagram for Problem 2.10.

    d. Transformer reactance referred to the low-voltage side in ohms
    e. Transformer reactance referred to the low-voltage side in per units

8. Verify the following equations:
    a. Equation 2.92
    b. Equation 2.93
    c. Equation 2.94
    d. Equation 2.96

9. Verify the following equations:
    a. Equation 2.100
    b. Equation 2.102

10. Consider the one-line diagram given in Figure 2.9. Assume that the three-phase transformer T_1 has nameplate ratings of 15,000 kVA, 7.97/13.8Y – 69Δ kV with leakage impedance of 0.01 + j0.08 pu based on its ratings, and that the three-phase transformer T2 has nameplate ratings of 1500 kVA, 7.97Δ kV – 277/480Y V with leakage impedance of 0.01 + j0.05 pu based on its ratings. Assume that the three-phase generator G_1 is rated 10/12.5 MW/MVA, 7.97/13.8Y kV with an impedance of 0 + j1.10 pu based on its ratings, and that three-phase generator G_2 is rated 4/5 MW/MVA, 7.62/13.2Y kV with an impedance of 0 + j0.90 pu based on its ratings. The transmission line TL_{23} has a length of 50 mi and is composed of 4/0 ACSR (aluminum conductor steel reinforced) conductors with an equivalent spacing (D_m) of 8 ft and has an impedance of 0.445 + j0.976 Ω/mi. Its shunt susceptance is given as 5.78 μS/mi. The line connects buses 2 and 3. Bus 3 is assumed to be an infinite bus, that is, the magnitude of its voltage remains constant at given values and its phase position is unchanged regardless of the power and power factor demands that may be put on it. Furthermore, it is assumed to have a constant frequency equal to the nominal frequency of the system studied. Transmission line TL_{14} connects buses 1 and 4. It has a line length of 2 mi and an impedance of 0.80 + j0.80 Ω/mi.

Because of the line length, its shunt susceptance is assumed to be negligible. The load that is connected to bus 1 has a current magnitude $|I_1|$ of 523 A and a lagging power factor of 0.707. The load that is connected to bus 5 is given as 8000 + j6000 kVA. Use the arbitrarily selected 5000 kVA as the three-phase kVA base and 39.84/69.00 kV as the line-to-neutral and line-to-line voltage base and determine the following:

Table 2.3
Table for Problem 10

Quantity	Nominally 69-kV Circuits	Nominally 13-kV Circuits	Nominally 480-V Circuits
$kVA_{B(3\phi)}$	5000 kVA	5000 kVA	5000 kVA
$kV_{B(L-L)}$	69 kV		
$kV_{B(L-N)}$	39.84 kV		
$I_{B(L)}$			
$I_{B(\phi)}$			
Z_B			
Y_B			

a. Complete Table 2.3 for the indicated values. Note the I_L means line current and I_ϕ means phase currents in delta-connected apparatus.

b. Draw a single-line positive-sequence network of this simple power system. Use the nominal π circuit to represent the 69-kV line. Show the values of all impedances and susceptances in per units on the chosen bases. Show all loads in per unit $P + jQ$.

11. Assume that a 500 + j200-kVA load is connected to a load bus that has a voltage of 1.0∠0° pu. If the power base is 1,000 kVA, determine the per-unit R and X of the load:

a. When load is represented by parallel connection

b. When load is represented by series connection

3 Steady-State Performance of Transmission Lines

3.1 INTRODUCTION

The primary purpose of an overhead three-phase electric power transmission line is to transmit bulk power to load centers and large industrial users that lie beyond the primary distribution lines [65–67]. The modeling of circuits has been extensively explored in numerous books from the 1960s and 1970s, representing significant scholarly progress in the field of transmission circuits and their modeling [43, 59, 64, 68]. A transmission system encompasses all land, conversion structures, and equipment required to provide power from a primary supply source, including lines, switching, and conversion stations that connect a generating or receiving point to a load center or wholesale point. It includes all lines and equipment designed to enhance, integrate, or interconnect power supply sources.

The decision to construct a transmission system is the result of system planning studies that aim to identify the best approach to meeting system requirements. At this stage, the following factors must be taken into consideration and established:

1. Voltage level
2. Conductor type and size
3. Line regulation and voltage control
4. Corona and losses
5. Proper load flow and system stability
6. System protection
7. Grounding
8. Insulation coordination
9. Mechanical design
   - Sag and stress calculations
   - Conductor composition
   - Conductor spacing
   - Insulator and conductor hardware selection
10. Structural design
    - Structure types
    - Stress calculations

The choice of a basic configuration for a transmission line depends on a range of interconnected factors, such as aesthetic considerations, economic viability, performance criteria, company policies and practices, line profile, right-of-way restrictions, preferred materials, and construction techniques. Figure 3.1 depicts typical compact configurations [69], while Figures 3.2–3.5 illustrate typical structures utilized in extra-high-voltage (EHV) transmission systems [62].

3.2 CONDUCTOR SIZE

Conductor sizes are based on the circular mil. A circular mil (cmil) is the area of a circle that has a diameter of 1 mil. A mil is equal to 1×10^{-3} in. The cross-sectional area of a wire in square inches equals its area in circular mils multiplied by 0.7854×10^{-6}. For the smaller conductors, up

DOI: 10.1201/9781003129769-3

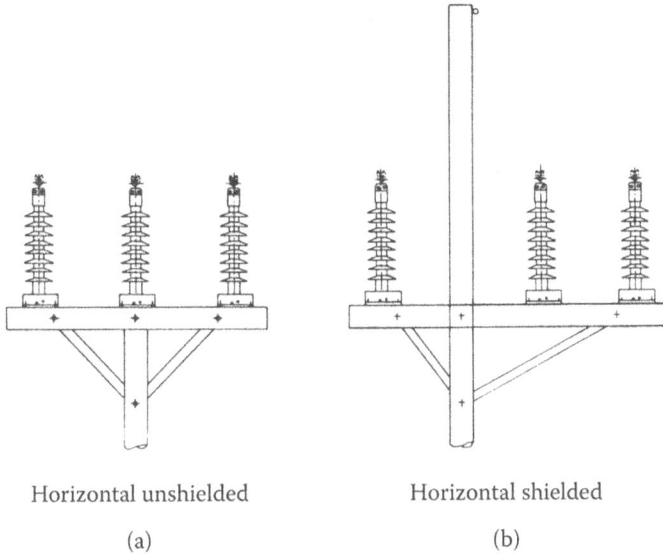

Horizontal unshielded   Horizontal shielded

(a)   (b)

Figure 3.1 Typical compact configurations. (From Elgerd, O. I., *Electric Energy Systems Theory: An Introduction*, McGraw-Hill, New York, 1971.)

to 211,600 cmil, the size is usually given by a gage number according to the American Wire Gauge (AWG) standard, formerly known as the Brown and Sharpe Wire Gauge (B&S).

In the AWG standard, gage sizes decrease as the wire increases in size. (The larger the gage size, the smaller the wire.) These numbers start at 40, the smallest, which is assigned to a wire with a diameter of 3.145 mil. The largest size is number 0000, written as 4/0 and read as "four odds." Above 4/0, the size is determined by cross-sectional area in circular mils. In summary

$$1 \text{ linear mil} = 0.001 \text{ in} = 0.0254 \text{ mm}$$

$$1 \text{ circular mil} = \text{area of circle 1 linear mil in diameter}$$

$$= \frac{\pi}{4} \text{mil}^2$$

$$= \frac{\pi}{4} \times 10^{-6} = 0.7854 \times 10^{-6} \text{in}^2$$

One thousand circular mils is often used as a unit, for example, a size given as 250 kcmil (or MCM) refers to 250,000 cmil. A given conductor may consist of a single strand, or several strands. If a single strand, it is solid; if of more than one strand, it is stranded. A solid conductor is often called a *wire*, whereas a stranded conductor is called a *cable*. A general formula for the total number of strands in concentrically stranded cables is

$$\text{Number of strands} = 3n^2 - 3n + 1$$

where n is the number of layers, including the single center strand.

In general, distribution conductors larger than 2 AWG are stranded. Insulated conductors for underground distribution or aerial cable lines are classified as cables and usually are stranded. Tables in appendices gives standard conductor sizes [45].

Conductors may be selected on the basis of Kelvin's law. According to Kelvin's law,* "the most economical area of conductor is that for which the annual cost of the energy wasted is equal to the interest on that portion of the capital expense which may be considered as proportional to the weight of the conductor" [61]. Therefore,

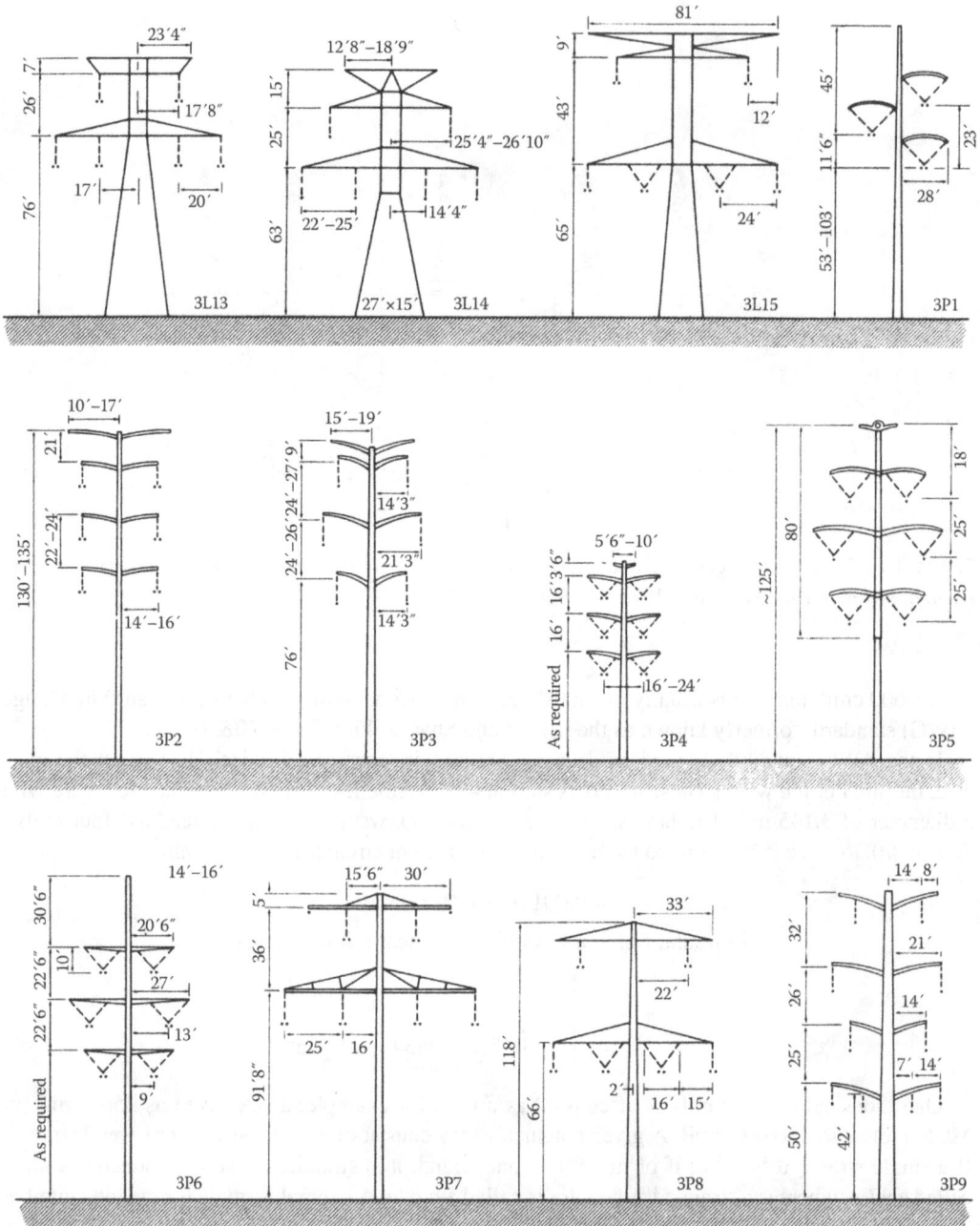

Figure 3.2 Typical pole and lattice structures for 345-kV transmission systems. (From Electric Power Research Institute, "Transmission Line Reference Book: 345kv and Above," 2nd ed., EPRI, Palo Alto, California, 1982.)

* → Expressed by Sir William Thomson (Lord Kelvin) in 1881.

$$\text{Annual cost} = \frac{3CI^2R}{1000} + \frac{p \times w \times a}{1000} \qquad (3.1)$$

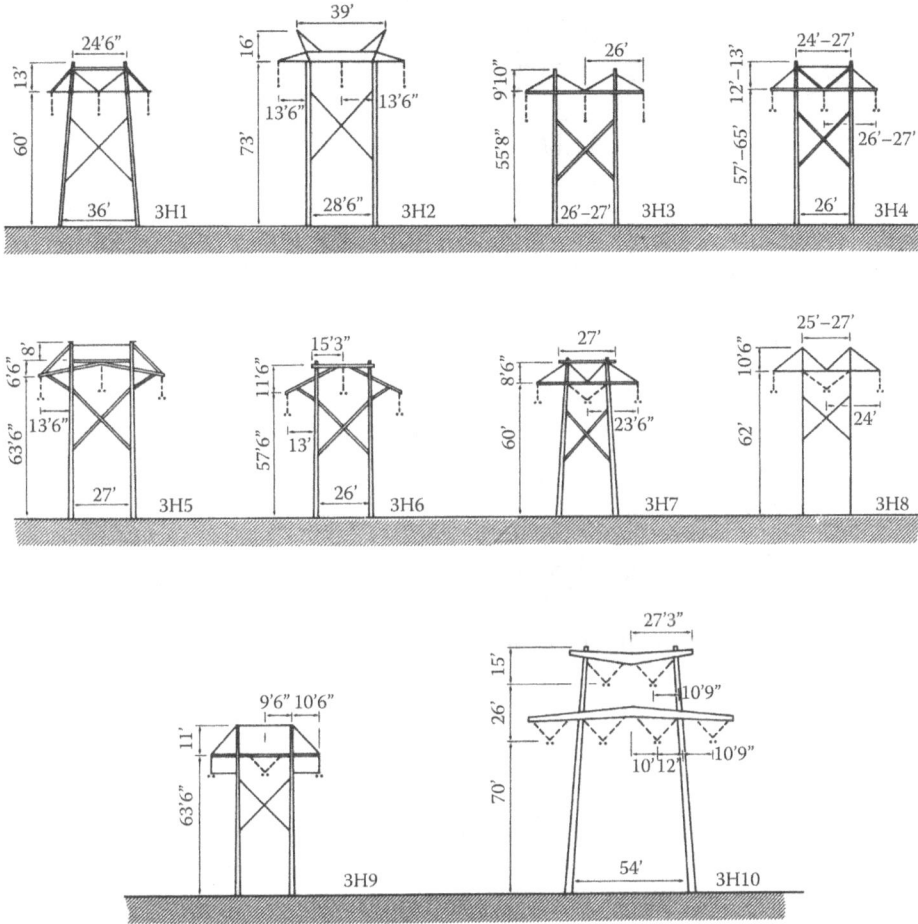

Figure 3.3 Typical wood H-frame structures for 345 kV. (From Electric Power Research Institute, "Transmission Line Reference Book: 345kv and Above," 2nd ed., EPRI, Palo Alto, California, 1982.)

where
    C = cost of energy wasted in dollars per kilowatt-year
    I = current per wire
    R = resistance per mile per conductor
    p = cost per pound conductor
    w = weight per mile of all conductors
    a = percent annual cost of money

Here, the minimum cost is obtained when

$$\frac{3 \times C \times I^2 \times R}{1000} = \frac{p \times w \times a}{1000}$$

However, in practice, Kelvin's law is seldom used. Instead, the I^2R losses are calculated for the total time horizon [70–72]. The conductors used in modern overhead power transmission lines are bare aluminum conductors, which are classified as follows:

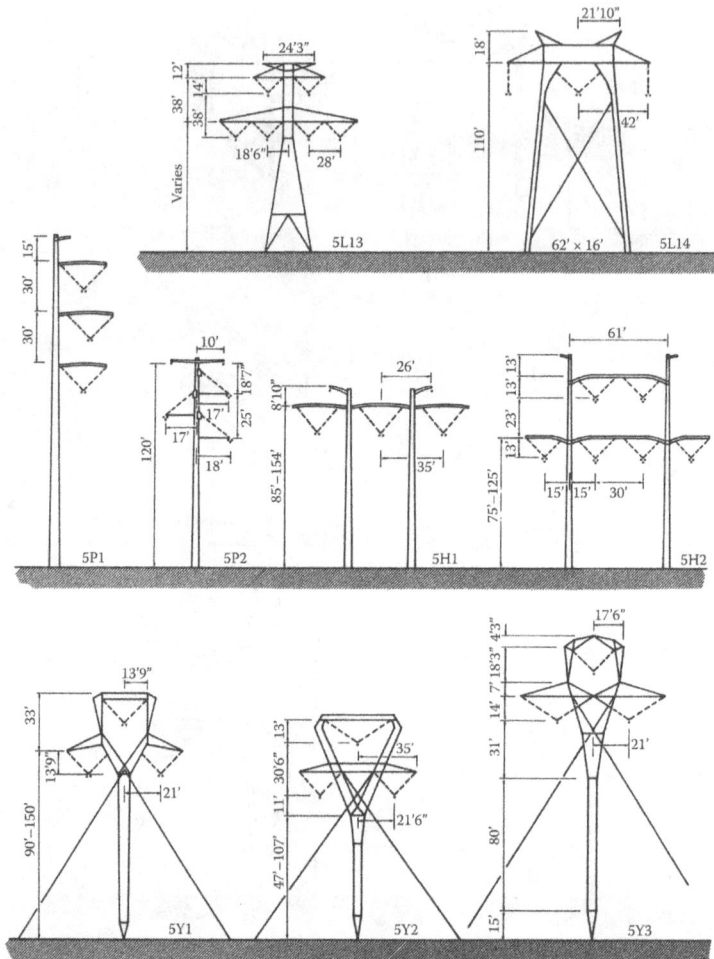

Figure 3.4 Typical 500-kV lattice, pole, H-frame, and guyed Y-type structures. (From Electric Power Research Institute, "Transmission Line Reference Book: 345kv and Above," 2nd ed., EPRI, Palo Alto, California, 1982.)

AAC: all-aluminum conductor
AAAC: all-aluminum-alloy conductor
ACSR: aluminum conductor steel reinforced
ACAR: aluminum conductor alloy reinforced

3.3 TRANSMISSION LINE CONSTANTS

For the purpose of system analysis, a given transmission line can be represented by its resistance, inductance or inductive reactance, capacitance or capacitive reactance, and leakage resistance (which is usually negligible).

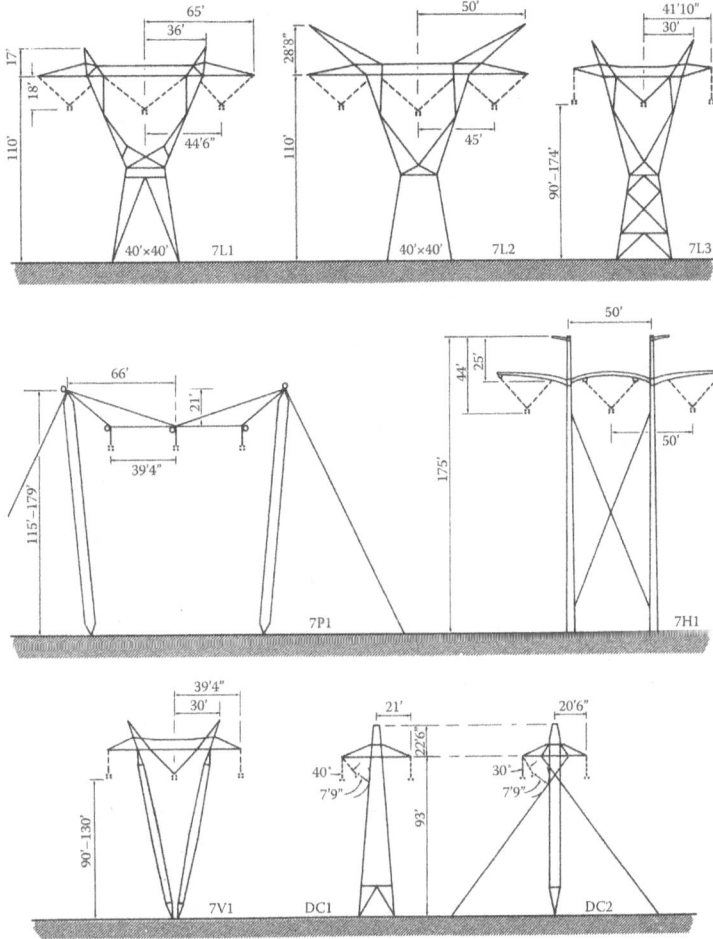

Figure 3.5 Typical dc structures (DC1 and DC2) and 735-800 kV ac designs. (From Electric Power Research Institute, "Transmission Line Reference Book: 345kv and Above," 2nd ed., EPRI, Palo Alto, California, 1982.)

3.4 RESISTANCE

The direct current (dc) resistance of a conductor is

$$R_{dc} = \frac{\rho \times l}{A} \, \Omega$$

where
ρ = conductor resistivity
l = conductor length
A = conductor cross-sectional area

In practice, several different sets of units are used in the calculation of the resistance. For example, in the International System of Units (SI units), l is in meters, A is in square meters, and ρ is in ohm meters. Whereas in power systems in the United States, ρ is in ohm circular mils per foot, l is in feet, and A is in circular mils. The resistance of a conductor at any temperature may be determined by

$$\frac{R_2}{R_1} = \frac{T_0 + t_2}{T_0 + t_1}$$

where
R_1 = conductor resistance at temperature t_1
R_2 = conductor resistance at temperature t_2
t_1, t_2 = conductor temperatures in degrees Celsius
T_0 = constant varying with conductor material
    = 234.5 for annealed copper
    = 241 for hard-drawn copper
    = 228 for hard-drawn aluminum

The phenomenon by which alternating current (ac) tends to flow in the outer layer of a conductor is called the *skin effect*. The skin effect is a function of conductor size, frequency, and the relative resistance of the conductor material.

Tables given in Appendix A provide the dc and ac resistance values for various conductors. The resistances to be used in the positive- and negative-sequence networks are the ac resistances of the conductors.

3.5 INDUCTANCE AND INDUCTIVE REACTANCE

3.5.1 SINGLE-PHASE OVERHEAD LINES

Figure 3.6 shows a single-phase overhead line. Assume that a current flows out in conductor *a* and returns in conductor *b*. These currents cause magnetic field lines that link between the conductors. A change in current causes a change in flux, which in turn results in an induced voltage in the circuit. In an ac circuit, this induced voltage is called the *IX* drop. In going around the loop, if *R* is the resistance of each conductor, the total loss in voltage due to resistance is $2IR$. Thus, the voltage

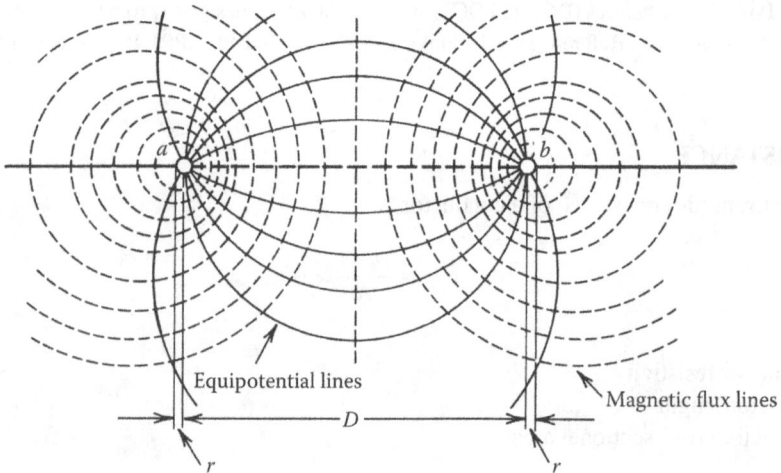

Figure 3.6 Magnetic field of single-phase line.

drop in the single-phase line due to loop impedance at 60 Hz is

$$VD = 2l \left(R + j0.2794 \log_{10} \frac{D_m}{D_s} \right) I \tag{3.2}$$

where
 VD = voltage drop due to line impedance in volts
 l = line length in miles
 R = resistance of each conductor in ohms per mile
 D_m = equivalent or geometric mean distance (GMD) between conductor centers in inches
 D_s = geometric mean radius (GMR) or self-GMD of one conductor in inches, = 0.7788r for cylindrical conductor
 r = radius of cylindrical conductor in inches (see Figure 3.6)
 I = phase current in amperes

Thus, the inductance of the conductor is expressed as

$$L = 2 \times 10^{-7} \frac{D_m}{D_s} \text{ H/m} \tag{3.3}$$

or

$$L = 0.7411 \log_{10} \frac{D_m}{D_s} \text{ mH/mi} \tag{3.4}$$

With the inductance known, the inductive reactance[1] can be found as

$$X_L = 2\pi f L = 2.02 \times 10^{-3} f \ln \frac{D_m}{D_s} \tag{3.5}$$

or

$$X_L = 4.657 \times 10^{-3} f \log_{10} \frac{D_m}{D_s} \tag{3.6}$$

or, at 60 Hz,

$$X_L = 0.2794 \log_{10} \frac{D_m}{D_s} \ \Omega/\text{mi} \tag{3.7}$$

$$X_L = 0.1213 \ln \frac{D_m}{D_s} \ \Omega/\text{mi} \tag{3.8}$$

By using the GMR of a conductor, D_s, the calculation of inductance and inductive reactance can be done easily. Tables give the GMR of various conductors readily.

3.5.2 THREE-PHASE OVERHEAD LINES

In general, the spacings D_{ab}, D_{bc}, and D_{ca} between the conductors of three-phase transmission lines are *not* equal. For any given conductor configuration, the average values of inductance and capacitance can be found by representing the system by one with equivalent equilateral spacing. The "equivalent spacing" is calculated as

$$D_{eq} \triangleq D_m = (D_{ab} \times D_{bc} \times D_{ca})^{1/3} \tag{3.9}$$

In practice, the conductors of a transmission line are transposed, as shown in Figure 3.7. The transposition operation, that is, exchanging the conductor positions, is usually carried out at switching stations.

---

[1] It is also the same as the positive- and negative-sequence of a line.

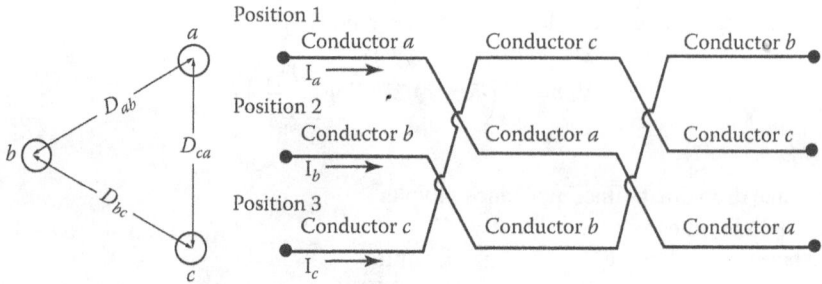

Figure 3.7 Complete transposition cycle of three-phase line.

Therefore, the average inductance per phase is

$$L_a 2 \times 10^{-7} \ln \frac{D_{eq}}{D_s} \text{ H/m} \tag{3.10}$$

or

$$L_a = 0.7411 \log_{10} \frac{D_{eq}}{D_s} \text{mH/mi} \tag{3.11}$$

and the inductive reactance is

$$X_L = 0.1213 \ln \frac{D_m}{D_s} \ \Omega/\text{mi} \tag{3.12}$$

or

$$X_L = 0.2794 \log_{10} \frac{D_m}{D_s} \ \Omega/\text{mi} \tag{3.13}$$

3.6 CAPACITANCE AND CAPACITIVE REACTANCE

3.6.1 SINGLE-PHASE OVERHEAD LINES

Figure 3.8 shows a single-phase line with two identical parallel conductors a and b of radius r separated by a distance D, center to center, and with a potential difference of V_{ab} volts. Let conductors a and b carry charges of $+q_a$ and $-q_b$ farads per meter, respectively. The capacitance between conductors can be found as

$$
\begin{aligned}
C_{ab} &= \frac{q_a}{V_{ab}} \\
&= \frac{2\pi\varepsilon}{\ln\left(\frac{D^2}{r_a \times r_b}\right)} \text{F/m}
\end{aligned} \tag{3.14}
$$

If $r_a = r_b = r$,

$$C_{ab} = \frac{2\pi\varepsilon}{2\ln\frac{D}{r}} \text{F/m} \tag{3.15}$$

Since

$$\varepsilon = \varepsilon_0 \times \varepsilon_r$$

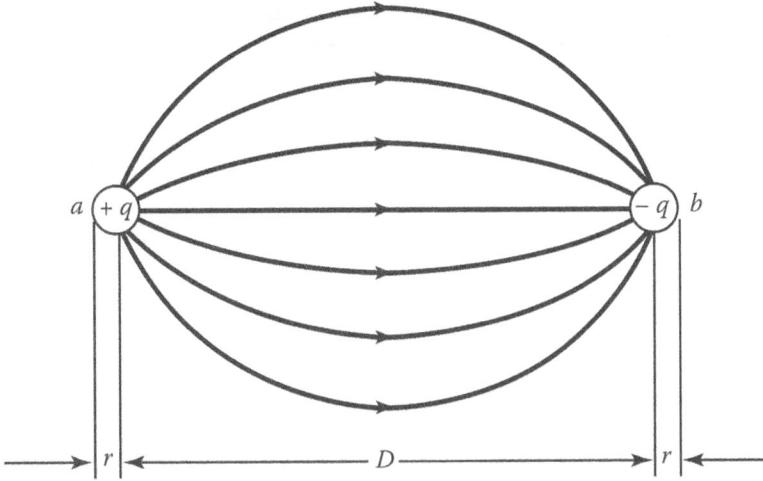

Figure 3.8　Capacitance of single-phase line.

where

$$\varepsilon_0 = \frac{1}{36\pi \times 10^9} = 8.85 \times 10^{-12} \ \text{F/m}$$

and

$$\varepsilon_r = 1 \ \text{for air}$$

Equation 3.15 becomes

$$C_{ab} = \frac{0.0388}{2 \times \log_{10}\left(\frac{D}{r}\right)} \mu\text{F/mi} \tag{3.16}$$

or

$$C_{ab} = \frac{0.0894}{2 \times \ln\left(\frac{D}{r}\right)} \mu\text{F/mi} \tag{3.17}$$

or

$$C_{ab} = \frac{0.0241}{2 \times \log_{10}\left(\frac{D}{r}\right)} \mu\text{F/mi} \tag{3.18}$$

As shown in Figures 3.9 and 3.10, Stevenson [61] explains that the capacitance to neutral or capacitance to ground for the two-wire line is twice the line-to-line capacitance or capacitance between conductors. Therefore, the line-to-neutral capacitance is

$$C_n = C_{an} = C_{bn} = \frac{0.0388}{\log_{10}\left(\frac{D}{r}\right)} [\mu\text{F/mi to neutral}] \tag{3.19}$$

This can then be verified since C_N must equal $2C_{ab}$ so that the capacitance between the conductors is expressed as follows:

$$C_{ab} = \frac{C_n \times C_n}{C_n + C_n}$$

$$= \frac{C_n}{2} \tag{3.20}$$

$$= C_{ab}$$

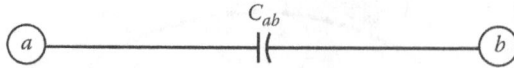

Figure 3.9 Line-to-line capacitance.

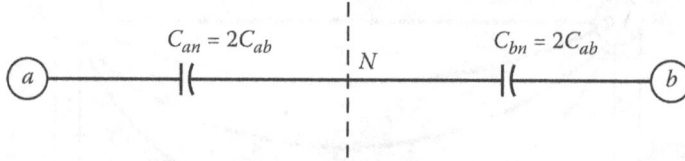

Figure 3.10 Line-to-neutral capacitance.

as demonstrated before. With the capacitance known, the capacitive reactance between one conductor and neutral can be found as

$$X_c = \frac{1}{2\pi \times f \times C_n} \tag{3.21}$$

or, for 60 Hz,

$$X_c = 0.06836 \log_{10} \frac{D}{r} [M\Omega \cdot mi \text{ to neutral}] \tag{3.22}$$

and the line-to-neutral susceptance is

$$b_c = \omega \times C_n$$

or equivalently,

$$b_c = \frac{1}{X_c} \tag{3.23}$$

By relating with reactance of Equation 3.22, the line-to-neutral susceptance is expressed as:

$$b_c = \frac{14.6272}{\log_{10}\left(\frac{D}{r}\right)} [m\Omega/mi \text{ to neutral}] \tag{3.24}$$

The *charging current of the line is*

$$\mathbf{I}_c = j\omega C_{ab} V_{ab} \text{ A/mi} \tag{3.25}$$

3.6.2   THREE-PHASE OVERHEAD LINES

Figure 3.11 shows the cross section of a three-phase line with equilateral spacing D. The line-to neutral capacitance can be found as

$$C_n = \frac{0.0388}{\log_{10}\left(\frac{D}{r}\right)} \mu F/mi \text{ to neutral} \tag{3.26}$$

which is identical to Equation 3.19.

On the other hand, if the spacings between the conductors of the three-phase line are not equal, the line-to-neutral capacitance is

$$C_n = \frac{0.0388}{\log_{10}\left(\frac{D_{eq}}{r}\right)} \mu F/mi \text{ to neutral} \tag{3.27}$$

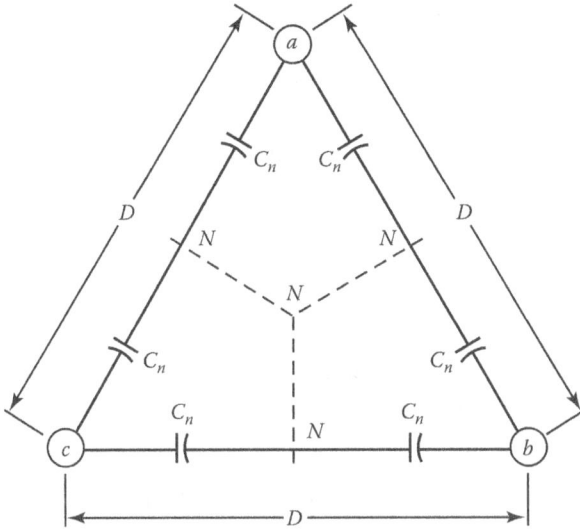

Figure 3.11 Three-phase line with equilateral spacing.

where

$$D_{eq} \triangleq D_m = (D_{ab} \times D_{bc} \times D_{ca})^{1/3}$$

The *charging current per phase* is

$$\mathbf{I}_c = j\omega C_n V_{an} \mathrm{A/mi} \tag{3.28}$$

3.7 TABLES OF LINE CONSTANTS

Tables provide the line constants directly without using equations for calculation. This concept was suggested by W. A. Lewis [57]. According to this concept, Equation 3.8 for inductive reactance at 60 Hz, that is,

$$X_L = 0.1213 \ln\frac{D_m}{D_s} \ \Omega/\mathrm{mi} \tag{3.29a}$$

can be broken down to

$$X_L = 0.1213 \ln\frac{1}{D_s} + 0.1213 \ln D_m \ \Omega/\mathrm{mi} \tag{3.29b}$$

where
D_s = GMR, which can be found from the tables for a given conductor
D_m = GMD between conductor centers

Therefore, Equation 3.29b can be rewritten as

$$X_L = x_a + x_d \ \Omega/\mathrm{mi} \tag{3.30}$$

where

x_a = inductive reactance at 1-ft spacing

$$= 0.1213 \ln\frac{1}{D_s} \; \Omega/\text{mi} \qquad (3.31)$$

x_d = inductive reactance spacing factor

$$= 0.1213 \, D_m \; \Omega/\text{mi} \qquad (3.32)$$

For a given frequency, the value of x_a depends only on the GMR, which is a function of the conductor type. However, x_d depends only on the spacing D_m. If the spacing is greater than 1 ft, x_d has a positive value that is added to x_a. On the other hand, if the spacing is less than 1 ft, x_d has a negative value that is subtracted from x_a. Tables given in Appendix A give x_a and x_d directly.

Similarly, Equation 3.22 for shunt capacitive reactance at 60 Hz, that is,

$$x_c = 0.06836 \log_{10}\frac{D_m}{r} \; \text{M}\Omega \times \text{mi} \qquad (3.33a)$$

can be split into

$$x_c = 0.06836 \log_{10}\frac{1}{r} + 0.06836 \log_{10}D_m \; \text{M}\Omega \times \text{mi} \qquad (3.33b)$$

or

$$x_c = x_a'' + x_a'' \; \text{M} \times \text{mi} \qquad (3.34)$$

where
x_a' = capacitive reactance at 1-ft spacing

$$= 0.06836 \log_{10}\frac{1}{r} \; \text{M}\Omega \times \text{mi} \qquad (3.35)$$

x_d' = capacitive reactance spacing factor

$$= 0.06836 \log_{10}D_m \; \text{M}\Omega \times \text{mi} \qquad (3.36)$$

Tables given in Appendix A provide x_a' and x_d' directly. The term x_d' is added or subtracted from x_a' depending on the magnitude of D_m.

Example 3.1: A three-phase, 60-Hz, transposed line has conductors that are made up of 4/0, seven-strand copper. At the pole top, the distances between conductors, center-to-center, are given as 6.8, 5.5, and 4 ft. The diameter of the conductor copper used is 0.1739 in. Determine the inductive reactances per mile per phase

a. By using Equation 3.29a
b. By using Tables and Equation 3.30

Solution:

a. First calculating the equivalent spacing for the pole top,

$$D_{eq} = D_m = (D_{ab} \times 5.5 \times 4)^{1/3}$$
$$= (6.8 \times 5.5 \times 4)^{1/3} = 5.3086 \text{ ft}$$

From Table A.1, $D_s = 0.01579$ ft for the conductor. Hence, its inductive reactance is

$$X_L = 0.1213 \ln \frac{5.3086 \text{ ft}}{0.01579 \text{ ft}}$$

$$= 0.1213 \ln \frac{5.3086 \text{ ft}}{0.01579 \text{ ft}}$$

$$= 0.705688 \ \Omega/\text{mi} \cong 0.7057 \ \Omega/\text{mi}$$

b. From Table A.1, $x_a = 0.503$ Ω/mi and from Table A.8, for $D_{eq} = 5.30086$ ft, by linear interpolation, $x_d = 0.2026$ Ω/mi. Thus, the inductive reactance is

$$X_L = x_a + x_d$$

$$= 0.503 + 0.2026$$

$$= 0.7056 \ \Omega/\text{mi}$$

Example 3.2: Consider the pole-top configuration given in Example 3.1. If the line length is 100 mi, determine the shunt capacitive reactance by using

a. Equation 3.33a
b. Using tables in Appendix B

Solution:

a. By using the Equation 3.33a,

$$X_c = 0.06836 \log_{10} \frac{D_m}{r}$$

$$= 0.06836 \log_{10} \frac{5.3086 \text{ ft}}{\left(\frac{0.522}{2 \times 12}\right) \text{ ft}}$$

$$= 0.06836 \log_{10} \frac{1}{\left(\frac{0.522}{2 \times 12}\right) \text{ ft}} + 0.06836 \log_{10}(5.3086 \text{ ft})$$

$$= 0.113651284 + 0.49559632$$

$$\cong 0.163211 \text{ M}\Omega \times \text{mi}$$

b. From Table A.1, $x'_a = 0.1136\text{M} \times \text{mi}$ and from Table A.9, $x'_d = 0.049543 \text{ M }\Omega \times \text{mi}$ Hence, from Equation 3.33b

$$X_c = X'_a + X'_d$$

$$= 0.1136 + 0.049543$$

$$= 0.163143 \text{ M}\Omega \times \text{mi}$$

c. The capacitive reactance of the 100-mi-long line is

$$X_c = \frac{x_c}{l}$$

$$= \frac{0.163143 \text{ M}\Omega \times \text{mi}}{100 \text{ mi}}$$

$$= 1.63143 \times 10^{-3} \text{ M}\Omega$$

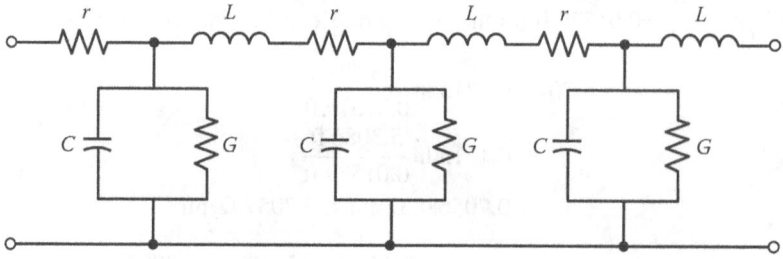

Figure 3.12 Distributed constant equivalent circuit of line.

3.8 EQUIVALENT CIRCUITS FOR TRANSMISSION LINES

A distributed constant circuit, as depicted in Figure 3.12, can be used to represent an overhead line or an underground cable. The circuit's resistance, inductance, capacitance, and leakage conductance are distributed uniformly throughout the line's length. In the figure, L denotes the inductance of a line conductor to neutral per unit length, r represents the AC resistance of a line conductor per unit length, C is the capacitance of a line conductor to neutral per unit length, and G represents the leakage conductance per unit length [58].

3.9 TRANSMISSION LINES

This section provides a brief overview of transmission system modeling. Transmission lines are modeled and classified based on their length as follows:

1. Short-transmission lines
2. Medium-length transmission lines
3. Long transmission lines

3.9.1 SHORT TRANSMISSION LINES (UP TO 50 MI OR 80 KM)

Transmission lines are categorized based on their length. Short transmission lines have lengths up to 50 miles or 80 kilometers, medium-length transmission lines have lengths up to 150 miles or 240 kilometers, and long transmission lines have lengths greater than 150 miles or 240 kilometers.

Modeling a short transmission line is relatively straightforward. Its shunt capacitance is so small that it can be ignored with minimal loss of accuracy. Since the current remains constant throughout the line, its shunt admittance is also neglected. This means that the capacitance and leakage resistance (or conductance) to the earth are generally ignored, as depicted in Figure 3.13. Therefore, the transmission line can be treated as a simple, lumped, and constant impedance circuit.

$$\begin{aligned}
\mathbf{Z} &= R + jX_L \\
&= \mathbf{z}l \\
&= rl + jxl \ \Omega
\end{aligned} \tag{3.37}$$

where
  $\mathbf{Z}$ = total series impedance per phase in ohms
  z = series impedance of one conductor in ohms per unit length
  X_L = total inductive reactance of one conductor in ohms
  x = inductive reactance of one conductor in ohms per unit length

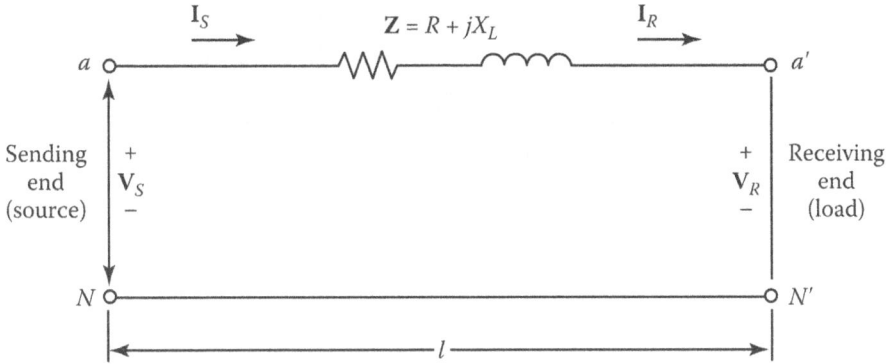

Figure 3.13   Equivalent circuit of short transmission line.

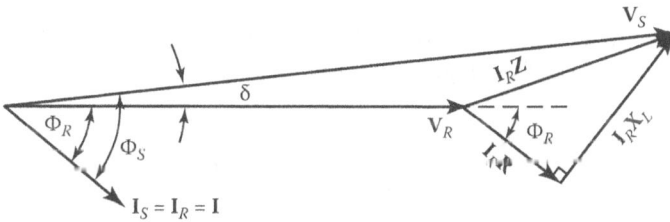

Figure 3.14   Phasor diagram of short transmission line to inductive load.

l = length of line

Vector (or phasor) diagrams for a short transmission line connected to an inductive load and a capacitive load are presented in Figures 3.14 and 3.15, respectively. These figures indicate that the current entering the line at the sending end is equal to the current leaving at the receiving end as follows:

$$\mathbf{V}_S = \mathbf{V}_R + \mathbf{I}_R\mathbf{Z} \tag{3.38}$$

$$\mathbf{I}_S = \mathbf{I}_R = \mathbf{I} \tag{3.39}$$

$$\mathbf{V}_R = \mathbf{V}_S - \mathbf{I}_R\mathbf{Z} \tag{3.40}$$

where
$\mathbf{V}_S$ = sending-end phase (line-to-neutral) voltage
$\mathbf{V}_R$ = receiving-end phase (line-to-neutral) voltage
$\mathbf{I}_S$ = sending-end phase current
$\mathbf{I}_R$ = receiving-end phase current
$\mathbf{Z}$ = total series impedance per phase

Thus, using $\mathbf{V}_R$ as the reference, Equation 3.38 can be written as

$$\mathbf{V}_S = \mathbf{V}_R + (I_R\cos\theta_R \pm I_R\sin\theta_R)(R + jX) \tag{3.41}$$

where the plus or minus sign is determined by θ_R, the power factor angle of the receiving end or load. If the power factor is lagging, the minus sign is employed. On the other hand, if it is leading, the plus sign is used.

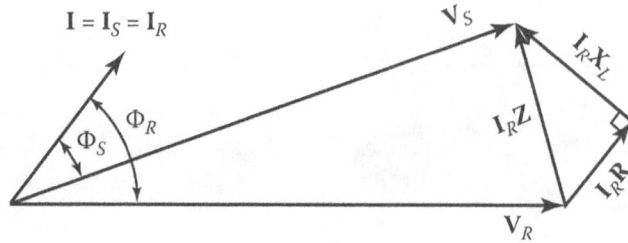

Figure 3.15 Phasor diagram of short transmission line connected to capacitive load.

However, if Equation 3.40 is used, it is convenient to use $\mathbf{V}_S$ as the reference. Therefore,

$$\mathbf{V}_R = \mathbf{V}_S - (I_R\cos\theta_R \pm I_R\sin\theta_R)(R + jX) \tag{3.42}$$

where θ_S is the sending-end power factor angle, that determines, as before, whether the plus or minus sign will be used. Also, from Figure 3.14, using $\mathbf{V}_R$ as the reference vector,

$$V_S = \sqrt{[(V_R + I \cdot R\cos\theta_R + I \cdot X\cos\theta_R)^2 + (I \cdot X\cos\theta_R \pm I \cdot R\sin\theta_R)^2]} \tag{3.43}$$

and load angle

$$\delta = \theta_S - \theta_R \tag{3.44}$$

or

$$\delta = \tan^{-1}\left(\frac{I \cdot X\cos\theta_R \pm I \cdot R\sin\theta_R}{V_R + I \cdot R\cos\theta_R + I \cdot X\cos\theta_R}\right) \tag{3.45}$$

The generalized constants, or **ABCD** parameters, can be determined by inspection of Figure 3.13. Since

$$\begin{bmatrix} \mathbf{V}_S \\ \mathbf{I}_S \end{bmatrix} = \begin{bmatrix} \mathbf{A} & \mathbf{B} \\ \mathbf{C} & \mathbf{D} \end{bmatrix}\begin{bmatrix} \mathbf{V}_R \\ \mathbf{I}_R \end{bmatrix} \tag{3.46}$$

and **AD – BC** = 1, where

$$\mathbf{A} = 1 \quad \mathbf{B} = \mathbf{Z} \quad \mathbf{C} = 0 \quad \mathbf{D} = 1 \tag{3.47}$$

then

$$\begin{bmatrix} \mathbf{V}_S \\ \mathbf{I}_S \end{bmatrix} = \begin{bmatrix} 1 & \mathbf{Z} \\ 0 & 1 \end{bmatrix}\begin{bmatrix} \mathbf{V}_R \\ \mathbf{I}_R \end{bmatrix} \tag{3.48}$$

and

$$\begin{bmatrix} \mathbf{V}_R \\ \mathbf{I}_R \end{bmatrix} = \begin{bmatrix} 1 & \mathbf{B} \\ 0 & 1 \end{bmatrix}^{-1}\begin{bmatrix} \mathbf{V}_S \\ \mathbf{I}_S \end{bmatrix} = \begin{bmatrix} 1 & -\mathbf{Z} \\ 0 & 1 \end{bmatrix}\begin{bmatrix} \mathbf{V}_S \\ \mathbf{I}_S \end{bmatrix}$$

The transmission efficiency of the short line can be expressed as

$$\begin{aligned} \eta &= \frac{\text{output}}{\text{input}} \\ &= \frac{\sqrt{3}V_R I \cos\theta_R}{\sqrt{3}V_S I \cos\theta_S} \\ &= \frac{V_R \cos\theta_R}{V_S \cos\theta_S} \end{aligned} \tag{3.49}$$

Equation 3.49 is applicable whenever the line is single phase. The transmission efficiency can also be expressed as

$$\eta = \frac{\text{output}}{\text{output} + \text{losses}}$$

For a single-phase line,

$$\eta = \frac{V_R I \cos \theta_R}{V_R I \cos \theta_R + 2I^2 R} \tag{3.50}$$

For a three-phase line,

$$\eta = \frac{\sqrt{3} V_R I \cos \theta_R}{\sqrt{3} V_R I \cos \theta_R + 3I^2 R} \tag{3.51}$$

3.9.2 STEADY-STATE POWER LIMIT

Suppose the impedance of a short transmission line is denoted as $\mathbf{Z} = Z \angle \theta$. Under this assumption, the real power delivered to the receiving end of the transmission line at steady-state can be represented as:

$$P_R = \frac{V_S \times V_R}{Z} \cos(\theta - \delta) - \frac{V_R^2}{Z} \cos \theta \tag{3.52}$$

and similarly, the reactive power delivered can be expressed as

$$Q_R = \frac{V_S \times V_R}{Z} \sin(\theta - \delta) - \frac{V_R^2}{Z} \sin \theta \tag{3.53}$$

Equations 3.52 and 3.53 provide values for P_R and Q_R per phase if V_S and V_R are the line-to-neutral voltages. Multiplying the resulting P_R and Q_R values by 3, or using the line-to-line values of V_S and V_R, will give the real and reactive power delivered to a balanced load at the receiving end of the line in three-phase form.

If Equation 3.52 is modified so that all variables are held constant except for δ, then P_R becomes a function of δ only. In this case, P_R will be at its maximum when $\delta = \theta$, and the maximum power that can be obtained at the receiving end for a given regulation can be expressed as the steady-state power limit.

$$P_{R,\max} = \frac{V_R^2}{Z^2} \left(\frac{V_S}{V_R} Z - R \right) \tag{3.54}$$

where V_S and V_R are the phase (line-to-neutral) voltages whether the system is single phase or three phase. The equation can also be expressed as

$$P_{R,\max} = \frac{V_S \times V_R}{Z} \frac{V_R^2 \times \cos \theta}{Z} \tag{3.55}$$

If $V_S = V_R$,

$$P_{R,\max} = \frac{V_R^2}{Z} (1 - \cos \theta) \tag{3.56}$$

or

$$P_{R,\max} = \left(\frac{V_R}{Z} \right)^2 (Z - R) \tag{3.57}$$

and similarly, the corresponding reactive power delivered to the load is given by

$$Q_{R,\max} = -\frac{V_R^2}{Z} \sin \theta \tag{3.58}$$

Both Equations 3.57 and 3.58 are independent of the V_S voltage, as can be observed. Equation 3.52 includes a negative sign indicating that the load is a sink of *leading VArs*[2] – that is, the load absorbs reactive power from the supply or produces reactive power that is lagging. The total three-phase power transmitted on the line is three times the power calculated using the above equations. If the voltage is given in volts, the power is expressed in watts or vars, and if the voltage is given in kilovolts, the power is expressed in megawatts or megavars.

The vars that are recognized, printed on varmeter scale plates, bought, and sold are only magnetizing vars. Therefore, in the following sections, the leading or lagging vars will be referred to as *magnetizing vars*. Similarly, the real and reactive powers for the sending end of a transmission line can be expressed as follows:

$$P_S = \frac{V_S^2}{Z}\cos\theta - \frac{V_S \times V_R}{Z}\cos(\theta + \delta) \tag{3.59}$$

and

$$Q_S = \frac{V_S^2}{Z}\sin\theta - \frac{V_S \times V_R}{Z}\sin(\theta + \delta) \tag{3.60}$$

If, in Equation 3.59, as before, all variables are kept constant with the exception of δ, so that the real power at the sending end, P_S, is a function of δ only, P_S is a maximum when

$$\theta + \delta = 180°$$

Therefore, the maximum power at the sending end, the maximum input power, can be expressed as

$$P_{S,\max} = \frac{V_S^2}{Z}\cos\theta + \frac{V_S \times V_R}{Z} \tag{3.61}$$

or

$$P_{S,\max} = \frac{V_S^2 \times R}{Z^2} + \frac{V_S \times V_R}{Z} \tag{3.62}$$

However, if $V_S = V_R$,

$$P_{S,\max} = \left(\frac{V_S}{Z}\right)^2 (Z + R) \tag{3.63}$$

and similarly, the corresponding reactive power at the sending end, the maximum input vars, is given by

$$Q_S = \frac{V_S^2}{Z}\sin\theta \tag{3.64}$$

As can be observed, both Equations 3.63 and 3.64 are independent of V_R voltage, and Equation 3.64 has a positive sign this time.

3.9.3 PERCENT VOLTAGE REGULATION

The voltage regulation of the line is defined by the increase in voltage when full load is removed, that is,

$$\text{Percentage of voltage regulation} = \frac{|\mathbf{V}_S| - |\mathbf{V}_R|}{|\mathbf{V}_R|} \times 100 \tag{3.65}$$

or

$$\text{Percentage of voltage regulation} = \frac{|\mathbf{V}_{R,\text{NL}}| - |\mathbf{V}_{R,\text{FL}}|}{|\mathbf{V}_{R,\text{FL}}|} \times 100 \tag{3.66}$$

---

[2]For many decades, the electrical utility industry has not recognized two different kinds of reactive power, *leading* and *lagging VArs*.

where

$|\mathbf{V}_S|$ = magnitude of sending-end phase (line-to-neutral) voltage at no load

$|\mathbf{V}_R|$ = magnitude of receiving-end phase (line-to-neutral) voltage at full load

$|\mathbf{V}_{R,\text{NL}}|$ = magnitude of receiving-end voltage at no load

$|\mathbf{V}_{R,\text{FL}}|$ = magnitude of receiving-end voltage at full load with constant $|\mathbf{V}_S|$

Therefore, if the load is connected at the receiving end of the line,

$$|\mathbf{V}_S| = |\mathbf{V}_{R,\text{NL}}|$$

and

$$|\mathbf{V}_R| = |\mathbf{V}_{R,\text{NL}}|$$

An approximate expression for percentage of voltage regulation is

$$\text{Percentage of voltage regulation} \cong I_R \left(\frac{R \cos\phi_R \pm X \sin\phi_R}{V_R} \right) \times 100 \qquad (3.67)$$

Example 3.3: A three-phase, 60-Hz overhead short transmission line has a line-to-line voltage of 23 kV at the receiving-end, a total impedance of $2.48 + j6.57$ Ω/phase, and a load of 9 MW with a receivingend lagging power factor of 0.85.

a. Calculate the line-to-neutral and line-to-line voltages at the sending end.
b. Calculate the load angle.

Solution:

METHOD 1. USING COMPLEX ALGEBRA:

a. The line-to-neutral reference voltage is

$$\mathbf{V}_{R(\text{L-N})} = \frac{\mathbf{V}_{R(\text{L-L})}}{\sqrt{3}}$$

$$= \frac{23 \times 10^3 \angle 0°}{\sqrt{3}} = 13,294 \angle 0° \text{ V}$$

The line current is

$$\mathbf{I} = \frac{9 \times 10^6}{\sqrt{3} \times 23 \times 10^3 \times 0.85} \times (0.85 - j0.527)$$

$$= 265.8(0.85 - j0.527)$$

$$= 225.83 - j140.08 \text{ A}$$

Therefore,

$$\mathbf{IZ} = (225.93 - j140.08)(2.48 + j6.57)$$

$$= (265.8\angle - 31.8°)(7.02\angle 69.32°)$$

$$= 1866.8\angle 37.52° \text{ V}$$

Thus, the line-to-neutral voltage at the sending end is

$$\mathbf{V}_{S(\text{L-N})} = \mathbf{V}_{R(\text{L-N})} + \mathbf{IZ}$$

$$= 14,803\angle 4.4° \text{ V}$$

The line-to-line voltage at the sending end is

$$V_{S(L\text{-}L)} = \sqrt{3}V_{S(L\text{-}N)}$$
$$= 25,640\angle 4.4° + 30° = 25,640\angle 34.4° \text{ V}$$

b. The load angle is 4.4°.

METHOD 2. USING THE CURRENT AS THE REFERENCE PHASOR:

a.

$$V_R \cos \theta_R + IR = 13,279.06 \times 0.85 + 265.8 \times 2.48 = 11,946 \text{ V}$$
$$V_R \sin \theta_R + IX = 13,294.8 \times 0.527 + 266.1 \times 6.57 = 8744 \text{ V}$$

then

$$V_{S(L\text{-}N)} = \sqrt{11,946.32^2 + 8,744^2} + 14,803 \text{ V/phase}$$

$$V_{S(L\text{-}L)} = 25,640 \text{ V}$$

b.

$$\theta_S = \theta_R + \delta = \tan^{-1}\left(\frac{8744}{11,946}\right) = 36.2°$$

$$\delta = \theta_S = \theta_R = 36.2 - 31.8 = 4.4°$$

METHOD 3. USING THE RECEIVING-END VOLTAGE AS THE REFERENCE PHASOR:

a.

$$V_{S(L\text{-}N)} = \sqrt{[(V_R + I \cdot R \cos \theta_R + I \cdot X \sin \theta_R)^2 + (I \cdot X \cos \theta_R + I \cdot R \sin \theta_R)^2]}$$

$$I \cdot R \cos \theta_R = 265.8 \times 2.48 \times 0.85 = 560.3$$
$$I \cdot R \sin \theta_R = 265.8 \times 2.48 \times 0.527 = 347.4$$
$$I \cdot X \cos \theta_R = 265.8 \times 6.57 \times 0.85 = 1484.4$$
$$I \cdot X \sin \theta_R = 265.8 \times 6.57 \times 0.527 = 920.3$$

Therefore,

$$V_{S(L\text{-}N)} = \sqrt{[(13,279 + 560.3 + 920.3)^2 + (1484.4 - 347.4)^2]}$$
$$= \sqrt{14,759.7^2 + 1137^2}$$
$$= 14,803 \text{ V}$$

b.

$$V_{S(L\text{-}L)} = \sqrt{3}V_{S(L\text{-}L)} = 25,640 \text{ V}$$

$$\delta = \tan^{-1}\left(\frac{1137}{14,759.7}\right) = 4.4°$$

METHOD 4. USING POWER RELATIONSHIPS:

The power loss in the line is

$$P_{\text{loss}} = 3I^2R$$
$$= 3 \times 265.8^2 \times 2.48 \times 10^{-6} = 0.526 \text{ MW}$$

The total input power to the line is

$$P_T = P + P_{\text{loss}}$$
$$= 9 + 0.526 = 9.526 \text{ MW}$$

The var loss in the line is

$$Q_{\text{loss}} = 3I^2X$$
$$= 3 \times 265.8^2 \times 6.57 \times 10^{-6} = 1.393 \text{ MVAr lagging}$$

The total megavar input to the line is

$$Q_T = \frac{P \sin \theta_R}{\cos \theta_R} + Q_{\text{loss}}$$
$$= \frac{9 \times 0.526}{0.85} + 1.393 = 6.973 \text{ MVAr lagging}$$

The total megavolt-ampere input to the line is

$$S_T = \sqrt{(P_T^2 + Q_T^2)}$$
$$= (9.526^2 + 6.973^2)^{1/2} = 11.81 \text{ MVA}$$

a.

$$V_{S(L\text{-}L)} = \frac{S_T}{\sqrt{3}I}$$
$$= \frac{11.81 \times 10^6}{\sqrt{3} \times 265.8} = 25,640 \text{ V}$$
$$V_{S(L\text{-}L)} = \frac{V_{S(L\text{-}L)}}{\sqrt{3}} = 14,803 \text{ V}$$

b.

$$\cos \theta_S = \frac{P_T}{S_T} = \frac{9,526}{11.81} = 0.807 \text{ lagging}$$

Therefore,

$$\theta_s = 36.2°$$
$$\delta = 36.2° - 31.8° = 4.4°$$

METHOD 5. TREATING THE THREE-PHASE LINE AS A SINGLE-PHASE LINE AND HAVING V_S AND v_R REPRESENT LINE-TO-LINE VOLTAGES, NOT LINE-TO-NEUTRAL VOLTAGES:

a. The power delivered is 4.5 MW

$$I_{\text{line}} = \frac{4.5 \times 10^6}{23 \times 10^3 \times 0.85} = 230.18 \text{ A}$$

$$R_{\text{loop}} = 2 \times 2.48 = 4.96 \ \Omega$$

$$X_{\text{loop}} = 2 \times 6.57 = 13.14 \ \Omega$$

$$V_R \cos \theta_R = 23 \times 10^3 \times 0.85 = 19{,}550 \text{ V}$$

$$V_R \sin \theta_R = 23 \times 10^3 \times 0.527 = 12{,}121 \text{ V}$$

$$I \cdot R = 230.18 \times 4.96 = 1141.7 \text{ V}$$

$$I \cdot X = 230.18 \times 13.14 = 3024.6 \text{ V}$$

Therefore,

$$\begin{aligned}
V_{S(\text{L-L})} &= \sqrt{(V_R \cos \theta_R + I \cdot R)^2 + (V_R \sin \theta_R + I \cdot X)^2} \\
&= \sqrt{(19{,}550 + 1{,}141.7)^2 + (12{,}121 + 3{,}024.6)^2} \\
&= \sqrt{20{,}691.7^2 + 15{,}145.6^2} \\
&= 25{,}640 \text{ V}
\end{aligned}$$

Thus,

$$V_{S(\text{L-N})} = \frac{V_{S(\text{L-N})}}{\sqrt{3}} = 14{,}803 \text{ V}$$

b.

$$\theta_s = \tan^{-1} \frac{15{,}145.6}{20{,}691.7} = 36.20°$$

and

$$\delta = 36.2° - 31.8° = 4.4°$$

Example 3.4: Calculate percentage of voltage regulation for the values given in Example 3.3

1. Using Equation 3.65
2. Using Equation 3.66

Solution:

1. Using Equation 3.67,

$$\begin{aligned}
\text{Percentage of voltage regulation} &= \frac{|\mathbf{V}_S| - |\mathbf{V}_R|}{|\mathbf{V}_R|} \times 100 \\
&= \frac{14{,}803 - 13{,}279.06}{13{,}279.06} \times 100 \\
&= 11.5
\end{aligned}$$

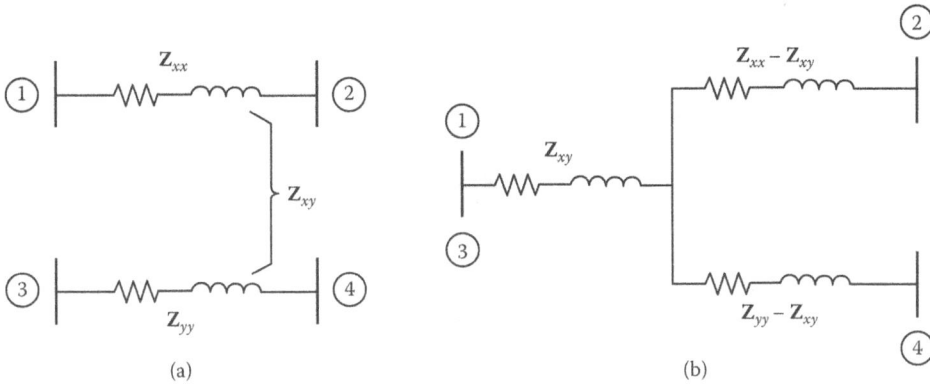

Figure 3.16   Representation of mutual impedance between two circuits.

2. Using Equation 3.67,

$$\text{Percentage of voltage regulation} \cong I_R \times \frac{(R\cos\theta_R \pm X\sin\theta_R)}{V_R} \times 100$$

$$= 265.8 \left(\frac{2.48 \times 0.85 + 6.57 \times 0.527}{13,279.06} \right) \times 100$$

$$= 11.1$$

3.9.4   REPRESENTATION OF MUTUAL IMPEDANCE OF SHORT LINES

Figure 3.16a shows a circuit of two lines, x and y, that have self-impedances of $\mathbf{Z}_{xx}$ and $\mathbf{Z}_{yy}$ and a mutual impedance of $\mathbf{Z}_{zy}$. Its equivalent circuit is shown in Figure 3.16b. Sometimes, it may be required to preserve the electrical identity of the two lines, as shown in Figure 3.17. The mutual impedance $\mathbf{Z}_{xy}$ can be in either line and transferred to the other by means of a transformer that has a 1:1 turns ratio. This technique is also applicable for three-phase lines.

Example 3.5: Assume that the mutual impedance between two parallel feeders is $0.09 + j0.3$ Ω/mi per phase. The self-impedances of the feeders are $0.604 \angle 50.4°$ and $0.567 \angle 52.9°$ Ω/mi per phase, respectively. Represent the mutual impedance between the feeders as shown in Figure 3.16b.

Solution:

$$\mathbf{Z}_{xy} = 0.09 + j0.3\ \Omega$$

$$\mathbf{Z}_{xx} = 0.604\angle 50.4° = 3.85 + j0.465\ \Omega$$

$$\mathbf{Z}_{yy} = 0.567\angle 52.9° = 0.342 + j0.452\ \Omega$$

Therefore,

$$\mathbf{Z}_{xx} - \mathbf{Z}_{xy} = 0.295 + j0.165\ \Omega$$
$$\mathbf{Z}_{yy} - \mathbf{Z}_{xy} = 0.252 + j0.152\ \Omega$$

Hence, the resulting equivalent circuit is shown in Figure 3.18.

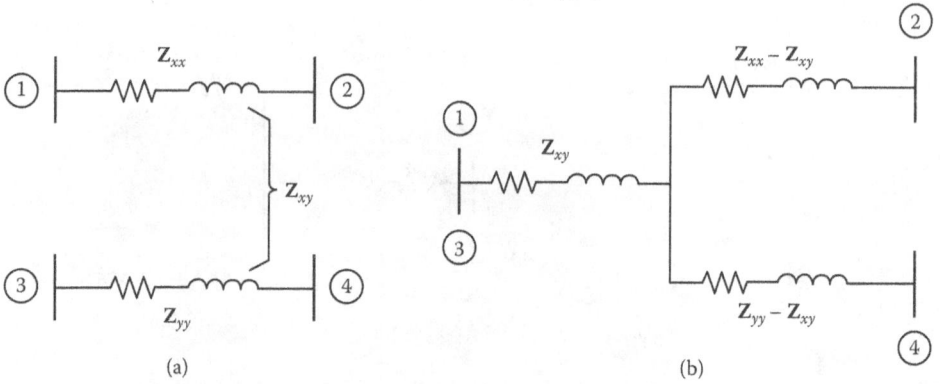

Figure 3.17 Representation of mutual impedance between two circuits by means of 1:1 transformer.

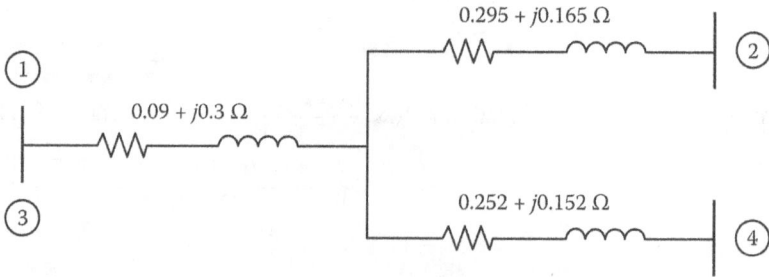

Figure 3.18 Resultant equivalent circuit for Example 3.5.

3.10   MEDIUM-LENGTH TRANSMISSION LINES (UP TO 150 MI OR 240 KM)

As the line length and voltage increase, the use of the formulas developed for the short transmission lines give inaccurate results. Thus, the effect of the current leaking through the capacitance must be taken into account for a better approximation. Thus, the shunt admittance is "lumped" at a few points along the line and represented by forming either a T or a π network, as shown in Figures 3.19 and 3.20. In the figures,

$$\mathbf{Z} = \mathbf{z} \cdot l$$

For the T circuit shown in Figure 3.19,

$$\mathbf{V}_S = \mathbf{I}_S \times \frac{1}{2}\mathbf{Z} + \mathbf{I}_R \times \frac{1}{2}\mathbf{Z} + \mathbf{V}_R$$

$$= \left[\mathbf{I}_R + \left(\mathbf{V}_R + \mathbf{I}_R \times \frac{1}{2}\mathbf{Z} \right) \mathbf{Y} \right] \frac{1}{2}\mathbf{Z} + \mathbf{V}_R + \mathbf{I}_R \frac{1}{2}\mathbf{Z}$$

or

$$\mathbf{V}_S = \underbrace{\left(1 + \frac{1}{2}\mathbf{ZY} \right)}_{A} \mathbf{V}_R + \underbrace{\left(\mathbf{Z} + \frac{1}{4}\mathbf{YZ}^2 \right)}_{B} \mathbf{I}_R \tag{3.68}$$

and

$$\mathbf{I}_S = \mathbf{I}_R + \left(\mathbf{V}_R + \mathbf{I}_R \times \frac{1}{2}\mathbf{Z} \right) \mathbf{Y}$$

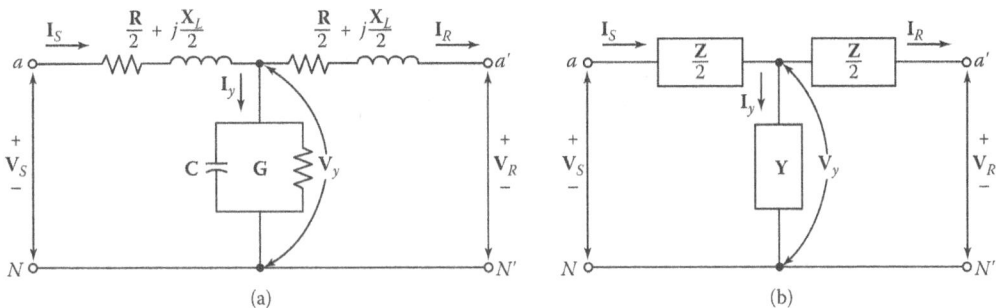

Figure 3.19 Nominal T circuit.

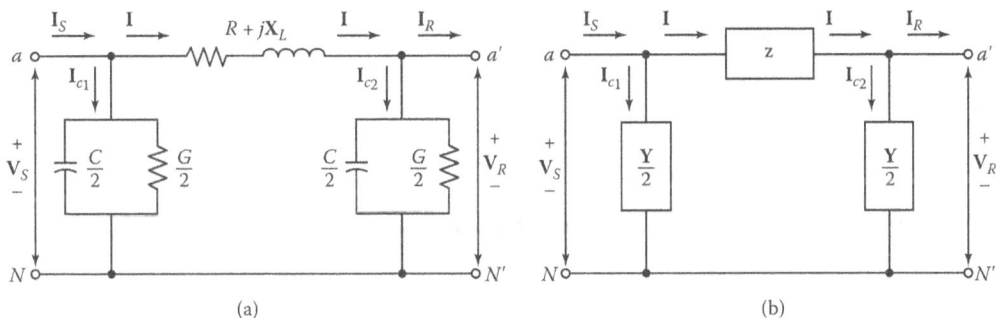

Figure 3.20 Nominal π circuit.

or

$$\mathbf{I}_S = \underbrace{\mathbf{Y}}_{\mathbf{C}} \times \mathbf{V}_R + \underbrace{\left(1 + \frac{1}{2}\mathbf{ZY}\right)}_{\mathbf{D}} \mathbf{I}_R \tag{3.69}$$

Alternatively, neglecting conductance so that

$$\mathbf{I}_C = \mathbf{I}_Y$$

and

$$\mathbf{V}_C = \mathbf{V}_Y$$

yields

$$\mathbf{I}_C = \mathbf{V}_C \times \mathbf{Y}$$

$$\mathbf{V}_C = \mathbf{V}_R + \mathbf{I}_R \times \frac{1}{2}\mathbf{Z}$$

Hence,

$$\mathbf{V}_S = \mathbf{V}_C + \mathbf{I}_S \times \frac{1}{2}\mathbf{Z}$$

$$= \mathbf{V}_R + \mathbf{I}_R \times \frac{1}{2}\mathbf{Z} + \left[\mathbf{V}_R\mathbf{Y} + \mathbf{I}_R\left(1 + \frac{1}{2}\mathbf{YZ}\right)\right]\left(\frac{1}{2}\mathbf{Z}\right)$$

or

$$V_S = \left(1 + \frac{1}{2}YZ\right)V_R + \underbrace{\left(Z + \frac{1}{4}YZ^2\right)}_{B}I_R \qquad (3.70)$$

where $\underbrace{\left(1 + \frac{1}{2}YZ\right)}_{A}$ and $\underbrace{\left(Z + \frac{1}{4}YZ^2\right)}_{B}$

Also,

$$\begin{aligned} I_S &= I_R + I_C \\ &= I_R + V_C \times Y \\ &= I_R + \left(V_R + I_R \times \frac{1}{2}Z\right)Y \end{aligned}$$

Again,

$$I_S = \underbrace{Y}_{C} \times V_R + \underbrace{\left(1 + \frac{1}{2}ZY\right)}_{D}I_R \qquad (3.71)$$

Since

$$A = 1 + \frac{1}{2}YZ \qquad (3.72)$$

$$B = Z + \frac{1}{4}YZ^2 \qquad (3.73)$$

$$C = Y \qquad (3.74)$$

$$D = 1 + \frac{1}{2}YZ \qquad (3.75)$$

for a nominal T circuit, the *general circuit parameter matrix*, or *transfer matrix*, becomes

$$\begin{bmatrix} A & B \\ C & D \end{bmatrix} = \begin{bmatrix} 1 + \frac{1}{2}YZ & Z + \frac{1}{4}YZ^2 \\ Y & 1 + \frac{1}{2}YZ \end{bmatrix}$$

Therefore,

$$\begin{bmatrix} V_S \\ I_S \end{bmatrix} = \begin{bmatrix} 1 + \frac{1}{2}YZ & Z + \frac{1}{4}YZ^2 \\ Y & 1 + \frac{1}{2}YZ \end{bmatrix}\begin{bmatrix} V_R \\ I_R \end{bmatrix} \qquad (3.76)$$

and

$$\begin{bmatrix} V_R \\ I_R \end{bmatrix} = \begin{bmatrix} 1 + \frac{1}{2}YZ & Z + \frac{1}{4}YZ^2 \\ Y & 1 + \frac{1}{2}YZ \end{bmatrix}\begin{bmatrix} V_S \\ I_S \end{bmatrix} \qquad (3.77)$$

For the π circuit shown in Figure 3.20,

$$V_S = \left(V_R \times \frac{1}{2}Y + I_R\right)Z + V_R$$

or

$$V_S = \underbrace{\left(1 + \frac{1}{2}YZ\right)}_{A}V_R + \underbrace{Z}_{B} \times I_R \qquad (3.78)$$

and

$$I_S = \frac{1}{2}Y \times V_S + \frac{1}{2}Y \times V_R + I_R \qquad (3.79)$$

By substituting Equation 3.78 into Equation 3.79,

$$\mathbf{I}_S = \left[\left(1 + \frac{1}{2}\mathbf{YZ}\right)\mathbf{V}_R + \mathbf{ZI}_R\right]\frac{1}{2}\mathbf{Y} + \frac{1}{2}\mathbf{Y} \times \mathbf{V}_R + \mathbf{I}_R$$

or

$$\mathbf{I}_S = \underbrace{\left(\mathbf{Y} + \frac{1}{4}\mathbf{Y}^2\mathbf{Z}\right)}_{\mathbf{C}}\mathbf{V}_R + \underbrace{\left(1 + \frac{1}{2}\mathbf{YZ}\right)}_{\mathbf{D}}\mathbf{I}_R \tag{3.80}$$

Alternatively, *neglecting conductance*,

$$\mathbf{I} = \mathbf{I}_{C2} + I_R$$

where

$$\mathbf{I}_{C2} = \frac{1}{2}\mathbf{Y} \times \mathbf{V}_R$$

yields

$$\mathbf{I} = \frac{1}{2}\mathbf{Y} \times \mathbf{V}_R + \mathbf{I}_R \tag{3.81}$$

Also,

$$\mathbf{V}_S = \mathbf{V}_R + \mathbf{IZ} \tag{3.82}$$

By substituting Equation 3.81 into Equation 3.82,

$$\mathbf{V}_S = \mathbf{V}_R + \left(\frac{1}{2}\mathbf{Y} \times \mathbf{V}_R + \mathbf{I}_R\right)\mathbf{Z}$$

or

$$\mathbf{V}_S = \underbrace{\left(1 + \frac{1}{2}\mathbf{YZ}\right)}_{\mathbf{A}}\mathbf{V}_R + \underbrace{\mathbf{Z}}_{\mathbf{B}} \times \mathbf{I}_R \tag{3.83}$$

and

$$\mathbf{I}_{C1} = \frac{1}{2}\mathbf{Y} \times \mathbf{V}_S \tag{3.84}$$

By substituting Equation 3.83 into Equation 3.84,

$$\mathbf{I}_{C1} = \frac{1}{2}\mathbf{Y} \times \left(1 + \frac{1}{2}\mathbf{YZ}\right)\mathbf{V}_R + \frac{1}{2}\mathbf{Y} \times \mathbf{ZI}_R \tag{3.85}$$

and since

$$\mathbf{I}_S = \mathbf{I} + \mathbf{I}_{C1} \tag{3.86}$$

by substituting Equation 3.81 into Equation 3.86,

$$\mathbf{I}_S = \frac{1}{2}\mathbf{YV}_R + \mathbf{I}_R + \frac{1}{2}\mathbf{Y}\left(1 + \frac{1}{2}\mathbf{YZ}\right)\mathbf{V}_R + \frac{1}{2}\mathbf{YZI}_R$$

or

$$\mathbf{I}_S = \underbrace{\left(\mathbf{Y} + \frac{1}{4}\mathbf{Y}^2\mathbf{Z}\right)}_{\mathbf{C}}\mathbf{V}_R + \underbrace{\left(1 + \frac{1}{2}\mathbf{YZ}\right)}_{\mathbf{D}}\mathbf{I}_R \tag{3.87}$$

Since

$$\mathbf{A} = 1 + \frac{1}{2}\mathbf{YZ} \tag{3.88}$$

$$\mathbf{B} = \mathbf{Z} \tag{3.89}$$

$$\mathbf{C} = \mathbf{Y} + \frac{1}{4}\mathbf{Y}^2\mathbf{Z} \tag{3.90}$$

$$\mathbf{D} = 1 + \frac{1}{2}\mathbf{Y}\mathbf{Z} \tag{3.91}$$

for a nominal π circuit, the general circuit parameter matrix becomes

$$\begin{bmatrix} \mathbf{A} & \mathbf{B} \\ \mathbf{C} & \mathbf{D} \end{bmatrix} = \begin{bmatrix} 1 + \frac{1}{2}\mathbf{Y}\mathbf{Z} & \mathbf{Z} \\ \mathbf{Y} + \frac{1}{4}\mathbf{Y}^2\mathbf{Z} & 1 + \frac{1}{2}\mathbf{Y}\mathbf{Z} \end{bmatrix} \tag{3.92}$$

Therefore,

$$\begin{bmatrix} \mathbf{V}_S \\ \mathbf{I}_S \end{bmatrix} = \begin{bmatrix} 1 + \frac{1}{2}\mathbf{Y}\mathbf{Z} & \mathbf{Z} \\ \mathbf{Y} + \frac{1}{4}\mathbf{Y}^2\mathbf{Z} & 1 + \frac{1}{2}\mathbf{Y}\mathbf{Z} \end{bmatrix} \begin{bmatrix} \mathbf{V}_R \\ \mathbf{I}_R \end{bmatrix} \tag{3.93}$$

and

$$\begin{bmatrix} \mathbf{V}_R \\ \mathbf{I}_R \end{bmatrix} = \begin{bmatrix} 1 + \frac{1}{2}\mathbf{Y}\mathbf{Z} & \mathbf{Z} \\ \mathbf{Y} + \frac{1}{4}\mathbf{Y}^2\mathbf{Z} & 1 + \frac{1}{2}\mathbf{Y}\mathbf{Z} \end{bmatrix} \begin{bmatrix} \mathbf{V}_S \\ \mathbf{I}_S \end{bmatrix} \tag{3.94}$$

As can be proved easily by using a delta-wye transformation, the nominal T and nominal π circuits are not equivalent to each other. This result is to be expected since two different approximations are made to the actual circuit, neither of which is absolutely correct. More accurate results can be obtained by splitting the line into several segments, each given by its nominal T or nominal π circuits and cascading the resulting segments.

Here, the power loss in the line is given as

$$P_{\text{loss}} = I^2 R \tag{3.95}$$

which varies approximately as the square of the through-line current. The reactive powers absorbed and supplied by the line are given as

$$Q_L = Q_{\text{absorbed}} = I^2 X_L \tag{3.96}$$

and

$$Q_C = Q_{\text{supplied}} = V^2 b \tag{3.97}$$

respectively. The Q_L varies approximately as the square of the through line current, whereas the Q_C varies approximately as the square of the mean line voltage. The result is that increasing transmission voltages decrease the reactive power absorbed by the line for heavy loads and increase the reactive power supplied by the line for light loads.

The percentage of voltage regulation for the medium-length transmission lines is given by Stevenson [61] as

$$\text{Percentage of voltage regulation} = \frac{\frac{|\mathbf{V}_{S,\text{LN}}|}{|\mathbf{A}|} - |\mathbf{V}_{R,\text{FL}}|}{|\mathbf{V}_{R,\text{FL}}|} \times 100 \tag{3.98}$$

where

$\quad$ $|\mathbf{V}_{S,\text{LN}}|$ = magnitude of sending-end phase (line-to-neutral) voltage

$\quad$ $|\mathbf{V}_{S,\text{FL}}|$ = magnitude of receiving-end phase (line-to-neutral) voltage at full load with constant $|\mathbf{V}_S|$

$\quad$ $|\mathbf{A}|$ = magnitude of line constant $\mathbf{A}$

Example 3.6: A three-phase 138-kV transmission line is connected to a 49-MW load at a 0.85 lagging power factor. The line constants of the 52-mi-long line are $\mathbf{Z} = 95\angle78°\Omega$ and $\mathbf{Y} = 0.001\angle90°S$. Using *nominal T circuit representation*, calculate the

a. $\mathbf{A}$, $\mathbf{B}$, $\mathbf{C}$, and $\mathbf{D}$ constants of the line
b. Sending-end voltage
c. Sending-end current
d. Sending-end power factor
e. Efficiency of transmission

Solution:

$$V_{R(L\text{-}N)} = \frac{138\text{kV}}{\sqrt{3}} = 79{,}624.3\text{V}$$

Using the receiving-end voltage as the reference,

$$\mathbf{V}_{R(L\text{-}N)} = 79{,}624.3\angle0°\text{V}$$

The receiving-end current is

$$I_R = \frac{49 \times 10^6}{\sqrt{3} \times 138 \times 10^3 \times 0.85}$$
$$= 241.18 \text{ A or } 241.18\angle -31.80° \text{ A}$$

a. The $\mathbf{A}$, $\mathbf{B}$, $\mathbf{C}$, and $\mathbf{D}$ constants for the nominal T circuit representation are

$$\mathbf{A} = 1 + \frac{1}{2}\mathbf{YZ}$$
$$= 1 + \frac{1}{2}(0.001\angle90°)(95\angle78°)$$
$$= 0.9535 + j0.0099$$
$$= 0.9536\angle0.6°$$

$$\mathbf{B} = \mathbf{Z} + \frac{1}{4}\mathbf{YZ}^2$$
$$= 95\angle78 + \frac{1}{4}(0.001\angle90°)(95\angle78°)^2$$
$$= 18.83 + j90.86$$
$$= 92.79\angle78.3°\Omega$$

$$\mathbf{C} = \mathbf{Y} = 0.001\angle90° \text{ S}$$

$$\mathbf{D} = 1 + \frac{1}{2}\mathbf{YZ} = \mathbf{A}$$
$$= 0.9536\angle0.6°$$

b.

$$\begin{bmatrix} \mathbf{V}_{S(L\text{-}N)} \\ \mathbf{I}_S \end{bmatrix} = \begin{bmatrix} 0.9536\angle0.6° & 92.79\angle78.3° \\ 0.001\angle90° & 0.9536\angle0.6° \end{bmatrix} \begin{bmatrix} 79.7674\angle0° \\ 241.46\angle -31.8° \end{bmatrix}$$

The sending-end voltage is

$$\mathbf{V}_{S(L\text{-}N)} = 0.9536\angle0.6° \times 79{,}674.8\angle0° + 92.79\angle78.3° \times 241.18\angle-31.8°$$
$$= 91{,}377 + j17{,}028.8 = 92{,}951.2\angle10.6° \text{ V}$$

or

$$\mathbf{V}_{S(L\text{-}L)} = 160{,}996.2\angle40.6° \text{ V}$$

c. The sending-end current is

$$\mathbf{I}_S = 0.001\angle90° \times 79{,}674.8\angle0° + 0.9536\angle0.6° \times 241.18\angle-31.8°$$
$$= 196.72 - j39.5 = 200.64\angle-11.3° \text{ A}$$

d. The sending-end power factor is

$$\theta_S = 10.6° + 11.3° = 21.9°$$

$$\cos\phi_S = 0.928$$

e. The efficiency of transmission is

$$\eta = \frac{\text{output}}{\text{input}}$$
$$= \frac{\sqrt{3}V_R I_R \cos\phi_R}{\sqrt{3}V_S I_S \cos\phi_S} \times 100$$
$$= \frac{138 \times 10^3 \times 241.18 \times 0.85}{160{,}996.2 \times 200.64 \times 0.928} \times 100$$
$$= 94.38\%$$

Example 3.7: Repeat Example 3.6 using the nominal π circuit representation.
Solution:

a. The **A, B, C,** and **D** constants for the nominal π circuit representation are

$$\mathbf{A} = 1 + \frac{1}{2}\mathbf{YZ}$$
$$= 0.9536\angle0.6°$$

$$\mathbf{B} = \mathbf{Z} = 95\angle78°$$

$$\mathbf{C} = \mathbf{Y} + \frac{1}{4}\mathbf{Y}^2\mathbf{Z}$$
$$= 0.001\angle90° + \frac{1}{4}(0.001\angle90°)^2(95\angle78°)$$
$$= -4.9379 \times 10^{-6} + j9.7677 \times 10^{-4} = 0.001\angle90.3° \text{ S}$$

$$\mathbf{D} = 1 + \frac{1}{2}\mathbf{YZ} = \mathbf{A}$$
$$= 0.9536\angle0.6°$$

b.

$$\begin{bmatrix} \mathbf{V}_{S(L\text{-}N)} \\ \mathbf{I}_S \end{bmatrix} = \begin{bmatrix} 0.9536\angle 0.6° & 95\angle 78° \\ 0.001\angle 90.3° & 0.9536\angle 0.6° \end{bmatrix} \begin{bmatrix} 79,674\angle 0° \\ 241.18\angle -31.8° \end{bmatrix}$$

Therefore,

$$\mathbf{V}_{S(L\text{-}N)} = 0.9536\angle 0.6° \times 79,674.8\angle 0° + 95\angle 78° \times 241.46\angle -31.8°$$
$$= 91,831.7 + j17,332.7 = 93,453.1\angle 10.7° \text{ V}$$

or

$$\mathbf{V}_{S(L\text{-}L)} = 161,865.5\angle 40.7° \text{ V}$$

c. The sending-end current is

$$\mathbf{I}_S = 0.001\angle 90.3° \times 79,674.3\angle 0° + 0.9536\angle 0.6° \times 241.18\angle -31.8°$$
$$= 196.31 - j39.47 = 200.24\angle -11.37° \text{ A}$$

d. The sending-end power factor is

$$\theta_S = 10.7° + 11.37° = 22.07°$$

and

$$\cos \theta_S = 0.927$$

e. The efficiency of transmission is

$$\eta = \frac{\text{output}}{\text{input}}$$
$$= \frac{\sqrt{3}V_R I_R \cos\phi_R}{\sqrt{3}V_S I_S \cos\phi_S} \times 100$$
$$= \frac{138 \times 10^3 \times 241.18 \times 0.85}{161,865.5 \times 200.24 \times 0.927} \times 100$$
$$= 94.16\%$$

The discrepancy between these results and the results of Example 4.4 is due to the fact that the nominal T and nominal π circuits of a medium-length line are not equivalent to each other. In fact, neither the nominal T nor the nominal π equivalent circuit exactly represents the actual line because the line is not uniformly distributed. However, it is possible to find the equivalent circuit of a long transmission line and to represent the line accurately.

3.11 LONG TRANSMISSION LINES (ABOVE 150 MI OR 240 KM)

Development of long transmission system modeling has been researched [73]. A more accurate analysis of the transmission lines require that the parameters of the lines are not lumped, as before, but are distributed uniformly throughout the length of the line.

Figure 3.21 shows a uniform long line with an incremental section dx at a distance x from the receiving end, its series impedance is $\mathbf{z}dx$, and its shunt admittance is $\mathbf{y}dx$, where $\mathbf{z}$ and $\mathbf{y}$ are the impedance and admittance per unit length, respectively. The voltage drop in the section is

$$d\mathbf{V}_x = (\mathbf{V}_x + d\mathbf{V}_x) - \mathbf{V}_x = d\mathbf{V}_x$$
$$= (\mathbf{I}_x + d\mathbf{I}_x)\mathbf{z}\, dx$$

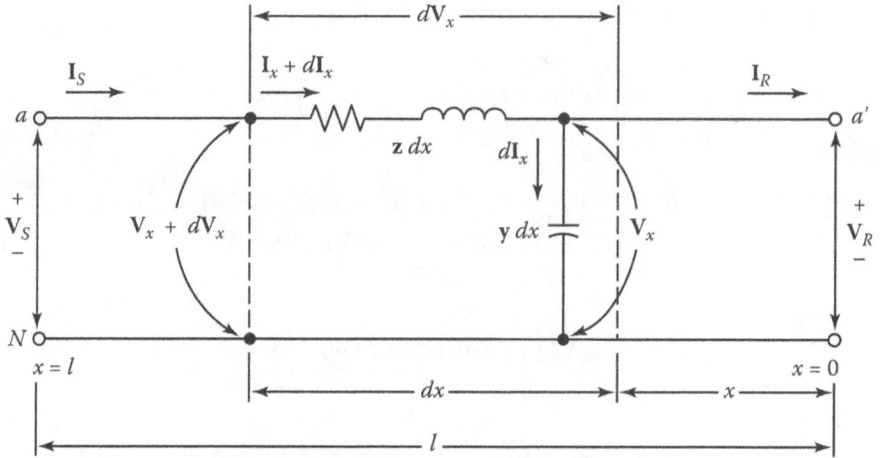

Figure 3.21 One phase and neutral connection of three-phase transmission line.

or

$$dV_x \cong I_x \, zd_x \qquad (3.99)$$

Similarly, the incremental charging current is

$$dI_x \cong V_x y \, dx \qquad (3.100)$$

Therefore,

$$\frac{dV_x}{dx} = zI_x \qquad (3.101)$$

and

$$\frac{dI_x}{dx} = yV_x \qquad (3.102)$$

Differentiating Equations 3.101 and 3.102 with respect to x,

$$\frac{d^2V_x}{dx^2} = z\frac{I_x}{dx} \qquad (3.103)$$

and

$$\frac{d^2I_x}{dx^2} = y\frac{V_x}{dx} \qquad (3.104)$$

Substituting the values of dI_x/dx and dV_x/dx from Equations 3.102 and 3.103 in Equations 3.105 and 3.106, respectively,

$$\frac{d^2V_x}{dx^2} = yzV_x \qquad (3.105)$$

and

$$\frac{d^2I_x}{dx^2} = yzI_x \qquad (3.106)$$

At x = 0, $V_x = V_R$ and $I_x = I_R$. Therefore, the solution of the ordinary second-order differential Equations 3.105 and 3.106 gives

$$V_{(x)} = \underbrace{(\cosh\sqrt{yz}x)}_{A} V_R + \underbrace{\left(\sqrt{\frac{z}{y}}\sinh\sqrt{yz}x\right)}_{B} I_R \qquad (3.107)$$

This application of hyperbolic functions in electrical engineering have been studied in 1925 [74]. Similarly,

$$\mathbf{I}_{(x)} = \underbrace{\left(\sqrt{\frac{\mathbf{y}}{\mathbf{z}}} \sinh\sqrt{\mathbf{yz}}x \right)}_{\mathbf{C}} \mathbf{V}_R + \underbrace{\left(\cosh\sqrt{\mathbf{yz}}x \right)}_{\mathbf{D}} \mathbf{I}_R \tag{3.108}$$

Equations 3.107 and 3.108 can be rewritten as

$$\mathbf{V}_{(x)} = (\cosh \gamma x)\mathbf{V}_R + (\mathbf{Z}_c \sinh \gamma x)\mathbf{I}_R \tag{3.109}$$

and

$$\mathbf{I}_{(x)} = (\mathbf{Y}_c \sinh \gamma x)\mathbf{V}_R + (\cosh \gamma x)\mathbf{I}_R \tag{3.110}$$

where
γ = propagation constant per unit length, = $\sqrt{\mathbf{yz}}$
$\mathbf{Z}_c$ = characteristic (or surge or natural) impedance of line per unit length, = $\sqrt{\mathbf{z}/\mathbf{y}}$
$\mathbf{Y}_c$ = characteristic (or surge or natural) admittance of line per unit length, = $\sqrt{\mathbf{y}/\mathbf{z}}$

Further,

$$\gamma = \alpha + j\beta \tag{3.111}$$

where
α = attenuation constant (measuring decrement in voltage and current per unit length in direction of travel) in nepers per unit length
β = phase (or phase change) constant in radians per unit length (i.e., change in phase angle between two voltages, or currents, at two points one per unit length apart on infinite line)

When x = l, Equations 3.109 and 3.110 become

$$\mathbf{V}_S = (\cosh \gamma l)\mathbf{V}_R + (\mathbf{Z}_c \sinh \gamma l)\mathbf{V}_R \tag{3.112}$$

and

$$\mathbf{I}_S = (\mathbf{Y}_c \sinh \gamma l)\mathbf{V}_R + (\cosh \gamma l)\mathbf{I}_R \tag{3.113}$$

Equations 3.112 and 3.113 can be written in matrix form as

$$\begin{bmatrix} \mathbf{V}_S \\ \mathbf{I}_S \end{bmatrix} = \begin{bmatrix} \cosh \gamma l & \mathbf{Z}_c \sinh \gamma l \\ \mathbf{Y}_c \sinh \gamma l & \cosh \gamma l \end{bmatrix} \begin{bmatrix} \mathbf{V}_R \\ \mathbf{I}_R \end{bmatrix} \tag{3.114}$$

and

$$\begin{bmatrix} \mathbf{V}_R \\ \mathbf{I}_R \end{bmatrix} = \begin{bmatrix} \cosh \gamma l & \mathbf{Z}_c \sinh \gamma l \\ \mathbf{Y}_c \sinh \gamma l & \cosh \gamma l \end{bmatrix} \begin{bmatrix} \mathbf{V}_S \\ \mathbf{I}_S \end{bmatrix} \tag{3.115}$$

or

$$\begin{bmatrix} \mathbf{V}_R \\ \mathbf{I}_R \end{bmatrix} = \begin{bmatrix} \cosh \gamma l & -\mathbf{Z}_c \sinh \gamma l \\ -\mathbf{Y}_c \sinh \gamma l & \cosh \gamma l \end{bmatrix} \begin{bmatrix} \mathbf{V}_S \\ \mathbf{I}_S \end{bmatrix} \tag{3.116}$$

Therefore,

$$\mathbf{V}_R = (\cosh \gamma l)\mathbf{V}_S + (\mathbf{Z}_c \sinh \gamma l)\mathbf{I}_S \tag{3.117}$$

and

$$\mathbf{I}_R = -(\mathbf{Y}_c \sinh \gamma l)\mathbf{V}_S + (\cosh \gamma l)\mathbf{I}_S \tag{3.118}$$

In terms of **ABCD** constants,

$$\begin{bmatrix} \mathbf{V}_S \\ \mathbf{I}_S \end{bmatrix} = \begin{bmatrix} \mathbf{A} & \mathbf{B} \\ \mathbf{C} & \mathbf{D} \end{bmatrix} \begin{bmatrix} \mathbf{V}_R \\ \mathbf{I}_R \end{bmatrix} = \begin{bmatrix} \mathbf{A} & \mathbf{B} \\ \mathbf{C} & \mathbf{A} \end{bmatrix} \begin{bmatrix} \mathbf{V}_R \\ \mathbf{I}_R \end{bmatrix} \tag{3.119}$$

and

$$\begin{bmatrix} \mathbf{V}_R \\ \mathbf{I}_R \end{bmatrix} = \begin{bmatrix} \mathbf{A} & -\mathbf{B} \\ -\mathbf{C} & \mathbf{D} \end{bmatrix} \begin{bmatrix} \mathbf{V}_S \\ \mathbf{I}_S \end{bmatrix} = \begin{bmatrix} \mathbf{A} & -\mathbf{B} \\ -\mathbf{C} & \mathbf{A} \end{bmatrix} \begin{bmatrix} \mathbf{V}_S \\ \mathbf{I}_S \end{bmatrix} \tag{3.120}$$

where

$$\mathbf{A} = (\cosh \gamma l) = \cosh \sqrt{\mathbf{YZ}} = \cosh \theta \tag{3.121}$$

$$\mathbf{B} = \mathbf{Z}_c \sinh \gamma l = \sqrt{\mathbf{Z/Y}} \sinh \sqrt{\mathbf{YZ}} = \mathbf{Z}_c \sinh \theta \tag{3.122}$$

$$\mathbf{C} = \mathbf{Y}_c \sinh \gamma l = \sqrt{\mathbf{Y/Z}} \sinh \sqrt{\mathbf{YZ}} = \mathbf{Y}_c \sinh \theta \tag{3.123}$$

$$\mathbf{D} = \mathbf{A} = \cosh \gamma l = \cosh \sqrt{\mathbf{YZ}} = \cosh \theta \tag{3.124}$$

$$\theta = \sqrt{\mathbf{YZ}} \tag{3.125}$$

$$\sinh \gamma l = \frac{1}{2}(e^{\gamma l} - e^{-\gamma l}) \tag{3.126}$$

$$\cosh \gamma l = \frac{1}{2}(e^{\gamma l} + e^{-\gamma l}) \tag{3.127}$$

Also,

$$\sinh(\alpha + j\beta) = \frac{e^{\alpha} e^{j\beta} - e^{-\alpha} e^{-j\beta}}{2} = \frac{1}{2}[e^{\alpha} \angle \beta - e^{-\alpha} \angle -\beta]$$

and

$$\cosh(\alpha + j\beta) = \frac{e^{\alpha} e^{j\beta} + e^{-\alpha} e^{-j\beta}}{2} = \frac{1}{2}[e^{\alpha} \angle \beta + e^{-\alpha} \angle -\beta]$$

Note that, the β in the above equations is the radian, and the radian is the unit found for β by computing the quadrature component of γ. Since 2π radians = 360°, 1 rad is 57.3°. Thus, the β is converted into degrees by multiplying its quantity by 57.3°. For a line length of l,

$$\sinh(\alpha l + j\beta l) = \frac{e^{\alpha l} e^{j\beta l} - e^{-\alpha l} e^{-j\beta l}}{2} = \frac{1}{2}[e^{\alpha l} \angle \beta l - e^{-\alpha l} \angle -\beta l]$$

and

$$\cosh(\alpha l + j\beta l) = \frac{e^{\alpha l} e^{j\beta l} + e^{-\alpha l} e^{-j\beta l}}{2} = \frac{1}{2}[e^{\alpha l} \angle \beta l + e^{-\alpha l} \angle -\beta l]$$

Equations 3.112 through 3.125 can be used if tables of complex hyperbolic functions or pocket calculators with complex hyperbolic functions are available.

Alternatively, the following expansions can be used:

$$\sinh \gamma l = \sinh(\alpha l + j\beta l) = \sinh \alpha l \cos \beta l + \cosh \alpha l \sin \beta l \tag{3.128}$$

and

$$\cosh \gamma l = \cosh(\alpha l + j\beta l) = \cosh \alpha l \cos \beta l + \sinh \alpha l \sin \beta l \tag{3.129}$$

The correct mathematical unit for βl is the radian, and the radian is the unit found for βl by computing the quadrature component of γl.

Furthermore, substituting for γl and $\mathbf{Z}_c$ in terms of $\mathbf{Y}$ and $\mathbf{Z}$, that is, the total line shunt admittance per phase and the total line series impedance per phase, in Equation 3.119 gives

$$\mathbf{V}_S = \left(\cosh \sqrt{\mathbf{YZ}} \right) \mathbf{V}_R + \left(\sqrt{\frac{\mathbf{Z}}{\mathbf{Y}}} \sinh \sqrt{\mathbf{YZ}} \right) \mathbf{I}_R \tag{3.130}$$

and

$$\mathbf{I}_S = \left(\sqrt{\frac{\mathbf{Z}}{\mathbf{Y}}}\sinh\sqrt{\mathbf{YZ}}\right)\mathbf{V}_R + \left(\cosh\sqrt{\mathbf{YZ}}\right)\mathbf{I}_R \qquad (3.131)$$

or, alternatively,

$$\mathbf{V}_S = \left(\cosh\sqrt{\mathbf{YZ}}\right)\mathbf{V}_R + \left(\frac{\sinh\sqrt{\mathbf{YZ}}}{\sqrt{\mathbf{YZ}}}\right)\mathbf{ZI}_R \qquad (3.132)$$

and

$$\mathbf{I}_S = \left(\frac{\sinh\sqrt{\mathbf{YZ}}}{\sqrt{\mathbf{YZ}}}\right)\mathbf{YV}_R + \left(\cosh\sqrt{\mathbf{YZ}}\right)\mathbf{I}_R \qquad (3.133)$$

The factors in parentheses in Equations 3.130 through 3.133 can readily be found by using Woodruff's charts, which are not included here but can be found in L. F. Woodruff, *Electric Power Transmission* (Wiley, NY, 1952).

The **ABCD** parameters in terms of infinite series can be expressed as

$$\mathbf{A} = 1 + \frac{\mathbf{YZ}}{2} + \frac{\mathbf{Y}^2\mathbf{Z}^2}{24} + \frac{\mathbf{Y}^3\mathbf{Z}^3}{720} + \frac{\mathbf{Y}^4\mathbf{Z}^4}{40,320} + \cdots \qquad (3.134)$$

$$\mathbf{B} = \mathbf{Z}\left(1 + \frac{\mathbf{YZ}}{6} + \frac{\mathbf{Y}^2\mathbf{Z}^2}{120} + \frac{\mathbf{Y}^3\mathbf{Z}^3}{5040} + \frac{\mathbf{Y}^4\mathbf{Z}^4}{362,880} + \cdots\right) \qquad (3.135)$$

$$\mathbf{A} = \mathbf{Y}\left(1 + \frac{\mathbf{YZ}}{6} + \frac{\mathbf{Y}^2\mathbf{Z}^2}{120} + \frac{\mathbf{Y}^3\mathbf{Z}^3}{5040} + \frac{\mathbf{Y}^4\mathbf{Z}^4}{362,880} + \cdots\right) \qquad (3.136)$$

where
 $\mathbf{Z}$ = total line series impedance per phase
  = $\mathbf{z}l$
  = $(r + jx_L)l$ Ω
 $\mathbf{Y}$ = total line shunt admittance per phase
  = $\mathbf{y}l$
  = $(g + jb)l$ S

In practice, usually not more than three terms are necessary in Equations 3.134 through 3.136. Weedy [75] suggests the following approximate values for the **ABCD** constants if the overhead transmission line is < 500 km in length:

$$\mathbf{A} = 1 + \frac{1}{2}\mathbf{YZ} \qquad (3.137)$$

$$\mathbf{B} = \mathbf{Z}\left(1 + \frac{1}{6}\mathbf{YZ}\right) \qquad (3.138)$$

$$\mathbf{C} = \mathbf{Y}\left(1 + \frac{1}{6}\mathbf{YZ}\right) \qquad (3.139)$$

However, the error involved may be too large to be ignored for certain applications.

Example 3.8: A single-circuit, 60-Hz, three-phase transmission line is 150 mi long. The line is connected to a load of 50 MVA at a lagging power factor of 0.85 at 138 kV. The line constants are given as $R = 0.1858$ Ω/mi, $L = 2.60$ mH/mi, and $C = 0.012$ μF/mi. Calculate the following:

a. **A**, **B**, **C**, and **D** constants of the line

b. Sending-end voltage
c. Sending-end current
d. Sending-end power factor
e. Sending-end power
f. Power loss in the line
g. Transmission line efficiency
h. Percentage of voltage regulation
i. Sending-end charging current at no load
j. Value of receiving-end voltage rise at no load if the sending-end voltage is held constant

Solution:

$$\mathbf{z} = 0.1858 + j2\pi \times 60 \times 2.6 \times 10^{-3}$$
$$= 0.1858 + j0.9802$$
$$= 0.9977\angle 79.27° \ \Omega/\text{mi}$$

and

$$\mathbf{y} = j2\pi \times 60 \times 0.012 \times 10^{-6}$$
$$= 4.5239 \times 10^{-6}\angle 90° \ \text{S/mi}$$

The propagation constant of the line is

$$\gamma = \sqrt{\mathbf{yz}}$$
$$= [(4.5239 \times 10^{-6}\angle 90°)(0.9977\angle 79.27°)]^{1/2}$$
$$= [4.5135 \times 10^{-6}]^{1/2}\angle \left(\frac{90° + 79.27°}{2}\right)$$
$$= 0.00214499\angle 84.63°$$
$$= 0.0002007 + j0.0021346$$

Thus,

$$\gamma l = \alpha l + j\beta l$$
$$= (0.0002007 + j0.0021346)150$$
$$\cong 0.0301 + j0.3202$$

The characteristic impedance of the line is

$$\mathbf{Z}_c = \sqrt{\frac{\mathbf{z}}{\mathbf{y}}} = \left(\frac{0.9977\angle 79.27°}{4.5239 \times 10^{-6}\angle 90°}\right)^{1/2}$$
$$= \left(\frac{0.9977 \times 10^6}{4.5239}\right)^{1/2} \angle \left(\frac{79.27° - 90°}{2}\right)$$
$$= 469.62\angle -5.37° \ \Omega$$

The receiving-end line-to-neutral voltage is

$$\mathbf{V}_{R(\text{L-N})} = \frac{138\text{kV}}{\sqrt{3}} = 79,674.34 \ \text{V}$$

Using the receiving-end voltage as the reference,

$$\mathbf{V}_{R(L-N)} = 79,674.34\angle 0° \text{ V}$$

The receiving-end current is

$$\mathbf{I}_R = \frac{50 \times 10^6}{\sqrt{3} \times 138 \times 10^3}$$
$$= 209.18 \text{ A} \quad \text{or} \quad 209.18\angle -31.8° \text{ A}$$

a. The **A**, **B**, **C**, and **D** constants of the line

$$\mathbf{A} = \cosh\gamma l$$
$$= \cosh(\alpha + j\beta)l$$
$$= \frac{e^{\alpha l}e^{j\beta l} + e^{\alpha l}e^{-j\beta l}}{2}$$
$$= \frac{e^{\alpha l}\angle \beta l + e^{\alpha l}\angle -\beta l}{2}$$

Therefore,

$$\mathbf{A} = \frac{e^{0.0301}e^{j0.3202} + e^{-0.0301}e^{-j0.3202}}{2}$$
$$= \frac{e^{0.0301}\angle 18.35° + e^{-0.0301}\angle -18.35°}{2}$$
$$= \frac{1.0306\angle 18.35° + 0.9703\angle -18.35°}{2}$$
$$= 0.9496 + j0.0095 = 0.9497\angle 0.57°$$

Note that, $e^{j0.3202}$ needs to be converted to degrees. Since 2π rad = 360°, 1 rad is 57.3°. Hence

$$(0.3202 \text{ rad})(57.3°/\text{rad}) = 18.35°$$

and

$$\mathbf{B} = \mathbf{Z}_C \sinh\gamma l = \mathbf{Z}_C \sinh(\alpha + j\beta)l$$
$$= \mathbf{Z}_C \left[\frac{e^{\alpha l}e^{j\beta l} - e^{\alpha l}e^{-j\beta l}}{2} \right]$$
$$= \mathbf{Z}_C \left[\frac{e^{\alpha l}\angle \beta l - e^{\alpha l}\angle -\beta l}{2} \right]$$
$$= (469.62\angle 5.37°) \left[\frac{e^{0.0301}e^{j0.3202} - e^{-0.0301}e^{-j0.3202}}{2} \right]$$
$$= 469.62\angle -5.37° \left[\frac{1.0306\angle 18.35° - 0.9703\angle -18.35°}{2} \right]$$
$$= 469.62\angle 5.37° \left(\frac{0.0572 + j0.63}{2} \right)$$
$$= 469.62\angle -5.37° \left(\frac{0.6326\angle 84.81°}{2} \right)$$
$$= 469.62\angle -5.37° (0.3163\angle 84.81°)$$
$$= 148.54\angle 79.44° \ \Omega$$

and

$$\mathbf{C} = \mathbf{Y}_C \sinh\gamma l = \frac{1}{\mathbf{Z}_C}\sinh\gamma l$$

$$= \frac{1}{469.62\angle-5.37°} \times \frac{0.63259}{2}\angle84.81°$$

$$= 0.00067\angle90.18° \text{ S}$$

and

$$\mathbf{D} = \mathbf{A} = \cos\gamma l = 0.9497\angle0.57°$$

b.

$$\begin{bmatrix} \mathbf{V}_{S(L\text{-}N)} \\ \mathbf{I}_S \end{bmatrix} = \begin{bmatrix} \mathbf{A} & \mathbf{B} \\ \mathbf{C} & \mathbf{D} \end{bmatrix}\begin{bmatrix} \mathbf{V}_{R(L\text{-}N)} \\ \mathbf{I}_R \end{bmatrix}$$

$$= \begin{bmatrix} 0.9497\angle0.57° & 148.54\angle79.44° \\ 0.00067\angle90.18° & 0.9497\angle0.57° \end{bmatrix}\begin{bmatrix} 79,674.34\angle0° \\ 209.18\angle-31.8° \end{bmatrix}$$

Thus, the sending-end voltage is

$$\mathbf{V}_{S(L\text{-}N)} = (0.9497\angle0.57°)(79,674.34\angle0°) + (148.54\angle79.44°)(209.18\angle-31.8°)$$
$$= 99,466.41\angle13.79° \text{ V}$$

and

$$\mathbf{V}_{S(L\text{-}L)} = \sqrt{3}\mathbf{V}_{S(L\text{-}N)}$$
$$= 172,280.87\angle13.79° + 30°$$
$$= 172,280.87\angle43.79° \text{ V}$$

Note that, an additional 30° is added to the angle since a line-to-line voltage is 30° ahead of its line-to-neutral voltage.

c. The sending-end current is

$$\mathbf{I}_S = (0.00067\angle90.18°)(79,674.34\angle0°) + (0.9497\angle0.57°)(209.18\angle-31.8°)$$
$$= 180.88\angle-16.3° \text{ A}$$

d. The sending-end power factor is

$$\theta_S = 13.79° + 16.3°$$
$$= 30.09°$$
$$\cos\theta_S = 0.9648$$

e. The sending-end power is

$$P_S = \sqrt{3}V_{S(L\text{-}L)}I_S\cos\theta_S$$
$$= \sqrt{3} \times 172,280.87 \times 180.88 \times 0.9948$$
$$\cong 45,652.79 \text{ kW}$$

f. The receiving-end power is

$$P_R = \sqrt{3}V_{R(L\text{-}L)}I_R\cos\theta_R$$
$$= \sqrt{3} \times 13810^3 \times 209.18 \times 0.85$$
$$= 42,499 \text{ kW}$$

Therefore, the power loss in the line is

$$P_L = P_S - P_R = 3,153.79 \text{ kW}$$

g. The transmission line efficiency is

$$\eta = \frac{P_R}{P_S} \times 100$$

$$= \frac{42,499}{45,652.79} \times 100$$

$$= 93.1\%$$

h. The percentage of voltage regulation is

$$\text{Percentage of voltage regulation} = \frac{99,470.05 - 79,674.34}{79,674.34} \times 100$$

$$= 24.8\%$$

i. The sending-end charging current at no load is

$$I_c = \frac{1}{2} Y V_{S(\text{L-N})}$$

$$= \frac{1}{2}(678.585 \times 10^{-6})(99,466\ 41)$$

$$= 33.75 \text{ A}$$

where

$$Y = y \times l$$

$$= (4.5239 \times 10^{-6} \text{ S/mi})(150 \text{ mi})$$

$$= 678.585 \times 10^{-6} \text{ S}$$

j. The receiving-end voltage rise at no load is

$$\mathbf{V}_{R(\text{L-N})} = \mathbf{V}_{S(\text{L-N})} - \mathbf{I}_c \mathbf{Z}$$

$$= 99,466.41\angle 13.79° - (33.75\angle 103.79°)(149.66\angle 79.27°)$$

$$= 104,433.09\angle 13.27° \text{ V}$$

Therefore, the line-to-line voltage at the receiving end is

$$\mathbf{V}_{R(\text{L-L})} = \sqrt{3}\mathbf{V}_{R(\text{L-N})} = 180,883.42\angle(13.27° + 30°)$$

$$= 180,883.42\angle 43.27° \text{ V}$$

Note that, in a well-designed transmission line, the voltage regulation and the line efficiency should be not greater than about 5%.

3.11.1 EQUIVALENT CIRCUIT OF LONG TRANSMISSION LINE

Using the values of the **ABCD** parameters obtained for a transmission line, it is possible to develop an exact π or an exact T, as shown in Figure 3.22. For the equivalent π circuit,

$$\mathbf{Z}_\pi = \mathbf{B} = \mathbf{Z}_c \sinh\theta \qquad\qquad (3.140)$$

$$= \mathbf{Z}_c \sinh\gamma l \qquad\qquad (3.141)$$

$$= \mathbf{Z}\left(\frac{\sinh\sqrt{\mathbf{YZ}}}{\sqrt{\mathbf{YZ}}}\right) \qquad\qquad (3.142)$$

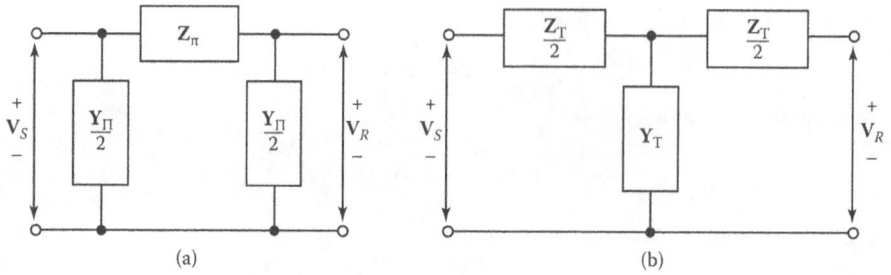

Figure 3.22 Equivalent π and T circuits for a long transmission line.

and

$$\frac{\mathbf{Y}_\pi}{2} = \frac{\mathbf{A}-1}{\mathbf{B}} = \frac{\cosh\theta-1}{\mathbf{Z}_c\sinh\theta} \tag{3.143}$$

or

$$\mathbf{Y}_\pi = \frac{2\tanh\left(\frac{\gamma l}{2}\right)}{\mathbf{Z}_c} \tag{3.144}$$

or

$$\frac{\mathbf{Y}_\pi}{2} = \frac{\mathbf{Y}}{2}\frac{2\tanh\left(\frac{\sqrt{\mathbf{YZ}}}{2}\right)}{\frac{\sqrt{\mathbf{YZ}}}{2}} \tag{3.145}$$

For the equivalent T circuit,

$$\frac{\mathbf{Z}_T}{2} = \frac{\mathbf{A}-1}{\mathbf{C}} = \frac{\cosh\theta-1}{\mathbf{Y}_c\sinh\theta} \tag{3.146}$$

or

$$\mathbf{Z}_T - 2\mathbf{Z}_c\tanh\frac{\gamma l}{2} \tag{3.147}$$

or

$$\frac{\mathbf{Z}_T}{2} = \frac{\mathbf{Z}}{2}\left(\frac{\tanh\frac{\sqrt{\mathbf{YZ}}}{2}}{\frac{\sqrt{\mathbf{YZ}}}{2}}\right) \tag{3.148}$$

and

$$\mathbf{Y}_T = \mathbf{C} = \mathbf{Y}_c\sinh\theta \tag{3.149}$$

or

$$\mathbf{Y}_T = \frac{\sinh\gamma l}{\mathbf{Z}_c} \tag{3.150}$$

or

$$\mathbf{Y}_T = \mathbf{Y}\frac{\sinh\sqrt{\mathbf{YZ}}}{\sqrt{\mathbf{YZ}}} \tag{3.151}$$

Example 3.9: Find the equivalent *pi* and the equivalent T circuits for the line described in Example 3.8 and compare them with the nominal π and the nominal T circuits.
Solution:
Figures 3.23 and 3.24 show the equivalent π and the nominal π circuits, respectively.

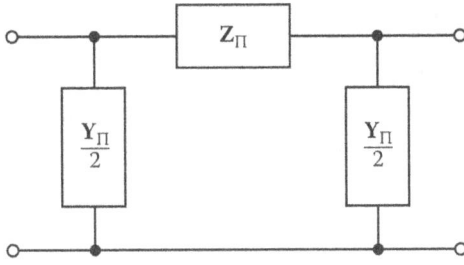

Figure 3.23 Equivalent π circuit.

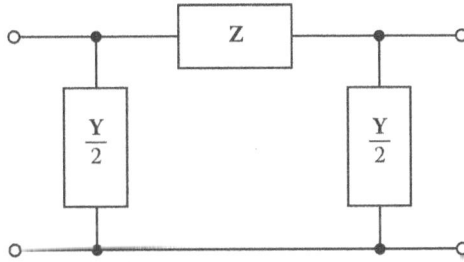

Figure 3.24 Nominal π circuit.

For the equivalent π circuit,

$$\mathbf{Z}_x = \mathbf{B} = 148.54\angle 79.44° \ \Omega$$

$$\frac{\mathbf{Y}_\pi}{2} = \frac{\mathbf{A} - 1}{\mathbf{B}}$$
$$= \frac{0.9497\angle 0.57° - 1}{148.54\angle 79.44}$$
$$= 0.000345\angle 89.93° \ \text{S}$$

For the nominal π circuit,

$$\mathbf{Z} = 150 \times 0.9977\angle 79.27°$$
$$= 149.655\angle 79.27° \ \Omega$$

$$\frac{\mathbf{Y}}{2} = \frac{150(4.5239 \times 10^{-6}\angle 90°)}{2}$$
$$= 0.000339\angle 90° \ \text{S}$$

Figure 3.25a and b shows the equivalent T and nominal T circuits, respectively.

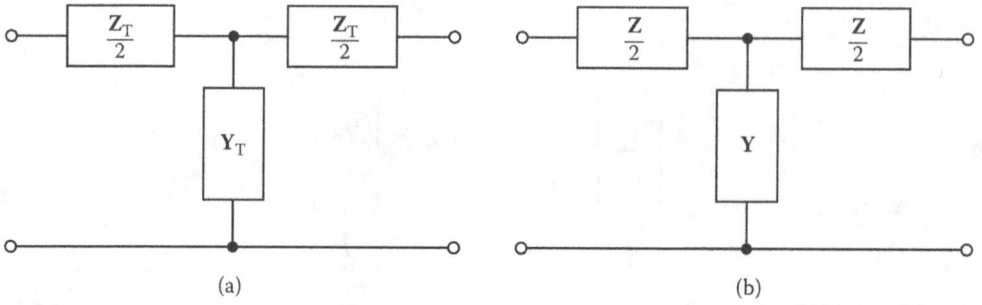

Figure 3.25 T circuits: (a) equivalent T; (b) nominal T.

For the equivalent T circuit,

$$
\begin{aligned}
\frac{\mathbf{Z}_T}{2} &= \frac{\mathbf{A} - 1}{\mathbf{C}} \\
&= \frac{0.9497\angle0.57° - 1}{0.00067\angle90.18°} \\
&= 76.46\angle79.19° \ \Omega
\end{aligned}
$$

$$\mathbf{Y}_T = \mathbf{C} = 0.00067\angle90.18° \ \text{S}$$

For the nominal T circuit,

$$
\begin{aligned}
\frac{\mathbf{Z}}{2} &= \frac{149.655\angle79.27°}{2} \\
&= 74.83\angle79.27°
\end{aligned}
$$

$$\mathbf{Y}_T = 0.000678\angle90° \ \text{S}$$

As can be observed from the results, the difference between the values for the equivalent and nominal circuits is very small for a 150-mi-long transmission line.

3.11.2   INCIDENT AND REFLECTED VOLTAGES OF LONG TRANSMISSION LINE

Previously, the propagation constant has been given as

$$\gamma = \alpha + j\beta \ \text{per-unit length} \tag{3.152}$$

and also

$$\cosh \gamma l = \frac{e^{\gamma l} + e^{-\gamma l}}{2} \tag{3.153}$$

$$\sinh \gamma l = \frac{e^{\gamma l} - e^{-\gamma l}}{2} \tag{3.154}$$

The sending-end voltage and current have been expressed as

$$\mathbf{V}_S = (\cosh \gamma l)\mathbf{V}_R + (\mathbf{Z}_c \sinh \gamma l)\mathbf{I}_R \tag{3.155}$$

and

$$\mathbf{I}_S = (\mathbf{Y}_c \sinh \gamma l)\mathbf{V}_R + (\cosh \gamma l)\mathbf{I}_R \qquad (3.156)$$

By substituting Equations 3.152 through 3.154 in Equations 3.112 and 3.113,

$$\mathbf{V}_S = \frac{1}{2}(\mathbf{V}_R + \mathbf{I}_R\mathbf{Z}_c)e^{\alpha l}e^{j\beta l} + \frac{1}{2}(\mathbf{V}_R - \mathbf{I}_R\mathbf{Z}_c)e^{-\alpha l}e^{-j\beta l} \qquad (3.157)$$

and

$$\mathbf{I}_S = \frac{1}{2}(\mathbf{V}_R\mathbf{Y}_c + \mathbf{I}_R)e^{\alpha l}e^{j\beta l} - \frac{1}{2}(\mathbf{V}_R\mathbf{Y}_c - \mathbf{I}_R)e^{-\alpha l}e^{-j\beta l} \qquad (3.158)$$

Equation 3.155 comprises two terms: the *incident voltage* and the *reflected voltage*, which propagate as *traveling waves* along the line with respect to its length l. The incident voltage increases in magnitude and phase as l increases from the receiving end and decreases as it moves towards the sending end. Conversely, the reflected voltage decreases in magnitude and phase as l increases from the receiving end towards the sending end. Thus, the voltage at any point along the line is the sum of the corresponding incident and reflected voltages. While the term $e^{\alpha l}$ varies as a function of l, the term $e^{j\beta l}$ has a constant magnitude of 1 and causes a phase shift of β radians per unit length of the line.

In Equation 3.155, when the two terms are 180° out of phase, a cancellation will occur. This happens when there is no load on the line, that is, when

$$\mathbf{I}_R = 0 \text{ and } \alpha = 0$$

and when $\beta x = \frac{\pi}{2}$ radians, or one-quarter wavelengths.

The wavelength A is defined as the distance l along a line between two points to develop a phase shift of 2π radians, or 360°, for the incident and reflected waves. If β is the phase shift in radians per mile, the wavelength in miles is

$$\lambda = \frac{2\pi}{\beta} \qquad (3.159)$$

Since the propagation velocity is

$$v = \lambda f \text{ mi/s} \qquad (3.160)$$

and is approximately equal to the speed of light, that is, 186,000 mil, at a frequency of 60 Hz, the wavelength is

$$\lambda = \frac{186,000 \text{ mi/s}}{60 \text{ Hz}} = 3100 \text{ mi}$$

On the other hand, at a frequency of 50 Hz, the wavelength is approximately 6000 km. If a finite line is terminated by its characteristic impedance $\mathbf{Z}_C$, that impedance could be imagined replaced by an infinite line. In this case, there is no reflected wave of either voltage or current since

$$\mathbf{V}_R = \mathbf{I}_R\mathbf{Z}_C$$

in Equations 3.155 and 3.156, and the line is called an *infinite* (or *flat*) line.

Stevenson [61] gives the typical values of $\mathbf{Z}_C$ as 400 ft for a single-circuit line and 200 Ω for two circuits in parallel. The phase angle of $\mathbf{Z}_C$ is usually between 0 and $-15°$ [63].

Example 3.10: Using the data given in Example 3.8, determine the following:

a. Attenuation constant and phase change constant per mile of the line
b. Wavelength and velocity of propagation
c. Incident and reflected voltages at the receiving end of the line
d. Line voltage at the receiving end of the line
e. Incident and reflected voltages at the sending end of the line
f. Line voltage at the sending end

Solution:

a. Since the propagation constant of the line is

$$\gamma = \sqrt{\mathbf{yz}}$$
$$= 0.0002 + j0.0021$$

then, the attenuation constant is 0.0002 Np/mi, and the phase change constant is 0.0021 rad/mi.
b. The wavelength of propagation is

$$\lambda = \frac{2\pi}{\beta}$$
$$= \frac{2\pi}{0.0021}$$
$$= 2991.99 \text{ mi}$$

and the velocity of propagation is

$$v = \lambda f = 2991.99 \times 60 = 179,519.58 \text{ mi/s}$$

c. From Equation 3.155,

$$\mathbf{V}_S = \frac{1}{2}(\mathbf{V}_R + \mathbf{I}_R\mathbf{Z}_c)e^{\alpha l}e^{j\beta l} + \frac{1}{2}(\mathbf{V}_R - \mathbf{I}_R\mathbf{Z}_c)e^{-\alpha l}e^{-j\beta l}$$

Since, at the receiving end, $I = 0$,

$$\mathbf{V}_S = \frac{1}{2}(\mathbf{V}_R + \mathbf{I}_R\mathbf{Z}_c) + \frac{1}{2}(\mathbf{V}_R - \mathbf{I}_R\mathbf{Z}_c)$$

Therefore, the incident and reflected voltage at the receiving end are

$$\mathbf{V}_{R(\text{incident})} = \frac{1}{2}(\mathbf{V}_R + \mathbf{I}_R\mathbf{Z}_c)$$
$$= \frac{1}{2}[79,674.34\angle 0° + (209.18\angle -31.8°)(469.62\angle -5.37°)]$$
$$= 84,367.77\angle -20.59° \text{ V}$$

and

$$\mathbf{V}_{R(\text{reflected})} = \frac{1}{2}(\mathbf{V}_R - \mathbf{I}_R\mathbf{Z}_c)$$
$$= \frac{1}{2}[79,674.34\angle 0° - (209.18\angle -31.8°)(469.62\angle -5.37°)]$$
$$= 29,684.15\angle -88.65° \text{ V}$$

d. The line-to-neutral voltage at the receiving end is

$$\mathbf{V}_{R(L\text{-}N)} = \mathbf{V}_{Rincident} + \mathbf{V}_{Rreflected}$$
$$= 79{,}674\angle 0° \text{ V}$$

Therefore, the line voltage at the receiving end is

$$\mathbf{V}_{R(L\text{-}L)} = \sqrt{3}V_{R(L\text{-}N)}$$
$$= 138{,}000 \text{ V}$$

e. At the sending end,

$$\mathbf{V}_{S(incident)} = \frac{1}{2}(\mathbf{V}_R + \mathbf{I}_R\mathbf{Z}_c)e^{\alpha l}e^{j\beta l}$$
$$= (84{,}367.77\angle -20.59°)e^{0.0301}\angle 18.35°$$
$$= 86{,}946\angle -2.24° \text{ V}$$

and

$$\mathbf{V}_{S(reflected)} = \frac{1}{2}(\mathbf{V}_R + \mathbf{I}_R\mathbf{Z}_c)e^{-\alpha l}e^{-j\beta l}$$
$$= (29{,}684.15\angle 88.65°)e^{-0.0301}\angle -18.5° = 28{,}802.5\angle 70.3° \text{ V}$$

f. The line-to-neutral voltage at the sending end is

$$\mathbf{V}_{S(L\text{-}N)} = \mathbf{V}_{S(incident)} + \mathbf{V}_{S(reflected)}$$
$$= 86{,}946\angle -2.24° + 28{,}802.5\angle -70.3° = 99{,}458.1\angle 13.8° \text{ V}$$

Therefore, the line voltage at the sending end is

$$\mathbf{V}_{S(L\text{-}L)} = \sqrt{3}\mathbf{V}_{S(L\text{-}N)} = 172{,}266.5 \text{ V}$$

3.11.3 SURGE IMPEDANCE LOADING OF TRANSMISSION LINE

In power systems, if the line is *lossless*,* the characteristic impedance Z_c of a line is sometimes called *surge impedance*. Therefore, for a loss-free line,

$$R = 0$$

and

$$\mathbf{Z}_L = jX_L$$

Thus,

$$Z_c = \sqrt{\frac{X_L}{Y_c}} \cong \sqrt{\frac{L}{C}} \ \Omega \qquad (3.161)$$

and its series resistance and shunt conductance are zero. It is a function of the line inductance and capacitance as shown and is independent of the line length.

The surge impedance loading (SIL) (or the *natural loading*) of a transmission line is defined as the power delivered by the line to a purely resistive load equal to its surge impedance. Therefore,

$$\text{SIL} = \frac{|kV_{R(L\text{-}L)}|^2}{Z_c^*} \text{ MW} \qquad (3.162)$$

* When dealing with high frequencies or with surges due to lightning, losses are often ignored [61].
or

$$\text{SIL} \cong \frac{|k\mathbf{V}_{R(\text{L-L})}|^2}{\sqrt{\frac{L}{C}}} \text{ MW} \qquad (3.163)$$

or

$$\text{SIL} = \sqrt{3}|\mathbf{V}_{R(\text{L-L})}||\mathbf{I}_L| \text{ W} \qquad (3.164)$$

where

$$\mathbf{I}_L = \frac{|\mathbf{V}_{R(\text{L-L})}|}{\sqrt{3} \times \sqrt{\frac{L}{C}}} \text{ A} \qquad (3.165)$$

where
    SIL = surge impedance loading in megawatts or watts
    $|k\mathbf{V}_{R(\text{L-L})}|$ = magnitude of line-to-line receiving-end voltage in kilovolts
    $|\mathbf{V}_{R(\text{L-L})}|$ = magnitude of line-to-line receiving-end voltage in volts
    Z_c = surge impedance in ohms $\cong \sqrt{L/C}$
    I_L = line current at SIL in amperes

The Safe Operating Limit (SOL) of a transmission line is often specified as a fraction of its surge impedance loading (SIL), which is used as a basis for comparing the load-carrying capacity of lines. However, SIL alone is not an indicator of the maximum power that a line can deliver, as other factors, such as line length, impedance of sending and receiving apparatus, and stability considerations must also be taken into account. Additionally, underground cables have a very low characteristic impedance, resulting in a much larger SIL (or natural load) than their rated load, effectively making them a source of lagging vars.

One way to increase the SIL of a line is to increase the voltage level, as seen in Equation 3.160, which shows that SIL increases with the square of the voltage. However, raising the voltage level can be costly. Instead, the surge impedance of the line can be reduced by adding capacitors or induction coils. There are four possible ways to alter line capacitance or inductance, as illustrated in Figures 3.26 and 3.27.

For a lossless line, the characteristic impedance and the propagation constant can be expressed as

$$Z_c = \sqrt{\frac{L}{C}} \qquad (3.166)$$

and

$$\gamma = \sqrt{LC} \qquad (3.167)$$

Hence, adding lumped inductances in series will increase the line inductance, which in turn increases the characteristic impedance and propagation constant and is not desirable. Adding incremental lumped inductance in parallel reduces the line capacitance, which decreases the propagation constant but increases the characteristic impedance, again not desirable. Adding capacitances in parallel increases the line capacitance, which decreases the characteristic impedance but increases the propagation constant, negatively affecting system stability. For short lines, this method may be effective.

The addition of capacitances in series, on the other hand, decreases the line inductance, reducing both the characteristic impedance and the propagation constant, which is desirable. Therefore, series capacitor compensation is used to improve stability limits and voltage regulation, provide a desired load division, and maximize the load-carrying capability of the system. However, during short circuits, full line current passing through the capacitors connected in series causes harmful overvoltages, making them problematic for line protective relaying. They introduce an impedance discontinuity (negative inductance) and subharmonic currents during fault conditions, and when the

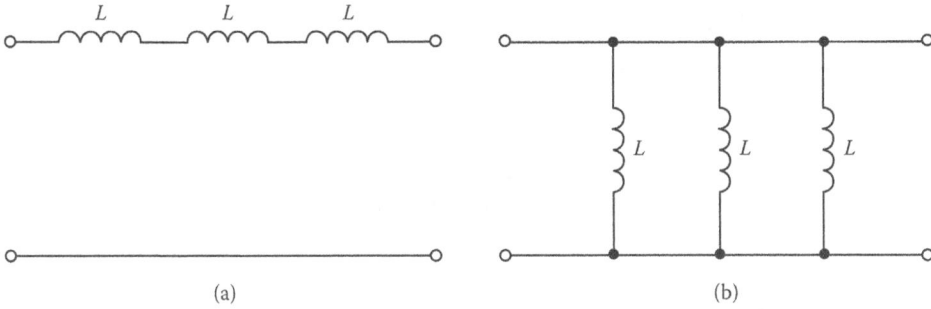

Figure 3.26 Transmission line compensation by adding lump inductances in (a) series or (b) parallel (i.e., shunt).

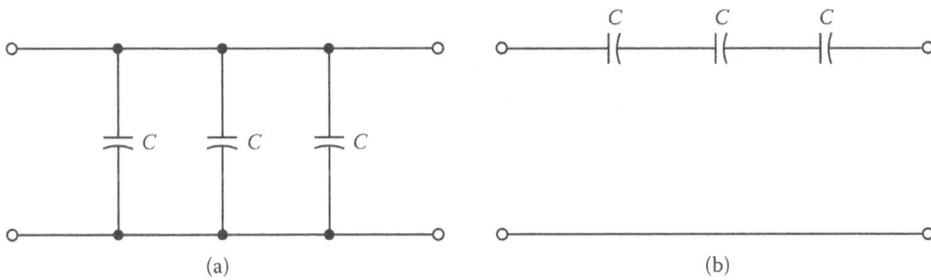

Figure 3.27 Transmission line compensation by adding capacitances in (a) parallel (i.e., shunt) or (b) series.

capacitor protective gap operates, they impress high-frequency currents and voltages on the system, causing incorrect operation of conventional relaying schemes. While series capacitance compensation has been attempted for distribution lines, it is not widely used.

Example 3.11: Determine the SIL of the transmission line given in Example 2.8.
Solution:
The approximate value of the surge impedance of the line is

$$Z_c \cong \sqrt{\frac{L}{C}}$$
$$= \left(\frac{2.6 \times 10^{-3}}{0.012 \times 10^{-6}} \right)^{1/2}$$
$$= 465.5 \ \Omega$$

Therefore,

* The occasional application of series compensation on new EHV lines has given rise to the problem of subsynchronous resonance, which can be defined as an oscillation that occurs due to the interaction between a series capacitor-compensated transmission system in electrical resonance and a turbine generator mechanical system in torsional mechanical resonance. This interaction results in the introduction of a negative resistance into the electric circuit by the turbine generator, which can increase oscillations until mechanical failures occur in the form of flexing or even breaking of the shaft. This event takes place when the electrical subsynchronous resonance frequency is equal

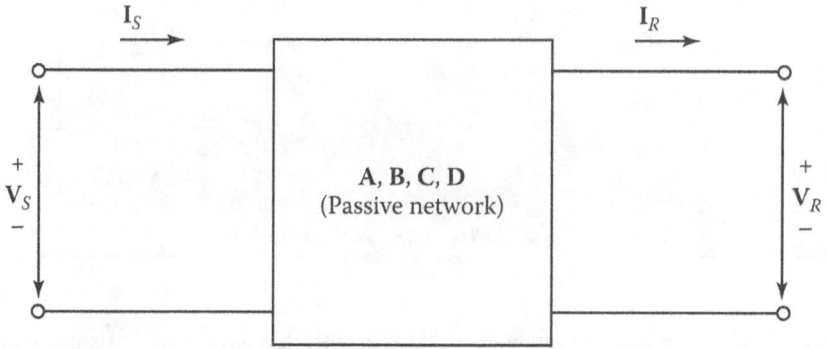

Figure 3.28 General two-port, four-terminal network.

or close to 60 Hz minus the frequency of one of the natural torsional modes of the turbine generator. The most well-known subsynchronous resonance problem occurred at Mojave Generating Station [76–79].

$$SIL \cong \frac{|k\mathbf{V}_{R(\text{L-L})}|^2}{\sqrt{\frac{L}{C}}}$$

$$= \frac{|138|^2}{469.5}$$

$$= 0.913 \text{ MW}$$

which is an approximate value of the SIL of the line. The exact value of the SIL of the line can be determined as

$$SIL \cong \frac{|k\mathbf{V}_{R(\text{L-L})}|^2}{Z_c}$$

$$= \frac{|138|^2}{469.62}$$

$$= 40.552 \text{ MW}$$

3.12   GENERAL CIRCUIT CONSTANTS

Figure 3.28 shows a general two-port, four-terminal network consisting of passive impedances connected in some fashion. From general network theory,

$$\mathbf{V}_S = \mathbf{A}\mathbf{V}_R + \mathbf{B}\mathbf{I}_R \tag{3.168}$$

and

$$\mathbf{I}_S = \mathbf{C}\mathbf{V}_R + \mathbf{D}\mathbf{I}_R \tag{3.169}$$

Also,

$$\mathbf{V}_S = \mathbf{D}\mathbf{V}_S - \mathbf{B}\mathbf{I}_S \tag{3.170}$$

and

$$\mathbf{I}_S = -\mathbf{C}\mathbf{V}_S - \mathbf{A}\mathbf{I}_S \tag{3.171}$$

It is always true that the determinant of Equations 3.166 and 3.167 or Equations 3.168 and 3.169 is always unity, that is,

$$\mathbf{A}\mathbf{D} - \mathbf{B}\mathbf{C} = 1 \tag{3.172}$$

In the above equations, **A**, **B**, **C**, and **D** are constants for a given network and are called general circuit constants. Their values depend on the parameters of the circuit concerned and the particular representation chosen. In general, they are complex numbers. For a network that has the symmetry of the uniform transmission line,

$$\mathbf{A} = \mathbf{D} \tag{3.173}$$

3.12.1 DETERMINATION OF A, B, C, AND D CONSTANTS

The **A**, **B**, **C**, and **D** constants can be calculated directly by network reduction. For example, when $\mathbf{I}_R = 0$, from Equation 3.167,

$$\mathbf{A} = \frac{\mathbf{V}_S}{\mathbf{V}_R} \tag{3.174}$$

and from Equation 3.167,

$$\mathbf{C} = \frac{\mathbf{I}_S}{\mathbf{V}_R} \tag{3.175}$$

Therefore, the **A** constant is the ratio of the sending- and receiving-end voltages, whereas the **C** constant is the ratio of sending-end current to receiving-end voltage when the receiving end is open circuited. When $\mathbf{V}_R = 0$, from Equation 3.166,

$$\mathbf{B} = \frac{\mathbf{V}_S}{\mathbf{I}_R} \tag{3.176}$$

When $\mathbf{V}_R = 0$, from Equation 3.167,

$$\mathbf{D} = \frac{\mathbf{I}_S}{\mathbf{I}_R} \tag{3.177}$$

The **B** constant represents the ratio of the sending-end voltage to the receiving-end current when the receiving end is short-circuited, while the **D** constant represents the ratio of the sending-end and receiving-end currents when the receiving end is short-circuited. The generalized circuit constants **A**, **B**, **C**, and **D** can also be indirectly calculated using knowledge of the system impedance parameters, as discussed in the previous sections. Table 3.1 provides the general circuit constants for different network types. It is worth noting that the dimensions of the **A** and **D** constants are numeric, while the dimension of the **B** constant is impedance in ohms, and the dimension of the **C** constant is admittance in siemens, as demonstrated in Equations 3.166 and 3.167.

3.12.2 MEASUREMENT OF ABCD PARAMETERS BY TEST

Since the transmission line is symmetrical, the measurement of the open circuit and short circuit is enough to determine the constants.

Short-circuit test: Short-circuiting one end of the line, the short-circuit impedance is measured at the other end of the line as

$$\mathbf{Z}_{sc} = \frac{\mathbf{B}}{\mathbf{A}} \tag{3.178}$$

since the short-circuited end voltage is zero, and using the reciprocity theorem for a symmetrical network,

$$\mathbf{A} = \mathbf{D} \tag{3.179}$$

then from Equation 3.176,

$$\mathbf{B} = \mathbf{A}\mathbf{Z}_{sc} \tag{3.180}$$

Open-circuit test: Open-circuiting one end of the line, the open-circuit impedance $\mathbf{Z}_{oc}$ is measured at the other end is zero, then

$$\mathbf{Z}_{oc} = \frac{\mathbf{A}}{\mathbf{C}} \tag{3.181}$$

Table 3.1
General Circuit Constants for Different Network Types

Network number	Type of network	Equations for general circuit constants in terms of constants of component networks			
		$A =$	$B =$	$C =$	$D =$
1	Series impedance	1	Z	0	1
2	Shunt admittance	1	0	Y	1
3	Transformer	$1 + \frac{Z_T Y_T}{2}$	$Z_T\left(1 + \frac{Z_T Y_T}{4}\right)$	Y_T	$1 + \frac{Z_T Y_T}{2}$
4	Transmission line	$\text{Cosh}\sqrt{ZY} = \left(1 + \frac{ZY}{2} + \frac{Z^2 Y^2}{24} + \cdots\right)$	$\sqrt{Z/Y}\,\text{Sinh}\sqrt{ZY} = Z\left(1 + \frac{ZY}{6} + \frac{Z^2 Y^2}{120} + \cdots\right)$	$\sqrt{Y/Z}\,\text{Sinh}\sqrt{ZY} = Y\left(1 + \frac{ZY}{6} + \frac{Z^2 Y^2}{120} + \cdots\right)$	Same as A
5	General network	A	B	C	D
6	General network and transformer impedance at receiving end	A_1	$B_1 + A_1 Z_{TR}$	C_1	$D_1 + C_1 Z_{TR}$
7	General network and transformer impedance at sending end	$A_1 + C_1 Z_{TS}$	$B_1 + D_1 Z_{TS}$	C_1	D_1
8	General network and transformer impedance at both ends–referred to high voltage	$A_1 + C_1 Z_{TS}$	$B_1 + A_1 Z_{TR} + D_1 Z_{TS} + C_1 Z_{TR} Z_{TS}$	C_1	$D_1 + C_1 Z_{TR}$
9	General network and transformer impedance at both ends–transformers having different ratios T_R and T_S referred to low voltage	$\frac{T_R}{T_S}(A_1 + C_1 Z_{TS})$	$\frac{1}{T_R T_S}(B_1 + A_1 Z_{TR} + D_1 Z_{TS} + C_1 Z_{TR} Z_{TS})$	$C_1 T_R T_S$	$\frac{T_S}{T_R}(D_1 + C_1 Z_{TR})$

(Continued on next page)

Table 3.1
General Circuit Constants for Different Network Types *Continued*.

Network number	Type of network	Equations for general circuit constants in terms of constants of component networks			
		$A =$	$B =$	$C =$	$D =$
10	General network and shunt admittance at receiving ends	$A_1 + B_1 Y_R$	B_1	$C_1 + D_1 Y_R$	D_1
11	General network and shunt admittance at sending end	A_1	B_1	$C_1 + A_1 Y_S$	$D_1 + B_1 Y_S$
12	General network and shunt admittance at both ends	$A_1 + B_1 Y_R$	B_1	$C_1 + A_1 Y_S + D_1 Y_R + B_1 Y_R Y_S$	$D_1 + B_1 Y_S$
13	Two general networks in series	$A_1 A_2 + C_1 B_2$	$B_1 A_2 + D_1 B_2$	$A_1 C_2 + C_1 D_2$	$B_1 C_2 + D_1 D_2$
14	Two general networks in series with intermediate impedance	$A_1 A_2 + C_1 B_2 + C_1 A_2 Z$	$B_1 A_2 + D_1 B_2 + D_1 A_2 Z$	$A_1 C_2 + C_1 D_2 + C_1 C_2 Z$	$B_1 C_2 + D_1 D_2 + D_1 C_2 Z$
15	Two general networks in series with intermediate shunt admittance	$A_1 A_2 + C_1 B_2 + A_1 B_2 Y$	$B_1 A_2 - D_1 B_2 + B_1 B_2 Y$	$A_1 C_2 + C_1 D_2 + A_1 D_2 Y$	$B_1 C_2 + D_1 D_2 + B_1 D_2 Y$
16	Three general networks in series	$A_3(A_1 A_2 + C_1 B_2) + B_3(A_1 C_2 + C_1 D_2)$	$A_3(B_1 A_2 + D_1 B_2) + B_3(B_1 C_2 + D_1 D_2)$	$C_3(A_1 A_2 + C_1 B_2) + D_3(A_1 C_2 + C_1 D_2)$	$C_3(B_1 A_2 + D_1 B_2) + D_3(B_1 C_2 + D_1 D_2)$
17	Two general networks in parallel	$\dfrac{A_1 B_2 + B_1 A_2}{B_1 + B_2}$	$\dfrac{B_1 B_2}{B_1 + B_2}$	$C_1 + C_2 + \dfrac{(A_1 - A_2)(D_2 - D_1)}{B_1 + B_2}$	$\dfrac{B_1 D_2 + D_1 B_2}{B_1 + B_2}$

since the current at the open-circuited end is zero, then

$$C = \frac{\mathbf{A}}{\mathbf{Z}_{oc}} \tag{3.182}$$

For a passive network the determinant is

$$\Delta = \begin{vmatrix} \mathbf{A} & \mathbf{B} \\ \mathbf{C} & \mathbf{D} \end{vmatrix} = \mathbf{AD} - \mathbf{BC} = 1 \tag{3.183}$$

By substituting Equation 3.177 into Equation 3.181,

$$\mathbf{A}^2 - \mathbf{BC} = 1 \tag{3.184}$$

Thus, by substituting Equation 3.178 and 3.180 into Equation 3.182,

$$\mathbf{A}^2 - \mathbf{A}\mathbf{Z}_{sc}\left(\frac{\mathbf{A}}{\mathbf{Z}_{sc}}\right) = 1 \tag{3.185}$$

or

$$\left(\frac{z_{oc} - z_{sc}}{z_{oc}}\right)\mathbf{A}^2 = 1 \tag{3.186}$$

Hence,

$$\mathbf{A} = \sqrt{\frac{\mathbf{Z}_{oc} - \mathbf{Z}_{sc}}{\mathbf{Z}_{oc}}} \tag{3.187}$$

Substituting Equation 3.185 into Equation 3.178,

$$\mathbf{B} = \mathbf{Z}_{sc}\sqrt{\frac{\mathbf{Z}_{oc} - \mathbf{Z}_{sc}}{\mathbf{Z}_{oc}}} \tag{3.188}$$

and substituting Equation 3.185 into Equation 3.180,

$$\mathbf{C} = \frac{1}{\sqrt{\mathbf{Z}_{oc}(\mathbf{Z}_{oc} - \mathbf{Z}_{sc})}} \tag{3.189}$$

3.12.3 ABCD CONSTANTS OF TRANSFORMER

Figure 3.29 shows the equivalent circuit of a transformer at no load. Neglecting its series impedance,

$$\begin{bmatrix} \mathbf{V}_S \\ \mathbf{I}_S \end{bmatrix} = \begin{bmatrix} \mathbf{A} & \mathbf{B} \\ \mathbf{C} & \mathbf{D} \end{bmatrix} \begin{bmatrix} \mathbf{V}_R \\ \mathbf{I}_R \end{bmatrix} \tag{3.190}$$

where the transfer matrix is

$$\begin{bmatrix} \mathbf{A} & \mathbf{B} \\ \mathbf{C} & \mathbf{D} \end{bmatrix} = \begin{bmatrix} 1 & 0 \\ \mathbf{Y}_T & 1 \end{bmatrix} \tag{3.191}$$

since

$$\mathbf{V}_S = \mathbf{V}_R \tag{3.192}$$

and

$$\mathbf{I}_S = \mathbf{Y}_I \mathbf{V}_R + \mathbf{I}_R \tag{3.193}$$

and where $\mathbf{Y}_T$ is the magnetizing admittance of the transformer.

Figure 3.30 shows the equivalent circuit of a transformer at full load that has a transfer matrix of

$$\begin{bmatrix} \mathbf{A} & \mathbf{B} \\ \mathbf{C} & \mathbf{D} \end{bmatrix} = \begin{bmatrix} 1 + \frac{\mathbf{Z}_T\mathbf{Y}_T}{2} & \mathbf{Z}_T\left(1 + \frac{\mathbf{Z}_T\mathbf{Y}_T}{4}\right) \\ \mathbf{Y}_T & 1 + \frac{\mathbf{Z}_T\mathbf{Y}_T}{2} \end{bmatrix} \tag{3.194}$$

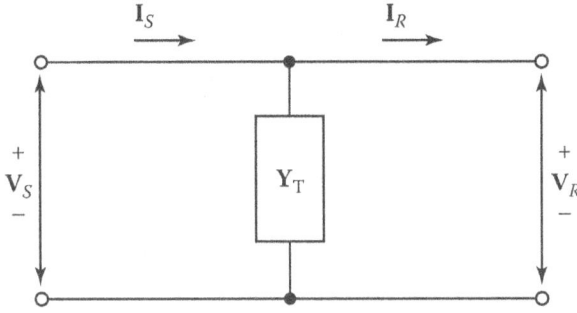

Figure 3.29 Transformer equivalent circuit at no load.

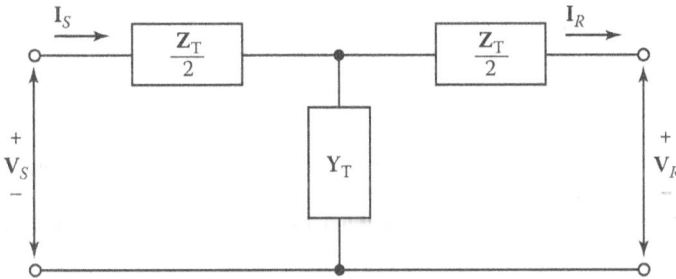

Figure 3.30 Transformer equivalent circuit at full load.

since

$$\mathbf{V}_S = \left(1 + \frac{\mathbf{Z}_T\mathbf{Y}_T}{2}\right)\mathbf{V}_R + \mathbf{Z}_T\left(1 + \frac{\mathbf{Z}_T\mathbf{Y}_T}{4}\right)\mathbf{I}_R \tag{3.195}$$

and

$$\mathbf{I}_S = (\mathbf{Y}_T)\mathbf{V}_R + \left(1 + \frac{\mathbf{Z}_T\mathbf{Y}_T}{2}\right)\mathbf{I}_R \tag{3.196}$$

where $\mathbf{Z}_T$ is the total equivalent series impedance of the transformer.

3.12.4   ASYMMETRICAL π AND T NETWORKS

Figure 3.31 shows an asymmetrical π network that can be thought of as a series (or *cascade*, or *tandem*) connection of a shunt admittance, a series impedance, and a shunt admittance.

The equivalent transfer matrix can be found by multiplying together the transfer matrices of individual components. Thus,

$$\begin{aligned}
\begin{bmatrix} \mathbf{A} & \mathbf{B} \\ \mathbf{C} & \mathbf{D} \end{bmatrix} &= \begin{bmatrix} 1 & 0 \\ \mathbf{Y}_1 & 1 \end{bmatrix}\begin{bmatrix} 1 & \mathbf{Z} \\ 0 & 1 \end{bmatrix}\begin{bmatrix} 1 & 0 \\ \mathbf{Y}_2 & 1 \end{bmatrix} \\
&= \begin{bmatrix} 1 + \mathbf{Z}\mathbf{Y}_1 & \mathbf{Z} \\ \mathbf{Y}_1 + \mathbf{Y}_2\mathbf{Z}\mathbf{Y}_1\mathbf{Y}_2 & 1 + \mathbf{Z}\mathbf{Y}_1 \end{bmatrix}
\end{aligned} \tag{3.197}$$

When the π network is symmetrical,

$$\mathbf{Y}_1 = \mathbf{Y}_2 = \frac{\mathbf{Y}}{2}$$

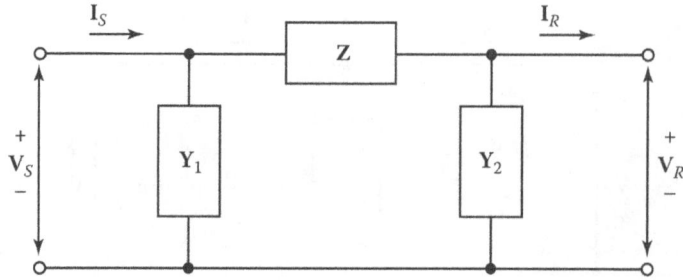

Figure 3.31 Asymmetrical π network.

and the transfer matrix becomes

$$\begin{bmatrix} \mathbf{A} & \mathbf{B} \\ \mathbf{C} & \mathbf{D} \end{bmatrix} = \begin{bmatrix} 1 + \frac{\mathbf{ZY}}{2} & \mathbf{Z} \\ \mathbf{Y} + \frac{\mathbf{ZY}^2}{4} & 1 + \frac{\mathbf{ZY}}{2} \end{bmatrix} \tag{3.198}$$

which is the same as Equation 3.56 for a nominal π circuit of a medium-length transmission line.

Figure 3.32 shows an asymmetrical T network that can be thought of as a cascade connection of a series impedance, a shunt admittance, and a series impedance. Again, the equivalent transfer matrix can be found by multiplying together the transfer matrices of individual components. Thus,

$$\begin{bmatrix} \mathbf{A} & \mathbf{B} \\ \mathbf{C} & \mathbf{D} \end{bmatrix} = \begin{bmatrix} 1 & \mathbf{Z}_1 \\ 0 & 1 \end{bmatrix} \begin{bmatrix} 1 & 0 \\ \mathbf{Y} & 1 \end{bmatrix} \begin{bmatrix} 1 & \mathbf{Z}_2 \\ 0 & 1 \end{bmatrix}$$

$$= \begin{bmatrix} 1 + \mathbf{Z}_1\mathbf{Y} & \mathbf{Z}_1 + \mathbf{Z}_2 + \mathbf{Z}_1\mathbf{Z}_2\mathbf{Y} \\ \mathbf{Y} & 1 + \mathbf{Z}_2\mathbf{Y} \end{bmatrix} \tag{3.199}$$

When the T network is symmetrical,

$$\mathbf{Z}_1 = \mathbf{Z}_2 = \frac{\mathbf{Z}}{2}$$

and the transfer matrix becomes

$$\begin{bmatrix} \mathbf{A} & \mathbf{B} \\ \mathbf{C} & \mathbf{D} \end{bmatrix} = \begin{bmatrix} 1 + \frac{\mathbf{ZY}}{2} & \mathbf{Z} + \frac{\mathbf{ZY}^2}{4} \\ \mathbf{Y} & 1 + \frac{\mathbf{ZY}}{2} \end{bmatrix} \tag{3.200}$$

which is the same as the equation for a nominal T circuit of a medium-length transmission line.

3.12.5 NETWORKS CONNECTED IN SERIES

Two four-terminal transmission networks may be connected in series, as shown in Figure 3.33, to form a new four-terminal transmission network. For the first four-terminal network,

$$\begin{bmatrix} \mathbf{V}_S \\ \mathbf{I}_S \end{bmatrix} = \begin{bmatrix} \mathbf{A}_1 & \mathbf{B}_1 \\ \mathbf{C}_1 & \mathbf{D}_1 \end{bmatrix} \begin{bmatrix} \mathbf{V} \\ \mathbf{I} \end{bmatrix} \tag{3.201}$$

and for the second four-terminal network,

$$\begin{bmatrix} \mathbf{V} \\ \mathbf{I} \end{bmatrix} = \begin{bmatrix} \mathbf{A}_2 & \mathbf{B}_2 \\ \mathbf{C}_2 & \mathbf{D}_2 \end{bmatrix} \begin{bmatrix} \mathbf{V}_R \\ \mathbf{I}_R \end{bmatrix} \tag{3.202}$$

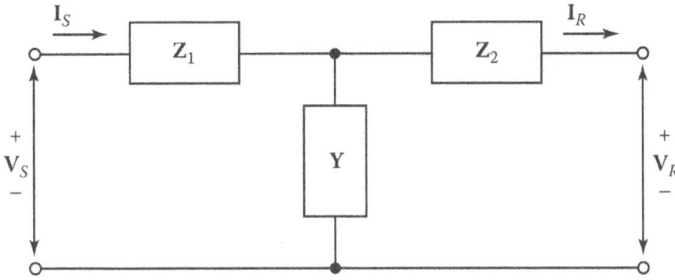

Figure 3.32 Asymmetrical T network.

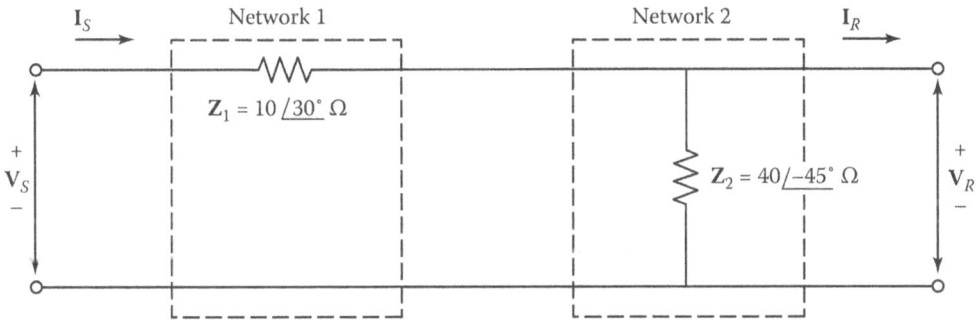

Figure 3.33 Two four-terminal transmission networks.

By substituting Equation 3.200 into Equation 3.199,

$$
\begin{bmatrix} \mathbf{V}_S \\ \mathbf{I}_S \end{bmatrix} = \begin{bmatrix} \mathbf{A}_1 & \mathbf{B}_1 \\ \mathbf{C}_1 & \mathbf{D}_1 \end{bmatrix} \begin{bmatrix} \mathbf{A}_2 & \mathbf{B}_2 \\ \mathbf{C}_2 & \mathbf{D}_2 \end{bmatrix} \begin{bmatrix} \mathbf{V}_R \\ \mathbf{I}_R \end{bmatrix}
$$
$$
= \begin{bmatrix} \mathbf{A}_1\mathbf{A}_2 + \mathbf{B}_1\mathbf{C}_2 & \mathbf{A}_1\mathbf{B}_2 + \mathbf{B}_1\mathbf{D}_2 \\ \mathbf{C}_1\mathbf{A}_2 + \mathbf{D}_1\mathbf{C}_2 & \mathbf{C}_1\mathbf{B}_2 + \mathbf{D}_1\mathbf{D}_2 \end{bmatrix} \begin{bmatrix} \mathbf{V}_R \\ \mathbf{I}_R \end{bmatrix}
$$

(3.203)

Thus, the equivalent **A**, **B**, **C**, and **D** constants for two networks connected in series are

$$\mathbf{A}_{eq} = \mathbf{A}_1\mathbf{A}_2 + \mathbf{B}_1\mathbf{C}_2 \tag{3.204}$$

$$\mathbf{B}_{eq} = \mathbf{A}_1\mathbf{B}_2 + \mathbf{B}_1\mathbf{D}_2 \tag{3.205}$$

$$\mathbf{C}_{eq} = \mathbf{C}_1\mathbf{A}_2 + \mathbf{D}_1\mathbf{C}_2 \tag{3.206}$$

$$\mathbf{D}_{eq} = \mathbf{C}_1\mathbf{B}_2 + \mathbf{D}_1\mathbf{D}_2 \tag{3.207}$$

Example 3.12: Figure 3.34 shows two networks connected in cascade. Determine the equivalent **A**, **B**, **C**, and **D** constants.

Solution:

For network 1,

$$
\begin{bmatrix} \mathbf{A}_1 & \mathbf{B}_1 \\ \mathbf{C}_1 & \mathbf{D}_1 \end{bmatrix} = \begin{bmatrix} 1 & 10\angle 30° \\ 0 & 1 \end{bmatrix}
$$

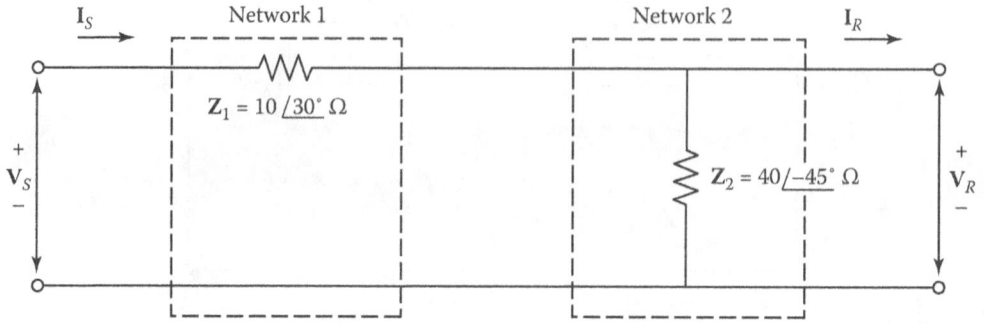

Figure 3.34 Network configurations for Example 3.12.

For network 2,

$$\mathbf{Y}_2 = \frac{1}{\mathbf{Z}_2} = \frac{1}{40 - \angle 45°} = 0.025 \angle 45° \text{ S}$$

Then

$$\begin{bmatrix} \mathbf{A}_2 & \mathbf{B}_2 \\ \mathbf{C}_2 & \mathbf{D}_2 \end{bmatrix} = \begin{bmatrix} 1 & 0 \\ 0.025 \angle 45° & 1 \end{bmatrix}$$

Therefore,

$$\begin{bmatrix} \mathbf{A}_{EQ} & \mathbf{B}_{EQ} \\ \mathbf{C}_{EQ} & \mathbf{D}_{EQ} \end{bmatrix} = \begin{bmatrix} 1 & 10 \angle 30° \\ 0 & 1 \end{bmatrix} \begin{bmatrix} 1 & 0 \\ 0.025 \angle 45° & 1 \end{bmatrix}$$

$$= \begin{bmatrix} 1.09 \angle 12.8° & 10 \angle 30° \\ 0.025 \angle 45° & 1 \end{bmatrix}$$

3.12.6 NETWORKS CONNECTED IN PARALLEL

Two four-terminal transmission networks may be connected in parallel, as shown in Figure 3.35, to form a new four-terminal transmission network.

Since

$$\mathbf{V}_S = \mathbf{V}_{S1} + \mathbf{V}_{S2} \tag{3.208}$$

$$\mathbf{V}_R = \mathbf{V}_{R1} + \mathbf{V}_{R2} \tag{3.209}$$

and

$$\mathbf{I}_S = \mathbf{I}_{S1} + \mathbf{I}_{S2} \tag{3.210}$$

$$\mathbf{I}_R = \mathbf{I}_{R1} + \mathbf{I}_{R2} \tag{3.211}$$

for the equivalent four-terminal network,

$$\begin{bmatrix} \mathbf{V}_S \\ \mathbf{I}_S \end{bmatrix} = \begin{bmatrix} \frac{\mathbf{A}_1\mathbf{B}_2+\mathbf{A}_2\mathbf{B}_1}{\mathbf{B}_1+\mathbf{B}_2} & \frac{\mathbf{B}_1\mathbf{B}_2}{\mathbf{B}_1+\mathbf{B}_2} \\ \mathbf{C}_1 + \mathbf{C}_2 + \frac{(\mathbf{A}_1-\mathbf{A}_2)(\mathbf{D}_2-\mathbf{D}_1)}{\mathbf{B}_1+\mathbf{B}_2} & \frac{\mathbf{D}_1\mathbf{B}_2+\mathbf{D}_2\mathbf{B}_1}{\mathbf{B}_1+\mathbf{B}_2} \end{bmatrix} \begin{bmatrix} \mathbf{V}_R \\ \mathbf{I}_R \end{bmatrix} \tag{3.212}$$

where the equivalent **A**, **B**, **C**, and **D** constants are

$$\mathbf{A}_{eq} = \frac{\mathbf{A}_1\mathbf{B}_2 + \mathbf{A}_2\mathbf{B}_1}{\mathbf{B}_1 + \mathbf{B}_2} \tag{3.213}$$

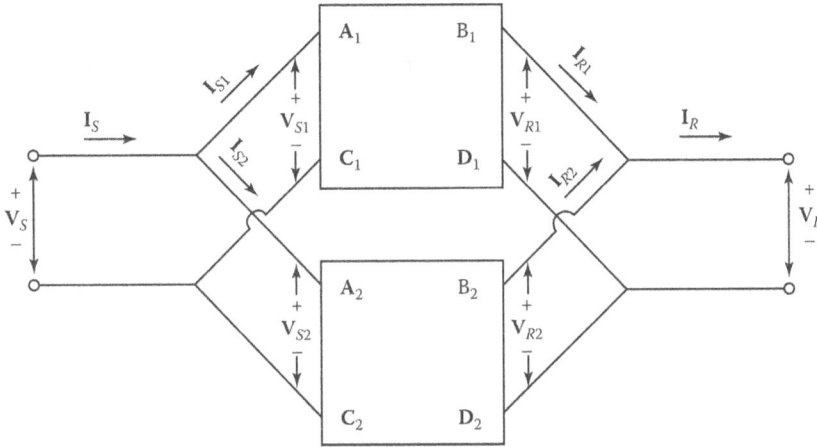

Figure 3.35 Transmission networks in parallel.

$$\mathbf{B}_{eq} = \frac{\mathbf{B}_1 \mathbf{B}_2}{\mathbf{B}_1 + \mathbf{B}_2} \tag{3.214}$$

$$\mathbf{C}_{eq} = \mathbf{C}_1 + \mathbf{C}_2 + \frac{(\mathbf{A}_1 - \mathbf{A}_2)(\mathbf{D}_2 - \mathbf{D}_1)}{\mathbf{B}_1 + \mathbf{B}_2} \tag{3.215}$$

$$\mathbf{D}_{eq} = \frac{\mathbf{D}_1 \mathbf{B}_2 + \mathbf{D}_2 \mathbf{B}_1}{\mathbf{B}_1 + \mathbf{B}_2} \tag{3.216}$$

Example 3.13: Assume that the two networks given in Example 3.13 are connected in parallel, as shown in Figure 3.12. Determine the equivalent **A**, **B**, **C**, and **D** constants (Figure 3.36).
Solution:
Using the **A**, **B**, **C**, and **D** parameters found previously for networks 1 and 2, that is,

$$\begin{bmatrix} \mathbf{A}_1 & \mathbf{B}_1 \\ \mathbf{C}_1 & \mathbf{D}_1 \end{bmatrix} = \begin{bmatrix} 1 & 10\angle 30° \\ 0 & 1 \end{bmatrix}$$

and

$$\begin{bmatrix} \mathbf{A}_2 & \mathbf{B}_2 \\ \mathbf{C}_2 & \mathbf{D}_2 \end{bmatrix} = \begin{bmatrix} 1 & 0 \\ 0.025\angle 45° & 1 \end{bmatrix}$$

the equivalent **A**, **B**, **C**, and **D** constants can be calculated as

$$\begin{aligned} \mathbf{A}_{eq} &= \frac{\mathbf{A}_1 \mathbf{B}_2 + \mathbf{A}_2 \mathbf{B}_1}{\mathbf{B}_1 + \mathbf{B}_2} \\ &= \frac{1 \times 0 + 1 \times 10\angle 30°}{10\angle 30° + 0} \\ &= 1 \end{aligned}$$

$$\begin{aligned} \mathbf{B}_{eq} &= \frac{\mathbf{B}_1 \mathbf{B}_2}{\mathbf{B}_1 + \mathbf{B}_2} \\ &= \frac{(10\angle 30°) \times 0}{(10\angle 30°) + 0} \\ &= 0 \end{aligned}$$

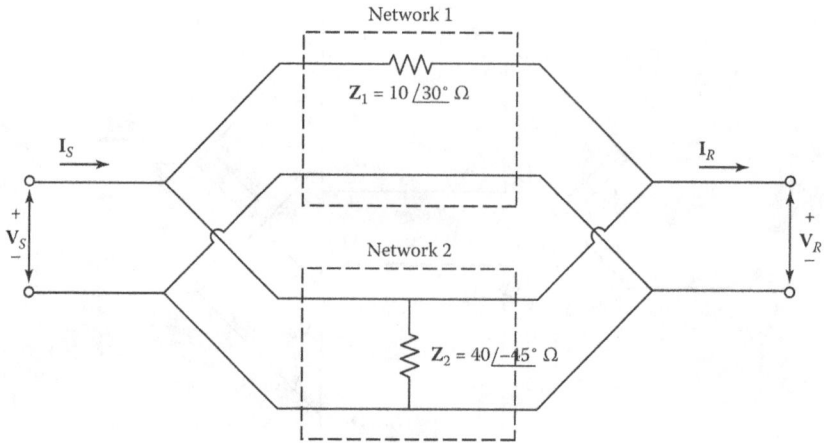

Figure 3.36 Transmission networks in parallel for Example 3.13.

$$\mathbf{C}_{eq} = \mathbf{C}_1 + \mathbf{C}_2 + \frac{(\mathbf{A}_1 - \mathbf{A}_2)(\mathbf{D}_2 - \mathbf{D}_1)}{\mathbf{B}_1 + \mathbf{B}_2}$$

$$= 0 + 0.025\angle 45^\circ + \frac{(1-1)(1-1)}{10\angle 30^\circ - 0}$$

$$= 0.025\angle 45^\circ$$

$$\mathbf{D}_{eq} = \frac{\mathbf{D}_1\mathbf{B}_2 + \mathbf{D}_2\mathbf{B}_1}{\mathbf{B}_1 + \mathbf{B}_2}$$

$$= \frac{1 \times 0 + 1 \times 10\angle 30^\circ}{10\angle 30^\circ + 0}$$

$$= 1$$

Therefore,

$$\begin{bmatrix} \mathbf{A}_{eq} & \mathbf{B}_{eq} \\ \mathbf{C}_{eq} & \mathbf{D}_{eq} \end{bmatrix} = \begin{bmatrix} 1 & 0 \\ 0.025\angle 45^\circ & 1 \end{bmatrix}$$

3.12.7 TERMINATED TRANSMISSION LINE

Figure 3.37 shows a four-terminal transmission network connected to (i.e., terminated by) a load $\mathbf{Z}_L$. For the given network,

$$\begin{bmatrix} \mathbf{V}_S \\ \mathbf{I}_S \end{bmatrix} = \begin{bmatrix} \mathbf{A} & \mathbf{B} \\ \mathbf{C} & \mathbf{D} \end{bmatrix} \begin{bmatrix} \mathbf{V}_R \\ \mathbf{I}_R \end{bmatrix} \tag{3.217}$$

$$\mathbf{V}_S = \mathbf{A}\mathbf{V}_R + \mathbf{B}\mathbf{I}_R \tag{3.218}$$

and

$$\mathbf{I}_S = \mathbf{C}\mathbf{V}_R + \mathbf{D}\mathbf{I}_R \tag{3.219}$$

and also

$$\mathbf{V}_R = \mathbf{Z}_L\mathbf{I}_R \tag{3.220}$$

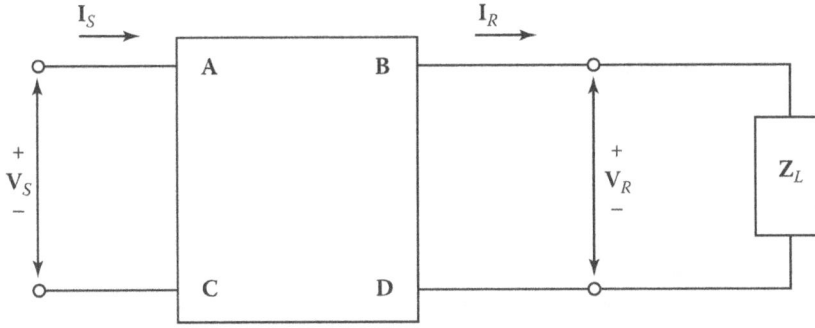

Figure 3.37 Terminated transmission line.

Therefore, the input impedance is

$$\mathbf{Z}_{\text{in}} = \mathbf{V}_S \mathbf{I}_S$$
$$= \frac{\mathbf{A}\mathbf{V}_R + \mathbf{B}\mathbf{I}_R}{\mathbf{C}\mathbf{V}_R + \mathbf{D}\mathbf{I}_R} \tag{3.221}$$

or by substituting Equation 3.169 into Equation 3.170,

$$\mathbf{Z}_{\text{in}} = \frac{\mathbf{A}\mathbf{Z}_L + \mathbf{B}}{\mathbf{C}\mathbf{Z}_L + \mathbf{D}} \tag{3.222}$$

Since for the symmetrical and long transmission line,

$$\mathbf{A} = \cosh\sqrt{\mathbf{YZ}} = \cosh\theta$$

$$\mathbf{B} = \sqrt{\frac{\mathbf{Z}}{\mathbf{Y}}}\sinh\sqrt{\mathbf{YZ}} = \mathbf{Z}_c\sinh\theta$$

$$\mathbf{C} = \sqrt{\frac{\mathbf{Y}}{\mathbf{Z}}}\sinh\sqrt{\mathbf{YZ}} = \mathbf{Y}_c\sinh\theta$$

$$\mathbf{D} = \mathbf{A} = \cosh\sqrt{\mathbf{YZ}} = \cosh\theta$$

the input impedance, from Equation 3.171, becomes

$$\mathbf{Z}_{\text{in}} = \frac{\mathbf{Z}_L\cosh\theta + \mathbf{Z}_c\sinh\theta}{\mathbf{Z}_L\mathbf{Y}_c\sinh\theta + \cosh\theta} \tag{3.223}$$

or

$$\mathbf{Z}_{\text{in}} = \frac{\mathbf{Z}_L[(\mathbf{Z}_c/\mathbf{Z}_L)\sinh\theta + \cosh\theta]}{(\mathbf{Z}_L/\mathbf{Z}_c)\sinh\theta + \cos\theta} \tag{3.224}$$

If the load impedance is chosen to be equal to the characteristic impedance, that is,

$$\mathbf{Z}_L = \mathbf{Z}_c \tag{3.225}$$

the input impedance, from Equation 3.173, becomes

$$\mathbf{Z}_{\text{in}} = \mathbf{Z}_c \tag{3.226}$$

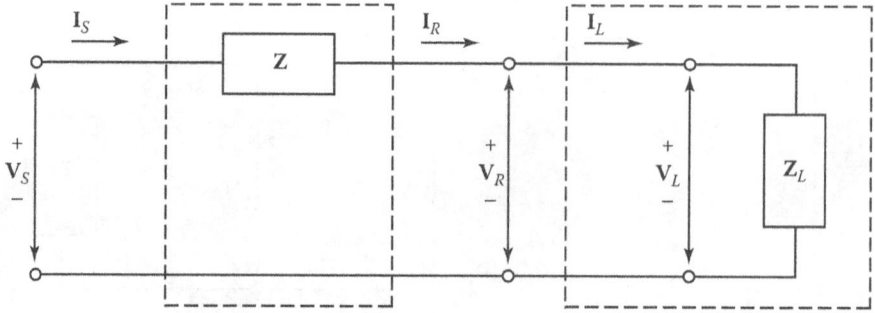

Figure 3.38 Transmission system for Example 3.14.

which is independent of θ and the line length. The value of the voltage is constant all along the line.

Example 3.14: Figure 3.38 shows a short transmission line that is terminated by a load of 200 kVA at a lagging power factor of 0.866 at 2.4 kV If the line impedance is 2.07 + j0.661 Ω, calculate

a. Sending-end current
b. Sending-end voltage
c. Input impedance
d. Real and reactive power loss in the line

Solution:

a. From Equation 3.166,

$$\begin{bmatrix} \mathbf{V}_S \\ \mathbf{I}_S \end{bmatrix} = \begin{bmatrix} \mathbf{A} & \mathbf{B} \\ \mathbf{C} & \mathbf{D} \end{bmatrix} \begin{bmatrix} \mathbf{V}_R \\ \mathbf{I}_R \end{bmatrix}$$
$$= \begin{bmatrix} 1 & \mathbf{Z} \\ 0 & 1 \end{bmatrix} \begin{bmatrix} \mathbf{V}_R \\ \mathbf{I}_R \end{bmatrix}$$

where

$$Z = 2.07 + j0.661 = 2.173\angle 17.7^\circ \ \Omega$$

$$\mathbf{I}_R = \mathbf{I}_S = \mathbf{I}_L$$

$$\mathbf{V}_R = \mathbf{Z}_L \mathbf{I}_R$$

Since

$$\mathbf{S}_R = 200\angle 30^\circ = 173.2 + j100 \ \text{kVA}$$

and

$$\mathbf{V}_R = 2.4\angle 0^\circ \text{kV}$$

then

$$\mathbf{I}_L^* = \frac{\mathbf{S}_R}{\mathbf{V}_L} = \frac{200\angle 30^\circ}{2.4\angle 0^\circ} = 83.33\angle 0^\circ \ \text{A}$$

or

$$\mathbf{I}_L = 83.33\angle -30° \text{ A}$$

hence,

$$\mathbf{I}_S = \mathbf{I}_R = \mathbf{I}_L = 83.33\angle -30° \text{ A}$$

b.

$$\mathbf{Z}_L = \frac{\mathbf{V}_L}{\mathbf{I}_L} = \frac{2.4 \times 10^3 \angle 0°}{83.33\angle -30°} = 28.8\angle 30° \text{ }\Omega$$

and

$$\mathbf{V}_R = \mathbf{Z}_L \mathbf{I}_R = 28.8\angle 30° \times 83.33\angle -30° = 2404\angle 0° \text{ kV}$$

thus,

$$\begin{aligned}
\mathbf{V}_S &= \mathbf{A}\mathbf{V}_R + \mathbf{B}\mathbf{I}_R \\
&= 2400\angle 0° + 2.173\angle 17.7° \times 83.33\angle -30° \\
&= 2576.9 - j38.58 \\
&= 2577.2\angle -0.9° \text{ V}
\end{aligned}$$

c. The input impedance is

$$\begin{aligned}
\mathbf{Z}_{\text{in}} &= \frac{\mathbf{V}_S}{\mathbf{I}_S} = \frac{\mathbf{A}\mathbf{V}_R + \mathbf{B}\mathbf{I}_R}{\mathbf{C}\mathbf{V}_R + \mathbf{D}\mathbf{I}_R} \\
&= \frac{2577.2\angle -0.9°}{83.33\angle -30°} = 30.93\angle 29.1° \text{ }\Omega
\end{aligned}$$

d. The real and reactive power loss in the line:

$$\mathbf{S}_L = \mathbf{S}_S - \mathbf{S}_R$$

where

$$\begin{aligned}
\mathbf{S}_S &= \mathbf{V}_S \mathbf{I}_S^* = 2577.2\angle -0.9° \times 83.33\angle 30° \\
&= 214,758\angle 29.1° \text{ VA}
\end{aligned}$$

or

$$\mathbf{S}_S = \mathbf{I}_S \times \mathbf{Z}_{\text{in}} \times \mathbf{I}_S^* = 214,758\angle 29.1° \text{ VA}$$

Thus,

$$\begin{aligned}
\mathbf{S}_L &= 214,758\angle 29.1° - 200,000\angle 30° \\
&= 14,444.5 - j4,444.4 \text{ VA}
\end{aligned}$$

that is, the active power loss is 14,444.5 W, and the reactive power loss is 4444.4 vars.

3.12.8 POWER RELATIONS USING A, B, C, AND D LINE CONSTANTS

For a given long transmission line, the complex power at the sending and receiving ends are

$$\mathbf{S}_S = P_S + jQ_S = \mathbf{V}_S\mathbf{I}_S^* \tag{3.227}$$

and

$$\mathbf{S}_R = P_R + jQ_R = \mathbf{V}_R\mathbf{I}_R^* \tag{3.228}$$

Also, the sending- and receiving-end voltages and currents can be expressed as

$$\mathbf{V}_S = \mathbf{A}\mathbf{V}_R + \mathbf{B}\mathbf{I}_R \tag{3.229}$$

$$\mathbf{I}_S = \mathbf{C}\mathbf{V}_R + \mathbf{D}\mathbf{I}_R \tag{3.230}$$

$$\mathbf{V}_R = \mathbf{A}\mathbf{V}_S - \mathbf{B}\mathbf{I}_S \tag{3.231}$$

$$\mathbf{I}_R = \mathbf{C}\mathbf{V}_S + \mathbf{D}\mathbf{I}_S \tag{3.232}$$

where

$$\mathbf{A} = A\angle\alpha = \cosh\sqrt{\mathbf{YZ}} \tag{3.233}$$

$$\mathbf{B} = A\angle\beta = \frac{\mathbf{Z}}{\mathbf{Y}}\sinh\sqrt{\mathbf{YZ}} \tag{3.234}$$

$$\mathbf{C} = C\angle\delta = \frac{\mathbf{Y}}{\mathbf{Z}}\sinh\sqrt{\mathbf{YZ}} \tag{3.235}$$

$$\mathbf{D} = \mathbf{A} = \cosh\sqrt{\mathbf{YZ}} \tag{3.236}$$

$$\mathbf{V}_R = V_R\angle 0° \tag{3.237}$$

$$\mathbf{V}_S = V_S\angle\delta \tag{3.238}$$

From Equation 3.215,

$$\mathbf{I}_S = \frac{\mathbf{A}}{\mathbf{B}}\mathbf{V}_S - \frac{\mathbf{V}_R}{\mathbf{B}} \tag{3.239}$$

or

$$\mathbf{I}_S = \frac{AV_S}{B}\angle\alpha + \delta - \beta - \frac{V_R\angle -\beta}{B} \tag{3.240}$$

and

$$\mathbf{I}_S^* = \frac{AV_S}{B}\angle -\alpha - \delta + \beta - \frac{V_R\angle\beta}{B} \tag{3.241}$$

and from Equation 3.225

$$\mathbf{I}_S = \frac{\mathbf{V}_S}{\mathbf{B}} - \frac{\mathbf{A}}{\mathbf{B}}\mathbf{V}_R \tag{3.242}$$

or

$$\mathbf{I}_R = \frac{V_S}{B}\angle\delta - \beta - \frac{AV_R}{B}\angle\alpha - \beta \tag{3.243}$$

and

$$\mathbf{I}_R^* = \frac{V_S}{B}\angle -\delta + \beta - \frac{AV_R}{B}\angle -\alpha + \beta \tag{3.244}$$

By substituting Equations 3.237 and 3.240 into Equations 3.223 and 3.224, respectively,

$$\mathbf{S}_S = P_S + jQ_S = \frac{AV_S^2}{B}\angle\beta - \alpha - \frac{V_SV_R}{B}\angle\beta + \alpha \tag{3.245}$$

and

$$\mathbf{S}_R = P_R + jQ_R = \frac{V_SV_R}{B}\angle\beta - \delta - \frac{AV_R^2}{B}\angle\beta - \alpha \tag{3.246}$$

Therefore, the real and reactive powers at the sending end are

$$\mathbf{P}_S = \frac{AV_S^2}{B}\cos(\beta - \alpha) - \frac{V_S V_R}{B}\cos(\beta + \alpha) \tag{3.247}$$

and

$$\mathbf{Q}_S = \frac{AV_S^2}{B}\sin(\beta - \delta) - \frac{V_S V_R}{B}\sin(\beta + \alpha) \tag{3.248}$$

and the real and reactive powers at the receiving end are

$$\mathbf{P}_R = \frac{V_S V_R}{B}\cos(\beta - \delta) - \frac{AV_R^2}{B}\cos(\beta - \alpha) \tag{3.249}$$

and

$$\mathbf{Q}_R = \frac{V_S V_R}{B}\sin(\beta - \delta) - \frac{AV_R^2}{B}\sin(\beta - \alpha) \tag{3.250}$$

For the constants V_S and V_R, for a given line, the only variable in Equations 3.245 through 3.248 is δ, the power angle. Therefore, treating P_S as a function of δ only in Equation 3.245, P_S is maximum when $\beta + \delta = 180°$. Therefore, the maximum power at the sending end, the maximum input power, can be expressed as

$$\mathbf{P}_{S,\mathrm{max}} = \frac{AV_S^2}{B}\cos(\beta - \alpha) + \frac{V_S V_R}{B} \tag{3.251}$$

and similarly the corresponding reactive power at the sending end, the maximum input vars, is

$$\mathbf{Q}_{S,\mathrm{max}} = \frac{AV_S^2}{B}\sin(\beta - \alpha) \tag{3.252}$$

On the other hand, P_R is maximum when $\delta = \beta$. Therefore, the maximum power obtainable (which is also called the *steady-state power limit*) at the receiving end can be expressed as

$$\mathbf{P}_{R,\mathrm{max}} = \frac{V_S V_R}{B} - \frac{AV_S^2}{B}\cos(\beta - \alpha) \tag{3.253}$$

and similarly, the corresponding reactive power delivered at the receiving end is

$$\mathbf{Q}_{R,\mathrm{max}} = -\frac{AV_S^2}{B}\sin(\beta - \alpha) \tag{3.254}$$

The equations above calculate the real and reactive powers delivered to the receiving end of a transmission line, where V_S and V_R represent the phase voltages (line-to-neutral) regardless of whether the system is single-phase or three-phase. Therefore, the total three-phase power transmitted on a three-phase line is three times the power calculated using the equations. If voltage values are given in volts, the power is expressed in watts or vars, while if they are given in kilovolts, the power is expressed in megawatts or megavars.

To calculate the power loss P_L in a long transmission line for a given value of γ, simply subtract the receiving-end real power from the sending-end real power:

$$P_S = P_S - P_R \tag{3.255}$$

and the lagging vars loss is

$$Q_L = Q_S - Q_R \tag{3.256}$$

Example 3.15: Figure 3.39 depicts a 345-kV AC transmission line with bundled conductors connecting two voltage-regulated buses. Suppose that both series capacitor and shunt reactor compensations are considered. The bundled conductor line consists of two 795-kcmil ACSR conductors per

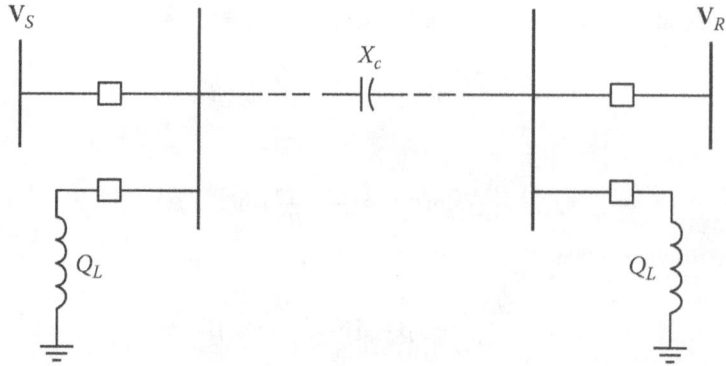

Figure 3.39 Compensated line for Example 3.15.

phase, with subconductors separated by 18 inches and a phase spacing of 24, 24, and 48 feet in a flat configuration. The resistance, inductive reactance, and susceptance of the line per phase per mile are 0.059 Ω, 0.588 Ω, and 7.20×10^{-6} S phase to neutral per phase per mile, respectively. The total length of the line is 200 miles, and the line resistance can be ignored for simplicity. Assume that there is no compensation in use, meaning that both reactors are disconnected and the series capacitor is bypassed. Determine the following:

a. The total three-phase SIL of the line in megavolt-amperes
b. The maximum three-phase theoretical steady-state power flow limit in megawatts
c. The total three-phase magnetizing vars generated by line capacitance
d. The open-circuit receiving-end voltage if the line is open at the receiving end

Solution:

a. The surge impedance of the line is

$$Z_c = \sqrt{(x_L \times x_c)} \tag{3.257}$$

where

$$x_c = \frac{1}{b_c} \ \Omega/\text{mi/phase} \tag{3.258}$$

Thus,

$$Z_c = \sqrt{\left(\frac{x_L}{b_c}\right)}$$

$$= \sqrt{\left(\frac{0.588}{7.20 \times 10^{-6}}\right)} \tag{3.259}$$

$$= 285.77 \ \Omega/\text{mi/phase}$$

Thus, the total three-phase SIL of the line is

$$\text{SIL} = \frac{|kV_{R(\text{L-L})}|^2}{Z_c}$$

$$= \frac{345^2}{285.77}$$

$$= 416.5 \ \text{MVA/mi}$$

b. Neglecting the line resistance,

$$P = P_S = P_R$$

or

$$P = \frac{V_S V_R}{X_L} \sin\delta \qquad (3.260)$$

When $\delta = 90°$, the maximum three-phase theoretical steady-state power flow limit is

$$
\begin{aligned}
P_{max} &= \frac{V_S V_R}{X_L} \\
&= \frac{(345 \text{ kV})^2}{117.6} \qquad (3.261) \\
&= 1012.1 \text{ MW}
\end{aligned}
$$

c. Using a nominal π circuit representation, the total three-phase magnetizing var generated by the line capacitance can be expressed as

$$
\begin{aligned}
Q_c &= V_S^2 \frac{b_c l}{2} + V_R^2 \frac{b_c l}{2} \\
&= V_S^2 \frac{B_c}{2} + V_R^2 \frac{B_c}{2} \qquad (3.262)
\end{aligned}
$$

Hence,

$$Q_c = (345 \times 10^3)^2 \left(\frac{1}{2}(7.20 \times 10^{-6}) \cdot 200\right) + (345 \times 10^3)^2 \left(\frac{1}{2}(7.20 \times 10^{-6}) \cdot 200\right)$$

$$= 171.4 \text{ MVAr}$$

d. If the line is open at the receiving end, the open-circuit receiving-end voltage can be expressed as

$$\mathbf{V}_S = \mathbf{V}_{R(oc)} \cosh\gamma l \qquad (3.263)$$

or

$$\mathbf{V}_{R(oc)} = \frac{\mathbf{V}_S}{\cosh\gamma l} \qquad (3.264)$$

where

$$
\begin{aligned}
\gamma &= j\omega\sqrt{LC} \\
&= j\omega \left(\frac{x_L}{\omega} \frac{1}{\omega x_c}\right)^{1/2} \\
&= j\left(\frac{x_L}{x_c}\right)^{1/2} \qquad (3.265) \\
&= j[(0.588)(7.20 \times 10^{-6})]^{1/2} = j0.0021 \text{ rad/mi}
\end{aligned}
$$

and

$$\gamma l = j(0.0021)(200) = j0.4115 \text{ rad}$$

thus,

$$
\begin{aligned}
\cosh\gamma l &= \cosh(0 + j0.4115) \\
&= \cosh(0)\cos(0.4115) + j\sinh(0)\sin(0.4115) \\
&= 0.9165
\end{aligned}
$$

therefore,

$$V_{R(oc)} = \frac{345 \text{ kV}}{0.9165}$$

$$= 376.43 \text{ kV}$$

Alternatively,

$$V_{R(oc)} = V_s \frac{X_c}{X_c + X_L}$$

$$= (345 \text{ kV})\left(\frac{-1388.9}{-1388.9 + 117.6}\right)$$

$$= 376.74 \text{ kV}$$

Example 3.16: Given the information provided in Example 3.13, consider the scenario where shunt compensation is utilized, and two shunt reactors are connected to absorb 60% of the total three-phase magnetizing var generation caused by line capacitance, where half of the total reactor capacity is placed at each end of the line. Determine the following quantities:

a. Total three-phase SIL of the line in megavolt-amperes
b. Maximum three-phase theoretical steady-state power flow limit in megawatts
c. Three-phase megavolt-ampere rating of each shunt reactor
d. Cost of each reactor at $10/kVA
e. Open-circuit receiving-end voltage if the line is open at the receiving end

Solution:

a. SIL = 416.5, as before, in Example 3.13.
b. P_{max} = 1012.1 MW, as before
c. The three-phase megavolt-ampere rating of each shunt reactor is

$$\frac{1}{2}Q_L = \frac{1}{2}0.60Q_c$$

$$= \frac{1}{2}0.60(171.4)$$

$$= 51.42 \text{ MVA}$$

d. The cost of each reactor at $10/kVA is

$$(51.42 \text{ MVA/reactor})(\$10/\text{kVA}) = \$514,200$$

e. Since

$$\gamma l = j0.260 \text{ rad}$$

and

$$\cosh \gamma l = 0.9663$$

then

$$V_{R(OC)} = \frac{345 \text{ kV}}{0.9663}$$

$$= 357.03 \text{ kV}$$

Alternatively,

$$V_{R(OC)} = V_S \frac{X_C}{X_C + X_L}$$
$$= (345 \text{ kV}) \left(\frac{-13{,}472}{-13{,}472 + 117.6} \right) \qquad (3.266)$$
$$= 357.1 \text{ kV}$$

Therefore, the inclusion of the shunt reactor causes the receiving-end open-circuit voltage to decrease.

3.13 EHV UNDERGROUND CABLE TRANSMISSION

As previously explained, the inductive reactance of an overhead high-voltage AC line is considerably larger than its capacitive reactance. However, in the case of an underground high-voltage AC cable, the capacitive reactance is much greater than the inductive reactance since the three-phase conductors are located close to one another within the same cable. The resultant vars generated by AC cables operating at 132, 220, and 400 kV phase-to-phase voltages are approximately 2000, 5000, and 15,000 kVA/mi, respectively, due to the capacitive charging currents. This var generation limits the practical non-interrupted length of an underground AC cable.

To address this issue, appropriate inductive shunt reactors can be installed along the cable line. The "critical length" of the cable is defined as the cable length having a three-phase charging reactive power equal in magnitude to the thermal rating of the cable line. For example, the approximate critical lengths for AC cables operating at phase-to-phase voltages of 132, 200, and 400 kV are 40, 25, and 15 mi, respectively.

Schifreen and Marble conducted a study [80] that highlights the operational limitations of high-voltage AC cable lines resulting from the charging current. For instance, as shown in Figure 3.40, an increase in cable length results in a reduction of the maximum allowable power output [81]. Additionally, Figure 3.41 illustrates that increasing cable length can only transmit full-rated current (1.0 pu) if the load power factor is decreased to resolve lagging values. It is worth noting that the critical length is used as the reference length in the figures. Table 3.4 [41] provides the characteristics of a 345-kV pipe-type cable, and Figure 3.42 shows the acceptable variation in per-unit vars delivered to the power system at each cable terminal for a given power transmission.

Example 3.17: Consider a high-voltage open-circuit three-phase insulated power cable of length l shown in Figure 3.43. Assume that a fixed sending-end voltage is to be supplied; the receiving-end voltage floats,

and it is an overvoltage. Furthermore, assume that at some critical length $(l = l_0)$, the sending-end current $\mathbf{I}_s$ is equal to the ampacity of the cable circuit, $\mathbf{I}_{l0}$. Therefore, if the cable length is l_0, no load, whatever of 1.0 or the leading power factor can be supplied without overloading the sending end of the cable. Use the general long-transmission-line equations, which are valid for steady-state sinusoidal operation, and verify that the approximate critical length can be expressed as

$$l_0 \cong \frac{I_{l0}}{V_S b}$$

Solution:
The long-transmission-line equations can be expressed as

$$\mathbf{V}_S = \mathbf{V}_R \cosh \gamma l + \mathbf{I}_R \mathbf{Z}_c \sinh \gamma l \qquad (3.267)$$

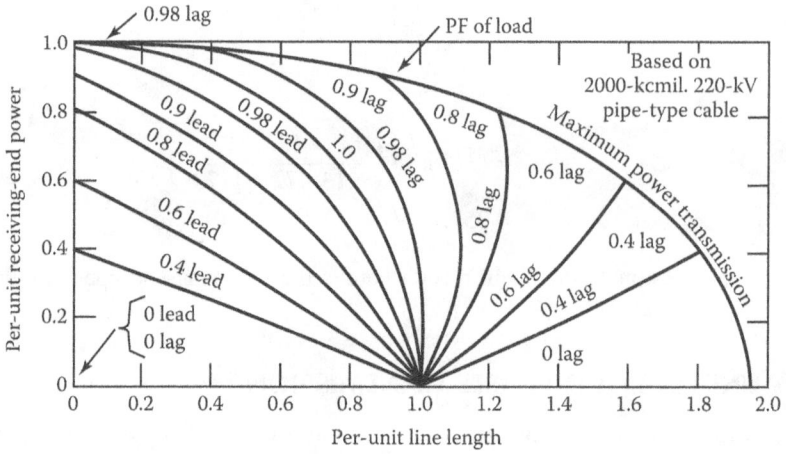

Figure 3.40 Power transmission limits of high-voltage ac cable lines. Curved lines: Sending-end current equal to rated or base current of cable. Horizontal lines: Receiving-end current equal to rated or base current of cable. (From Wiseman, R. T., *Trans. Am. Inst. Electr. Eng.*, 26, 803 1956 IEEE.)

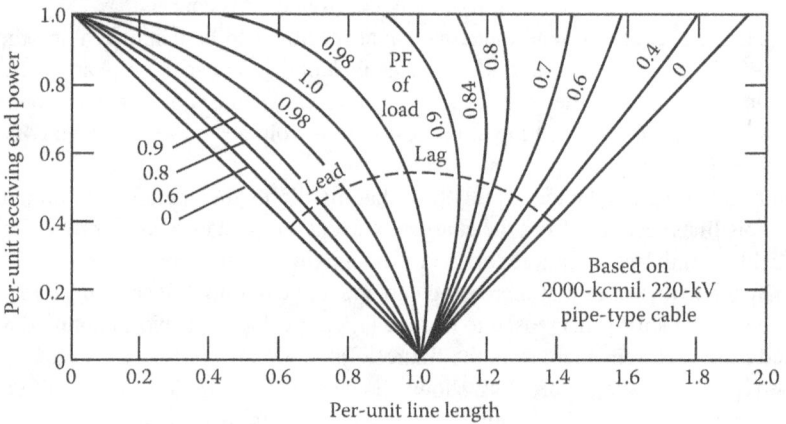

Figure 3.41 Receiving-end current limits of high-voltage ac cable lines. Curved lines: Sending-end current equal to rated or base current of cable. (From Wiseman, R. T., *Trans. Am. Inst. Electr. Eng.*, 26, 803 1956 IEEE. Used with permission.)

and

$$\mathbf{I}_S = \mathbf{I}_R \cosh \gamma I + \mathbf{V}_R \mathbf{Y}_c \sinh \gamma I \tag{3.268}$$

Since at critical length, $I = I_0$ and

$$\mathbf{I}_R = 0$$

and

$$\mathbf{I}_S = \mathbf{I}_{/0}$$

from Equation 3.264, the sending-end current can be expressed as

$$\mathbf{I}_{/0} = \mathbf{V}_R \mathbf{Y}_c \sinh \gamma I_0 \tag{3.269}$$

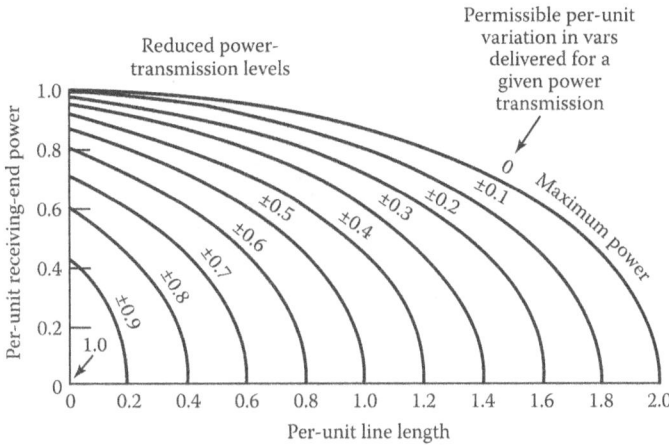

Figure 3.42 Permissible variations in per-unit vars delivered to electric system at each terminal of ac cable line for given power transmission. (From Wiseman, R. T., *Trans. Am. Inst. Electr. Eng.*, 26, 803 1956 IEEE.)

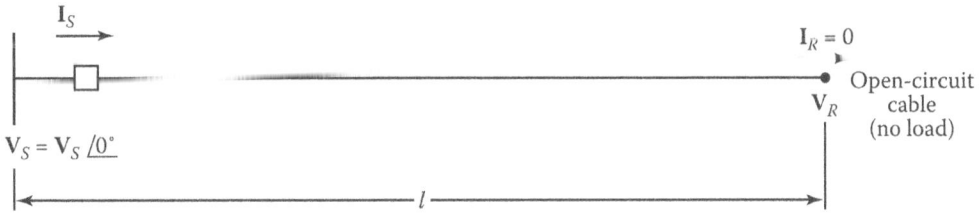

Figure 3.43 Cable system for Example 3.17.

or

$$\mathbf{I}_{/0} = \mathbf{V}_R \mathbf{Y}_c \left(\frac{e^{\gamma l_0} - e^{\gamma l_0}}{2} \right) \tag{3.270}$$

$$\mathbf{I}_{/0} = \mathbf{V}_R \mathbf{Y}_c \left(\frac{[1 + \gamma l_0 + (\gamma l_0)^2/2! + \cdots] - [1 - \gamma l_0 + (\gamma l_0)^2/2! - \cdots]}{2} \right)$$

or

$$\mathbf{I}_{/0} = \mathbf{V}_R \mathbf{Y}_c \left(\gamma l_0 + \frac{(\gamma l_0)^3}{3!} + \cdots \right) \tag{3.271}$$

Neglecting $(\gamma l_0)3/3!$ and higher powers of γl_0,

$$\mathbf{I}_{/0} = \mathbf{V}_R \mathbf{Y}_c \gamma l_0 \tag{3.272}$$

Similarly, from Equation 3.263, the sending-end voltage for the critical length can be expressed as

$$\mathbf{V}_S = \mathbf{V}_R \cosh \gamma l_0 \tag{3.273}$$

or

$$\mathbf{V}_S = \mathbf{V}_R \left(\frac{e^{\gamma l_0} + e^{\gamma l_0}}{2} \right) \tag{3.274}$$

or

$$\mathbf{V}_S = \mathbf{V}_R \left(\frac{[1 + \gamma I_0 + (\gamma I_0)^2/2! + \cdots] + [1 - \gamma I_0 + (\gamma I_0)^2/2! - \cdots]}{2} \right)$$

or

$$\mathbf{V}_S = \mathbf{V}_R \left(1 + \frac{(\gamma I_0)^2}{2!} + \cdots \right) \tag{3.275}$$

Neglecting higher powers of γI_0,

$$\mathbf{V}_S \cong \mathbf{V}_R \left(1 + \frac{(\gamma I_0)^2}{2!} \right) \tag{3.276}$$

Thus,

$$\mathbf{V}_R = \frac{\mathbf{V}_S}{1 + \frac{(\gamma I_0)^2}{2!} + \cdots} \tag{3.277}$$

Substituting Equation 3.274 into Equation 3.269,

$$\mathbf{I}_{/0} = \left(\frac{\mathbf{V}_S}{1 + \frac{(\gamma I_0)^2}{2!} + \cdots} \right) \mathbf{Y}_c \gamma I_0 \tag{3.278}$$

or

$$\mathbf{I}_{/0} = \mathbf{V}_S \mathbf{Y}_c \gamma I_0 \left(1 + \frac{(\gamma I_0)^2}{2!} \right)^{-1}$$
$$= \mathbf{V}_S \mathbf{Y}_c \gamma I_0 \left(1 - \frac{(\gamma I_0)^2}{2!} + \cdots \right) \tag{3.279}$$

or

$$\mathbf{I}_{/0} = \mathbf{V}_S \mathbf{Y}_c \gamma I_0 - \frac{\mathbf{V}_S \mathbf{Y}_c (\gamma I_0)^3}{2!} \tag{3.280}$$

Neglecting the second term,

$$\mathbf{I}_{/0} \cong \mathbf{V}_S \mathbf{Y}_c \gamma I_0 \tag{3.281}$$

Therefore, the critical length can be expressed as

$$\mathbf{I}_0 \cong \frac{\mathbf{I}_{/0}}{\mathbf{V}_S \mathbf{Y}_c \gamma} \tag{3.282}$$

where

$$\mathbf{Y}_c = \sqrt{\frac{\mathbf{y}}{\mathbf{z}}}$$

$$\gamma = \sqrt{\mathbf{z} \times \mathbf{y}}$$

Thus,

$$\mathbf{y} = \mathbf{Y}_c \gamma \tag{3.283}$$

or

$$\mathbf{y} = g + jb \tag{3.284}$$

Therefore, the critical length can be expressed as

$$I_0 \cong \frac{\mathbf{I}_{/0}}{\mathbf{V}_S \times y} \qquad (3.285)$$

or

$$I_0 \cong \frac{\mathbf{I}_{/0}}{\mathbf{V}_S \times g + jb} \qquad (3.286)$$

Since, for cables, $g \ll b$,

$$y \cong b \angle 90° $$

and assuming

$$\mathbf{I}_{/0} \cong \mathbf{I}_{/0} \angle 90° $$

from Equation 3.231, the critical length can be expressed as

$$I_0 \cong \frac{\mathbf{I}_{/0}}{\mathbf{V}_S \times b} \qquad (3.287)$$

Example 3.18: Figure 3.44a depicts an open-circuit high-voltage insulated AC underground cable circuit. The critical length of an uncompensated cable is I_0, where $\mathbf{I}_S = \mathbf{I}_0$ is equal to the cable ampacity rating. It is noteworthy that $Q_0 = 3V_S I_0$, where the sending-end voltage $\mathbf{V}_S$ is controlled and the receiving-end voltage $\mathbf{V}_R$ floats. The difference between $|\mathbf{V}_R|$ and $|\mathbf{V}_S|$ is minimal due to the low series inductive reactance of the cables. Based on the available information, the system's performance should be analyzed with $\mathbf{I}_R = 0$ (i.e., *zero load*).

a. Assume that one shunt inductive reactor sized to absorb Q_0 magnetizing vars is to be purchased and installed as shown in Figure 3.44b. Locate the reactor by specifying I_1 and I_2 in terms of I_0. Place arrowheads on the four short lines, indicated by a solid line, to show the directions of magnetizing var flows. Also show on each line the amounts of var flow, expressed in terms of Q_0.
b. Assume that one reactor size $2Q_0$ can be afforded and repeat part (a) on a new diagram.
c. Assume that two shunt reactors, each of size $2Q_0$, are to be installed, as shown in Figure 3.44c, hoping, as usual, to extend the feasible length of cable. Repeat part (a).

Solution:
The answers for parts (a), (b), and (c) are given in Figure 3.45a, b and c, respectively.

3.14 GAS-INSULATED TRANSMISSION LINES

The gas-insulated transmission line (GIL) is a system designed for transmitting electric power at bulk power ratings over long distances. Its first application was in 1974, connecting the electric generator of a hydro pump storage plant in Germany through a tunnel built in a mountain for a 420-kV overhead line. Currently, over 100 km of GILs have been constructed globally at high-voltage levels ranging from 135 to 550 kV, including power plants with adverse environmental conditions. In situations where overhead lines are not feasible, the GIL may be a viable alternative since it provides a transmission line that is not affected by environmental conditions, sealed completely inside a metallic enclosure, maintaining transmission capacity.

The applications of GIL include connecting high-voltage transformers with high-voltage switchgear within power plants, connecting high-voltage transformers inside the cavern power

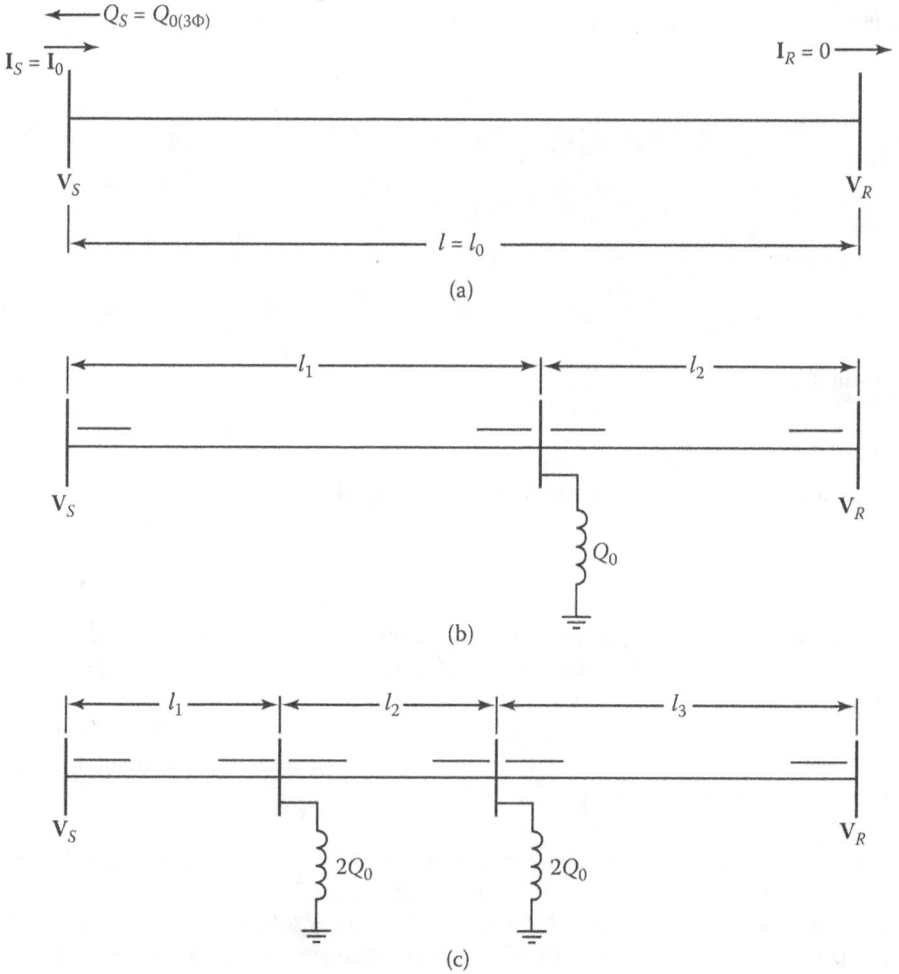

Figure 3.44 Insulated HV underground cable circuit for Example 3.18.

plants to overhead lines on the outside, connecting gas-insulated switchgear (GIS) with overhead lines, and serving as a bus duct within GIS.

Initially, GIL systems were only utilized in special applications because of their high cost. However, the second-generation GIL system is now used for high-power transmission over long distances due to significant cost reductions. These cost reductions are due to the use of an N2 − SF6 gas mixture for electrical insulation. The GIL system has several advantages, including low losses, low magnetic field emissions, high reliability with high transmission capacity, no adverse impact on the environment or the landscape, and underground laying with a transmission capacity equivalent to an overhead transmission line.

Example 3.19: A power utility company is mandated to construct a 500 kV power line to supply electricity to a nearby town. There are two possible routes to build the necessary power line. Route A is 80 miles long and circumvents a lake. The cost of constructing the overhead transmission line is estimated to be $1 million per mile, with an annual maintenance cost of $500 per mile per year. At the end of 40 years, its salvage value will be $2000 per mile.

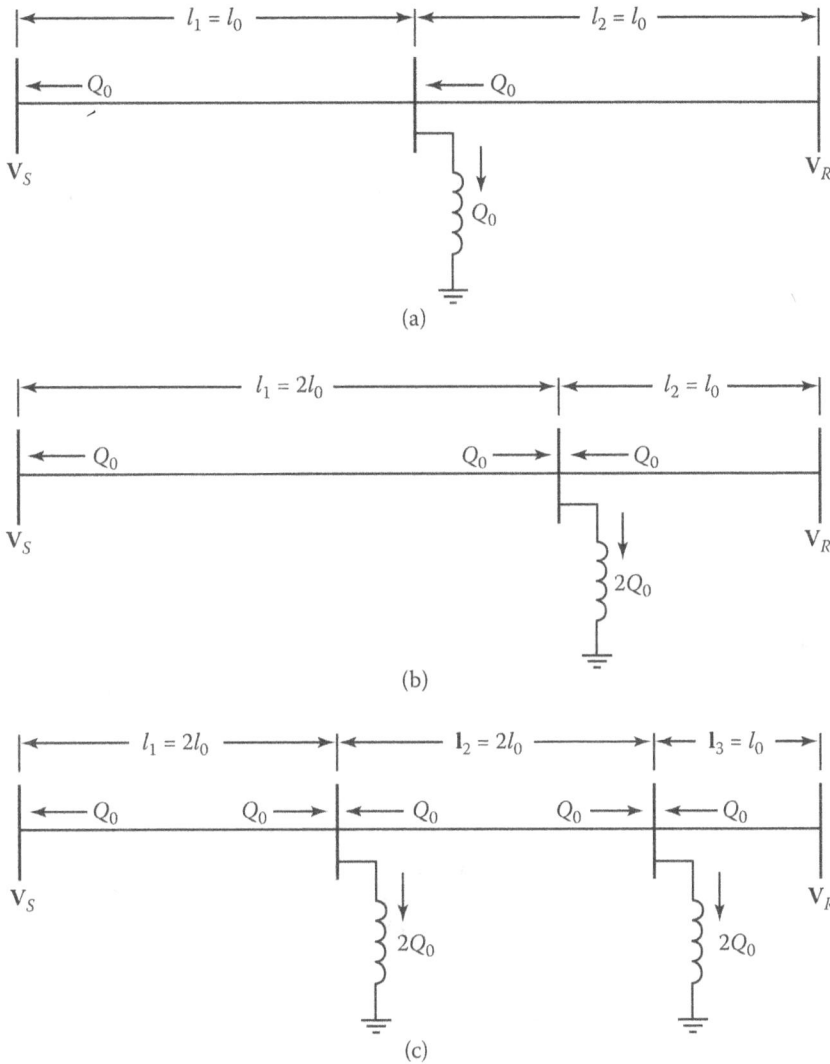

Figure 3.45 Solution for Example 3.16.

On the other hand, route B is 50 miles long and consists of an underwater (submarine) line that runs across the lake. The cost of constructing the underwater line using submarine power cables is estimated to be $4 million per mile, with an annual maintenance cost of $1500 per mile per year. At the end of 40 years, its salvage value will be $6000 per mile.

Alternatively, GIL can be used in route C, which crosses the lake, spanning a length of 30 miles. The cost of constructing the required GIL transmission is estimated to be $7.6 million per mile, with an annual maintenance cost of $200 per mile. At the end of 40 years, its salvage value will be $1000 per mile. The GIL alternative is expected to yield relative savings of $17.5 $\times$ 10^6 per year in power losses compared to the other alternatives.

Assuming a fixed charge rate of 10% and annual ad valorem taxes of 3% of the first costs of each alternative, with the energy cost at $0.10 per kWh, the economically preferable alternative must be determined using engineering economy interest tables*.

Solution:

OVERHEAD TRANSMISSION:

The first cost of the 500 kV overhead transmission line is

$$P = (\$1,000,000/\text{mi})(80 \text{ mi}) = \$80,000,000$$

and its estimated salvage value is

$$F = (\$2,000/\text{mi})(80 \text{ mi}) = \$160,000$$

The annual equivalent cost of capital invested in the line is

$$A_1 = \$80,000,000(A/P)_{40}^{10\%} - \$100,000(A/F)_{40}^{10\%}$$
$$= \$80,000,000(0.10226) - \$100,000(0.00226)$$
$$= 8,180,800 - \$266 = \$8,180,534$$

The annual equivalent cost of the tax and maintenance is

$$A_2 = \$80,000,000(0.03) + (\$500/\text{mi})(80 \text{ mi}) = \$2,440,000$$

The total annual equivalent cost of the overhead transmission line is

$$A = A_1 + A_2 = \$8,180,534 + \$2,440,000$$
$$= \$10,620,534$$

SUBMARINE TRANSMISSION:

The first cost of the 500 kV submarine power transmission line is

$$P = (\$4,000,000/\text{mi})(50 \text{ mi}) = \$200,000,000$$

and its estimated salvage value is

$$F = (\$6000/\text{mi})(50 \text{ mi}) = \$300,000$$

* For example, see *Engineering Economy for Engineering Managers*, T. Gönen, Wiley, 1990.
The annual equivalent cost of capital invested in the line

$$A_1 = \$200,000,000(A/P)_{40}^{10\%} - \$300,000(A/F)_{40}^{10\%}$$
$$= \$200,000,000(0.10296) - \$300,000(0.00296)$$
$$= \$20,591,112$$

The annual equivalent cost of tax and maintenance is

$$A_2 = (\$200,000,000)(0.03) + (\$1500/\text{mi})(50 \text{ mi}) = \$6,075,000$$

The total annual equivalent cost of the overhead transmission line is

$$A = A_1 + A_2 = \$20,591,112 + \$6,075,000$$
$$= \$26,666,112$$

GIL TRANSMISSION:

The first cost of the 500-kV GIL transmission line is

$$P = (\$7,600,000/\text{mi})(30 \text{ mi}) = \$228,000,000$$

and its estimated salvage value is

$$F = (\$1000/\text{mi})(30 \text{ mi}) = \$30,000$$

The annual equivalent cost of capital invested in the GIL line is

$$
\begin{aligned}
A_1 &= \$228,000,000(A/P)_{40}^{10\%} - \$30,000(A/F)_{40}^{10\%} \\
&= \$228,000,000(0.10226) - \$30,000(0.00226) \\
&= \$23,315,280 - \$67.5 = \$23,315,212.5
\end{aligned}
$$

The annual equivalent cost of the tax and maintenance is

$$A_2 = (\$228,000,000)(0.03) + (\$200/\text{mi})(30 \text{ mi}) = \$6,846,000$$

The total annual equivalent cost of the GIL transmission line is

$$
\begin{aligned}
A - A_1 + A_2 &= \$23,315,212.5 + \$6,846,000 \\
&= \$30,221,212.5
\end{aligned}
$$

Since the relative savings in power losses is $17,500,000, then the total net annual equivalent cost of the GIL transmission is

$$
\begin{aligned}
A_{\text{net}} &= \$30,221,212.5 - \$17,500,000 \\
&= \$12,721,212.5
\end{aligned}
$$

The results show that the use of overhead transmission for this application is the best choice. The next best alternative is the GIL transmission. However, the above example is only a rough and very simplistic estimate. In real applications, there are many other cost factors that need to be included in such comparisons.

Example 3.20: Consider transmitting 2,100-MVA electric power across 30 km by using an overhead transmission line (OH) versus by using a GIL. The resulting power losses at peak load are 820 and 254 kW/km for the overhead transmission and the GIL, respectively. Assume that the annual load factor and the annual power loss factor are the same and are equal to 0.7 for both alternatives. Also, assume that the cost of electric energy is $0.10 per kWh. Determine the following:

a. The power loss of the overhead line at peak load
b. The power loss of the GIL
c. The total annual energy loss of the overhead transmission line at peak load
d. The total annual energy loss of the gas insulated transmission line at peak load
e. The average energy loss of the overhead transmission line
f. The average energy loss of the GIL at peak load
g. The average annual cost of losses of the overhead transmission line
h. The average annual cost of losses of the GIL
i. The annual resultant savings in losses using the GIL
j. Find the breakeven (or payback) period when the gas-insulated line alternative is selected, if the investment cost of the gas-insulated line is $200,000,000

Solution:

a. The power loss of the overhead transmission line at peak load is

$$(\text{Power loss})_{\text{OH line}} = (829 \text{ kW/km})30 \text{ km} = 24,870 \text{ kW}$$

b. The power loss of the GIL transmission line at peak load is

$$(\text{Power loss})_{\text{GILL line}} = (254 \text{ kW/km})30 \text{ km} = 7,620 \text{ kW}$$

c. The total annual energy loss of the overhead transmission line at peak load is

$$(\text{Total annual energy loss})_{\text{at peak}} = (24,870 \text{ kW/km})(8,760\text{h/hr})$$
$$= 21,786 \times 10^4 \text{ kWh/yr}$$

d. The total annual energy loss of the gas-insulated line at peak load is

$$(\text{Total annual energy loss})_{\text{at peak}} = (7,620 \text{ kW/km})(8,760\text{h/hr})$$
$$= 6,675.2 \times 10^4 \text{ kWh/yr}$$

e. The total annual energy loss of the overhead transmission line at peak load is

$$(\text{Average annual energy loss})_{\text{OH line}} = 0.7(21,786 \times 10^4\text{kWh/yr})$$
$$= 15,250.2 \times 10^4\text{kWh/yr}$$

f. The average energy loss of the gas-insulated line at peak load is

$$(\text{Average annual energy loss})_{\text{GIL line}} = 0.7(6,675\text{kWh/yr})$$
$$= 4,672.64 \times 10^4\text{kWh/yr}$$

g. The average annual cost of losses of the overhead transmission line is

$$(\text{Average annual cost of loss})_{\text{OH line}} = (\$0.10/\text{kWh})(15,250.2 \times 10^4\text{kWh/yr})$$
$$= \$1,525.02 \times 10^3/\text{yr}$$

h. The average annual cost of losses of the GIL is

$$(\text{Average annual cost of losses})_{\text{GIL line}} = (\$0.10/\text{kWh})(4,672.64 \times 10^4\text{kWh/yr})$$
$$= \$467.264 \times 10^3/\text{yr}$$

i. The annual resultant savings in power losses using the GIL is

$$(\text{Annual savings in losses}) = (\text{Annual cost of losses})_{\text{OH line}} - (\text{Annual cost of losses})_{\text{GIL line}}$$
$$= \$1,525 \times 10^3 - \$467.264 \times 10^3$$
$$= \$1,057.736 \times 10^3/\text{yr}$$

j. If the GIL alternative is selected

$$(\text{Breakeven period}) = \frac{\text{Total investment cost}}{\text{Savings per year}}$$
$$= \frac{\$200,000,000}{\$1,057.736 \times 10^3} \cong 189 \text{ year}$$

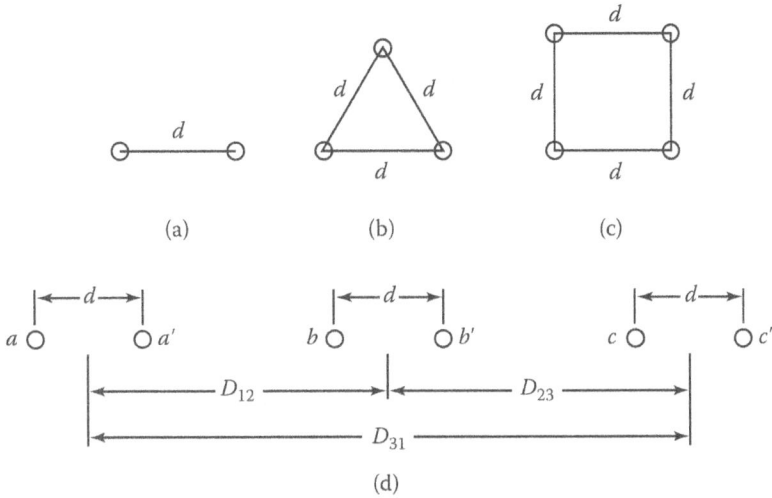

Figure 3.46 Bundle arrangements: (a) two-conductor bundle; (b) three-conductor bundle; (c) fourconductor bundle; (d) cross section of bundled-conductor three-phase line with horizontal tower configuration.

3.15 BUNDLED CONDUCTORS

Bundled conductors are utilized at or above 345 kV. Rather than a single large conductor per phase, two or more conductors with a comparable total cross-section are suspended from each insulator string. This arrangement significantly reduces the voltage gradient at the conductor surface by having two or more conductors per phase in close proximity, compared to the spacing between phases. The bundles used in the extra-high-voltage range typically have two, three, or four subconductors, as illustrated in Figure 3.46. In the ultrahigh-voltage range, bundles with eight, twelve, or even sixteen conductors may also be employed.

Bundle conductors are sometimes referred to as duplex, triplex, and so on, conductors, indicating the number of subconductors and are also known as grouped or multiple conductors. The use of bundled conductors has several advantages over single conductors per phase, including reduced line inductive reactance, reduced voltage gradient, increased corona critical voltage, less corona power loss, audible noise, and radio interference, greater power capacity per unit mass of the conductor, and a reduction in high-frequency vibration amplitude and duration. However, the disadvantages of bundled conductors include increased wind and ice loading, more complex suspension requiring duplex or quadruple insulator strings, increased tendency to gallop, higher costs, increased clearance requirements at structures, and increased charging kilovolt-amperes.

If the subconductors of a bundle are transposed, the current is evenly distributed among the conductors in the bundle. The GMRs of bundled conductors composed of two, three, and four subconductors can be expressed, respectively, as follows:

$$D_s^b = (D_s \times d)^{1/2} \tag{3.288}$$

$$D_s^b = (D_s \times d^2)^{1/3} \tag{3.289}$$

$$D_s^b = 1.09(D_s \times d_3)^{1/4} \tag{3.290}$$

where
D_s^b = GMR of bundled conductor
D_s = GMR of subconductors

d = distance between two subconductors

Therefore, the *average* inductance per phase is

$$L_a = 2 \times 10^{-7} \ln \frac{D_{eq}}{D_s^b} \ \text{H/m} \tag{3.291}$$

and the inductive reactance is

$$x_L = 0.1213 \ln \frac{D_{eq}}{D_s^b} \tag{3.292}$$

where

$$D_{eq} \triangleq D_m (D_{12} \times D_{23} \times D_{31})^{1/3} \tag{3.293}$$

The modified GMRs (to be used in capacitance calculations) of bundled conductors made up of two, three, and four subconductors can be expressed, respectively, as

$$D_{sC}^b = (r \times d)^{1/2} \tag{3.294}$$

$$D_{sC}^b = (r \times d)^{1/3} \tag{3.295}$$

$$D_{sC}^b = 1.09(r \times d^3)^{1/4} \tag{3.296}$$

where
D_{sC}^b = modified GMR of bundled conductor
r = outside radius of subconductors
d = distance between two subconductors

Therefore, the line-to-neutral capacitance can be expressed as

$$C_n = \frac{2\pi \times 8.8538 \times 10^{-12}}{\ln \left(\frac{D_{eq}}{D_{sC}^b} \right)} \ \text{F/m} \tag{3.297}$$

or

$$C_n = \frac{55.63 \times 10^{-12}}{\ln \left(\frac{D_{eq}}{D_{sC}^b} \right)} \ \text{F/m} \tag{3.298}$$

For a two-conductor bundle, the maximum voltage gradient at the surface of a subconductor can be expressed as

$$E_0 = \frac{V_0 \left(1 + \frac{2r}{d}\right)}{2r\ln \left(\frac{D}{\sqrt{r \times d}} \right)} \tag{3.299}$$

Example 3.21: Consider the bundled-conductor three-phase 200-km line shown in Figure 3.46d. Assume that the power base is 100 MVA and the voltage base is 345 kV. The conductor used is a 1113 kcmil ACSR, and the distance between two subconductors is 12 in. Assume that the distances D_{12}, D_{23}, and D_{31} are 26, 26, and 52 ft, respectively, and determine the following:

a. Average inductance per phase in henries per meter
b. Inductive reactance per phase in ohms per kilometer and ohms per mile
c. Series reactance of line in per units
d. Line-to-neutral capacitance of line in farads per meter
e. Capacitive reactance to neutral of line in ohm per kilometers and ohm per miles.

Solution:

a. From Table A.3 in Appendix A, Ds is 0.0435 ft; therefore,

$$D_s^b = (D_s \times d)^{1/2}$$
$$= (0.0435 \times 0.3048 \times 12 \times 0.0254)^{1/2}$$
$$= 0.0636 \text{ m}$$

$$D_{eq} = (D_{12} \times D_{23} \times D_{31})^{1/3}$$
$$= (26 \times 26 \times 52 \times 0.3048^3)^{1/3}$$
$$= 9.9846 \text{ m}$$

Thus, from Equation 3.219,

$$L_a = 2 \times 10^{-7} \ln \frac{D_{eq}}{D_s^b}$$
$$= 2 \times 10^{-7} \ln \left(\frac{9.9846}{0.0636} \right)$$
$$= 1.0112 \mu H/m$$

b.

$$X_L = 2\pi f L_a$$
$$= 2\pi \times 60 \times 1.0112 \times 10^{-6} \times 10^3$$
$$= 0.3812 \text{ } \Omega/km$$

and

$$X_L = 0.3812 \times 1.609$$
$$= 0.6134 \text{ } \Omega/mi$$

c.

$$Z_B = \frac{345^2}{100}$$
$$= 1190.25 \text{ } \Omega$$

$$X_L = \frac{0.3812 \times 200}{1190.25}$$
$$= 0.0641 \text{ pu}$$

d. From Table A.3, the outside diameter of the subconductor is 1.293 in; therefore, its radius is

$$r = \frac{1.293 \times 0.3048}{2 \times 12}$$
$$= 0.0164 \text{ m}$$

$$D_{sC}^b = (r \times d)^{1/2}$$
$$= (0.0164 \times 12 \times 0.0254)^{1/2}$$
$$= 0.0707 \text{ m}$$

Thus, the line-to-neutral capacitance of the line is

$$C_n = \frac{55.63 \times 10^{-12}}{\ln(D_{eq}/D_{sC}^b)}$$

$$= \frac{55.63 \times 10^{-12}}{\ln(9.9846/0.0707)}$$

$$= 11.238 \times 10^{-12} \text{ F/m}$$

e. The capacitive reactance to the neutral of the line is

$$X_c = \frac{1}{2\pi f C_n}$$

$$= \frac{10^{12} \times 10^{-3}}{2\pi \times 60 \times 11.238}$$

$$= 0.236 \times 10^6 \ \Omega - \text{km}$$

and

$$X_c = \frac{0.236 \times 10^6}{1.609}$$

$$= 0.147 \times 10^6 \ \Omega\text{mi}$$

3.16   EFFECT OF GROUND ON CAPACITANCE OF THREE-PHASE LINES

Figure 3.47 depicts three-phase line conductors and their images beneath the ground surface. The line is assumed to be transposed, and conductors a, b, and c have charges q_a, q_b, and q_c, respectively, while their images have charges $-q_a$, $-q_b$, and $-q_c$. The line-to-neutral capacitance can be described as [5,57,63].

$$C_n = \frac{2\pi \times 8.8538 \times 10^{-12}}{\ln\left(\frac{D_{eq}}{r}\right) - \ln\left(\frac{l_{12}l_{23}l_{31}}{h_{11}h_{22}h_{33}}\right)^{1/3}} \text{ F/m} \qquad (3.300)$$

If the effect of the ground is not taken into account, the line-to-neutral capacitance is

$$C_n = \frac{2\pi \times 8.8538 \times 10^{-12}}{\ln\left(\frac{D_{eq}}{r}\right)} \text{ F/m} \qquad (3.301)$$

It is apparent that the ground has the effect of increasing the line capacitance. Nonetheless, given that the conductor heights are significantly greater than the distances between them, the impact of the ground is frequently disregarded for three-phase lines.

3.17   ENVIRONMENTAL EFFECTS OF OVERHEAD TRANSMISSION LINES

In recent times, minimizing the environmental impacts of overhead transmission lines has gained significant importance, especially with the increasing use of extra-high and ultrahigh-voltage levels. It is crucial to predict, analyze, and measure the magnitude and impact of radio noise, television interference, audible noise, electric field, and magnetic fields not only during the line design stage but also in the measurement stage. Measurements related to corona phenomena should include radio and television station signal strengths and radio, television, and audible noise levels.

To determine the effects of transmission lines on these quantities, measurements should be taken at three different times: (1) prior to the construction of the line, (2) after construction but before

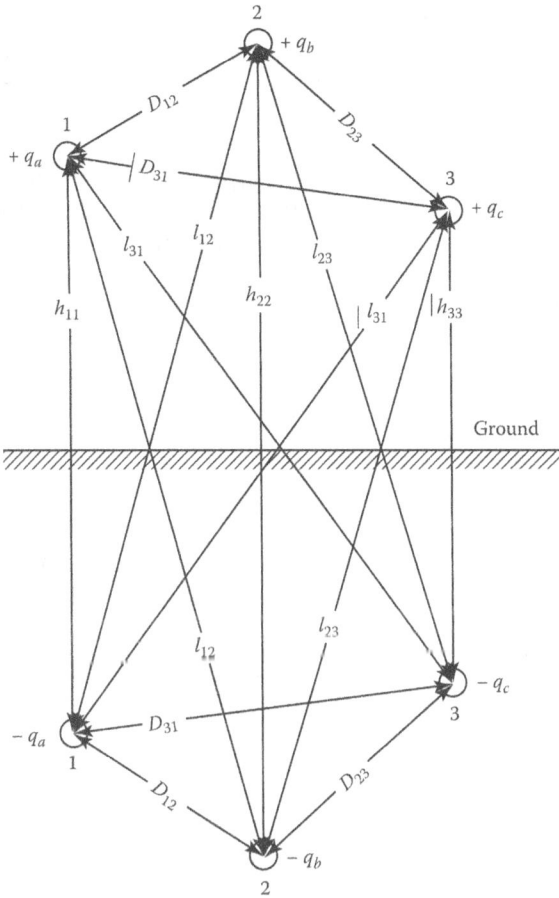

Figure 3.47 Three-phase line conductors and their images.

energization, and (3) after energization of the line. Noise measurements must be conducted at several locations along the transmission line. Additionally, measurements may be taken at several points of interest at each location, such as the point of maximum noise, the edge of the right of way, and the point 50 ft from the outermost conductor.

Overhead transmission lines and stations also produce electric and magnetic fields, which must be considered during the design process. The study of field effects, such as induced voltages and currents in conducting bodies, is becoming increasingly crucial as transmission line operating voltage levels increase due to economic and operational benefits. Presently, the study of such effects at the ultrahigh-voltage level involves the following:

1. Calculation and measurement techniques for electric and magnetic fields
2. Calculation and measurement of induced currents and voltages on objects of various shapes for all line voltages and design configurations
3. Calculation and measurement of currents and voltages induced in people as result of various induction mechanisms
4. Investigation of sensitivity of people to various field effects
5. Study of conditions resulting in fuel ignition, corona from grounded objects, and other possible field effects [41]

In order to determine the profile of the electric field, measurements must be taken laterally at midspan and should extend to the edges of the right of way. Additionally, electric field effects on related objects such as vehicles and fences should also be considered. Magnetic field effects are not as much of a concern as electric field effects for extra-high and ultrahigh-voltage transmission because magnetic field levels for normal load current values are generally low. The impact of currents induced in the human body by magnetic effects is considerably less than those caused by electric induction. For instance, the induced current densities in the human body are less than one-tenth those caused by electric field induction. It should be noted that most environmental measurements are highly influenced by the prevailing weather conditions and transmission line geometry, including factors such as temperature, humidity, barometric pressure, precipitation levels, and wind velocity.

PROBLEMS

1. Redraw the phasor diagram shown in Figure 3.14 by using $\mathbf{I}$ as the reference vector and derive formulas to calculate the
    (a) Sending-end phase voltage, V_s
    (b) Sending-end power-factor angle, ϕ_s
2. A three-phase, 60-Hz, a 20-mi-long short transmission line provides 12 MW at a lagging power factor of 0.85 at a line-to-line voltage of 34.5 kV. The line conductors are made of 26-strand 397.5-kcmil ACSR conductors that operate at 50° C and are equilaterally spaced 6 ft apart. Calculate the following:
    (a) Source voltage
    (b) Sending-end power factor
    (c) Transmission efficiency
    (d) Regulation of line
3. Repeat Problem 2 assuming the receiving-end power factor of 0.8 lagging.
4. Repeat Problem 2 assuming the receiving-end power factor of 0.8 leading.
5. A single-phase load is supplied by a 24-kV feeder whose impedance is $60 + j310 \, \Omega$ and a 24/2.4-kV transformer whose equivalent impedance is $0.25 + j'1.00 \, \Omega$ referred to its low voltage side. The load is 210 kW at a leading power factor of 0.9 and 2.3 kV. Calculate the following:
    (a) Sending-end voltage of feeder
    (b) Primary-terminal voltage of transformer
    (c) Real and reactive-power input at sending end of feeder
6. A short three-phase line has the series reactance of $151 \, \Omega$ per phase. Neglect its series resistance. The load at the receiving end of the transmission line is 15 MW per phase and 12 MVAr lagging per phase. Assume that the receiving-end voltage is given as $115 + j0$ kV per phase and calculate
    (a) Sending-end voltage
    (b) Sending-end current
7. A short 40-mi-long three-phase transmission line has a series line impedance of $0.6 + j0.95 \, \Omega$ /mi per phase. The receiving-end line-to-line voltage is 69 kV. It has a full-load receiving end current of $300\angle-30°$ A. Do the following:
    (a) Calculate the percentage of voltage regulation.
    (b) Calculate the **ABCD** constants of the line.
    (c) Draw the phasor diagram of $\mathbf{V}_S$, $\mathbf{V}_R$, and $\mathbf{I}$.
8. Repeat Problem 7 assuming the receiving-end current of $300\angle-45°$ A.
9. A three-phase, 60-Hz, 12-MW load at a lagging power factor of 0.85 is supplied by a threephase, 138-kV transmission line of 40 mi. Each line conductor has a resistance of $41 \, \Omega$/mi and an inductance of 14 mH/mi. Calculate
    (a) Sending-end line-to-line voltage

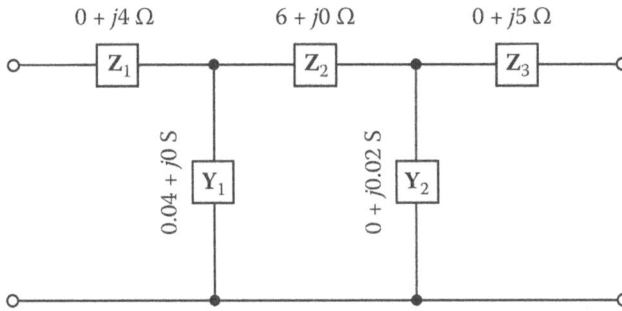

Figure 3.48 Network for Problem 15.

(b) Loss of power in transmission line

(c) Amount of reduction in line power loss if load-power factor were improved to unity

10. A three-phase, 60-Hz transmission line has sending-end voltage of 39 kV and receiving end voltage of 34.5 kV. If the line impedance per phase is $18 + j57\ \Omega$, compute the maximum power receivable at the receiving end of the line.

11. A three-phase, 60-Hz, 45-mi-long short line provides 20 MVA at a lagging power factor of 0.85 at a line-to-line voltage of 161 kV. The line conductors are made of 19-strand 4/0 copper conductors that operate at 50○ C. The conductors are equilaterally spaced with 4 ft spacing between them

   (a) Determine the percentage of voltage regulation of the line.

   (b) Determine the sending-end power factor.

   (c) Determine the transmission line efficiency if the line is single phase, assuming the use of the same conductors.

   (d) Repeat part (c) if the line is three phase.

12. A three-phase, 60-Hz, 15-MW load at a lagging power factor of 0.9 is supplied by two parallel connected transmission lines. The sending-end voltage is 71 kV, and the receiving-end voltage on a full load is 69 kV. Assume that the total transmission line efficiency is 98%. If the line length is 10 mi and the impedance of one of the lines is $0.7 + j1.2\ \Omega/\text{mi}$, compute the total impedance per phase of the second line.

13. Verify that $(\cosh \gamma l - 1)/\sinh \gamma l = \tanh(1/2)\ \gamma l$.

14. Derive Equations 3.78 and 3.79 from Equations 3.76 and 3.77.

15. Find the general circuit parameters for the network shown in Figure 3.48.

16. Find a T equivalent of the circuit shown in Figure 3.48.

17. Assume that the line is a 200-mi-long transmission line and repeat Example 3.6. Use the Y and Z given in Example 3.6 as if they are for the whole 200-mi-long line given in this problem. Use the long-line model.

18. Assume that the line in Example 3.8 is 75 mi long and the load is 100 MVA, and repeat the example. (Use T-model for the medium line.)

19. Develop the equivalent transfer matrix for the network shown in Figure 3.49 by using matrix manipulation.

20. Develop the equivalent transfer matrix for the network shown in Figure 3.50 by using matrix manipulation.

21. Verify Equations 3.202 through 3.205 without using matrix methods.

22. Verify Equations 3.209 through 3.212 without using matrix methods.

23. Assume that the line given in Example 3.6 is a 200-mi-long transmission line. Use the other data given in Example 3.6 accordingly and repeat Example 3.10.

24. Use the data from Problem 23 and repeat Example 3.11.

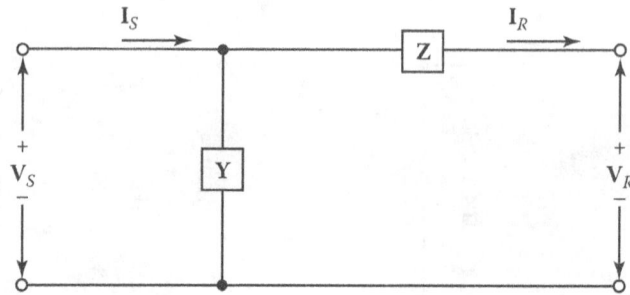

Figure 3.49 Network for Problem 19.

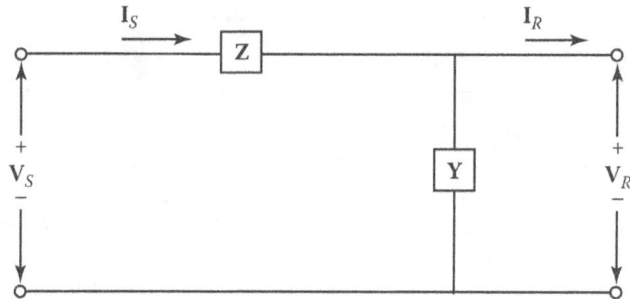

Figure 3.50 Network for Problem 20.

25. Assume that the shunt compensation of Example 3.16 is to be retained and now 60% series compensation is to be used, that is, the X_c is equal to 60% of the total series inductive reactance per phase of the transmission line. Determine the following:
    (a) Total three-phase SIL of line in megavolt-amperes
    (b) Maximum three-phase theoretical steady-state power flow limit in megawatts

26. Assume that the line given in Problem 25 is designed to carry a contingency peak load of 2 × SIL and that each phase of the series capacitor bank is to be of series and parallel groups of two-bushing, 12-kV, 150-kvar shunt power factor correction capacitors.
    (a) Specify the necessary series-parallel arrangements of capacitors for each phase.
    (b) Such capacitors may cost about $1.50/kvar. Estimate the cost of the capacitors in the entire three-phase series capacitor bank. (Take note that the structure and the switching and protective equipment associated with the capacitor bank will add a great deal more cost.)

27. Use Table 3.4 for a 345-kV, pipe-type, three-phase, 1000-kcmil cable. Assume that the percent power factor cable is 0.5 and maximum electric stress is 300 V/mil and that

$$\mathbf{V}_s = \frac{345,000}{\sqrt{3}} \angle 0° \ \mathbf{V}$$

Use I_T = 585A for the cable and calculate the following:
    (a) Susceptance b of cable
    (b) Critical length of cable and compare to value given in Table 4.3

28. Consider the cable given in Problem 27 and use Table 3.4 for the relevant data; determine the value of

$$\mathbf{I}_{lo} = \frac{\mathbf{V}_S}{\mathbf{Z}_c} \tanh \gamma l_0$$

accurately and compare it with the given value of cable ampacity in Table 3.3. (*Hint*: Use the exponential form of the tanh γl_0 function.)

29. Consider Equation 3.52 and verify that the maximum power obtainable (i.e., the steadystate power limit) at the receiving end can be expressed as

$$P_{R,\max} = \frac{|\mathbf{V}_S||\mathbf{V}_R|}{|X|}\sin\gamma$$

30. Repeat Problem 8 assuming that the given power is the sending-end power instead of the receiving-end power.

31. Assume that a three-phase transmission line is constructed of 700 kcmil, 37-strand copper conductors, and the line length is 100 mi. The conductors are spaced horizontally with $D_{ab} = 10\ \Omega$, $D_{bc} = 8\ \Omega$, and $D_{ca} = 18\ \Omega$. Use 60 Hz and 25° C, and determine the following line constants from tables in terms of

    (a) Inductive reactance in ohms per mile
    (b) Capacitive reactance in ohms per mile
    (c) Total line resistance in ohms
    (d) Total inductive reactance in ohms
    (e) Total capacitive reactance in ohms

32. A 60-Hz, single-circuit, three-phase transmission line is 150 mi long. The line is connected to a load of 50 MVA at a logging power factor of 0.85 at 138 kV. The line impedance and admittance are $\mathbf{z} = 0.7688 \angle 77.4°\ \Omega/\text{mi}$ and $\mathbf{y} = 4.5239 \times 10^{-6} \angle 90°$ S/mi, respectively. Use the long-line model and determine the following:

    (a) Propagation constant of the line
    (b) Attenuation constant and phase-change constant, per mile, of the line
    (c) Characteristic impedance of the line
    (d) SIL of the line
    (e) Receiving-end current
    (f) Incident voltage at the sending end
    (g) Reflected voltage at the sending end

33. Consider a three-phase transmission and assume that the following values are given:

$$\mathbf{V}_{R(L-N)} = 79,134\angle 0°\,\text{V}, \mathbf{I}_R = 209.18\angle -31.8°\ \text{A}, \mathbf{Z}_c = 469.62\angle 5.37°\Omega \text{ and} \gamma l$$
$$= 0.0301 + j0.3202$$

Determine the following:

    (a) Incident and reflected voltages at the receiving end of the line
    (b) Incident and reflected voltages at the sending end of the line
    (c) Line voltage at the sending end of the line

34. Repeat Example 3.17 but assume that the conductor used is 1431-kcmil ACSR and that the distance between two subconductors is 18 in. Also, assume that the distances D_{12}, D_{23}, and D_{31} are 25, 25, and 50 ft, respectively.

4 Disturbance of Normal and Abnormal Operating Conditions

4.1 INTRODUCTION

A fault in a power system can cause abnormal currents to flow through an abnormal path due to partial or complete insulation failure, which can disrupt normal system operation [82]. Insulation failure, also known as a short circuit or fault, occurs when one or more energized conductors come into contact with each other or ground, and it is not always necessary for the conductors to be in actual contact. Current can flow through an ionized path, such as air or another normally insulating substance, causing a short circuit [83–93]. When a fault occurs, the voltage between the two parts is either reduced to zero in the case of metal-to-metal contacts or to a very low value if the short-circuit path is through an arc. This results in currents of abnormally high magnitude flowing through the system to the fault point.

An unusually high electrical current can occur when the voltage in the system is abnormally high, either due to lightning strikes or sudden changes in power flow, which can cause the insulation to break down or create a flashover. This, in turn, can cause damage to the insulation or create an ionization effect that leads to a power arc. Additionally, even during normal voltage conditions, the presence of moisture, dirt, or salt on the insulator can also result in flashover.

Switching surges are caused by sudden changes in current flow that can happen when turning on or off electrical lines or equipment. These operations can generate traveling waves that may cause flashing on lines or equipment and create stress on weak insulation points. When current is started or stopped, line and apparatus insulation may experience transient overvoltages, which are a part of the "recovery" voltages. The most intense switching surges are observed when interrupting current that lags or leads the applied voltage by 90°, such as fault current or line-charging current.

When an unloaded line is dropped on a grounded system, the line voltage may reach the crest line-to-neutral voltage during the first interruption. With each subsequent half-cycle, the arc may restrike, leading to even higher voltages of three times the initial value on the first restrike, five times the initial value on the second restrike, and so on [94–96]. These switching surges can be significantly higher for systems that are not solidly grounded, which can cause severe overvoltages on the system's insulation, especially when interrupting line-charging current with breaker recovery voltages that are even higher.

Overhead transmission lines, which are primarily made up of bare conductors, are among the most vulnerable components in power systems [97]. These lines experience a significant number of faults, which may result from various causes such as wind, sleet, conductor clashing from conductor galloping, small animals (including birds, snakes, and squirrels), trees, cranes, airplanes, vandalism, vehicle collisions with poles or towers, line breaks due to excessive ice loading, or damage to supporting structures.

It is important to distinguish between a fault and an overload. An "overload" simply means that the system is carrying a load that exceeds its normal capacity. While the voltage at the overload point may drop to a low value, it will not drop to zero, which is what would happen in the case of a fault. This may result in an "undervoltage" condition that can spread to points far from the fault [98]. Faults can be classified as either "permanent" or "temporary." Permanent faults typically

DOI: 10.1201/9781003129769-4

cause equipment damage, which may occur violently due to the sudden flow of energy into the short circuit causing an explosion. Such a fault can burn or melt conductors.

The majority of faults that occur on overhead transmission lines are typically of a temporary or transient nature. As a result, service can be restored by quickly isolating the faulty section of the line and then reclosing it with enough speed to prevent restriking, but allowing sufficient time for the air to be deionized after the fault arc has been extinguished. This technique is known as *high-speed reclosing.*

Substation buses are engineered to withstand the highest expected faults, which can generate significant mechanical forces due to heavy short-circuit currents. Transformer failures can result from insulation deterioration caused by aging or overvoltages resulting from lightning or switching transients. Although faults may occur externally to the transformer, they can still generate significant internal mechanical forces. These forces can cause the movement of windings or other components, damaging insulation or even resulting in structural failure.

Generator failures can be categorized into three types: (1) failures that occur in the exciter circuit or control equipment, which account for about 50% of generator faults, (2) failures that occur in auxiliary equipment such as cooling systems, which make up around 40%, and (3) failures that encompass (a) stator faults due to conductor insulation breakdown caused by overvoltage or overheating due to unbalanced currents, (b) rotor faults resulting from damaged rotor windings due to ground faults, (c) overspeed caused by sudden loss of load, (d) loss of synchronism due to interphase faults, incorrect switching, or loss of field, (e) loss of field due to pilot-exciter failure, main-exciter failure, or field breaker failure, and (f) bearing failure resulting from cooling or oil supply failure.

When two line conductors carry current in parallel, they experience a force that is either attractive or repulsive, depending on the direction of the current. The force affecting each conductor can be quantified as follows:

$$F \propto \frac{I^2}{d} \qquad (4.1)$$

where
  I = current in each conductor
  d = distance between conductors

When the current flow in two parallel conductors is in the same direction, it creates an attractive force. Conversely, if the current flows in opposite directions, it generates a repulsive force. According to a study conducted by the Electric Power Research Institute (EPRI) [43], these forces can be significant enough to cause conductor movement, particularly when the conductors are closely spaced, such as in extra-high-voltage conductor bundles or adjacent phases of a compact line, during short-circuit currents. Naturally, the actual movement of a conductor, taking into account inertia, is determined by the current's magnitude and duration and is therefore dependent on circuit breaker interrupting time. Other issues during normal operating conditions may include corona, corona losses, and problems related to radio noise or radio interference (RI).

4.2 STATE OF THE ART

Extensive research has been conducted on transmission system modeling, with a focus on finite elements, lightning performance, and system characteristics [99–102]. Detecting a fault in a specific section of a transmission line is the reversal of the process used to locate the fault. The earliest research on fault detection systems for multiple terminals within a single transmission line can be traced back to previous studies [103]. Subsequent research has focused on using global positioning system (GPS) waveforms to locate electrical faults based on anomalous measurements that demonstrate lightning strike transients [104, 105]. The GPS is equipped with a time clock that captures timestamps designed to synchronize time across all devices within the various geographical regions of a waveform control area. As development has progressed, microprocessor-based relays with advanced communication infrastructure have replaced electromechanical ones gradually [106].

Digital relays have become an essential component of the power grid's cyberinfrastructure. Protection relays are designed to prevent electrical faults from damaging equipment. Numerous fault location methods have been developed in recent years for multiple terminals within a transmission line [107, 108], including wavelet traversal and directional sensing [109, 110], and analysis of transient behavior [111]. In the early 2000s, methods focused on synchronized measurements [112, 113] and the improvement of algorithms with single-ended detection sensors [114, 115]. Traveling wave methods involve superposition theorem with steady-state and post-fault signals [116, 117] and can enhance machine learning [118]. The discrete wavelet transformation has been used to decompose current samples to determine the fault zone, and an accurate non-iterative algorithm has been developed for identifying faults between two terminals of substations without time-synchronized measurements [119]. Later, efforts were made to correct unsynchronized measurements for the two-terminal fault-location problem [120–122]. A traveling wave algorithm has been implemented to detect transients in three-phase power systems by incorporating Park's transformation [123]. A data-driven approach based on an impedance-based network perspective is proposed to identify potential fault locations in a transmission network [124]. Matrix pencils, which are part of numerical linear algebra, are used in traveling-wave detection techniques to capture short arrival times [125]. The proposed method demonstrates the algorithm's robustness, which is not affected by measurement noise.

Traveling wave phenomena have been extensively studied mathematically, and recent contributions have focused on phase fault-launched waves and relevant parameters [126, 127]. The study also considers Clarke's transformation as a reference to phase A, and the results demonstrate the effectiveness of the methods, especially for aerial traveling waves. The analysis of reflection and refraction matrices for line terminations at the faulted location has been studied as well. Recent breakthroughs have extended these contributions to handle multiple terminals of transmission lines with modified algorithms for distributed fault location [128]. Researchers have also attempted fault distance prediction without line models and time synchronization to increase accuracy, which can be extremely useful for field crews [108]. Testbeds have been established for evaluating traveling wave fault locators (TWFLs) [129]. Many improvements have been made in the literature recently, including adaptability of threshold values and applications [130], stronger arrival information [131], generalized schemes for locating concurrent faults [132], and more precise methods using single-ended data from substations [115].

4.3 FAULT ANALYSIS AND FAULT TYPES

This section elaborates on the analysis of hypothetical faults. The analysis, also known as a short-circuit study or analysis, aims to calculate the maximum and minimum fault currents and voltages at different locations in the power system for various types of faults. This information is used to select the appropriate protective schemes, relays, and circuit breakers to quickly restore the system to normal conditions.

In practice, the following simplifying assumptions are typically made when performing fault analysis:

1. Neglecting normal loads, shunt capacitances, shunt elements in transformer equivalent circuits, and other shunt connections to the ground.
2. Assuming all generated system voltages are equal in magnitude and in phase.
3. Neglecting series resistances of lines and transformers if they are considered small compared to their reactances.
4. Assuming all transformers are at their nominal taps.

5. Representing the generator with a constant-voltage source in series with an appropriate reactance, which may be subtransient ($X''d$), transient ($X'd$), or synchronous at steady-state (X_d) reactance. Typically, the subtransient reactance (X''_d) is selected for the positive-sequence reactance, which is sufficient to calculate fault current magnitudes for the first three to four cycles after a fault occurs.

The initial assumption is based on the fact that the fault circuit has a significantly lower impedance than the shunt impedances. Although this assumption may lead to a slight loss in accuracy, the computational effort saved justifies its use.

The second assumption follows from the first, as the power system network becomes open-circuited, leading to the neglect of normal load currents (i.e., prefault currents), resulting in all prefault bus voltages having the same magnitude and phase angle. Therefore, in per-unit analysis, prefault bus voltages are set equal to $1.0 \angle 0°$ pu. If necessary, taking into account load current can be done using the superposition theorem.

The third assumption is typically used for hand calculations and educational purposes, simplifying the power system network to contain only reactances, allowing for a simplified reactance diagram. However, it is unnecessary if a digital computer is used for computation.

The fourth assumption neglects transformer tapings to allow for fault analysis in a per-unit system, removing transformers from the circuit. Usually, the subtransient reactance ($X''d$) is selected as the positive-sequence reactance, with the negative-sequence reactance being slightly different from the positive-sequence reactance for salient-type machines. For non-salient generators, selecting $X''d$ for the positive-sequence reactance results in the negative-sequence reactance becoming identical to the positive-sequence reactance.

To calculate maximum fault currents, it is common practice to assume a "bolted" fault, with no fault impedance ($\mathbf{Z}_f = 0$) due to fault arc, simplifying calculations and providing a safety factor.

In the case of a short-circuit path that is not metallic but through an arc or the ground, non-linear impedances can inject harmonics into the current and/or voltage. Fault resistance has two components: the resistance of the arc and the resistance of the ground. When the fault is between phases such as line to line, the fault only includes the arc, and the fault arc resistance is given by Warrington [133]:

$$R_{\mathrm{arc}} = \frac{8750 \times l}{I1.4} \Omega \tag{4.2}$$

where
l = length of arc in still air in feet
I = fault current in amperes

If time is involved, the arc resistance is calculated from

$$R_{\mathrm{arc}} = \frac{8750(d \times 3vt)}{I^1.4} \Omega \tag{4.3}$$

where
d = conductor spacing in feet (from Figure 4.1)
v = wind velocity in miles per hour
t = duration in seconds

The relation between the fault current and the arc voltage is given as

$$R_{\mathrm{arc}} = \frac{8750(d \times 3vt)}{I^{0.4}} \Omega \tag{4.4}$$

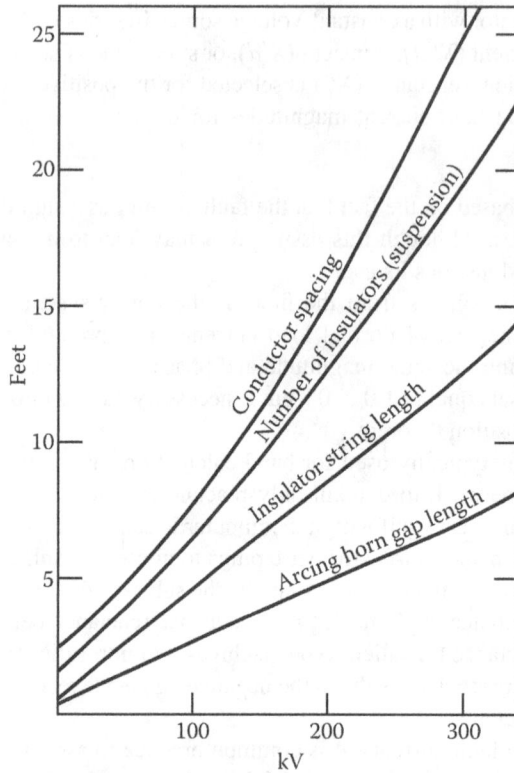

Figure 4.1 Minimum arcing distances on overhead lines (based on average tower dimensions in the United Kingdom and the United States). (From Warrington, A. R. van C., *Protective Relays: Their Theory and Practice*, vol. 2. Chapman & Hall, London, 1969.)

If a high-resistance line-to-ground fault occurs, the important impedances in the fault circuit are the contact to ground and the path through it. Warrington [133] has shown that the resistance of the ground fault is somewhat nonlinear partly because there are small arcs between conducting particles and partly due to the compounds of silicon, carbon, etc., which have nonlinear resistance (see Figure 4.1). It is interesting to note that, in practice, such *fault resistance* is erroneously called the *fault impedance*, and it is assumed to include a fictitious reactive component.

In general, the fault types that may occur in a three-phase power system can be categorized as follows.

A. Shunt faults:
    1. Balanced (also called symmetrical) three-phase faults
        a. Three-phase direct (L–L–L) faults
        b. Three-phase faults through a fault impedance to ground (L–L–L–G)
    2. Unbalanced (also called unsymmetrical) faults
        a. Single line-to-ground (SLG) faults
        b. Line-to-line (L–L) faults
        c. Double line-to-ground (DLG) faults
B. Series faults:
    1. One line open (OLO)
    2. Two lines open (TLO)
    3. Unbalanced series impedance condition

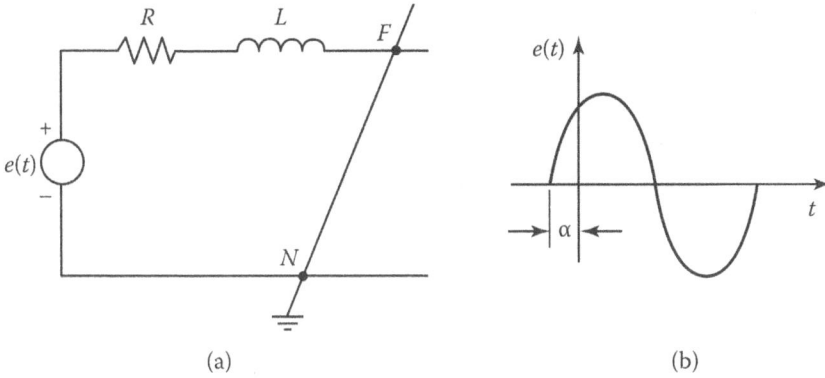

Figure 4.2 Synchronous generator with balanced fault: (a) per-phase representation of generator; (b) voltage waveform.

C. Simultaneous faults:
 1. A shunt fault at one fault point and a shunt fault at the other
 2. A shunt fault at one fault point and a series fault at the other
 3. A series fault at one fault point and a series fault at the other
 4. A series fault at one fault point and a shunt fault at the other

Shunt faults are more severe compared to series faults, while balanced faults are simpler to compute than unbalanced faults. Although simultaneous faults, which involve two or more faults happening simultaneously, are regarded as the most challenging fault analysis problem, this chapter only covers balanced and series faults, while unbalanced faults will be discussed in Chapter 6. The probability of having a simultaneous fault is much lower than a shunt fault, and therefore, the topic is beyond the scope of this book. However, for readers who wish to learn more about simultaneous faults, it is recommended to refer to Anderson's book [134].

4.4 BALANCED THREE-PHASE FAULTS AT NO LOAD

Three-phase transmission circuits have been designed and implemented in the generation of transmission expansion across the country [46, 64, 135–137]. Consider the per-phase representation of a synchronous generator, as shown in Figure 4.2. Assume that there is a balanced three-phase fault between the points F and N. If the generator voltage is $e(t) = V_m\sin(\omega t + \alpha)$ and the fault occurs at $t = 0$, it can be shown that there will be a transient current $i(t)$ that can be expressed as

$$i(t) = \frac{V_m}{Z}[\sin(\omega t + \alpha - \theta) - \sin(\alpha - \theta)e^{-\frac{Rt}{L}}]$$ (4.5)

where

$$Z = (R^2 + \omega^2 L^2)^{1/2} \quad \theta = \tan^{-1}\left(\frac{\omega L}{R}\right)$$

It can be seen in Equation 4.5 that the first term is a sinusoidal term and its value changes with time and that the second term is a nonperiodic term and its value decreases exponentially with time. The second term is a *unidirectional offset* and is also called the *direct current (dc) component* of the fault current. It will in general exist, and its initial magnitude (i.e., at $t = 0$) can be as large as the magnitude of the steady-state current term, as shown in Figure 4.3a. If the fault occurs at $t = 0$ when

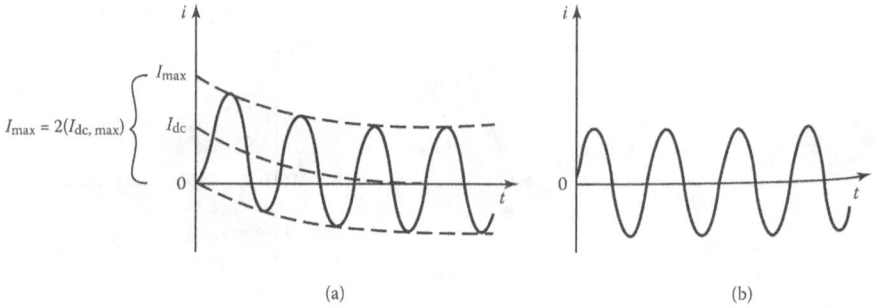

(a)   (b)

Figure 4.3   Balanced fault current wave shapes: (a) $\alpha - \theta = -90°$; (b) $\alpha = \theta$.

the angle $\alpha - \theta = -90°$, the value of the transient current becomes twice the steady-state maximum value and can be expressed as

$$i(t) = \frac{V_m}{2}\left(-\cos\omega t + e^{-\frac{Rt}{L}}\right) \tag{4.6}$$

and is shown in Figure 4.3a. The associated value of α is obtained from

$$\tan\alpha = -\frac{R}{\omega L} \tag{4.7}$$

On the other hand, if $\alpha = 0$, at $t = 0$, the dc offset does not exist, as shown in Figure 4.3b, and the value of the transient current can be expressed as

$$i(t) = \frac{V_m}{Z}\sin \omega t \tag{4.8}$$

Obviously, if a $\alpha - \theta = \pi$ at $t = 0$, the dc offset current again cannot exist. Thus, the value of the transient current depends on the angle α of the voltage wave. However, the time of the fault cannot be predicted in practice, and therefore the value of cannot be known ahead of time. However, the dc component diminishes very fast, usually in 8–10 cycles.

Furthermore, since the voltages generated in the phases of a three-phase synchronous generator are 120° apart from each other, each phase will have, in general, a different offset.

Note that, in the aforementioned discussions, the value of L for the generator is assumed to be constant. In reality, however, the reactance of a synchronous machine varies with time immediately after the occurrence of the fault. Thus, it is customary to represent a synchronous generator by a constant driving voltage in series with impedance that varies with time.

This varying impedance consists primarily of reactance since $X \gg R_a$, that is, X is much larger than the armature resistance. Hence, the value of impedance is approximately equal to its reactance. For the purpose of fault current calculations, the variable reactances of a synchronous machine can be represented, as shown in Figure 4.4b, by the following three reactance values:

X_d'' = *subtransient reactance*: determines the fault current during the first cycle after the fault occurs. In about 0.05–0.1 s, this reactance increases to
X_d' = *transient reactance*: determines the fault current after several cycles at 60 Hz. In about 0.2–2 s, it reactance increases to
$X_d = X_s$ = *synchronous reactance*: determines the fault current after a steady-state condition is reached

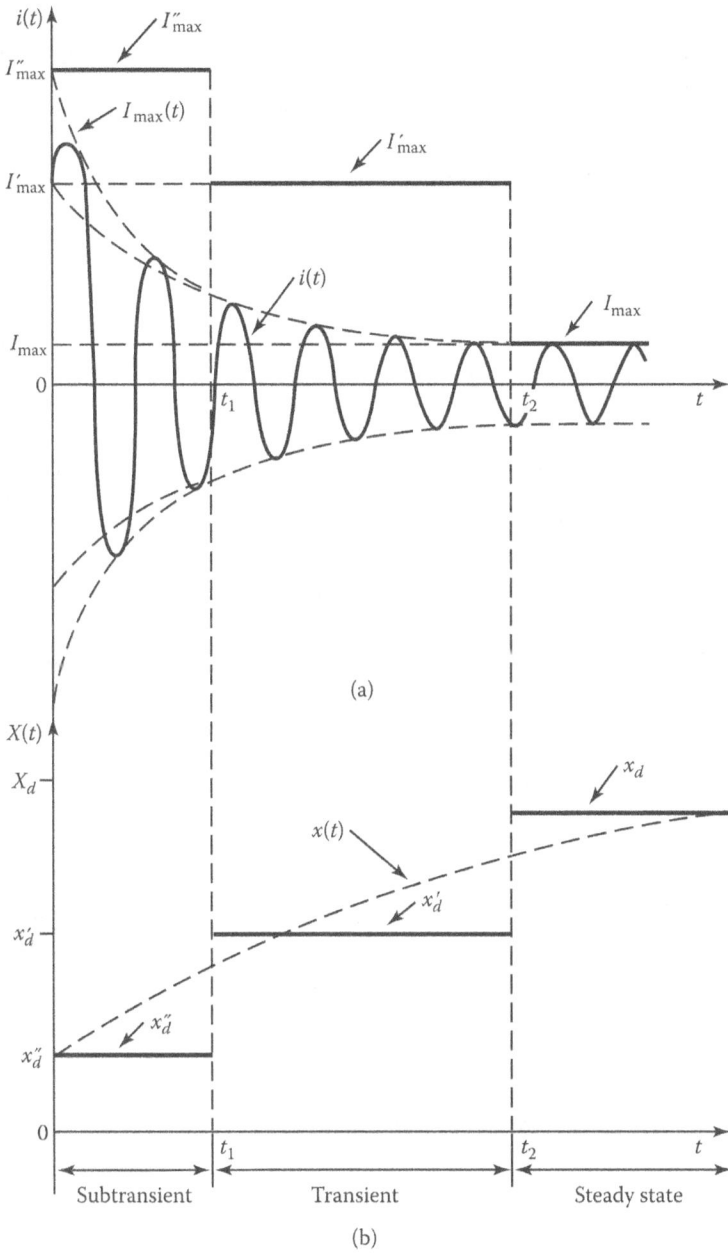

Figure 4.4 Balanced fault current and reactance for one phase of synchronous machine: (a) balanced instantaneous fault current without dc offset; (b) reactance $X(t)$ vs. time with stepped approximation.

This representation of the machine reactance by three different reactances is due to the fact that the flux across the air gap of the machine is much greater at the instant the fault occurs than it is a few cycles later. Thus, when a fault occurs at the terminals of a synchronous machine, time is necessary for the decrease in flux across the air gap. As the flux lessens, the armature current lessens since the voltage produced by the air gap flux regulates the current.

Thus, the subtransient reactance X_d'' includes the leakage reactance of the stator and rotor windings of the generator, the influences of damper windings and of the solid parts of the rotor body being included in the rotor leakage. The subtransient reactance is also called the *initial reactance*. The transient reactance X_d'' includes the leakage reactance of the stator and excitation windings of the generator. It is usually larger than the subtransient reactance.

If, however, the rotor has laminated poles and yokes and no damper windings, the transient reactance is the same as the subtransient reactance. If, however, the rotor has laminated poles and yokes and no damper windings,[*][*] the transient reactance is the same as the subtransient reactance.

The synchronous reactance X_d is the total reactance of the armature winding, which includes the stator leakage reactance and the armature reaction reactance of the generator. It is much larger than the transient reactance X_d''. Note that, all three reactances are considered to be the positive-sequence reactance of the synchronous machine.

In a *salient-pole* machine, the index d means that the reactances refer to a position of the rotor such that the axis of the rotor winding coincides with the axis of the stator winding so that the flux flows directly into the pole face. Therefore, it is called the *direct axis*, and thus the three reactances are also known as the *direct-axis reactances*.

In addition to these reactances, the generator also has reactances in the corresponding *quadrature axis*, that is, X_q'', X_q' and X_q, due to the flux path between poles, that is, midway between the field poles. The quadrature axis is 90 electrical degrees apart from the adjacent direct axes. However, the quadrature axis reactances are not relevant to the fault calculations.

Note that in a *nonsalient-pole* machine (i.e., *cylindrical-rotor machine*), values of X_d and X_q are basically equal. Therefore, there is no need to differentiate X_d from X_q but only call the synchronous reactance X_s. For the sake of simplification, in this book, all synchronous machines are assumed to have cylindrical rotors.

If the generator is operating at no load before the occurrence of a three-phase fault at its terminals, then the continuously varying symmetrical maximum current, $I_{max(t)}$, and reactance can be approximated with the discrete current levels, as shown in Figure 4.4, so that

$$X_d'' = \frac{E_{max}}{I_{max}''} \tag{4.9}$$

[*] They are located in the pole faces of a generator and are used to reduce the effects of hunting.

$$X_d' = \frac{E_{max}}{I_{max}'} \tag{4.10}$$

$$X_d = \frac{E_{max}}{I_{max}} \tag{4.11}$$

where E_{max} is the no-load line-to-neutral maximum voltage of the generator. Alternatively, using the root mean square (rms) values,

$$I'' = \frac{E_g}{X_d''} \tag{4.12}$$

$$I' = \frac{E_g}{X_d'} \tag{4.13}$$

$$I = \frac{E_g}{X_d} \tag{4.14}$$

where
E_g = no-load line-to-neutral rms voltage
I'' = subtransient current,[*] rms value without dc offset
I' = transient current, rms value without dc offset
I = steady-state current, rms value

Note that, the importance of the reactances given by Equations 4.9 through 4.11 depends on what percentage they represent of the short-circuit impedance. For example, if the fault occurs right at the terminals of the generator, they are very important; however, if the fault is remote from the generator, their importance is smaller.

The fault current will be lagging in power in a system where X ≫ R. Table 4.1 gives the typical values of the reactances for synchronous machines.

It is interesting to observe in Figure 4.4a that the total alternating component of armature current consists of the steady-state value and the two components that decay with time constants T_d' and T_d'. It can be expressed as

$$I_{ac} = (I'' - I')\exp\left(-\frac{t}{T_d''}\right) + (I' - I)\exp\left(-\frac{t}{T'}\right) + I \qquad (4.15)$$

where all quantities are in rms values and are equal but displaced 120 electrical degrees in the three phases.

* It is also called the *initial symmetrical rms current*.

4.5 FAULT INTERRUPTION

The fault-protective devices, such as fuses and circuit breakers, typically operate before steady-state conditions are achieved. As a result, the generator synchronous reactance is almost never used for calculating fault currents for these devices. To determine the initial symmetrical RMS current, subtransient reactances of synchronous generators and motors are used, as discussed earlier.

However, the circuit breaker's interrupting capacity is determined using the subtransient reactance for generators and the transient reactance for synchronous motors, while ignoring the effects of induction motors. The use of subtransient reactance for synchronous motors is appropriate if fast-acting circuit breakers are used. For instance, modern air blast circuit breakers can operate in 2.5 cycles of 60 Hz, whereas older circuit breakers and those used on lower voltages may take more than eight cycles to operate.

Note that the discussion so far has excluded the dc offset (or unidirectional current component). With fast-acting circuit breakers, the actual current to be interrupted increases by the dc component of the fault current, and the initial symmetrical RMS current value increases by a specific factor depending on the circuit breaker's speed. For example, if the circuit breaker opening time is eight, three, or two cycles, the corresponding multiplying factor is 1.0, 1.2, or 1.4, respectively. Therefore, the circuit breaker's interrupting capacity or rating is expressed as:

$$S_{\text{interrupting}} = \sqrt{3}(V_{\text{prefault}})(I'')\zeta \times 10^{-6} \text{ MVA} \qquad (4.16)$$

where

V_{prefault} = prefault voltage at point of fault in volts
I'' = initial symmetrical rms current in amperes
ζ = multiplying factor

Equation 4.16 only considers the AC component. The types of reactances and corresponding multiplying factors are provided in Table 4.1 [138]. As discussed previously, the asymmetrical current waveform gradually decays into a symmetrical current waveform, and the rate of decay of the DC component depends on the L/R of the current source. The time constant for the decay of the DC component can be calculated using the following formula:

$$T_{\text{dc}} = \text{circuit } L/R \text{ s} \qquad (4.17a)$$

Table 4.1

Reactance Quantities and Multiplying Factors for Application of Circuit Breakers

		Reactance Quantity for Use in X_1		
	Multiplying Factor	Synchronous Generators and Condensers	Synchronous Motors	Induction Machines
A. Circuit Breaker Interrupting Duty				
1. General case:				
Eight-cycle or slower circuit breakers[a]	1	Subtransient[b]	Transient	Neglect
Five-cycle circuit breakers	1.1			
Three-cycle circuit breakers	1.2			
Two-cycle circuit breakers	1.4			
2. Special case for circuit breakers at generator voltage only. For short-circuit calculations of more than 500,000 kVA (before the application of any multiplying factor) fed predominantly direct from generators, or through current-limiting reactors only:		Subtransient[b]	Transient	Neglect
Eight-cycle or slower circuit breakersa	1.1			
Five-cycle circuit breakers	1.2			
Three-cycle circuit breakers	1.3			
Two-cycle circuit breakers	1.5			
3. Air circuit breakers rated 600 V and less	1.25	Subtransient	Subtransient	Subtransient
B. Mechanical Stress and Momentary Duty of Circuits Breakers				
1. General case	1.6	Subtransient	Subtransient	Subtransient
2. At 500 V and below, unless current is fed predominantly by directly connected synchronous machines or through reactors	1.5	Subtransient	Subtransient	Subtransient

Source: Westinghouse Electric Corporation, *Electrical Transmission and Distribution Reference Book. WEC, Pittsburgh, 1964.*

[a] As old circuit breakers are slower that modern ones, it might be expected that a low multiplier could be used with old circuit breakers. However, modern circuit breakers are likely to be more effective than their slower predecessors, and therefore, the application procedure with the older circuit breakers should be more conservative than with modern circuit breakers. Also, there is no assurance that a short circuit will not change its character and initiate a higher current flow through a circuit breaker while it is opening. Consequently, the factors to be used with older and slower circuit breakers well may be the same as for modern eight-cycle circuit breakers.

[b] This is based on the condition that any hydroelectric generators involved have amortisseur (damper) windings. For hydroelectric generators without amortisseur windings, a value of 75% of the transient reactance should be used for this calculation rather than the subtransient value.

or

$$T_{dc} = \frac{\text{circuit } L/R}{2\pi} \quad \text{cycles} \tag{4.17b}$$

The momentary duty (or rating) of a circuit breaker is expressed as

$$S_{\text{momentary}} = \sqrt{3}(V_{\text{prefault}})(I'')(1.6) \times 10^{-6} \text{ MVA} \tag{4.18}$$

This equation includes the dc component. Thus, the rms momentary current can be expressed as
* Note that, the fault megavolt-ampere is often referred to as the *fault level*.

$$I_{\text{momentary}} = 1.6 \times I''A \tag{4.19}$$

Here, the Imomentary current is the total rms current that includes ac and dc components, and it is used for oil circuit breakers of 115 kV and above. The circuit breaker must be able to withstand this rms current during the first half-cycle after the fault occurs. Note that, if the I'' is measured in peak amperes, then the peak momentary current is expressed as

$$I_{\text{momentary}} = 2.7 \times I''A \tag{4.20}$$

In the United States, the ratings of circuit breakers are given in the American National Standards Institute (ANSI) standards [139] based on symmetrical current,[†] in terms of nominal voltage, rated maximum voltage, rated voltage range factor K, rated continuous current, and rated short-circuit current. The required symmetrical current-interrupting capability is defined as

$$(\text{Current-interrupting capability}) = (\text{Rated short-circuit current}) \left(\frac{\text{Rated maximum voltage}}{\text{operating voltage}}\right)$$

According to the standard, if the operating voltages are below 1/K times the rated maximum voltage, then the required symmetrical current-interrupting capability of the circuit breaker must be equal to K times the rated short-circuit current. Table 4.2 provides ratings for outdoor circuit breakers based on symmetrical current. It is important to note that the rated voltage factor K is defined as the ratio of the rated maximum voltage to the lower limit of the range of operating voltage where the required symmetrical and asymmetrical interrupting capabilities are inversely proportional to the operating voltage [140]. Therefore, the general expressions for the rms and peak momentary currents, respectively, taking into account the rated voltage range factor K, are as follows:
[†] The ratings of the circuit breakers can also be based on total current, which includes the DC component.

$$I_{\text{momentary}} = 1.6 \times K \times I'' \text{ A} \tag{4.21}$$

and

$$I_{\text{momentary}} = 2.7 \times K \times I'' \text{ A} \tag{4.22}$$

Notice that, in Table 4.2, the factor K is 1 for the nominal voltages of 115 kV and above. Therefore, Equations 4.21 and 4.22 become the same as Equations 4.19 and 4.20, respectively.

Example 4.1: A circuit breaker has a rated maximum rms voltage of 38 kV and is being operated at 34.5 kV. Determine the following:

(a) The highest symmetrical current-interrupting capability
(b) The operating voltage at the highest symmetrical current capability

Table 4.2
Outdoor Circuit Breaker Ratings Based on Symmetrical Current

Nominal rms Voltage Class (kV)	Rated Maximum rms Voltage (kV)	Rated Voltage Range Factor, K	Rated Continuous rms Current (kA)	Rated Short-Circuit rms Current (at Rated Maximum kV) (kA)	Rated Interrupting Time (Cycles)	Rated Maximum rms Voltage Divided by K (kV)	Maximum rms Symmetrical Interrupting Capability[a] (kA)
14.4	15.5	2.67	0.6	8.9	5	5.8	24
14.4	25.5	1.29	1.2	18	5	12	23
23	25.8	2.15	1.2	11	5	12	24
34.5	38	1.65	1.2	22	5	23	36
46	48.3	1.21	1.2	17	5	40	21
69	72.5	1.21	1.2	19	5	60	23
115	121	1	1.2	20	3	121	20
115	121	1	1.6	40	3	121	40
...	...	...	...	...	...	...	...
115	121	1	3	63	3	121	63
138	145	1	1.2	20	3	145	20
138	145	1	1.6	40	3	145	40
138	145	1	2	40	3	145	40
...	...	...	...	...	...	...	...
138	145	1	3	80	3	145	80
161	169	1	1.2	16	3	169	16
161	169	1	1.6	31.5	3	169	31.5
...	...	...	...	...	...	...	...
161	169	1	2	50	3	169	50
230	242	1	1.6	31.5	3	242	31.5
230	242	1	2	31.5	3	242	31.5
...	...	...	...	...	...	...	...
230	242	1	3	63	3	242	63
345	362	1	2	40	3	362	40
345	362	1	3	40	3	362	40
500	550	1	2	40	2	550	40
500	550	1	3	40	2	550	40
700	765	1	2	40	2	765	40
700	765	1	3	40	2	765	40

[a] Equal to K times the rated short-circuit rms current.

(c) The associated rms momentary current rating
(d) The associated peak momentary current rating

Solution:

(a) From Table 4.2, the rated voltage range factor K is 1.65, the rated continuous rms current is 1200 A, and the rated short-circuit rms symmetrical current at the rated maximum rms voltage of 38 kV is 22,000 A. However, since the circuit breaker is used at 34.5 kV, its symmetrical current-interrupting capability is

$$(22{,}000 \text{ A}) \left(\frac{38 \text{ kV}}{34.5 \text{ kV}} \right) = 24{,}232 \text{ A}$$

The highest symmetrical current-interrupting capability is

$$(22{,}000 \text{ A})K = 22{,}000 \times 1.65 \cong 36{,}000 \text{ A}$$

(b) Which is possible when the operating voltage is
(c) Note that, at lower operating voltages, the highest symmetrical current-interrupting capability of 36,000 A cannot be exceeded. Hence, the associated rms momentary current is

$$\frac{38 \text{ kV}}{K} = \frac{38}{1.65} \cong 23 \text{ kV}$$

(d) The associated peak momentary current rating is

$$\begin{aligned} I_{\text{momentary}} &= 1.6 \times K \times I'' \\ &= 1.6(36{,}000 \text{ A}) \\ &= 57{,}600 \text{ A rms} \end{aligned}$$

A simplified procedure for determining the symmetrical fault current is known as the "E/X method" and is described in Section 5.3.1 of ANSI C37.010. This method [63, 138–140] gives results approximating those obtained by more rigorous methods. In using this method, it is necessary first to make an E/X calculation. The method then corrects this calculation to take into account both the dc and ac decay of the current, depending on circuit parameters X/R. The approximation basically results owing to the use of curves.

Example 4.2: Consider the system shown in Figure 4.5 and assume that the generator is unloaded and running at the rated voltage with the circuit breaker open at bus 3. Assume that the reactance values of the generator are given as $X_d'' = X_1 = X_2 = 0.14$ pu and $X_0 = 0.08$ pu based on its ratings. The transformer impedances are $Z_1 = Z_2 = Z_0 = j0.05$ pu based on its ratings. The transmission line TL$_{23}$ has $Z_1 = Z_2 = j0.04$ pu and $Z_0 = j0.10$ pu. Assume that the fault point is located on bus 1, and determine the subtransient fault current for a three-phase fault in per units and amperes. Select 25 MVA as the megavolt-ampere base and 8.5 and 138 kV as the low-voltage and high-voltage bases.
Solution:

$$\begin{aligned} I_f'' &= \frac{E_g}{X_d''} \\ &= \frac{1.0\angle 0°}{j0.14} \\ &= j7.143 \text{ pu} \end{aligned}$$

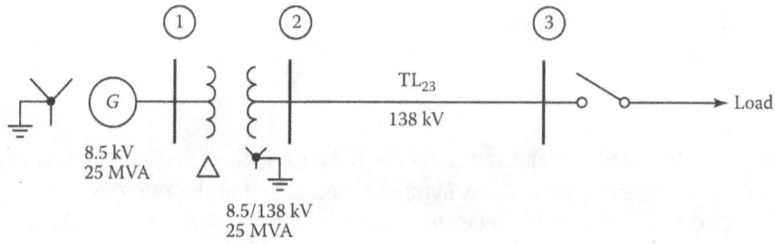

Figure 4.5 Transmission system for Example 4.2.

The current base for the low-voltage side is

$$
\begin{aligned}
I_{B(\text{LV})} &= \frac{S_B}{\sqrt{3}V_{B(\text{LV})}} \\
&= \frac{25,000 \text{ kVA}}{\sqrt{3}(8.5 \text{ kV})} \\
&= 1698.1 \text{ A}
\end{aligned}
$$

Therefore,

$$
\begin{aligned}
I_f'' = |\mathbf{I}_f''| &= (7.143)(1698.1) \\
&= 12,129.52 \text{ A}
\end{aligned}
$$

Example 4.3: Use the results of Example 4.2 and determine the following:

(a) The possible value of the maximum of the dc current component
(b) Total maximum instantaneous current
(c) Momentary current
(d) Interrupting capacity of the three-cycle circuit breaker if located at bus 1
(e) Momentary interrupting capacity of the three-cycle circuit breaker if located at bus 1

Solution:

(a) Let

$$
I_{\text{dc,max}} = \text{peak-to-peak amplitude}
$$

then

$$
\begin{aligned}
I_{\text{dc,max}} &= \sqrt{2}(I_f'') \\
&= \sqrt{2}(I_{\text{ms}}) \\
&= \sqrt{2}(7.143 \text{ pu}) \\
&= 10.1 \text{ pu}
\end{aligned}
$$

(b) From Figure 4.3a,

$$I''_{max} = 2I_{dc,max}$$
$$= 2(\sqrt{2}I''_f)$$
$$= 20.2 \text{ pu}$$

(c) $I_{momentary}$ current represents the summation of I_{ac} and I_{dc}, where I_{dc} is about 50% of I_{ac}. Hence,

$$I_{momentary} = 1.6 \times I_f$$
$$= 1.6(7.143 \text{ pu})$$
$$= 11.43 \text{ pu}$$

(d)

$$I_{interrupting} = \sqrt{3}V_f I''_f \zeta \times 10^{-6}$$
$$= \sqrt{3}(8500)(12,129.52)(1.2) \times 10^{-6}$$
$$= 214.3 \text{ MVA}$$

(e)

$$I_{interrupting} = \sqrt{3}V_f I''_f (1.6) \times 10^{-6}$$
$$= \sqrt{3}(8500)(12,129.52)(1.6) \times 10^{-6}$$
$$= 285.7 \text{ MVA}$$

4.6 BALANCED THREE-PHASE FAULTS AT FULL LOAD

The preceding section covered balanced three-phase faults, assuming that the generator supplied a no-load current. However, in this section, it will be presumed that the generator had been supplying a load current before the fault occurred, which was of substantial magnitude and hence could not be disregarded. To calculate the balanced three-phase fault, any of the following two techniques may be employed:

Method 1: Using Machine Internal Voltage Behind Subtransient Reactance

Consider the one-line diagram shown in Figure 4.6a and assume that the three-phase synchronous generator feeds the three-phase synchronous motor. The load current flowing before the fault, at point F, is $\mathbf{I}_L$ and the voltage at the fault point is $\mathbf{V}_F$. The line impedance between the generator and the motor is $\mathbf{Z}_{12}$. Therefore, the terminal voltage $\mathbf{V}_t$ of the generator and the motor is not the same as shown in Figure 4.6b. Since there is a prefault load current $\mathbf{I}_L$, the internal voltages of the synchronous generator can be expressed as

$$\mathbf{E}''_g = \mathbf{V}_t + \mathbf{I}_L X''_d \tag{4.23}$$
$$\mathbf{E}'_g = \mathbf{V}_t + \mathbf{I}_L X'_d \tag{4.24}$$
$$\mathbf{E}_g = \mathbf{V}_t + \mathbf{I}_L X_d \tag{4.25}$$

where
$\mathbf{E}''_g$ = voltage behind the subtransient reactance X''_d (also called *subtransient internal voltage*)

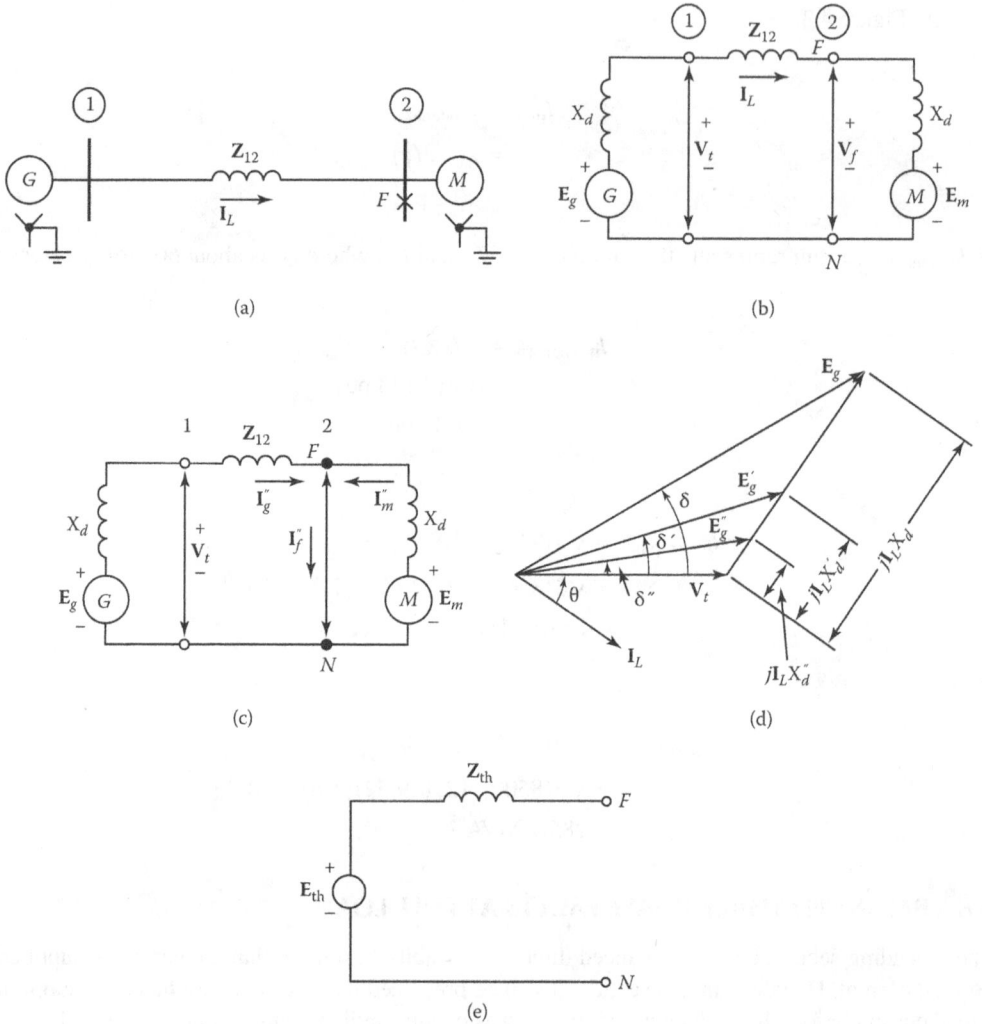

Figure 4.6 Calculation of balanced three-phase fault at full load: (a) one-line diagram of system under study; (b) equivalent circuit before fault; (c) equivalent circuit after fault; (d) phasor diagram for generator; (e) Thévenin equivalent of circuit given in (b).

E'_g = voltage behind the transient reactance X'_d (also called *transient internal voltage*)
E_g = voltage behind the synchronous reactance X_d (also called *steady-state internal voltage*)

It is evident that when the generator is running without a load, the internal reactances of all three phases are identical, therefore.

$$E''_g = E'_g + E_g \tag{4.26}$$

as used in Equations 4.12 through 4.14. Figure 4.6d shows a phasor diagram depicting the relations between the internal voltages and the terminal voltage V_t for the generator.

Similarly, the internal voltages of the synchronous motor can be expressed as

$$E''_m = V_t + I_L X''_d \tag{4.27}$$

$$E'_m = V_t + I_L X'_d \tag{4.28}$$

$$\mathbf{E}_m = \mathbf{V}_t + \mathbf{I}_L X_d \tag{4.29}$$

Choosing the appropriate internal voltage relies on the duration that has elapsed since the fault occurred. For instance, if the fault current immediately after the fault is needed, the subtransient internal voltage, denoted as $\mathbf{E}''_g$, is utilized.

It should be noted that if a synchronous motor is implicated in the fault, its field remains energized despite the fact that the motor does not receive any power from the line. Due to the inertia of its rotor and connected mechanical load, it continues to rotate for a certain period. Hence, the motor starts to operate as a generator and contributes to the fault current, as depicted in Figure 4.6c. Consequently, the fault current produced by the generator and the motor can be expressed as:

$$\mathbf{I}''_g = \frac{E''_g}{X''_d + Z_{12}} \tag{4.30}$$

and

$$\mathbf{I}''_m = \frac{E''_m}{X''_d} \tag{4.31}$$

Therefore, the total current, which includes both fault and load currents, can be found from

$$\mathbf{I}''_f = \mathbf{I}''_g + \mathbf{I}''_m \tag{4.32}$$

Method 2: Using the Thévenin Voltage and Impedance at the Fault Point

The subtransient fault current can be determined by using Thévenin's theorem, as shown in Figure 4.6e. Note that, the Thévenin voltage, $\mathbf{E}_{th}$, is the same as the V_f voltage before the fault, as shown in Figure 4.6b, where the Thévenin impedance is

$$\mathbf{Z}_{th} = \frac{(X''_d + X_{12})X''_d}{(X''_d + X_{12}) + X''_d} \tag{4.33}$$

Thus, the subtransient fault current at the fault point F is

$$\begin{aligned}\mathbf{I}''_f &= \frac{E_{th}}{Z_{th}} \\ &= \frac{V_f}{Z_{th}}\end{aligned} \tag{4.34}$$

By using the current division, the contributions of the generator and motor to the fault current can be found as

$$\mathbf{I}''_{f(g)} = \left[\frac{X''_d}{(X''_d + X_{12}) + X''_d}\right]\mathbf{I}''_f \tag{4.35}$$

and

$$\mathbf{I}''_{f(m)} = \left[\frac{(X''_d + X_{12})}{(X''_d + X_{12}) + X''_d}\right]\mathbf{I}''_f \tag{4.36}$$

To calculate the total current contributions from both the generator and the motor, it is necessary to take into account the prefault load current, IL. As a result, the subtransient currents in both the generator and the motor can be fully expressed as:

$$\mathbf{I}''_g = \mathbf{I}''_f(g) + \mathbf{I}_L \tag{4.37}$$

and

$$\mathbf{I}_m'' = \mathbf{I}_f''(m) + \mathbf{I}_L \qquad (4.38)$$

and hence, the total fault current can be expressed as

$$\mathbf{I}_f'' = \mathbf{I}_{(g)}'' + \mathbf{I}_m'' \qquad (4.39)$$

The corresponding fault (or short-circuit) megavolt-ampere can be found as

$$S_f = \sqrt{3}(V_{\text{prefault}})(I_f)10^{-6} \text{ MVA} \qquad (4.40)$$

where the prefault voltage is the nominal voltage in kilovolts.

In the event that the nominal prefault voltage and fault current are known for a given fault point, the corresponding Thévenin equivalent system reactance can be found as

$$X_{\text{th}} = \frac{V_{\text{prefault}} \times 1000}{\sqrt{3}I_f} \; \Omega \qquad (4.41)$$

Alternatively, finding I_f from Equation 4.40 and substituting into Equation 4.41,

$$X_{\text{th}} = \frac{(V_{\text{prefault}})^2}{S_f} \; \Omega \qquad (4.42)$$

where the prefault voltage is in kilovolts and the fault megavolt-ampere is in megavolt-amperes.

Furthermore, in the event that base voltage and prefault voltage are the same, then the Thévenin equivalent system reactance can be expressed as

$$X_{\text{th}} = \frac{S_B}{S_f} \text{ pu} \qquad (4.43)$$

or

$$X_{\text{th}} = \frac{I_B}{I_f} \text{ pu} \qquad (4.44)$$

Example 4.4: Consider the system shown in Figure 4.6 and assume that the prefault voltage at the terminals of the motor is 1.0 ∠ 0° pu. The subtransient reactances of the generator and the motor are both 0.07 pu. The load current I_L is 0.8 + j0.5 pu. The line reactance X_{12} is 0.09 pu. Use the first method given in Section 4.4 and determine the following:

(a) Subtransient internal voltage of the generator
(b) Subtransient internal voltage of the motor
(c) Fault current contribution of the generator
(d) Fault current contribution of the motor
(e) Total fault current

Solution:

(a)

$$\mathbf{I}_L = 0.8 + j0.50 = 0.9434\angle 32° \text{ pu}$$

Thus, the prefault terminal voltage of the generator is

$$\begin{aligned} \mathbf{V}_t &= \mathbf{V}_f + \mathbf{I}_L \mathbf{X}_{12} \\ &= 1.0\angle 0° + (0.8 + j0.5)(0.09\angle 90°) \\ &= 0.955 + j0.072 = 0.9577\angle 4.3° \text{ pu} \end{aligned}$$

$$\mathbf{E}_g'' = \mathbf{V}_f + \mathbf{I}_L \mathbf{X}_d''$$
$$= (0.955 + j0.072) + (0.9434\angle 32°)(0.07\angle 90°)$$
$$= 0.920 + j0.128 \text{ pu}$$

(b)

$$\mathbf{V}_t + \mathbf{V}_f = 1.0\angle 0° \text{ pu}$$

Hence,

$$\mathbf{E}_m'' = \mathbf{V}_t + \mathbf{I}_L \mathbf{X}_d''$$
$$= 1.0\angle 0° - (0.9434\angle 32°)(0.07\angle 90°)$$
$$= 1.035 + j0.056 \text{ pu}$$

(c)

$$\mathbf{I}_g'' = \frac{\mathbf{E}_g''}{X_d'' + X_{12}}$$
$$= \frac{0.920 + j0.128}{j0.07 + j0.09}$$
$$= 0.8 - j5.7502 \text{ pu}$$

(d)

$$\mathbf{I}_m'' = \frac{\mathbf{E}_m''}{X_d''}$$
$$= \frac{1.035 + j0.056}{j0.07}$$
$$= -0.8 - j14.7855 \text{ pu}$$

(e)

$$\mathbf{I}_f'' = \mathbf{I}_g'' + \mathbf{I}_m''$$
$$= -j20.5357 \text{ pu}$$

Example 4.5: Use the data given in Example 4.4 and determine

(a) Thévenin's impedance at the fault point
(b) Subtransient fault at the fault point
(c) Subtransient fault contribution of the generator
(d) Subtransient fault contribution of the motor
(e) Total current contribution of the generator
(f) Total current contribution of the motor
(g) Total fault current at the fault point

Solution:

(a)

$$\mathbf{Z}_{th} = \frac{(X_d'' + X_{12})X_d''}{(X_d'' + X_{12}) + X_d''}$$
$$= \frac{(j0.16)j0.07}{j0.16 + j0.07}$$
$$= j0.0487 \text{ pu}$$

(b)

$$\mathbf{I}''_f = \frac{\mathbf{E}_{th}}{\mathbf{Z}_{th}} = \frac{\mathbf{V}_f}{\mathbf{Z}_{th}}$$

$$= \frac{1.0\angle 0^\circ}{j0.0487}$$

$$= 20.5357\angle -90^\circ \text{ pu}$$

(c)

$$\mathbf{I}''_{f(g)} = \left[\frac{X''_d}{(X''_d + X_{12}) + X''_d} \right] \mathbf{I}''_f$$

$$= \left[\frac{j0.07}{j0.16 + j0.07} \right] \times 20.5357\angle -90^\circ$$

$$= -j6.25 \text{ pu}$$

(d)

$$\mathbf{I}''_{f(m)} = \left[\frac{(X''_d + X_{12}}{(X''_d + X_{12}) + X''_d} \right] \mathbf{I}''_f$$

$$= \left[\frac{j0.16}{j0.16 + j0.07} \right] \times 20.5357\angle -90^\circ$$

$$= -j14.2857 \text{ pu}$$

(e)

$$\mathbf{I}''_g = \mathbf{I}''_{f(g)} + \mathbf{I}_L$$

$$= j6.25 + 0.8 + j0.5$$

$$= 0.8 - j5.75 \text{ pu}$$

(f)

$$\mathbf{I}''_f = \mathbf{I}''_{f(m)} + \mathbf{I}_L$$

$$= -j14.2557 - 0.8 + j0.5$$

$$= 0.8 - j14.7857 \text{ pu}$$

(g)

$$\mathbf{I}''_f = \mathbf{I}''_g + \mathbf{I}''_m$$

$$= -(0.8 - j5.57) - 0.8 + j14.7857$$

$$= -j20.5357 \text{ pu}$$

It checks with the previous results.

4.7 APPLICATION OF CURRENT-LIMITING REACTORS

If a fault occurs, the current is limited by the system reactance, which includes the impedance of generators, transformers, lines, and other components of the system. The reactance of modern generators is typically large enough to constrain the short-circuit megavoltampere to a value that circuit breakers can interrupt. However, if the system is large or some generators are old, additional impedance may be necessary, and reactors can provide it.

Current-limiting reactors are coils used to restrict current during fault conditions, and it is crucial that magnetic saturation at high current does not reduce the coil reactance to accomplish this task. Reactors can be either air-cored or iron-cored. Air-cored reactors do not experience magnetic saturation, and their reactances are therefore independent of current. They come in two types: oil-immersed and dry-type. Oil-immersed reactors can be cooled by any of the means used to cool power transformers, while dry-type reactors are typically cooled by natural ventilation but can also be designed with forced-air and heat-exchanger auxiliaries.

Reactors are typically built as single-phase units. If dry-type reactors are positioned near metal objects such as I-beams, plates, and channels, magnetic shielding is necessary to prevent the reactor flux from inducing eddy currents in those objects. Otherwise, the proximity of metal objects will increase the power loss of the reactor and change its reactance.

These reactors are limited to 34.5 kV as a maximum insulation class due to the required clearances and construction details necessary to minimize corona. However, oil-immersed reactors can be used at any voltage level, both indoors and outdoors. They provide high safety against flashover, high thermal capacity, and have no magnetic field outside the tank to cause heating or magnetic forces in surrounding metal objects during short circuits.

Air-cored reactors are designed to withstand great mechanical stresses that occur under short-circuit conditions and thermally for not more than $33^{1/3}$ times the normal full-load current for 5 s under short-circuit conditions. Iron-cored reactors are typically built as oil-immersed reactors. Their reactance-to-resistance ratio is much greater than for the air-cored type. They can be designed for any voltage level and are more expensive than air-cored reactors.

In general, the effective resistance of a reactor is negligible (its R/X ratio is about 0.03). The inductive reactance, X, of a reactor can be determined from

$$X_r = \frac{(\%X_r)V_r}{100(\sqrt{3}) \times I_r} \ \Omega/\text{phase} \tag{4.45}$$

or

$$X_r = \frac{(\%X_r)V_{2r}}{100P_r} \ \Omega/\text{phase} \tag{4.46}$$

where
$\%X_r$ = reactor reactance in percent
V_r = reactor-rated voltage in kilovolts
I_r = reactor-rated current in kiloamperes
P_r = reactor power rating in megavolt-amperes

As power systems become more interconnected, fault levels can increase, making it necessary to increase the system reactance. This can be achieved by strategically placing reactors at certain points in the system. These reactors can be located in various ways, including (1) in series with generators, (2) in series with lines or feeders, (3) between buses, (4) in a tie-bus arrangement, or (5) in a ring arrangement.

Reactor placement in series with generators is not common because modern generators typically have sufficient leakage reactance, and the arrangement can cause large voltage drops and power losses under steady-state conditions. Additionally, if a fault occurs on or near the buses, the generators may lose synchronism due to the resulting bus undervoltage conditions. This method is only used to protect older generators with low leakage reactances.

Reactor placement in series with lines is more frequently used because it results in a large voltage drop in the associated reactor in the event of a fault, but only a small reduction in the bus voltage.

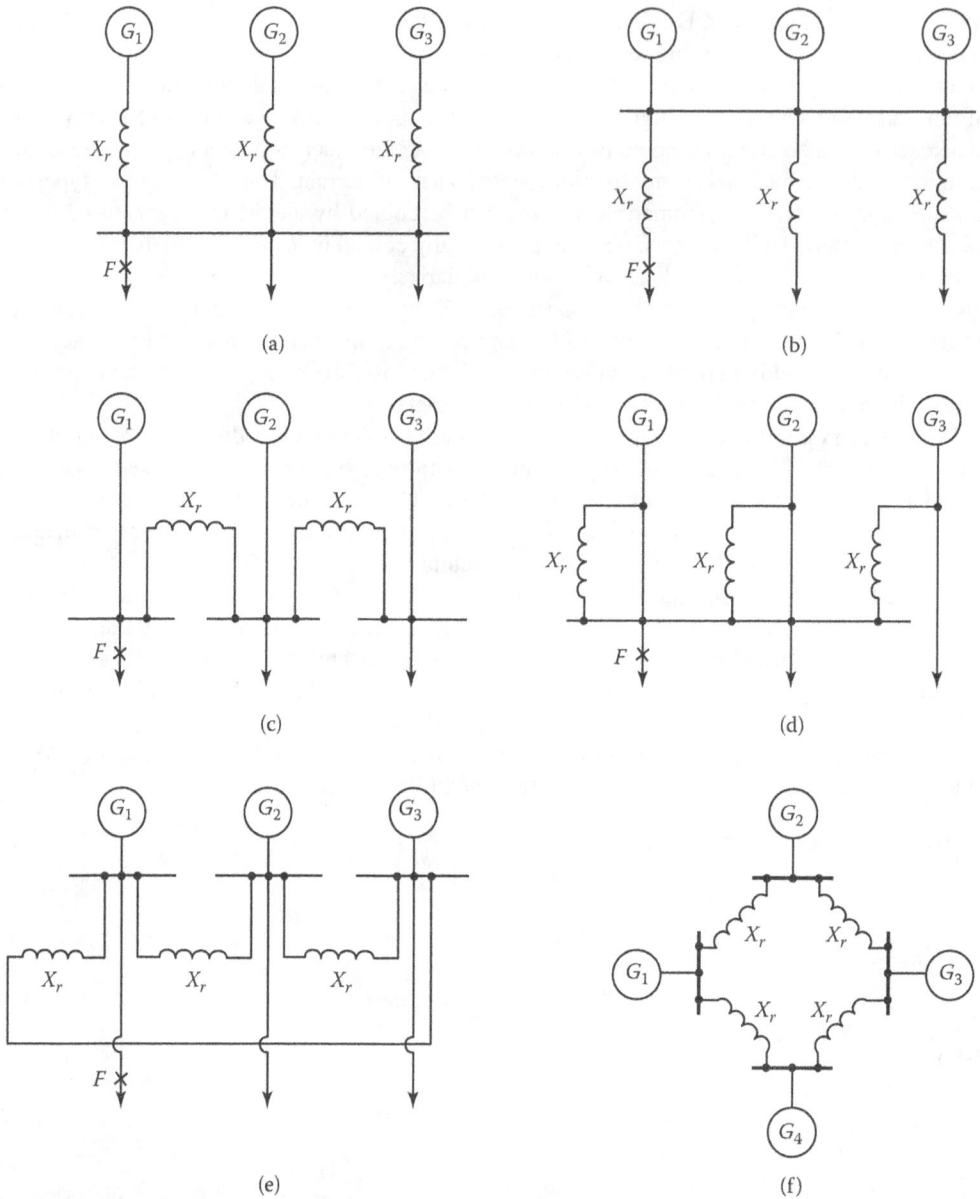

Figure 4.7 Various reactor connections: (a) series with generators; (b) series with lines; (c) between buses; (d) tie-bus system; (e) ring system; (f) ring system.

As a result, the synchronism of the generators will be preserved, and the fault can be isolated by disconnecting the faulted line. The arrangements shown in Figure 4.7c and d help to limit and localize disturbances to the faulted bus and generator.

In the ring arrangements shown in Figure 4.7e and f, current between two sections flows through two paths in parallel, while in the tie-bus arrangement, current flows through two reactors in series. Therefore, the reactors in the tie-bus arrangement have only one-third of the reactance of ring reactors. However, they carry twice as much current as the ring reactors.

Example 4.6: Three 20-kV solidly grounded generators are connected to three reactors in a tie–bus arrangement as shown in Figure 4.7d. The reactances of generators and the reactors are 0.2 and 0.1 pu, respectively, based on a 50-MVA base. If there is a symmetrical three-phase fault at the fault point F, determine the following:

(a) Short-circuit megavolt-amperes
(b) Fault current distribution in system

Solution:

(a) Figure 4.8a shows the equivalent circuit after the fault. The Thévenin equivalent reactance can be calculated from Figure 4.8b as

$$X_{th} = \frac{(j0.2)(j0.25)}{(j0.2)(j0.25)}$$
$$= j0.1111 \text{ pu}$$

Therefore, the short-circuit megavolt-amperes can be found as

$$S_f = \frac{S_{B(3\phi)}}{|X_{th}|}$$
$$= \frac{50 \text{ MVA}}{0.1111}$$
$$= 45,005 \text{ MVA}$$

(b) Since the short-circuit current at the fault point F is

$$\mathbf{I}_f = \frac{V_f}{Z_{th}}$$
$$= \frac{1.0\angle 0°}{j0.1111}$$
$$= 9\angle -90° \text{ pu}$$

and

$$I_B = \frac{50 \text{ MVA}}{\sqrt{3}(20 \text{ kV})}$$
$$= 1443.4 \text{ A}$$

then

$$\mathbf{I}_f = (9\angle -90°)1443.4$$
$$= 12,990.6\angle -90° \text{ A}$$

Assume that generator G_1 supplies some $\mathbf{I}_{f_1}$ amount of the short-circuit current $\mathbf{I}_f$. Then, from Figure 4.8b,

$$0.2\mathbf{I}_{f_1} = 0.25(12,990.6\angle -90°)$$

Thus,

$$\mathbf{I}_{f_1} = 16,238.25\angle -90° \text{ A}$$

The remaining short-circuit current distribution is shown in Figure 4.8c.

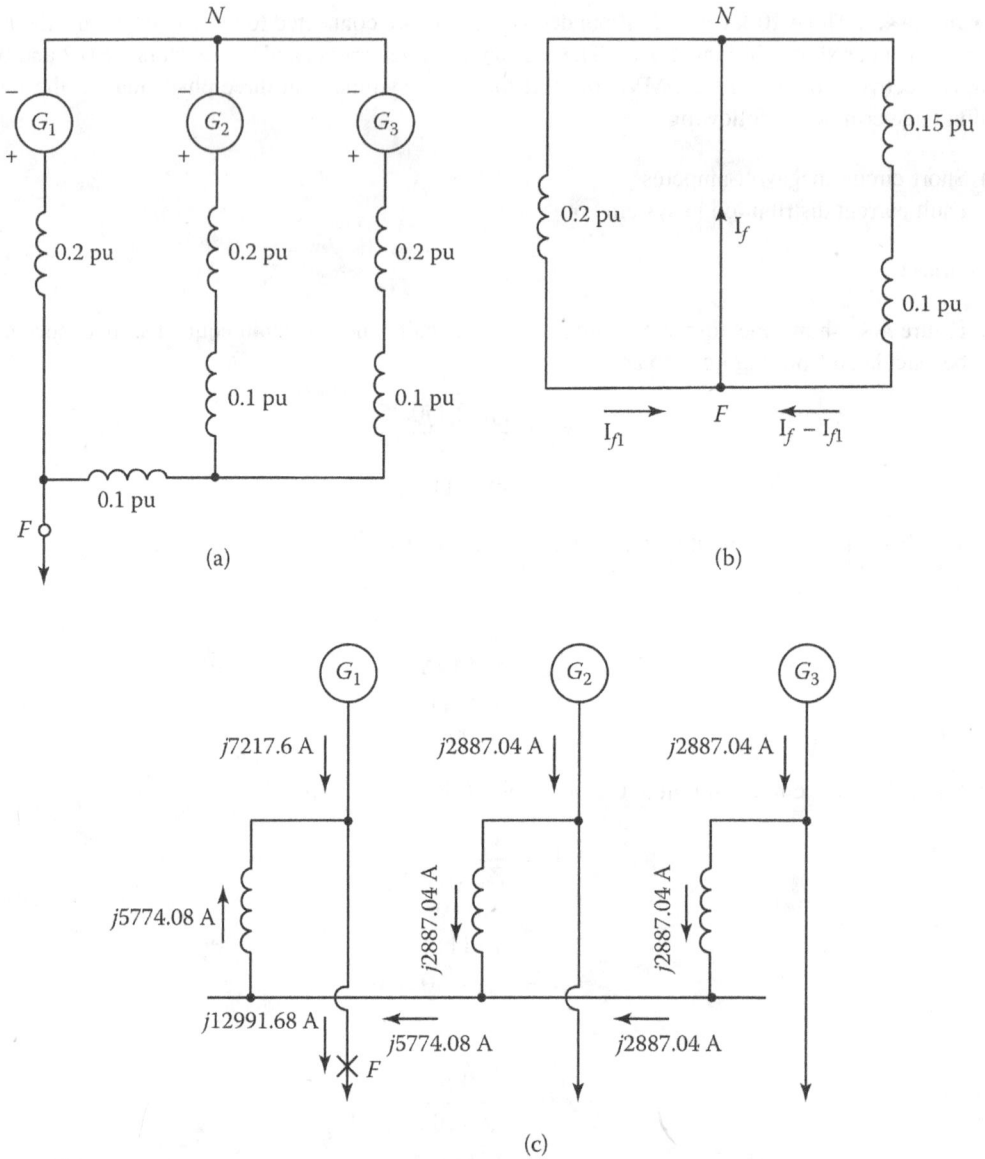

(a)

(b)

(c)

Figure 4.8 Three reactors in tie-bus arrangement for Example 4.5.

4.8 INSULATORS

4.8.1 TYPES OF INSULATORS

An *insulator* is a material that prevents the flow of an electric current and can be used to support electrical conductors. The function of an insulator is to provide for the necessary clearances between the line conductors, between conductors and ground, and between conductors and the pole or tower. Insulators are made of porcelain, glass, and fiberglass treated with epoxy resins. However, porcelain is still the most common material used for insulators.

The basic types of insulators include (1) pin-type insulators, (2) suspension insulators, and (3) strain insulators. The pin insulator gets its name from the fact that it is supported on a pin. The pin

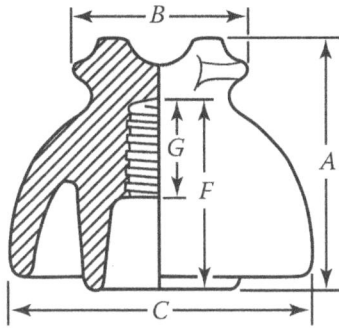

Figure 4.9 Typical pin-insulator, i.e., one-piece pin insulator.

holds the insulator, and the insulator has the conductor tied to it. They may be made in one piece for voltages below 23 kV, in two pieces for voltages from 23 to 46 kV, in three pieces for voltages from 46 to 69 kV, and in four pieces for voltages from 69 to 88 kV. Pin insulators are seldom used on transmission lines having voltages above 44 kV, although some 88-kV lines using pin insulators are in operation.

The glass pin insulator is mainly used on low-voltage circuits. The porcelain pin insulator is used on secondary mains and services, as well as on primary mains, feeders, and transmission lines. Figure 4.11 shows typical pin-type porcelain insulators. A modified version of the pin-insulator is known as the post insulator. The post-type insulators are used on distribution, subtransmission, and transmission lines and are installed on wood, concrete, and steel poles. The line post insulators are constructed for vertical or horizontal mounting. The line post insulators are usually made as one-piece solid porcelain units. Figure 4.12 shows a typical post-type porcelain insulator. Suspension insulators consist of a string of interlinking separate disks made of porcelain. A string may consist of many disks depending on the line voltage.* For example, on average, for 115-kV lines usually seven disks are used; however, for 345-kV lines usually 18 disks are used.

The assembly of suspension units arranged to dead-end the conductor of such a structure is called a *dead-end*, or *strain*, insulator. In such an arrangement, suspension insulators are used as strain insulators. The dead-end string is usually protected against damage from arcs by using one to three additional units and installing or rings, as shown in Figure 4.11. Such devices are designed to ensure that an arc (e.g., due to lightning impulses) will hold free of the insulator string.

The protect the insulator string by providing a shorter path for the arc, as shown in Figure 4.11a. The effectiveness of the *arcing ring* (or *grading shield*), shown in Figure 4.11b, is due to its tendency to equalize the voltage gradient over the insulator, causing a more uniform field. Therefore, protection of the insulator is not dependent on simply providing a shorter arcing path, as is the case with horns. Figure 4.11c shows a control ring developed by Ohio Brass company that can be used to "control" the voltage stress at the line end of the insulator strings.

* In average practice, the number of units used in an insulator string is approximately proportional to the line voltage, with a slight increase for the highest voltages and with some allowances for the length of the insulator unit. For example, 4 or 5 units have generally been used at 69 kV, 7 or 8 at 115 kV, 8–10 at 138 kV, 9–11 at 161 kV, 14–20 at 230 kV, 15–18 at 345 kV, 24–35 at 500 kV (with the 35-unit insulator strings used at high altitudes), 33–35 at 735 kV (Hydro-Quebec), and 30–35 at 765 kV.

It has been shown that their use can also reduce the corona formation on the line hardware. Control rings are used on single-conductor high-voltage transmission lines operating above 250 kV. Transmission lines with bundled conductors do not require the use of and rings nor control rings, provided that the bundle is not made up of two conductors one above the other.

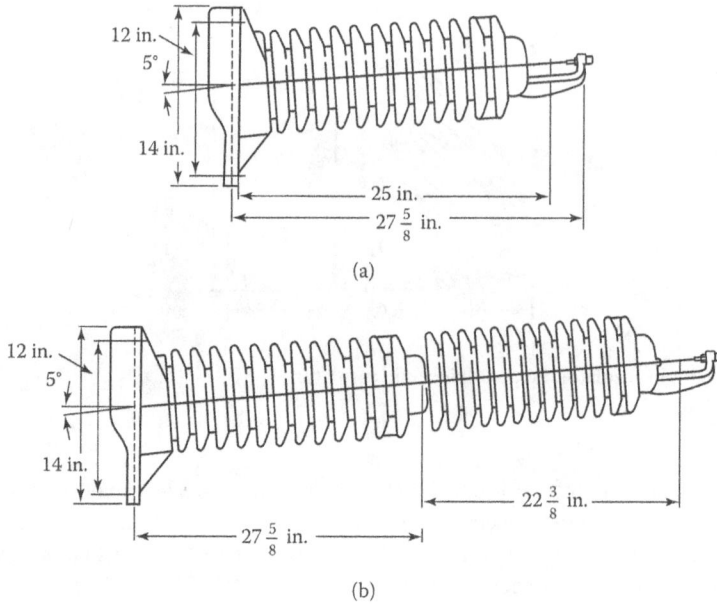

(a)

(b)

Figure 4.10 Typical (side) post-type porcelain insulators used in: (a) 69 kV; (b) 138 kV.

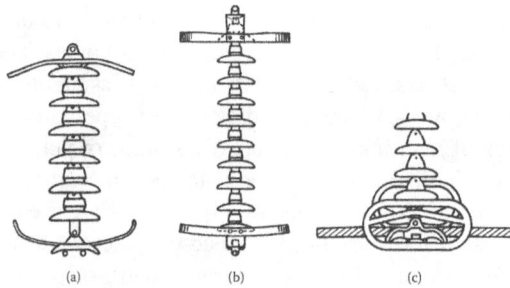

Figure 4.11 Devices used to protect insulator strings: (a) suspension string with ; (b) suspension string with grading shields (or arcing rings); (c) suspension string with control ring. (Courtesy of Ohio Brass Company.)

4.8.2 TESTING OF INSULATORS

The operating performance of a transmission line depends largely on the insulation. Therefore, experience has shown that for a satisfactory operation, the dry flashover voltage of the assembled insulator must be equal to three to five times the nominal operating voltage, and its leakage path must be about twice the shortest air gap distance. Thus, insulators used on overhead lines are subjected to tests that can generally be classified as (1) design tests, (2) performance tests, and (3) routine tests.

The *design tests* include the dry flashover test, pollution flashover test, wet flashover test, pollution flashover test, and impulse test. The *flashover voltage* is defined as the voltage at which the insulator surface breaks down (by ionization of the air surrounding the insulator), allowing current to flow on the outside of the insulator between the conductor and the cross-arm.

Whether an insulator breaks down depends not only on the magnitude of the applied voltage but also on the rate at which the voltage increases. Since insulations have to withstand steep-fronted lightning and switching surges when they are in use, their design must provide the flashover voltage*

Table 4.3

Flashover Characteristics of Suspension Insulator Strings and Air Gaps

Impulse Air Gap		Impulse Flashover, Positive Critical	No. of Insulator	Wet 60-Hz Flashover	Wet 60-Hz Air Gap	
in.	mm	(kV)	Units[a]	(kV)	mm	in.
8	203	150	1	50	254	10
14	356	255	2	90	305	12
21	533	355	3	130	406	16
26	660	440	4	170	508	20
32	813	525	5	215	660	26
38	965	610	6	255	762	30
43	1092	695	7	295	889	35
49	1245	780	8	335	991	39
55	1397	860	9	375	1118	44
60	1524	945	10	415	1245	49
66	1676	1025	11	455	1346	53
71	1803	1105	12	490	1473	58
77	1956	1185	13	525	1575	62
82	2083	1265	14	565	1676	66
88	2235	1345	15	600	1778	70
93	2362	1425	16	630	1880	74
99	2515	1505	17	660	1981	78
104	2642	1585	18	690	2083	82
110	2794	1665	19	720	2184	86
115	2921	1745	20	750	2286	90
121	3073	1825	21	780	2388	94
126	3200	1905	22	810	2464	97
132	3353	1985	23	840	2565	101
137	3480	2065	24	870	2692	106
143	3632	2145	25	900	2794	110
148	3759	2225	26	930	2921	115
154	3912	2305	27	960	3023	119
159	4039	2385	28	990	3124	123
165	4191	2465	29	1020	3251	128
171	4343	2550	30	1050	3353	132

Source : Transmission Line Design Manual, U.S. Department of the Interior, Denver, Colorado, 1980.

[a] Insulator units are $146 \times 254\ mm(5\frac{3}{4} \times 10\ in)$ or $146 \times 267\ mm(5\frac{3}{4} \times 10\ in)$

on a steep-fronted impulse waveform that is greater than that on a normal system waveform. The ratio of these voltages is defined as the *impulse ratio*. Thus,

$$\text{impulse ratio} = \frac{\text{impulse flashover voltage}}{\text{power frequency flashover voltage}} \qquad (4.47)$$

Table 4.3 gives flashover characteristics of suspension insulator strings and air gaps [141]. The performance tests include the puncture test, mechanical test, temperature test, porosity test, and electromechanical test (for suspension insulators only). The event that takes place when the dielectric of the insulator breaks down and allows current to flow inside the insulator between the conductor and the cross-arm is called the *puncture*. Thus, the design must facilitate the occurrence of a flashover at a voltage that is lower than the voltage for puncture. An insulator may survive a flashover without damage but must be replaced when punctured. The test of the glaze on porcelain insulators is called the *porosity test*. The routine tests include the proof-load test, corrosion test, and high-voltage test (for pin insulators only).*

4.8.3 VOLTAGE DISTRIBUTION OVER A STRING OF SUSPENSION INSULATORS

Figure 4.12 shows the voltage distribution along the surface of a single clean insulator disk (known as the cap-and-pin insulator unit) used in suspension insulators. Note that, the highest voltage gradient takes place close to the cap and the pin (which are made up of metal), whereas much lower voltage gradients take place along most of the remaining surfaces.

The underside (i.e., the *inner skirt*) of the insulator has been given the shape, as shown in Figure 4.12, to minimize the effects of moisture and contamination and to provide the longest path possible for the leakage currents that might flow on the surface of the insulator. In the figure, the voltage drop between the cap and the pin has been taken as 100% of the total voltage.

Thus, approximately 24% of this voltage is distributed along the surface of the insulator from the cap to point 1 and only 6% from point 1 to point 9. The remaining 70% of this voltage is distributed between point 9 and the pin. The main problem with suspension insulators having a string of identical insulator disks is the nonuniform distribution voltage over the string.

Each insulator disk with its hardware (i.e., cap and pin) constitutes a capacitor, the hardware acting as the plates or electrodes and the porcelain as the dielectric. Figure 4.13 shows the typical voltage distribution on the surfaces of three clean cap-and-pin insulator units connected in series [42]. The figure clearly illustrates that when several units are connected in series, (1) the voltage on each insulator over the string is not the same, (2) the location of the unit within the insulator string dictates the voltage distribution, and (3) the maximum voltage gradient takes place at the pin of the insulator unit nearest to the line conductor.

As shown in Figure 4.14a, when several insulator units are placed in series, two sets of capacitances take place; the series capacitance C_1 (i.e., the capacitance of each insulator unit) and the shunt capacitance to ground, C_2. Note that, all the charging current I for the series and shunt capacitances flows through the first (with respect to the conductor) of the series capacitance C_1.

The I_1 portion of this current flows through the first shunt capacitance C_2, leaving the remaining $I-I_1$ portion of the current to flow through the second series capacitance, and so on. Therefore, this diminishing current flowing through the series capacitance C_1 results in a diminishing voltage (drop) distribution through them from the conductor end to the ground end (i.e., cross-arm), as illustrated in Figure 4.14b. Thus,

$$V_5 > V_4 > V_3 > V_2 > V_1$$

In summary, the voltage distribution over a string of identical suspension insulator units is not uniform owing to the capacitances formed in the air between each cap/pin junction and the grounded (metal) tower. However, other exist between metal parts at different potentials. For example, there are between the cap–pin junction of each unit and the line conductor. Figure 4.15 shows the resulting equivalent circuit for the voltage distribution along a clean eight-unit insulator string. The voltage distribution on such a string can be expressed as

$$V_k = \frac{V_n}{\beta^2 \sinh \beta n}\left[\frac{C_2}{C_1}\right]\sinh \beta k + \left[\frac{C_3}{C_1}\right]\sinh\beta(k-n)+\left[\frac{C_3}{C_1}\right]\sinh \beta n \qquad (4.48)$$

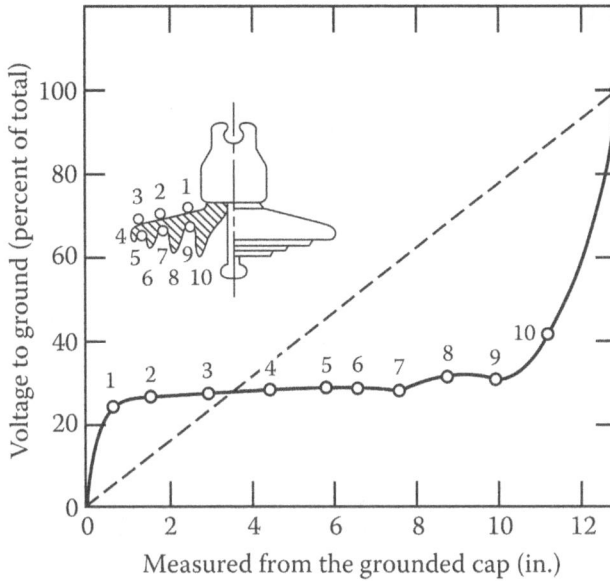

Figure 4.12 Voltage distribution along surface of single dean cap-and-pin suspension insulator. (From Edison Electric Institute, *EHV Transmission Line Reference Book*, EEI, New York, 1968.) [41]

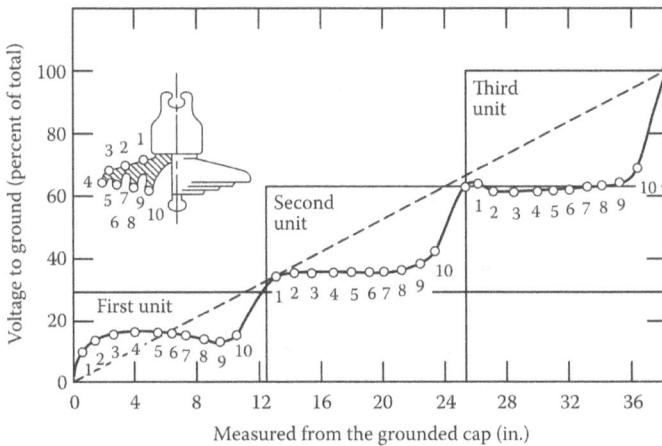

Figure 4.13 Typical voltage distribution on surfaces of three clean cap-and-pin suspension insulator units in series. (From Edison Electric Institute, *EHV Transmission Line Reference Book*. EEI, New York, 1968.)

where
  V_k = voltage across k units from ground end
  V_n = voltage across all n units (i.e., applied line-to-ground voltage in volts)

$$\beta = \text{a constant} = \sqrt{\left(\frac{C_2 + C_1}{2}\right)} \tag{4.49}$$

  C_1 = capacitance between cap and pin of each unit
  C_2 = capacitance of one unit to ground

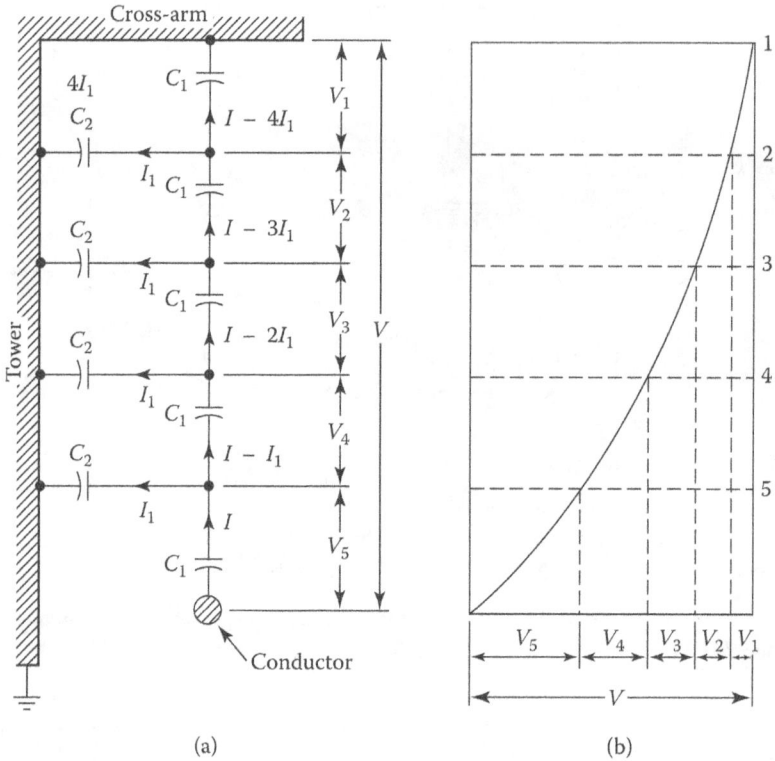

Figure 4.14 Voltage distribution among suspension insulator units [42].

C_3 = capacitance of one unit to line conductor

However, the capacitance C_3 is usually very small, and thus its effect on the voltage distribution can be neglected. Hence, Equation 4.48 can be reexpressed as

$$V_k = V_n \left(\frac{\sinh \alpha k}{\sinh \alpha n} \right) \tag{4.50}$$

where

$$\alpha = \text{a constant} = \sqrt{\left(\frac{C_2}{C_1} \right)} \tag{4.51}$$

Figure 4.16 shows how the voltage changes along the eight-unit string of insulators when the ratio C_2/C_1 is about .083333 and the ratio C_3/C_1 is about zero (i.e., $C_3 = 0$). It is interesting to note that a calculation based on Equation 4.48 gives almost the same result. The ratio C_2/C_1 is usually somewhere between 0.1 and 0.2.

Furthermore, there is the that exists between the conductor and the tower. However, it has no effect on the voltage distribution over the insulator string, and therefore it can be neglected.

Note that, this method of calculating the voltage distribution across the string is based on the assumption that the insulator units involved are clean and dry, and therefore they act as a purely capacitive voltage divider. In reality, however, the insulator units may not be clean or dry. Thus, in the equivalent circuit of the insulator string, each capacitance C_1 should be shunted by a resistance R representing the leakage resistance.

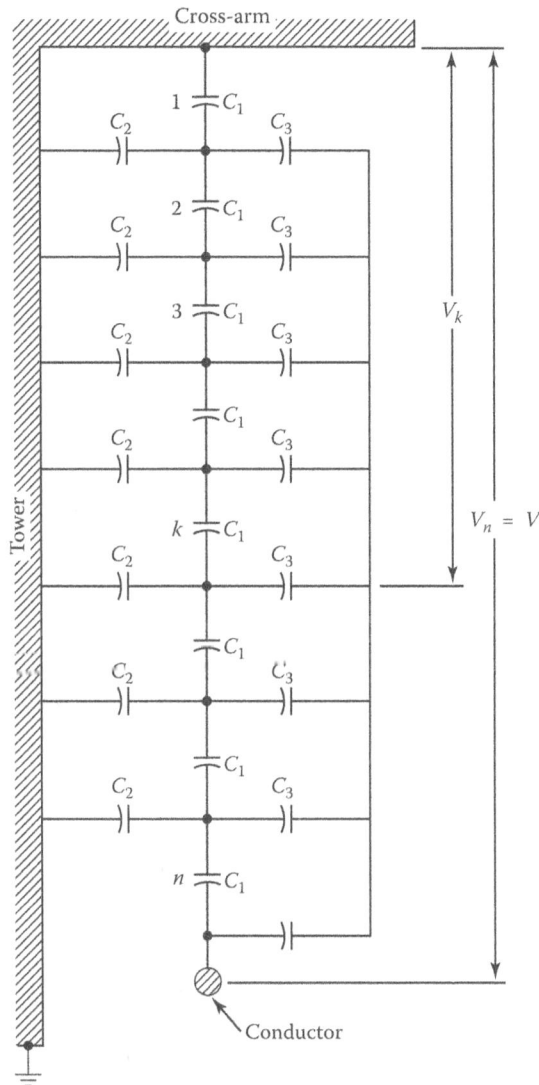

Figure 4.15 Equivalent circuit for voltage distribution along clean eight-unit insulator string. (Adopted from Edison Electric Institute, *EHV Transmission Line Reference Book*. EEI, New York, 1968.)

Such resistance depends on the presence of contamination (i.e., pollution) on the insulator surfaces and is considerably modified by rain and fog. If, however, the units are badly contaminated, the surface leakage (resistance) currents could be greater than the capacitance currents, and the extent of the contamination could vary from unit to unit, causing an unpredictable voltage distribution. It is also interesting to note that if the insulator unit nearest to the line conductor is electrically stressed to its safe operating value, then all the other units are electrically understressed, and consequently, the insulator string as a whole is being inefficiently used. Therefore, the string efficiency (in per units) for an insulator string made up of n series units can be defined as

$$\text{string efficiency} = \frac{\text{voltage across string}}{n(\text{voltage across unit adjacent to line conductor})} \tag{4.52}$$

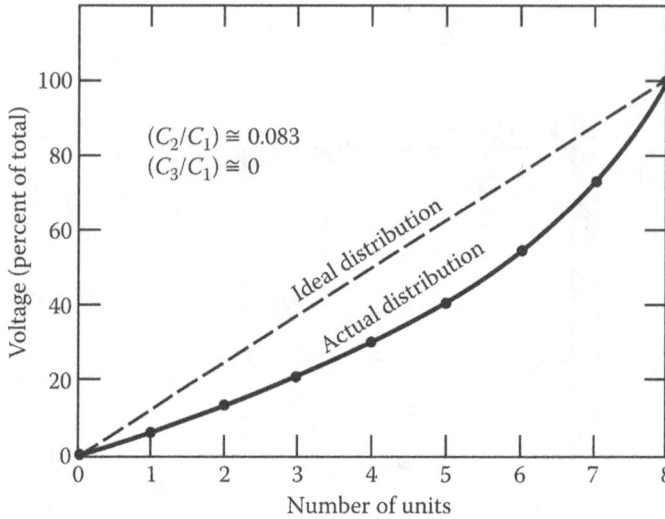

$(C_2/C_1) \cong 0.083$
$(C_3/C_1) \cong 0$

Ideal distribution

Actual distribution

Figure 4.16 Voltage distribution along clean eight-unit cap-and-pin insulator string. (From Edison Electric Institute, *EHV Transmission Line Reference Book*. EEI, New York, 1968.)

If the unit adjacent to the line conductor is about to flash over, then the whole string is about to flash over. Therefore, the string efficiency can be reexpressed as

$$\text{string efficiency} = \frac{\text{flashover voltage of string}}{n(\text{flahover voltage of one unit})} \tag{4.53}$$

Note that, the string efficiency decreases as the number of units increases. The methods to improve the *string efficiency (grading)* include the following:

1. By grading the insulators so that the top unit has the minimum series capacitance C_1, whereas the bottom unit has the maximum capacitance. This may be done by using different sizes of disks and hardware or by putting metal caps on the disks or by a combination of both methods.* However, this is a rarely used method since it would involve stocking spares of different types of units, which is contrary to the present practice of the utilities to standardize on as few types as possible.
2. By installing a large circular or oval grading shield ring (i.e., an arcing ring) at the line end of the insulator string [142]. This method introduces a capacitance C_3, as shown in Figure 4.15, from the ring to the insulator hardware to neutralize the capacitance C_2 from the hardware to the tower. This method substantially improves the string efficiency. However, it is not usually possible in practice to achieve completely uniform voltage distribution by using the grading shield, especially if the string has a large number of units.
3. By reducing the air (shunt) capacitance C_3, between each unit and the tower (i.e., the ground), by increasing the length of the cross-arms. However, this method is restricted in practice owing to the reduction in cross-arm rigidity and the increase in tower cost.
4. By using a semiconducting (or stabilizing) high-resistance glaze on the insulator units to achieve a resistor voltage divider effect. This method is based on the fact that the string efficiency increases owing to the increase in surface leakage resistance current when the units are wet. Thus, the leakage resistance current becomes the same for all units, and the voltage distribution improves since it does not depend on the capacitance currents only. However, this method is restricted by the risk of thermal instability.

4.8.4 INSULATOR FLASHOVER DUE TO CONTAMINATION

An insulator must be capable of enduring extreme and sudden temperature changes such as ice, sleet, and rain as well as environmental contaminants such as smoke, dust, salt, fogs, saltwater sprays, and chemical fumes without deterioration from chemical action, breakage from mechanical stresses, or electrical failure. Further, the insulating material must be thick enough to resist puncture by the combined working voltage of the line and any probable transient whose time lag to spark over is great.

If this thickness is greater than the desirable amount, then two or more pieces are used to achieve the proper thickness. The thickness of a porcelain part must be so related to the distance around it that it will flash over before it will puncture. The ratio of puncture strength to flashover voltage is called the "safety factor" of the part or of the insulator against puncture. This ratio should be high enough to provide sufficient protection for the insulator from puncture by the transients.

The insulating materials mainly used for line insulators are (1) wet-process porcelain, (2) dry-process porcelain, and (3) glass. However, wet-process porcelain is used much more than dry-process porcelain. One of the reasons for this is that wet-process porcelain has greater resistance to impact and is practically incapable of being penetrated by moisture without glazing, whereas dry-process porcelain is not.

However, in general, dry-process porcelain has a somewhat higher crushing strength. Dry-process porcelain is only used for the lowest voltage lines. As a result of recent developments in the technology of glass manufacturing, glass insulators, which are very tough and have low internal resistance, can be produced. Thus, usage of glass insulators is increasing.

To select insulators properly for a given overhead line design, not only the aforementioned factors but also the geographic location of the line needs to be considered. For example, the overhead lines that will be built along the seashore, especially in California, will be subjected to winds blowing in from the ocean, which carry a fine salt vapor that deposits salt crystals on the windward side of the insulator.

On the order hand, if the line is built in areas where rain is seasonal, the insulator surface leakage resistance may become so low during the dry seasons that insulators flash over without warning. Another example is that if the overhead line is going to be built near gravel pits, cement mills, and refineries, its insulators may become so contaminated that extra insulation is required. Contamination flashovers on transmission systems are initiated by airborne particles deposited on the insulators. These particles may be of natural origin or they may be generated by pollution that is mostly a result of industrial, agricultural, or construction activities. Thus, when line insulators are contaminated, many insulator flashovers occur during light fogs unless arcing rings protect the insulators or special fog-type insulators are used.

Table 4.4 lists the types of contaminants causing contamination flashover [143]. The mixed contamination condition is the most common, caused by the combination of industrial pollution and sea salt or by the combination of several industrial pollutions. Table 4.4 also presents the prevailing weather conditions at the time of flashover. Note that fog, dew, drizzle, and mist are common weather conditions, accounting for 72% of the total. In general, a combination of dew and fog is considered as the most severe wetting condition, even though fog is not necessary for the wetting process.

Note that, the surface leakage resistance of an insulator is unaffected by deposits of dry dirt. However, when these contamination deposits become moist or wet, they constitute continuous conducting layers. Leakage current starts to flow in these layers along the surface of the insulators. This leakage current heats the wet contamination, and the water starts to evaporate from those areas where the product of current density and surface resistivity is greater, causing the surface resistivity to further increase. Therefore, the current continues to flow around such a dry spot, causing the current density in the neighboring regions to increase. This, in turn, produces more heat, which evaporates the moisture in these surrounding regions, causing the formation of circular patterns

Table 4.4

Numbers of Flashovers Caused by Various Contaminant, Weather, and Atmospheric Conditions

Type of Contaminant	Weather and Atmospheric Conditions								
	Fog	Dew	Drizzle, Mist	Ice	Rain	No Wind	High Wind	Wet Snow	Fair
Sea salt	14	11	22	1	12	3	12	3	—
Cement	12		16	2	11	4	1	4	—
Fertilizer	7	5	8	—	1	1	—	4	—
Fly ash	11	6	19	1	6	3	1	3	1
Road salt	8	2	6	—	4	2	—	6	—
Potash	3		3	—	—	—	—	—	—
Cooling tower	2	2	2	—	2	—	—	—	—
Chemicals	9	5	7	1	1	—	—	1	1
Gypsum	2	1	2	—	2	—	—	2	—
Mixed contamination	32	19	37	—	13	1	—	1	—
Limestone	2	1	2	—	4	—	2	2	—
Phosphate and sulfate	4	1	4	—	3	—	—	—	—
Paint	1		1	—	—	1	—	—	—
Paper mill	2	2	4	—	2	—	—	1	—
Drink milk	1	1	1	—	—	1	—	1	—
Acid exhaust	2		3	—	—	—	—	1	—
Bird droppings	2	2	3	—	1	2	—	—	2
Zinc industry	2	1	2	—	1	—	—	1	—
Carbon	5	4	5	—	—	4	3	3	—
Soap	2	2	1	—	—	1	—	—	—
Steel works	6	5	3	2	2	—	—	1	—
Carbide residue	2	1	1	1	—	—	—	1	—
Sulfur	3	2	2	—	—	1	—	1	—
Copper and nickel salt	2	2	2	—	—	2	—	1	—
Wood fiber	1	1	1	—	1	—	—	1	—
Bulldozing dust	2	1	1	—	—	2	—	—	—
Aluminum plant	2	2	1	—	1	—	—	—	—
Sodium plant	1		1	—	—	—	—	—	—
Active dump	1	1	1	—	—	—	—	—	—
Rock crusher	3	3	5	—	1	—	—	—	—

Source: Electric Power Research Institute, *Transmission Line Reference Book: 345 kV and Above* , 2nd ed. EPRI, Palo Alto, CA, 1982.

known as "dry bands" until the leakage current is decreased to a value insufficient to sustain further evaporation, and the voltage builds up across the dry bands.

Further wetting results in further reduction of the resistance, and small flashovers take place on the dry bands on which moisture droplets fall. Since many dry bands on the insulator are in about the same condition, the arcs extend rapidly over the whole surface, forcing all dry bands to discharge in a rapid cascade known as the "flashover" of the insulator.

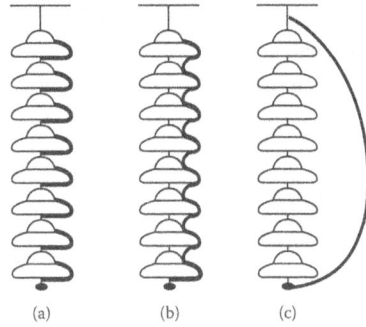

Figure 4.17 Changes in channel position of contaminated flashover. (From Edison Electric Institute, *EHV Transmission Line Reference Book*. EEI, New York, 1968.)

Figure 4.17 illustrates the phenomenon of insulator flashover due to contamination. Severe contamination may reduce the 60-Hz flashover voltage from approximately 50 kV rms per unit to as low as 6 to 9 kV rms per unit. The condition for such flashover may be developed during the melting of contaminated ice on the insulator by leakage currents. An insulator flashover due to contamination is easily distinguished from other types of flashover due to the fact that the arc always begins close to the surface of each insulator unit, as shown in Figure 4.17a. As shown in Figure 4.17c, only in the final stage does the flashover resemble an air strike. Furthermore, since the insulator unit at the conductor end has the greatest voltage, the flashover phenomenon usually starts at that insulator unit.

To prevent insulator flashovers, the insulators of an overhead transmission line may be cleaned simply by washing them, a process that can be done basically either by conventional techniques or by a new technique. In the conventional techniques, the line is deenergized, and its conductors are grounded at each pole or tower where the members of an insulator cleaning crew wash and wipe the insulators by hand.

In the new technique, the line is kept energized while the insulators may be cleaned by high-pressure water jets produced by a truck-mounted high-pressure pump that forces water through a nozzle at 500–850 psi, developing a round solid stream. The water jet strikes the insulator with a high velocity, literally tearing the dirt and other contaminants from the insulator surface. The cost of insulator cleaning per unit is very low by this technique.

Certain lines may need insulator cleaning as often as three times a year. To overcome the problem of surface contamination, some insulators may be covered with a thin film of silicone grease that absorbs the dirt and makes the surface water form into droplets rather than a thin film. This technique is especially effective for spot contamination where maintenance is possible, and it is also used against sea salt contamination.

Finally, specially built semiconducting glazed insulators having a resistive coating are used. The heat produced by the resistive coating keeps the surface dry and provides for a relatively linear potential distribution.

4.8.5 INSULATOR FLASHOVER ON OVERHEAD HIGH-VOLTAGE DC LINES

Even though mechanical considerations are similar for ac and dc lines, electrical characteristics of insulators on dc lines are significantly different from those on ac lines. For example, when conventional ac insulators are used on dc lines, flashover takes place much more frequently than on an ac line of equivalent voltage. This is caused partly by the electrostatic forces of the steady dc field, which increases the deposit of pollution on the insulator surface.

Further, arcs tend to develop into flashovers more readily in the absence of voltage zero. To improve the operating performance and reduce the construction cost of overhead high-voltage dc (HVDC) lines by using new insulating materials and new insulator configurations particularly suited to dc voltages stress, more compact line designs can be produced, therefore saving money on towers and rights of way.

For example, to improve the operating performance and reduce the construction cost of overhead HVDC lines, the EPRI has sponsored the development of a new insulator. One of the more popular designs, the composite insulator, uses a fiberglass rod for mechanical and electrical strength and flexible skirts made of organic materials for improved flashover performance. The composite insulator appears to be especially attractive for use on HVDC lines because it is better able to withstand flashover in all types of contaminated environments, particularly in areas of light and medium contamination.

Furthermore, there are various design measures that may be taken into account to prevent contamination flashovers, for example, overinsulation, installment of V-string insulators, and installment of horizontal string insulators. Over insulation may be applicable in the areas of heavy contamination.

Up to 345 kV, overinsulation is often achieved by increasing the number of insulators. However, severe contamination may dictate the use of very large leakage distances that may be as large as double the nominal requirements. Thus, electrical, mechanical, and economic restrictions may limit the use of this design measure.

The use of the V-string insulators can prevent the insulation contamination substantially. They self-clean more effectively in rain than vertical string insulators since both sides of each insulator disk are somewhat exposed to rain. They can be used in heavy contamination areas very effectively.

The installment of horizontal insulator strings is the most effective design measure that can be used to prevent contamination flashovers in the very heavy contamination areas. The contaminants are most effectively washed away on such strings. However, they may require a strain tower support depending on the tower type.

Other techniques used include the installation of specially designed and built insulators. For example, the use of fog-type insulators has shown that the contamination flashover can be effectively reduced since most of the flashovers occur in conditions where there is mist, dew, and fog.

4.9 GROUNDING

4.9.1 ELECTRIC SHOCK AND ITS EFFECTS ON HUMANS

To properly design a grounding* for the high-voltage lines and/or substations, it is important to understand the electrical characteristics of the most important part of the circuit, the human body [144–146]. In general, shock currents are classified according to the degree of the severity of the shock they cause. For example, currents that produce direct physiological harm are called *primary shock currents*. However, currents that cannot produce direct physiological harm but may cause involuntary muscular reactions are called *secondary shock currents*. These shock currents can be either steady state or transient in nature. In ac power systems, steady-state currents are sustained currents of 60 Hz or its harmonics [147, 148].

The transient currents, on the other hand, are capacitive discharge currents whose magnitudes diminish rapidly with time. Table 4.5 gives the possible effects of electrical shock currents on humans. Note that, the threshold value for a normally healthy person to be able to feel a current is about 1 mA.[†] This is the value of current at which a person is just able to detect a slight tingling sensation on the hands or fingers due to current flow [149].

Currents of approximately 10–30 mA can cause lack of muscular control. In most humans, a current of 100 mA will cause ventricular fibrillation. Currents of higher magnitudes can stop the heart completely or cause severe electrical burns.

Table 4.5

Typical Effects of Electrical Shock Current on Humans

60 Hz Current	Effect
0–1 mA	No sensation (not felt)
0–3 mA	Perceptible, mild
3–5 mA	Annoyance, pain, or surprise
5–10 mA	Painful shock
10–15 mA	Local muscle contractions, sufficient to cause "freezing" to circuit for 2.5% of population
15–30 mA	Local muscle contractions, sufficient to cause "freezing" to circuit for 50% of population
30–50 mA	Difficulty in breathing, can cause loss of consciousness
50–100 mA	Possible ventricular fibrillation of the heart
100–200 mA	Certain ventricular fibrillation of the heart
>200 mA	Severe burns and muscular contractions; heart more apt to stop than fibrillated
Over a few amperes	Irreparable damage to body tissue

Ventricular fibrillation is a condition where the heart beats in an abnormal and ineffective manner, with fatal results. Thus, its threshold is the main concern in grounding design. It is defined as very rapid uncoordinated contractions of the ventricles of the heart, resulting in loss of synchronization between heartbeat and pulse beat. Institute of Electrical and Electronics Engineers (IEEE) Standard (Std.) 80-2000 gives the following equation to find the non-fibrillating current of magnitude I_b at a duration ranging from 0.03 to 3.0 s in relation to the energy absorbed by the body as [150]

$$S_b = (I_b)^2 \times t_s \tag{4.54}$$

where

I_b = rms magnitude of the current through the body in amperes

I_s = duration of the current exposure in seconds

S_b = empirical constant related to the electric shock energy tolerated by a certain percent of a given population

A human heart can be seen as a muscle operating rhythmically due to a nerve pulse that provides the heartbeat. Therefore, when a false signal (i.e., the electrical shock) is injected into the heart, it could upset the rhythmic flow of operation of values and other components of the heart. This causes a condition known as ventricular fibrillation. Once this "out-of-phase" rhythm is established, it is extremely difficult to stop. It usually requires the injection of another shock to stop the fibrillation and reestablish the normal rhythm. Obviously, if it takes place in the field or at a remote location, the time delay before medical defibrillation may be too long, causing a fatality.

A fatality may also occur owing to a coronary arrest, that is, the stopping of the heartbeat. Furthermore, the current passing through the body may temporarily paralyze either the nerves or the area of the brain that controls respiration. This may also lead to death by causing cessation of respiration (asphyxia) if the victim has grasped a live conductor and cannot let go.

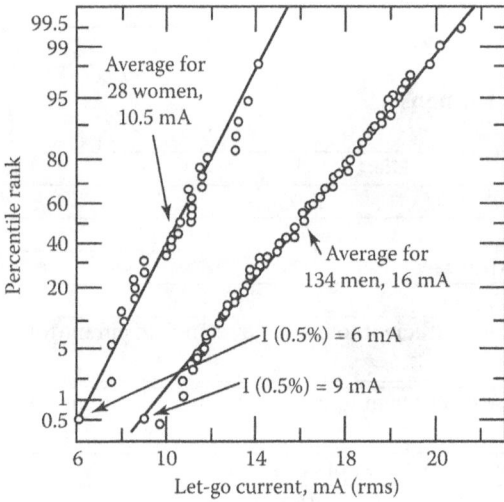

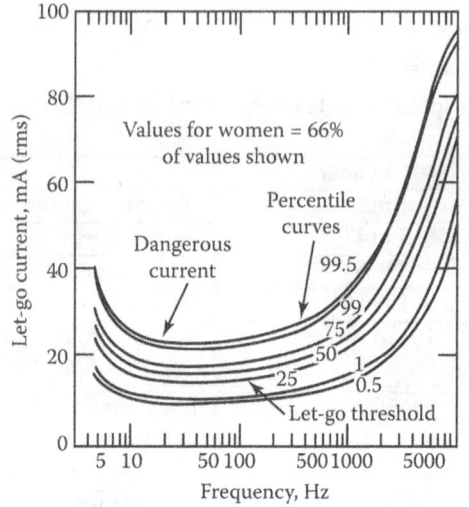

(a)   (b)

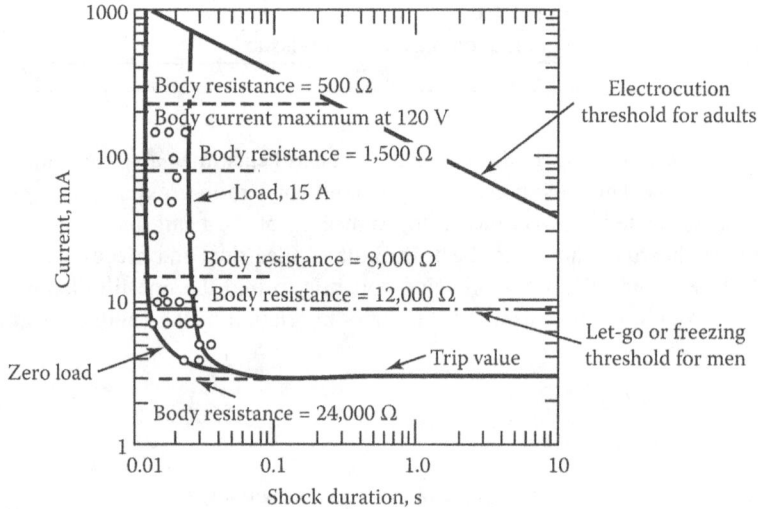

(c)

Figure 4.18 Effects of shock currents on humans: (a) 60-Hz let-go current distribution curves for 134 men and 28 women; (b) let-go currents vs. frequency; (c) trip current vs. shock duration. (From Dalziel, C. F., and Lee, W. R., IEEE Spectrum 6, 44–50 1972 IEEE.)

Currents of 1 mA or greater but less than 6 mA are often defined as the *secondary shock currents (let-go currents)*. The let-go current is the maximum current level at which a human holding an energized conductor can control his muscles enough to release it. Dalziel's classic experiment [151,152], with 28 women and 134 men, provides data indicating an average let-go current of 10.5 rnA for women and 16 rnA for men, with 6 and 9 rnA as the respective threshold values, as shown in Figure 4.21a. According to Dalziel, not only the individual's physiological development but also psychological factors can play an important role in limiting both the minimum and maximum values.* In general, currents with magnitudes of 6 mA or greater are known as *primary shock currents*.

Note that, it is virtually impossible to produce primary shock currents with less than 25 V owing to normal body resistance. Among the possible consequences of primary shock current is ventricular fibrillation. On the basis of the electrocution formula developed by Dalziel [151, 152], the 60-Hz minimum required body current leading to possible fatality through ventricular fibrillation can be expressed as

$$I_b = \frac{0.116}{\sqrt{t}} \text{ A} \tag{4.55}$$

where t is in seconds, in the range from approximately 8.3 ms to 5 s.

Table 4.6 presents a summary of quantitative effects of electric current on humans based on a report by the IEEE Working Group on Electrostatic Effects of Transmission Lines. It is important to be aware that such tables are developed on the assumption that the individuals involved are 100% fit. However, in reality, not all individuals are 100% fit. Therefore, they are more susceptible to shock hazards [153]. Thus, in these days of the Occupational Safety and Health Administration and lawsuits, utilities have to be very cautious when it comes to grounding. The effects of an electric current passing through the vital parts of a human body depend on the duration, magnitude, and frequency of this current. Experiments have shown that the heart requires about 5 min to return to normal after experiencing a severe shock [151]. Thus, two or more closely spaced shocks (such as those that would take place in systems with automatic reclosing) would tend to have a cumulative effect. Present industry practice considers two closely spaced shocks to be equivalent to a single shock whose duration is the sum of the intervals of the individual shocks. Experiments have also shown that humans are very vulnerable to the effects of electric current at frequencies of 50–60 Hz [154–158].

As shown in Figure 4.18b, the human body can tolerate slightly larger currents at 25 Hz and about five times larger at dc current. Similarly, at frequencies of 1000 or 10,000 Hz, even larger currents can be tolerated. In the case of lighting surges, the human body appears able to tolerate very high currents, perhaps on the order of several hundreds of amperes [159]. Figure 4.18c shows the relation between trip current and shock duration for a typical ground fault interrupter (used at 120/240 V level), with electrocution threshold and let-go threshold for adults indicated to provide perspective.

When the human body becomes a part of the electric circuit, the current that passes through it can be found by applying Thévenin's theorem and Kirchhoff's current law, as illustrated in Figure 4.19. For dc and ac currents at 60 Hz, the human body can be substituted by a resistance in the equivalent circuits. The body resistance considered is usually between two extremities, either from one hand to both feet or from one foot to the other one.

Experiments have shown that the body can tolerate much more current flowing from one leg to the other than it can when current flows from one hand to the legs. Figure 4.19a shows a touch contact with current flowing from hand to feet. On the other hand, Figure 4.19b shows a step contact where current flows from one foot to the other. Note that, in each case, the body current I_b is driven by the potential difference between points A and B.

Currents of 1 mA or greater but less than 6 mA are often defined as secondary shock currents (let-go currents). The let-go current is the maximum current level at which a human holding an energized conductor can control his muscles enough to release it. For 99.5% of the population, the 60-Hz minimum required body current, I_b, leading to possible fatality through ventricular fibrillation can be expressed as

$$I_b = \frac{0.116}{\sqrt{t_s}} \text{ A} \quad \text{for 50 kg body weight} \tag{4.56a}$$

can be broken down to

$$I_b = \frac{0.157}{\sqrt{t_s}} \quad \text{A for 70 kg body weight} \tag{4.56b}$$

where t is in seconds in the range from approximately 8.3 ms to 5 s.

Table 4.6

Effect of Electric Current (mA) on Men and Women

	Direct Current		60 Hz rms	
Effect	Men	Women	Men	Women
1. No sensation on hand	1	0.6	0.4	0.3
2. Slight tingling; perception threshold	5.2	3.5	1.1	0.7
3. Shock—not painful but muscular control not lost	9	6	1.8	1.2
4. Painful shock—painful but muscular control not lost	62	41	9	6
5. Painful shock—let-go threshold[a]	76	51	16	10.5
6. Painful and severe shock, muscular contractions, breathing difficulty	90	60	23	15
7. Possible ventricular fibrillation from				
Short shocks:				
(a) Shock duration 0.03 s	1300	1300	1000	1000
(b) Shock duration 3.0 s	500	500	100	100
(c) Almost certain ventricular fibrillation (if shock duration over one heartbeat interval)	1375	1375	275	275

Source: IEEE Working Group Report, *IEEE Trans. Power Appar. Syst.* PAS-9, 422–426 1972
IEEE.

[a] Threshold for 50% of the males and females tested.

The effects of an electric current passing through the vital parts of a human body depend on the duration, magnitude, and frequency of this current. The body resistance considered is usually between two extremities, either from one hand to both feet or from one foot to the other one. Figure 4.20 show five basic situations involving a person and grounded facilities during fault.

Note that, in the figure, the *mesh voltage* is defined by the maximum touch voltage within a mesh of a ground grid. However, the *metal-to-metal touch voltage* defines the difference in potential between metallic objects or structures within the substation site that may be bridged by direct hand-to-hand or hand-to-feet contact. However, the *step voltage* represents the difference in surface potential experienced by a person bridging a distance of 1 m with the feet without contacting any other grounded object.

On the other hand, the *touch voltage* represents the potential difference between the ground potential rise (GPR) and the surface potential at the point where a person is standing while at the same time having a hand in contact with a grounded structure. The *transferred voltage* is a special case of the touch voltage where a voltage is transferred into or out of the substation from or to a remote point external to the substation site [160].

Finally, GPR is the maximum electrical potential that a substation grounding grid may have relative to a distant grounding point assumed to be at the potential of remote earth. This voltage, GPR, is equal to the maximum grid current times the grid resistance. Under normal conditions, the grounded electrical equipment operates at near-zero ground potential. That is, the potential of a grounded neutral conductor is nearly identical to the potential of remote earth. During a ground fault, the portion of fault current that is conducted by substation grounding grid into the earth causes the rise of the grid potential with respect to remote earth.

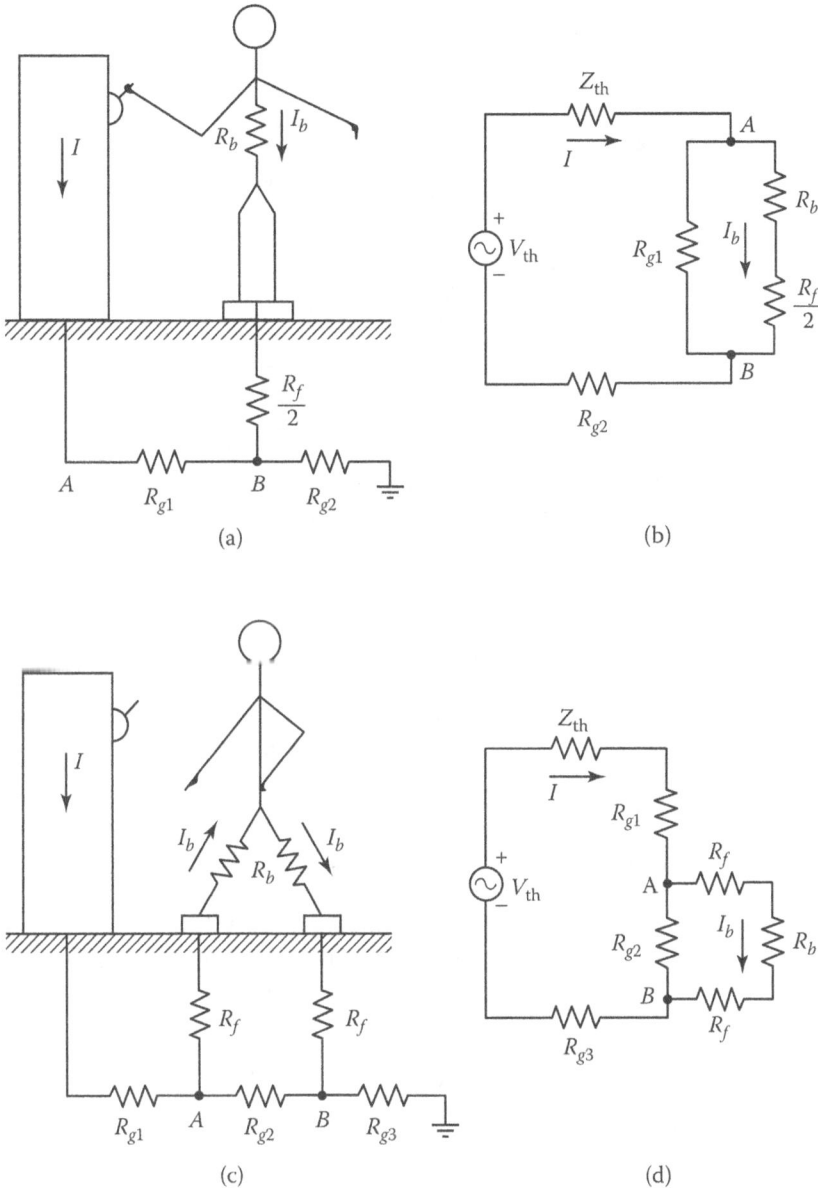

Figure 4.19 Typical electric shock hazard situations: (a) touch potential; (b) its equivalent circuit; (e) step potential; (d) its equivalent circuit.

Exposure to a touch potential normally poses a greater danger than exposure to a step potential. The step potentials are usually smaller in magnitude (due to the greater corresponding body resistance), and the allowable body current is higher than the touch contacts. In either case, the value of the body resistance is difficult to establish.

As said before, experiments have shown that the body can tolerate much more current flowing from one leg to the other than it can when current flows from one hand to the legs. Treating the foot as a circular plate electrode gives an approximate resistance of $3\rho_s$, where ρ_s is the soil resistivity.

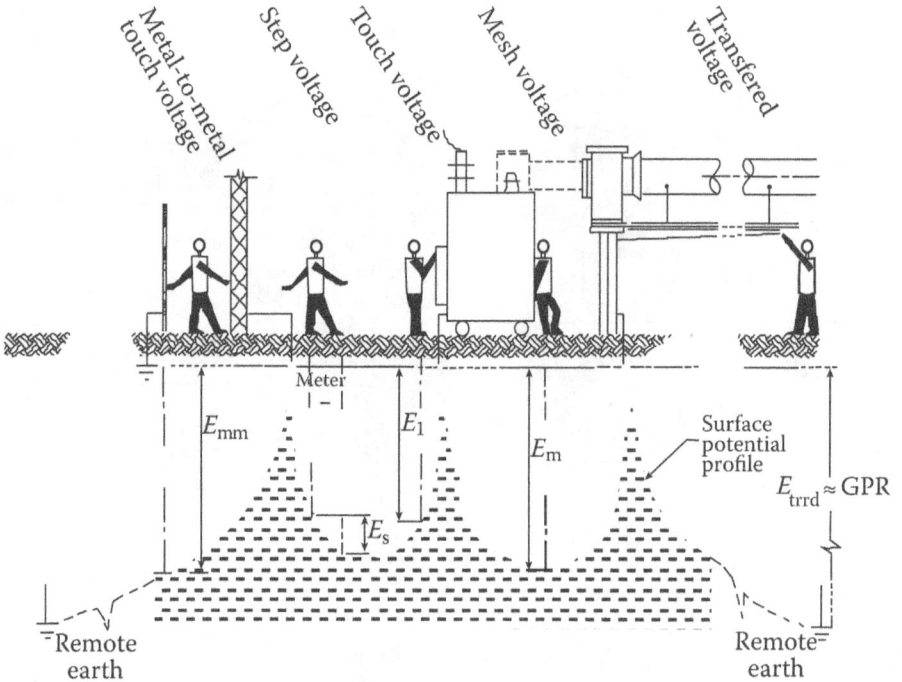

Figure 4.20 Possible basic shock situations. (From Keil, R. P., *Substation Grounding, in Electric Power Substations Engineering*, McDonald, J. D. (ed.), 2nd ed., CRC Press, Boca Raton, FL, 2007.)

The resistance of the body itself is usually used as about 2300 Ω hand to hand or 1100 Ω hand to foot.

However, IEEE Std. 80-2000 [160] recommends the use of 1000 Ω as a reasonable approximation for body resistance. Therefore, the total branch resistance, for hand-to-foot currents, can be expressed as

$$R_b = 1000 + 1.5\rho_s \ \Omega \text{ for touch voltage} \tag{4.57a}$$

and, for foot-to-foot currents,

$$R_b = 1000 + 6\rho_s \ \Omega \text{ for step voltage} \tag{4.57b}$$

where ρ_s is the soil resistivity in ohm meters. If the surface of the soil is covered with a layer of crushed rock or some other high-resistivity material, its resistivity should be used in Equations4.56 and 4.57. The touch voltage limit can be determined from

$$V_{\text{touch}} = \left(R_b + \frac{R_f}{2} \right) I_b \tag{4.58}$$

and

$$V_{\text{step}} = \left(R_b + 2R_f \right) I_b \tag{4.59}$$

where

$$R_f = 3C_s\rho_s \tag{4.60}$$

where

R_b = resistance of human body, typically 1000 Ω for 50 and 60 Hz

R_f = ground resistance of one foot

I_b = rms magnitude of the current going through the body in A, per Equations 4.56a and 4.56b

C_s = surface layer derating factor based on the thickness of the protective surface layer spread above the earth grade at the substation (per IEEE Std. 80-2000, if no protective layer is used, then $C_s = 1$)

Since it is much easier to calculate and measure the potential than the current, the fibrillation threshold, given by Equations 4.56a and 4.56b, are usually given in terms of voltage. Thus, for a person with a body weight of 50 or 70 kg, the maximum allowable (or tolerable) touch voltages, respectively, can be expressed as

$$V_{\text{touch } 50} = \frac{0.116(1000 + 1.5\rho_s)}{\sqrt{t_s}} \text{ V} \quad 50 \text{ kg body weight} \tag{4.61a}$$

and, for foot-to-foot currents,

$$V_{\text{touch } 70} = \frac{0.157(1000 + 1.5\rho_s)}{\sqrt{t_s}} \text{ V} \quad 70 \text{ kg body weight} \tag{4.61b}$$

Note that, the above equations are applicable only in the event of no protective surface layer is used. Hence, for the metal-to-metal touch in V, Equations 4.61a and 4.61b become

$$V_{\text{mm - touch } 50} = \frac{116}{\sqrt{t_s}} \text{ V} \quad 50 \text{ kg body weight} \tag{4.61c}$$

and

$$V_{\text{mm - touch } 70} = \frac{157}{\sqrt{t_s}} \text{ V} \quad 70 \text{ kg body weight} \tag{4.61d}$$

The maximum allowable (or tolerable) step voltages, for a person with a body weight of 50 or 70 kg, are given, respectively, as

$$V_{\text{step } 50} = \frac{0.116(1000 + 6C_s\rho_s)}{\sqrt{t_s}} \text{ V} \quad 50 \text{ kg body weight} \tag{4.62a}$$

and

$$V_{\text{step } 70} = \frac{0.157(1000 + 6C_s\rho_s)}{\sqrt{t_s}} \text{ V} \quad 70 \text{ kg body weight} \tag{4.62b}$$

$$V_{\text{touch } 50} = \frac{0.116(1000 + 1.5C_s\rho_s)}{\sqrt{t_s}} \text{ V} \quad 50 \text{ kg body weight} \tag{4.62c}$$

and

$$V_{\text{touch } 70} = \frac{0.157(1000 + 1.5C_s\rho_s)}{\sqrt{t_s}} \text{ V} \quad 70 \text{ kg body weight} \tag{4.62d}$$

The above equations are applicable only in the event that a protection surface layer is used. For metal-to-metal contacts, use $\rho_s = 0$ and $C_s = 1$. For more detailed applications, see IEEE Std. 2000 [160]. Also, it is important to note that in using the above equations, it is assumed that they are applicable to 99.5% of the population. There are always exceptions.

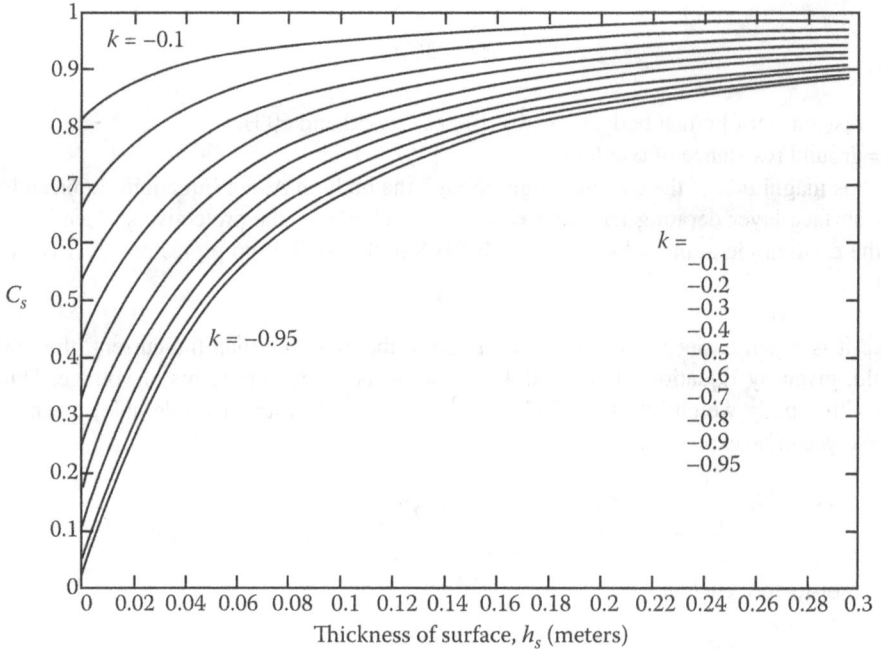

Figure 4.21 Surface layer derating factor vs. thickness of surface material in meters. (From Keil, R. P., *Substation Grounding, in Electric Power Substations Engineering*, McDonald, J. D. (ed.), 2nd ed., CRC Press, Boca Raton, FL, 2007.)

4.9.2   REDUCTION OF FACTOR C_S

Note that, according to IEEE Std. 80-2000, a thin layer of highly resistive protective surface material such as gravel spread across the earth at a substation greatly reduces the possibility of shock situation at that substation. IEEE Std. 80-2000 gives the required equations to determine the ground resistance of one foot on a thin layer of surface material as

$$C_s = 1 + \frac{1.6b}{\rho_s} \sum_{n=1}^{\infty} K^n R_{m(2nh_s)} \tag{4.63}$$

and

$$C_s = 1 - \frac{0.09\left(1 - \frac{\rho}{\rho_s}\right)}{2h_s + 0.09} \tag{4.64}$$

where

$$K = \frac{\rho - \rho_s}{\rho + \rho_s} \tag{4.65}$$

where
   C_s = surface layer derating factor (It can be considered as a corrective factor to compute the effective foot resistance in the presence of a finite thickness of surface material.) (See Figure 4.21)
   ρ_s = surface material resistivity, $\Omega \cdot m$
   K = reflection factor between different material resistivities
   ρ = resistivity of earth beneath the substation, $\Omega \cdot m$
   h_s = thickness of the surface material, m b = radius of circular metallic disc representing the foot, m

Table 4.7
Resistivity of Different Soils

Ground Type	Resistivity, ρ_S
Seawater	0.01–1.0
Wet organic soil	10
Moist soil (average earth)	100
Dry soil	1000
Bedrock	10^4
Pure slate	10^7
Sandstone	10^9
Crushed rock	1.5×10^8

$R_{m(2nh_s)}$ = mutual ground resistance between two similar, parallel, coaxial plates that are separated by a distance of $(2nh_s)$, $\Omega \cdot m$

Note that, Figure 4.21 gives the exact value of C_s instead of using the empirical Equation 4.64 for it. The empirical equation gives approximate values that are within 5% of the values that can be found with the equation.

Table 4.7 gives typical values for various ground types. However, the resistivity of ground also changes as a function of temperature, moisture, and chemical content. Therefore, in practical applications, the only way to determine the resistivity of soil is by measuring it.

Example 4.7: Assume that a human body is part of a 60-Hz electric power circuit for about 0.49 s and that the soil type is average earth. On the basis of IEEE Std. 80-2000, determine the following:

(a) Tolerable touch potential, for 50 kg body weight
(b) Tolerable step potential
(c) Tolerable touch voltage limit for metal-to-metal contact, if the person is 50 kg
(d) Tolerable touch voltage limit for metal-to-metal contact, if the person is 70 kg

Solution:

(a) Using Equation 4.61a, for 50 kg body weight

$$V_{touch\,50} = \frac{0.116(1000 + 1.5\rho_s)}{\sqrt{t_s}}$$
$$= \frac{0.116(1000 + 1.5 \times 100)}{\sqrt{0.49}}$$
$$\cong 191 \text{ V}$$

(b) Using Equation 4.61b

$$V_{step\,50} = \frac{0.116(1000 + 6\rho_s)}{\sqrt{t_s}}$$
$$= \frac{0.116(1000 + 6 \times 100)}{\sqrt{0.49}}$$
$$\cong 265 \text{ V}$$

Table 4.8
Effect of Moisture Content on Soil Resistivity

Moisture Content (wt.%)	Resistivity ($\Omega \cdot$cm)	
	Topsoil	Sandy Loam
0	>109	$>10^9$
2.5	250,000	150,000
5	165,000	43,000
10	53,000	18,500
15	19,000	10,500
20	12,000	6,300
30	6,400	4,200

(c) Since $\rho_s = 0$,

$$V_{\text{mm - touch 50}} = \frac{116}{\sqrt{t_s}} = \frac{116}{\sqrt{0.49}} = 165.7 \text{ V for 50 Kg body weight}$$

(d) Since $\rho_s = 0$,

$$V_{\text{mm - touch 70}} = \frac{157}{\sqrt{t_s}} = \frac{157}{\sqrt{0.49}} = 224.3 \text{ V for 70 Kg body weight}$$

4.9.3 GROUND POTENTIAL RISE (GPR) AND GROUND RESISTANCE

The GPR is a function of fault current magnitude, system voltage, and ground (system) resistance. The current through the ground system multiplied by its resistance measured from a point remote from the substation determines the GPR with respect to remote ground.

The ground resistance can be reduced by using electrodes buried in the ground. For example, metal rods or *counterpoise* (i.e., buried conductors) are used for the lines of the grid system are made of copper-stranded cable, on the other hand rods are used for the substations.

The grounding resistance of a buried electrode is a function of (1) the resistance of the electrode itself and the connections to it, (2) the contact resistance between the electrode and the surrounding soil, and (3) the resistance of the surrounding soil, from the electrode surface outward.

The first two resistances are very small with respect to soil resistance and therefore may be neglected in some applications. However, the third one is usually very large depending on the type of soil, chemical ingredients, moisture level, and temperature of the soil surrounding the electrode.

Table 4.8 presents data indicating the effect of moisture contents on soil resistivity. The resistance of the soil can be measured by using the three-electrode method or by using self-contained instruments such as the Biddle Megger ground resistance tester.

If the surface of the soil is covered with a layer of crushed rock or some other high-resistivity material, its resistivity should be used in the previous equations. Table 4.7 gives typical values for various ground types. However, the resistivity of ground also changes as a function of temperature, moisture, and chemical content. Thus, in practical applications, the only way to determine the resistivity of soil is by measuring it.

In general, soil resistivity investigations are required to determine the soil structure. Table 4.7 gives only very rough estimates. The soil resistivity can very substantially with changes in temperature, moisture, and chemical content. To determine the soil resistivity of a specific site, soil resistivity measurements are required to be taken. Since soil resistivity can change both horizontally and vertically, it is necessary to take more than one set of measurements. IEEE Std. 80-2000 [160] describes various measuring techniques in detail. There are commercially available computer programs that use the soil data and calculate the soil resistivity and provide a confidence level based on the test. There is also a graphical method that was developed by Sunde [161] to interpret the test results.

4.9.4 GROUND RESISTANCE

Ground is defined as a conducting connection, either intentional or accidental, by which an electric circuit or equipment becomes grounded. Thus, *grounded* means that a given electric system, circuit, or device is connected to the earth or to some other equivalent conducting body of relatively large extent, serving in the place of the former with the purpose of establishing and maintaining the potential of conductors connected to it approximately at the potential of the earth and allowing for conducting electric currents from and to the earth of its equivalent.

Thus, a safe grounding design should provide the following:

1. A means to carry and dissipate electric currents into ground under normal and fault conditions without exceeding any operating and equipment limits or adversely affecting the continuity of service
2. Assurance for such a degree of human safety so that a person working or walking in the vicinity of grounded facilities is not subjected to the danger of critical electrical shock

However, a low ground resistance is not, in itself, a guarantee of safety. For example, about three or four decades ago, a great many people assumed that any object grounded, however crudely, could be safely touched. This misconception probably contributed to many tragic accidents in the past. Thus, since there is no simple relation between the resistance of the ground system as a whole and the maximum shock current to which a person might be exposed, a system or system component (e.g., substation or tower) of relatively low ground resistance may be dangerous under some conditions, but another system component with very high ground resistance may still be safe or can be made safe by careful design.

Table 4.9 gives data showing the effect of temperature on soil resistivity.

Figure 4.22 shows a ground rod driven into the soil and conducting current in all directions. The resistance of the soil has been illustrated in terms of successive shells of the soil of equal thickness. With increased distance from the electrode, the soil shells have greater area and therefore lower resistance. Thus, the shell nearest the rod has the smallest cross section of the soil and therefore the highest resistance. Measurements have shown that 90% of the total resistance surrounding an electrode is usually with a radius of 6–10 ft. Table 4.10 gives formulas to determine resistance to ground of various types of electrodes [162].

The assumptions that have been made in deriving these formulas are that the soil is perfectly homogeneous and the resistivity is of the same known value throughout the soil surrounding the electrode. Of course, these assumptions are seldom true. The only way one can be sure of the resistivity of the soil is by actually measuring it at the actual location of the electrode and at the actual depth.

Figure 4.23 shows the variation of soil resistivity with depth for a soil having uniform moisture content at all depth. In reality, however, deeper soils have greater moisture content, and the advantage of depth is more visible. Some nonhomogeneous soils can also be modeled by using the two-layer method [161, 163–165].

The resistance of the soil can be measured by using the three-electrode method or by using self-contained instruments such as the Biddle Megger ground resistance tester. Figure 4.24 shows the approximate ground resistivity distribution in the United States.

Table 4.9
Effect of Temperature on Soil Resistivity[a]

Temperature (°C)	°F	Resistivity (Ω·cm)
20	68	7,200
10	50	9,900
0 (water)	32	13,800
0 (ice)	32	30,000
−5	23	79,000
−15	14	330,000

[a]Sandy loam with 15.2% moisture.

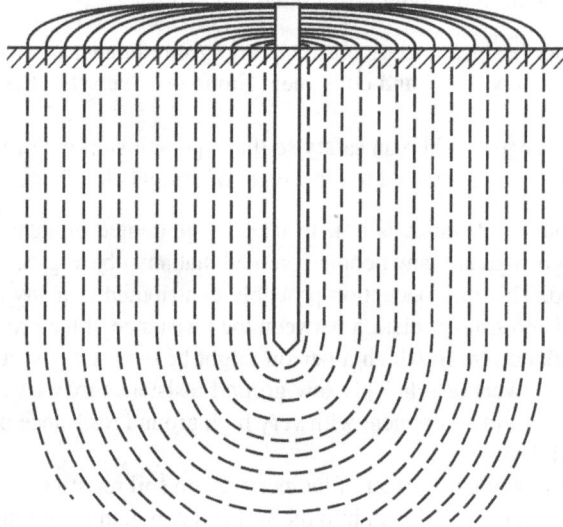

Figure 4.22 Resistance of earth surrounding an electrode.

4.9.5 SOIL RESISTIVITY MEASUREMENTS

Table 4.7 gives estimates on soil classification that are only an approximation of the actual resistivity of a given site. Actual resistivity tests therefore are crucial. They should be made at a number of places within the site. In general, substation sites where the soil has uniform resistivity throughout the entire area and to a considerable depth are seldom found.

4.9.5.1 Wenner Four-Pin Method

More often than not, there are several layers, each having a different resistivity. Furthermore, lateral changes also take place, but with respect to the vertical changes; these changes usually are more gradual. Hence, soil resistivity tests should be made to find out if there are any substantial changes in resistivity with depth. If the resistivity varies considerably with depth, it is often desirable to use an increased range of probe spacing to get an estimate of the resistivity of the deeper layers.

Table 4.10

Formulas for Calculations of Resistance to Ground

⬤	Hemisphere, radius a	$R = \frac{\rho}{2\pi a}$
●	One ground rod, length L, radius a	$R = \frac{\rho}{2\pi L}\left(In\frac{4L}{a} - 1\right)$
● ●	Two ground rod, $s > L$; spacing s	$R = \frac{\rho}{2\pi L}\left(In\frac{4L}{a} - 1\right) + \frac{\rho}{4\pi s}\left(1 - \frac{L^2}{3s^2} + \frac{2L^4}{5s^2}\cdots\right)$
● ●	Two ground rod, $s > L$; spacing s	$R = \frac{\rho}{2\pi L}\left(In\frac{4L}{a} + In\frac{4L}{s} - 2 + \frac{s}{2L} - \frac{s^2}{16L^2} + \frac{s^4}{512L^4}\cdots\right)$
—	Buried horizontal wire, length $2L$, depth $s/2$	$R = \frac{\rho}{2\pi L}\left(In\frac{4L}{a} + In\frac{4L}{s} - 2 + \frac{s}{2L} - \frac{s^2}{16L^2} + \frac{s^4}{512L^4}\cdots\right)$
L	Rigth-angle turn of wire, length of arm L, depth $s/2$	$R = \frac{\rho}{4\pi L}\left(In\frac{2L}{a} + In\frac{2L}{s} + 0.2373 - 0.2146\frac{s}{L}\right.$ $\left. + 0.1035\frac{s^2}{L^2} - 0.0424\frac{s^4}{L^4}\cdots\right)$
人	Three-point star, length of arm L, depth $s/2$	$R = \frac{\rho}{6\pi L}\left(In\frac{2L}{a} + In\frac{2L}{s} + 1.071 - 0.209\frac{s}{L} + 0.238\frac{s^2}{L^2} - 0.054\frac{s^4}{L^4}\cdots\right)$
+	Four-point star, length of arm L, depth $s/2$	$R = \frac{\rho}{8\pi L}\left(In\frac{2L}{a} + In\frac{2L}{s} + 2.912 - 1.071\frac{s}{L} + 0.645\frac{s^2}{L^2} - 0.145\frac{s^4}{L^4}\cdots\right)$
✳	Six-point star, length of arm L, depth $s/2$	$R = \frac{\rho}{12\pi L}\left(In\frac{2L}{a} + In\frac{2L}{s} + 6.851 - 3.128\frac{s}{L} + 1.758\frac{s^2}{L^2} - 0.490\frac{s^4}{L^4}\cdots\right)$
✴	Eight-point star, length of arm L, depth $s/2$	$R = \frac{\rho}{16\pi L}\left(In\frac{2L}{a} + In\frac{2L}{s} + 10.98 - 5.51\frac{s}{L} + 3.26\frac{s^2}{L^2} - 1.17\frac{s^4}{L^4}\cdots\right)$
○	Ring of wire, diameter of ring D, diameter of wire d, depth $s/2$	$R = \frac{\rho}{2\pi^2 D}\left(In\frac{8D}{d} + In\frac{4D}{s}\right)$
—	Buried horizontal strip, length $2L$, section a by b, depth $s/2$, $b < a/8$	$R = \frac{\rho}{4\pi L}\left(In\frac{4L}{a} + \frac{a^2 - \pi ab}{2(a+b)^2} + In\frac{4L}{s} - 1 + \frac{s}{2L} - \frac{s^2}{16L^2} + \frac{s^4}{512L^4}\cdots\right)$
◎	Buried vertical round plate, radius a, depth $s/2$	$R = \frac{\rho}{8a} + \frac{\rho}{4\pi s}\left(1 - \frac{7}{12}\frac{a^2}{s^2} + \frac{33}{40}\frac{a^4}{s^4}\cdots\right)$
	Buried vertical round plate, radius a, depth $s/2$	$R = \frac{\rho}{8a} + \frac{\rho}{4\pi s}\left(1 + \frac{7}{24}\frac{a^2}{s^2} + \frac{99}{320}\frac{a^4}{s^4}\cdots\right)$

Source: Dwight, H. B., *Electr. Eng.* (*Am. Inst. Electr. Eng.*) 55, 1319–1328 1936 IEEE.

[a] Approximate formulas, including effects of images. Dimensions must be in centimeters to give resistance in ohms. The symbol ρ is the resistivity of earth in ohm centimeters.

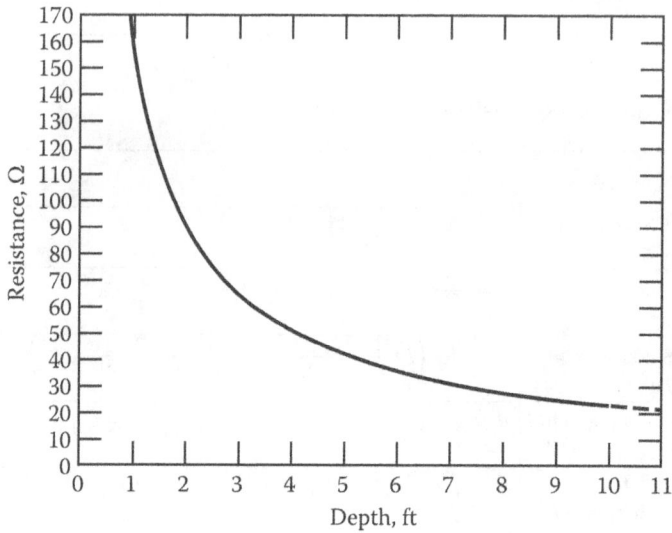

Figure 4.23 Variation of soil resistivity with depth for soil having uniform moisture content at all depths. (From National Bureau of Standards Technical Report 108.)

IEEE Std. 81-1983 describes a number of measuring techniques. The Wenner four-pin method is the most commonly used technique. Figure 4.25 illustrates this method. In this method, four probes (or pins) are driven into the earth along a straight line, at equal distances apart, driven to a depth b. The voltage between the two inner (i.e., potential) electrodes is then measured and divided by the current between the two outer (i.e., current) electrodes to give a value of resistance R. The apparent resistivity of soil is determined from

$$\rho_a = \frac{4\pi a R}{1 + \frac{2a}{\sqrt{a^2+4b^2}}} - \frac{a}{\sqrt{a^2+b^2}} \tag{4.66}$$

where
   ρ_a = apparent resistivity of the soil in ohm meters
   R = measured resistivity in ohms
   a = distance between adjacent electrodes in meters
   b = depth of the electrodes in meters

In the event that b is small in comparison to a, then

$$\rho_a = 2\pi a R \tag{4.67}$$

The current tends to flow near the surface for the small probe spacing, whereas more of the current penetrates deeper soils for large spacing. Because of this fact, the previous two equations can be used to determine the apparent resistivity ρ_a at a depth a.

The Wenner four-pin method obtains the soil resistivity data for deeper layers without driving the test pins to those layers. No heavy equipment is needed to do the four-pin test. The results are not greatly affected by the resistance of the test pins or the holes created in driving the test pins into the soil. Because of these advantages, the Wenner method is the most popular method.

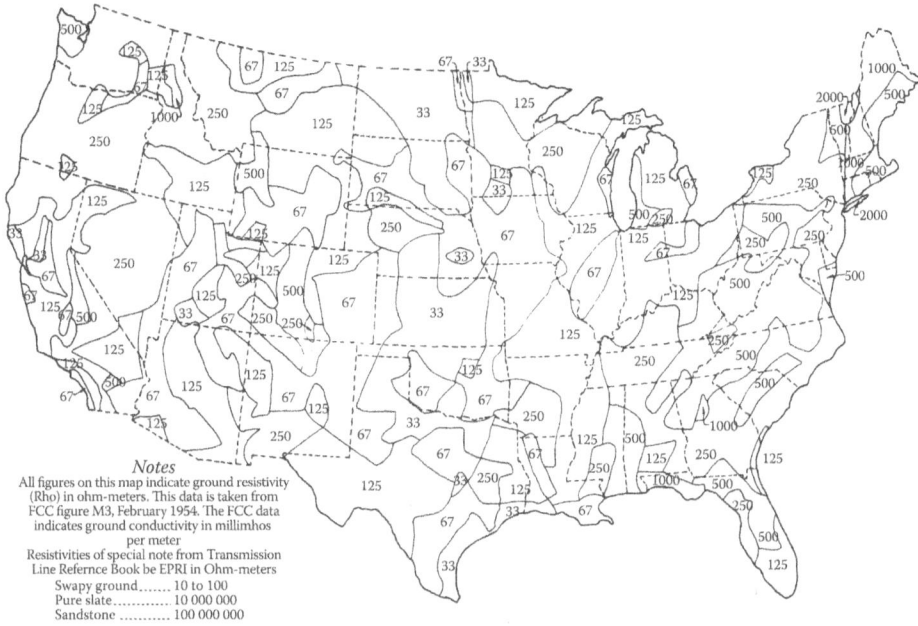

Notes

All figures on this map indicate ground resistivity (Rho) in ohm-meters. This data is taken from FCC figure M3, February 1954. The FCC data indicates ground conductivity in millimhos per meter

Resistivities of special note from Transmission Line Refernce Book be EPRI in Ohm-meters

Swapy ground....... 10 to 100
Pure slate............ 10 000 000
Sandstone 100 000 000

Figure 4.24 Approximate ground resistivity distribution in the United States. (From Farr, H. H., *Transmission Line Design Manual*. U.S. Department of the Interior, Water and Power Resources Service, Denver, 1980.)

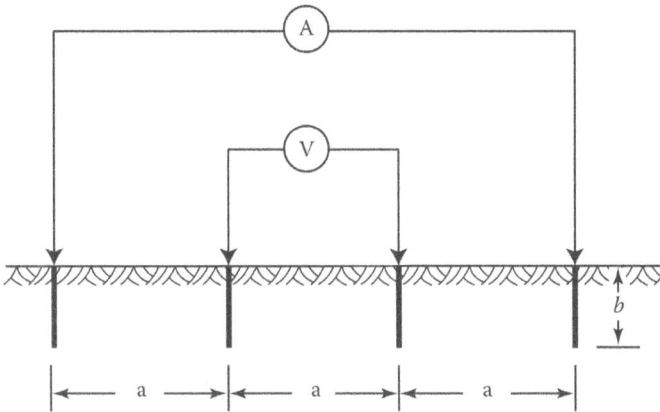

Figure 4.25 Wenner four-pin method. (From Gönen, T., *Electric Power Transmission System Engineering*, 2nd ed., CRC Press, Boca Raton, FL, 2009.) [166]

4.9.5.2 Pin or Driven-Ground Rod Method

IEEE Std. 81-1983 describes a second method of measuring soil resistivity. It is illustrated in Figure 4.26. In this method, the depth (L_r) of the driven rod located in the soil to be tested is varied. The other two rods are known as *reference rods*. They are driven to a shallow depth in a straight line. The location of the voltage rod is varied between the test rod and the current rod. Alternatively, the voltage rod can be placed on the other side of the driven rod. The apparent resistivity is found from

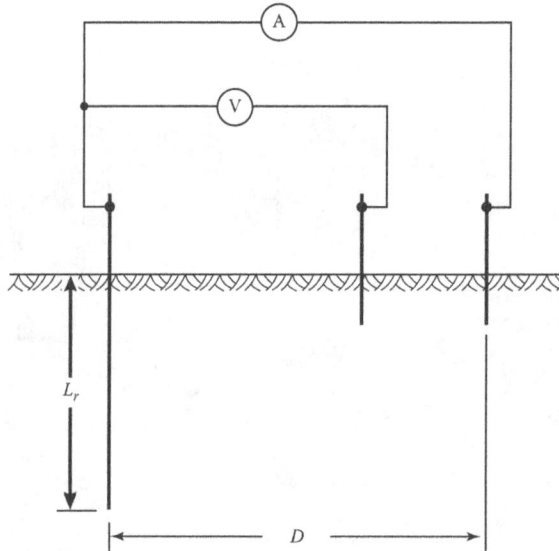

Figure 4.26 Circuit diagram for three-pin or driven-ground rod method. (From Gönen, T., *Electric Power Transmission System Engineering*, 2nd ed., CRC Press, Boca Raton, FL, 2009.)

$$\rho_a = \frac{2\pi L_r R}{\ln\left(\frac{8L_r}{d}\right) - 1} \tag{4.68}$$

where
  L_r = length of the driven rod in meters
  d = diameter of the rod in meters
  R = measured resistivity in ohms

A plot of the measured resistivity value ρ_a vs. the rod length (L_r) provides a visual aid for finding out earth resistivity variations with depth. An advantage of the driven-rod method, even though not related necessarily to the measurements, is the ability to determine to what depth the ground rods can be driven. This knowledge can save the need to redesign the ground grid. Because of hard layers in the soil such as rock and hard clay, it becomes practically impossible to drive the test rod any further, resulting in insufficient data.

A disadvantage of the driven-rod method is that when the test rod is driven deep in the ground, it usually losses contact with the soil owing to the vibration and the larger-diameter couplers resulting in higher measured resistance values. A ground grid designed with these higher soil resistivity values may be unnecessarily conservative. Thus, this method presents an uncertainty in the resistance value.

4.10 SUBSTATION GROUNDING

Grounding at the substation has paramount importance. Again, the purpose of such a grounding system includes the following:

1. To provide the ground connection for the grounded neutral for transformers, reactors, and capacitors
2. To provide the discharge path for lightning rods, arresters, gaps, and similar devices

3. To ensure safety to operating personnel by limiting potential differences that can exist in a substation
4. To provide a means of discharging and deenergizing equipment in order to proceed with the maintenance of the equipment
5. To provide a sufficiently low resistance path to ground to minimize the rise in ground potential with respect to remote ground

A multigrounded, common neutral conductor used for a primary distribution line is always connected to the substation grounding system where the circuit originates and to all grounds along the length of the circuit. If separate primary and secondary neutral conductors are used, the conductors have to be connected together provided the primary neutral conductor is effectively grounded.

The substation grounding system is connected to every individual equipment, structure, and installation so that it can provide the means by which grounding currents are connected to remote areas. It is extremely important that the substation ground has a low ground resistance, adequate current-carrying capacity, and safety features for personnel. It is crucial to have the substation ground resistance very low so that the total rise of the ground system potential will not reach values that are unsafe for human contact.*

The substation grounding system is normally made of buried horizontal conductors and driven ground rods interconnected (by clamping, welding, or brazing) to form a continuous grid (also called mat) network. A continuous cable (usually it is 4/0 bare copper cable buried 12–18 in. below the surface) surrounds the grid perimeter to enclose as much ground as possible and to prevent current concentration and thus high gradients at the ground cable terminals. Inside the grid, cables are buried in parallel lines and with uniform spacing (e.g., about 10 × 20 ft).

All substation equipment and structures are connected to the ground grid with large conductors to minimize the grounding resistance and limit the potential between equipment and the ground surface to a safe value under all conditions. All substation fences are built inside the ground grid and attached to the grid in short intervals to protect the public and personnel. The surface of the substation is usually covered with crushed rock or concrete to reduce the potential gradient when large currents are discharged to ground and to increase the contact resistance to the feet of the personnel in the substation.

IEEE Std. 80-1976 [167] provides a formula for a quick simple calculation of the grid resistance to ground after a minimum design has been completed. It is expressed as

$$R_{grid} = \frac{\rho_s}{4r} + \frac{\rho_s}{L_T} \tag{4.69}$$

where

ρ_s = soil resistivity in ohm meters
L = total length of grid conductors in meters
R = radius of circle with area equal to that of grid in meters

IEEE Std. 80-2000 provides the following equation to determine the grid resistance after a minimum design has been completed:

$$R_{grid} = \frac{\rho_s}{4}\sqrt{\frac{\pi}{A}} \tag{4.70}$$

Also, IEEE Std. 80-2000 provides the following equation to determine the upper limit for grid resistance to ground after a minimum design has been completed:

$$R_{grid} = \frac{\rho_s}{4}\sqrt{\frac{\pi}{A}} + \frac{\rho_s}{L_T} \tag{4.71}$$

where

R_{grid} = grid resistance in ohms
ρ = soil resistance in ohm meters
A = area of the ground in square meters
L_T = total buried length of conductors in meters

However, Equation 4.71 requires a uniform soil resistivity. Hence, a substantial engineering judgment is necessary for reviewing the soil resistivity measurements to decide the value of soil resistivity. However, it does provide a guideline for the uniform soil resistivity to be used in the ground grid design. Alternatively, Sverak et al. [168] provides the following formula for the grid resistance:

$$R_{grid} = \rho_s \left[\frac{1}{L_T} + \frac{1}{\sqrt{20A}} \left(1 + \frac{1}{1 + h\sqrt{\frac{20}{A}}} \right) \right] \tag{4.72}$$

where
R_{grid} = substation ground resistance, Ω
ρ_s = soil resistivity, $\Omega \cdot$m
A = area occupied by the ground grid, m^2
H = depth of the grid, m
L_T = total buried length of conductors, m

IEEE Std. 80-1976 also provides formulas to determine the effects of the grid geometry on the step and mesh voltage in volts. Mesh voltage is the worst possible value of a touch voltage to be found within a mesh of a ground grid if standing at or near the center of the mesh. They can be expressed as

$$E_{step} = \frac{\rho_s \times K_s \times K_i \times I_G}{L_s} \tag{4.73}$$

and

$$E_{mesh} = \frac{\rho_s \times K_m \times K_i \times I_G}{L_m} \tag{4.74}$$

where
ρ_s = average soil resistivity in ohm meters
K_s = step coefficient
K_m = mesh coefficient
K_i = irregularity coefficient
I_G = maximum rms current flowing between ground grid and earth in amperes
L_s = total length of buried conductors, including cross connections, and (optionally) the total effective length of ground rods in meters
L_m = total length of buried conductors, including cross connections, and (optionally) the combined length of ground rods in meters

Many utilities have computer programs for performing grounding grid studies. The number of tedious calculations that must be performed to develop an accurate and sophisticated model of a system is no longer a problem.

In general, in the event of a fault, overhead ground wires, neutral conductors, and directly buried metal pipes and cables conduct a portion of the ground fault current away from the substation ground grid and have to be taken into account when calculating the maximum value of the grid current. On the basis of the associated equivalent circuit and resultant current division, one can determine what portion of the total current flows into the earth and through other ground paths. It can be used to determine the approximate amount of current that did not use the ground as flow path. The fault current division factor (also known as the *split factor*) can be expressed as

$$E_{split} = \frac{I_{grid}}{3I_{a0}}$$ (4.75)

where

S_{split} = fault current division factor
I_{grid} = rms symmetrical grid current, A
I_{a0} = zero-sequence fault current, A

The *split factor* is used to determine the approximate amount of current that did not use the ground flow path. Computer programs can determine the split factor easily, but it is also possible to determine the split factor through graphs. With the Y ordinate representing the split factor and the X axis representing the grid resistance, it is obvious that the grid resistance has to be known to determine the split factor. As previously said, the split factor determines the approximate amount of current that uses the earth as a return path. The amount of current that does enter the earth is found from the following equation. Hence, the design value of the *maximum grid current* can be found from

$$I_G = D_f \times I_{grid}$$ (4.76)

where

I_G = maximum grid current in amperes
D_f = decrement factor for the entire fault duration of t_f, given in seconds
I_{grid} = rms symmetrical grid current in amperes

Here, Figure 4.27 illustrates the relation between asymmetrical fault current, dc decaying component, and symmetrical fault current, and the relation between the variables I_F, I_f, and D_f for the fault duration t_f.

The *decrement factor* is an adjustment factor that is used in conjunction with the symmetrical ground fault current parameter in safety-oriented grounding calculations. It determines the rms equivalent of the asymmetrical current wave for a given fault duration, accounting for the effect of initial dc offset and its attenuation during the fault. The decrement factor can be calculated from

$$D_f = \sqrt{1 + \frac{T_a}{I_f}\left(1 - e^{-\frac{2t_f}{T_a}}\right)}$$ (4.77)

where

t_f = time duration of fault in seconds
$T_a = \frac{X}{\omega R}$ = dc offset time constant in seconds

Here, t_f should be chosen as the fastest clearing time, and includes breaker and relay time for transmission substations. It is assumed here that the ac components do not decay with time.

The symmetrical grid current is defined as that portion of the symmetrical ground fault current that flows between the grounding grid and surrounding earth. It can be expressed as

$$I_{grid} = S_f \times I_f$$ (4.78)

where

I_f = rms value of symmetrical ground fault current in amperes
S_f = fault current division factor

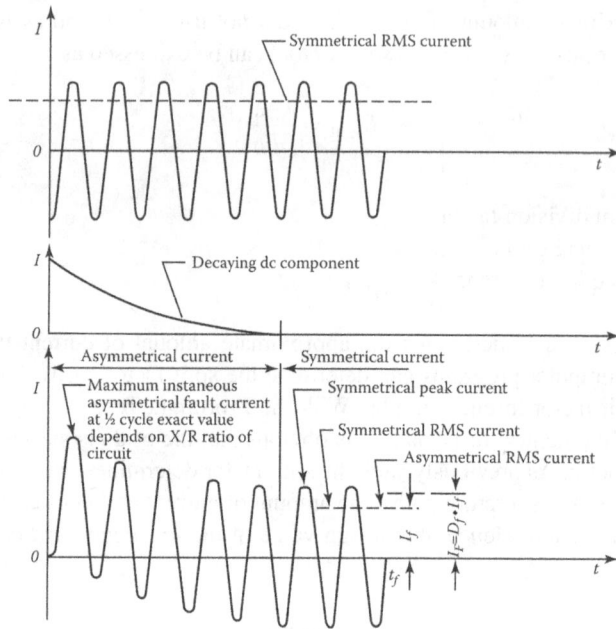

Figure 4.27 Relation between asymmetrical fault current, dc decaying component, and symmetrical fault current.

IEEE Std. 80-2000 provides a series of current based on computer simulations for various values of ground grid resistance and system conditions to determine the grid current. On the basis of those split-current curves, one can determine the maximum grid current.

4.11 GROUND CONDUCTOR SIZING FACTORS

The flow of excessive currents will be very dangerous if the right equipment is not used to help dissipate the excessive currents. Ground conductors are means of providing a path for excessive currents from the substation to ground grid. Hence, the ground grid than can spread the current into the ground, creating a zero potential between the substation and the ground. Table 4.11 gives the list of possible conductors that can be used for such conductors. In the United States, there are only two types of conductors, namely, copper and/or copper-clad steel conductors, that are used for this purpose. The copper one is mainly used because of its high conductivity and the high resistance to corrosion. The next step is to determine the size of ground conductor that needs to be buried underground [169, 170]. Thus, based on the *symmetrical conductor current*, the required conductor size can be found from

$$I_f = A_{mm^2} \left[\left(\frac{TCAP \times 10^{-4}}{t_c \times \alpha_r \times \rho_r} \right) \ln \left(\frac{K_0 + T_{max}}{K_0 + T_{amb}} \right) \right]^{1/2} \tag{4.79}$$

if the conductor size needs to be found in square millimeters, the conductor size can be found from

$$A_{mm^2} = \frac{I_f}{\left[\left(\frac{TCAP \times 10^{-4}}{t_c \times \alpha_r \times \rho_r} \right) \ln \left(\frac{K_0 + T_{max}}{K_0 + T_{amb}} \right) \right]^{1/2}} \tag{4.80}$$

Alternatively, in the event that the conductor size needs to be found in kilo-circular mils, since

$$A_{kcmil} = 1.974 \times A_{mm^2} \tag{4.81}$$

Table 4.11

Material Constants of the Typical Grounding Material Used

Description	K_f	T_m (°C)	α_r Factor at 20°C (1/°C)	ρ_r 20°C (mΩ·cm)	K_o at 0°C (0°C)	Fusing Temperature, T_m (0°C)	Material Conducting (%)	TCAP Thermal Capacity [J/cm^3 × °C]
Copper annealed soft-drawn	7	1083	0.0393	1.72	234	1083	100	3.42
Copper annealed hard-drawn	1084	1084	0.00381	1.78	242	1084	97	3.42
Copper-clad steel wire	1084	12.06	0.00378	5.86	245	1084	30	3.85
Stainless steel 304	1510	14.72	0.0013	15.86	749	1400	2.4	3.28
Zinc-coated steel rod	28.96	28.96	0.003	72	293	419	8.6	4.03

then Equation 4.72 can be expressed as

$$I_f = 5.07 \times 10^{-3} A_{\text{kcmil}} \left[\left(\frac{\text{TCAP} \times 10^{-4}}{t_c \times \alpha_r \times \rho_r} \right) \ln \left(\frac{K_0 + T_{\max}}{K_0 + T_{\text{amb}}} \right) \right]^{1/2} \quad (4.82)$$

Note that both α_r and ρ_r can be found at the same reference temperature of T_r (° C). Also, note that Equations 4.79 and 4.82 can also be used to determine the short-time temperature rise in a ground conductor. Thus, taking other required conversions into account, the conductor size in kilo-circular mils can be found from

$$A_{\text{kcmil}} = \frac{197.4 \times I_f}{\left[\left(\frac{\text{TCAP} \times 10^{-4}}{t_c \times \alpha_r \times \rho_r} \right) \ln \left(\frac{K_0 + T_{\max}}{K_0 + T_{\text{amb}}} \right) \right]^{1/2}} \quad (4.83)$$

where
I_f = rms current (without dc offset), kA
A_{mm^2} = conductor cross section, mm^2
A_{kcmil} = conductor cross section, kcmil
TCAP = thermal capacity per unit volume, J/(cm3.° C). (It is found from Table 4.11, per IEEE Std.80-2000.)
t_c = duration of current, s
α_r = thermal coefficient of resistivity at reference temperature Tr, 1/° C. (It is found from Table 4.11, per IEEE Std. 80-2000 for 20° C.)
ρ_r = resistivity of the ground conductor at reference temperature Tr, $\mu\Omega$·cm. (It is found from Table 4.11, per IEEE Std. 80-2000 for 20° C.)
$K_0 = 1/\alpha_0$ or $(1/\alpha_r) - $ Tr, °C
$T_{\max}$ = maximum allowable temperature, ° C
T_{amb} = ambient temperature, ° C
I_f = rms current (without dc offset), kA

A_{mm^2} = conductor cross section, mm^2
A_{kcmil} = conductor cross section, kcmil

For a given conductor material, once the TCAP is found from Table 4.11 or calculated from

$$\text{TCAP}[\text{J}/(\text{cm}^3 \times^\circ \text{C})] = 4.184(\text{J/cal}) \times \text{SH}[(\text{cal/g} \times^\circ \text{C}))] \times \text{SW}(\text{g/cm}^3) \qquad (4.84)$$

where
SH = specific heat, in cal/(g $\times$ °C) is related to the thermal capacity per unit volume in J/(cm^3 $\times$ °C)
SW = specific weight, in g/cm^3 is related to the thermal capacity per unit volume in J/(cm^3 $\times$ °C)

Thus, TCAP is defined by

$$\text{TCAP}[\text{J}/(\text{cm}^3 \times^\circ \text{C})] = 4.184(\text{J/cal}) \times \text{SH}[(\text{cal/g} \times^\circ \text{C}))] \times \text{SW}(\text{g/cm}^3) \qquad (4.85)$$

Asymmetrical fault currents consist of subtransient, transient, and steady-state ac components, and the dc offset current component. To find the asymmetrical fault current (i.e., if the effect of the dc offset is needed to be included in the fault current), the equivalent value of the asymmetrical current I_F is found from

$$I_F = D_f \times I_f \qquad (4.86)$$

where I_F represents the rms value of an asymmetrical current integrated over the entire fault duration, t_c, can be found as a function of X/R by using D_f, before using Equations 4.79 or 4.82, and where D_f is the decrement factor and is found from

$$D_f = \left[1 + \frac{T_a}{t_f} \left(1 - e^{\frac{-2t_f}{T_a}} \right) \right]^{1/2} \qquad (4.87)$$

where t_f is the time duration of fault in seconds, and T_a is the dc offset time constant in seconds. Note that,

$$T_a = \frac{X}{\omega R} \qquad (4.88)$$

and for 60 Hz,

$$T_a = \frac{X}{120\pi R} \qquad (4.89)$$

If the GPR value exceeds the tolerable touch and step voltages, it is necessary to perform the mesh voltage design calculations to determine whether the design of a substation is safe. If the design is again unsafe, conductors in the form of ground rods are added to the design until the design is considered safe. The mesh voltage is found from

The resulting I_F is always greater than I_f. However, if the X/R ratio is less than 5 and the fault duration is greater than 1 s, the effects of the dc offset are negligible.

4.12   MESH VOLTAGE DESIGN CALCULATIONS

If the GPR value exceeds the tolerable touch and step voltages, it is necessary to perform the mesh voltage design calculations to determine whether the design of a substation is safe. If the design is again unsafe, conductors in the form of ground rods are added to the design until the design is considered safe. The mesh voltage is found from

$$E_{mesh} = \frac{\rho \times K_m \times K_i \times I_G}{L_M} \tag{4.90}$$

where
ρ = soil resistivity, $\Omega \cdot$ m
K_m = mesh coefficient
K_i = correction factor for grid geometry
I_G = maximum grid current that flows between ground grid and surrounding earth, A
L_m = length of $L_c + L_R$ for mesh voltage, m
L_c = total length of grid conductor, m
L_R = total length of ground rods, m

The mesh coefficient K_m is determined from

$$K_M = \frac{1}{2\pi} \left[\ln \left\{ \frac{D^2}{16h \times d} + \frac{(D+2h)2}{8D \times d} - \frac{h}{4d} \right\} + \frac{K_{ii}}{K_h} \times \ln \left\{ \frac{8}{\pi(2n-1)} \right\} \right] \tag{4.91}$$

where
d = diameter of grid conductors, m
D = spacing between parallel conductors, m
K_{ii} = irregularity factor (*corrective weighting factor* that adjusts for the effects of inner conductors on the corner mesh)
K_h = corrective weighting factor that highlight for the effects of grid depth
n = geometric factor
h = depth of ground grid conductors, m

Note that, the value of K_{ii} depends on the following circumstances:

(a) For the grids with ground rods existing in grid corners as well as perimeter:

$$K_{ii} = 1 \tag{4.92}$$

(b) For the grids with no or few ground rods with none existing in corners or perimeter:

$$K_{ii} = \frac{1}{(2n)^{\frac{2}{n}}} \tag{4.93}$$

and

$$K_h = \sqrt{1 + \frac{h}{h_0}} \tag{4.94}$$

where h_0 = grid reference depth = 1 m.
The effective number of parallel conductors (n) given in a given grid are found from

$$n = n_a \times n_b \times n_c \times n_d \tag{4.95}$$

where

$$n = \frac{2L_c}{L_p}$$

$n_b = 1$, for square grids

$n_c = 1$, for square and rectangular grids
$n_d = 1$, for square, rectangular, and L-shaped grids

Otherwise, the following equations are used to determine the n_b, n_c, and n_d so that

$$n_a = \sqrt{\frac{L_p}{4\sqrt{A}}} \tag{4.96}$$

$$n_c = \left[\frac{L_X \times L_Y}{A}\right]^{\frac{0.7A}{L_X \times L_Y}} \tag{4.97}$$

$$n_d = \frac{D_m}{\sqrt{L_X^2 + L_Y^2}} \tag{4.98}$$

where
L_p = peripheral length of the grid, m
L_C = total length of the conductor in the horizontal grid, m
A = area of the grid, m^2
L_X = maximum length of the grid in the X direction, m
L_Y = maximum length of the grid in the Y direction, m
d = diameter of grid conductors, m
D = spacing between parallel conductors, m
D_m = maximum distance between any two points on the grid, m
h = depth of ground grid conductors, m

Note that, the irregularity factor is determined from

$$K_{ii} = 0.644 + 0.148n \tag{4.99}$$

The effective buried length (L_M) for grids:

(a) With little or no ground rods but none located in the corners or along the perimeter of the grid:

$$L_M = L_C + L_R \tag{4.100}$$

where
L_R = total length of all ground rods, m
L_C = total length of the conductor in the horizontal grid, m
(b) With ground rods in corners and along the perimeter and throughout the grid:

$$L_M = L_C + \left[1.55 + 1.22\left(\frac{L_R}{\sqrt{L_X^2 + L_Y^2}}\right)\right] L_R \tag{4.101}$$

where L_R = length of each ground rod, m.

4.13   STEP VOLTAGE DESIGN CALCULATIONS

According to IEEE STD. 80-2000, in order for the ground system to be safe, step voltage has to be less than the tolerable step voltage. Furthermore, step voltages within the grid system designed for safe mesh voltages will be well within the tolerable limits, the reason for this is both feet and legs are in series rather than in parallel and the current takes the path from one leg to the other rather than through vital organs. The step voltage is determined from

$$E_{step} = \frac{\rho \times K_s \times K_i \times I_G}{L_S} \tag{4.102}$$

where

K_s = step coefficient
L_S = buried conductor length, m

Again, for grids with or without ground rods,

$$L_S = 0.75 L_C + 0.85 L_R \tag{4.103}$$

so that the step coefficient can be found from

$$K_S = \frac{1}{\pi}\left[\frac{1}{2h} + \frac{2}{D+h} + \frac{1}{D}(1 - 0.5^{n-2})\right] \tag{4.104}$$

where h = depth of ground grid conductors in meters, usually between 0.25 m ¡ h ¡ 2.5 m.

4.14 TYPES OF GROUND FAULTS

In general, it is difficult to determine which fault type and location will result in the greatest flow of current between the ground grid and the surrounding earth because no simple rule applies. IEEE Std. 80-2000 recommends not to consider multiple simultaneous faults since their probability of occurrence is negligibly small. Instead, it recommends investigating single-line-to-ground and line-to-line-to-ground faults.

4.14.1 LINE-TO-LINE-TO-GROUND FAULT

For a line-to-line-to-ground (i.e., double line-to-ground) fault, IEEE Std. 80-2000 gives the following equation to calculate the zero-sequence fault current:

$$I_{a0} = \frac{E(R_2 + jX_2)}{(R_1 + jX_1)[R_0 + R_2 + 3R_f + j(X_0 + X_2)] + (R_2 + jX_2)(R_0 + 3R_f + jX_0)} \tag{4.105}$$

where

I_{a0} = symmetrical rms value of zero-sequence fault current, A
E = phase-to-neutral voltage, V
R_f = estimated resistance of the fault, Ω (normally it is assumed $R_f = 0$)
R_1 = positive-sequence system resistance, Ω
R_2 = negative-sequence system resistance, Ω
R_0 = zero-sequence system resistance, Ω
X_1 = positive-sequence system reactance (subtransient), Ω
X_2 = negative-sequence system reactance, Ω
X_0 = zero-sequence system reactance, Ω

The values of R_0, R_1, R_2, and X_0, X_1, X_2 are determined by looking into the system from the point of fault. In other words, they are determined from the Thévenin equivalent impedance at the fault point for each sequence.* Often, however, the resistance quantities given in the above equation is negligibly small. Hence,

$$I_{a0} = \frac{E \times X_2}{X_1(X_0 + X_2)(X_0 + X_2)} \tag{4.106}$$

4.14.2 SINGLE-LINE-TO-GROUND FAULT

For a single-line-to-ground fault, IEEE Std. 80-2000 gives the following equation to calculate the zero-sequence fault current,

$$I_{a0} = \frac{E}{3R_f + R_0 + R_1 + R_2 + j(X_0 + X_1 + X_2)} \tag{4.107}$$

Often, however, the resistance quantities in the above equation are negligibly small. Hence,

$$I_{a0} = \frac{E}{X_0 + X_1 + X_2} \tag{4.108}$$

4.15 GROUND POTENTIAL RISE

As said before in Section 4.8.2, the GPR is a function of fault-current magnitude, system voltage, and ground-system resistance. The GPR with respect to remote ground is determined by multiplying the current flowing through the ground system by its resistance measured from a point remote from the substation. Here, the current flowing through the grid is usually taken as the maximum available line-to-ground fault current.

The GPR is a function of fault current magnitude, system voltage, and ground (system) resistance. The current through the ground system multiplied by its resistance measured from a point remote from the substation determines the GPR with respect to remote ground. Hence, GPR can be found from
where

$$V_{\text{GPR}} = I_G \times R_g \tag{4.109}$$

V_{GPR} = ground potential rise, V
R_g = ground grid resistance, Ω

For example, if a ground fault current of 20,000 A is flowing into a substation ground grid due to a line-to-ground fault and the ground grid system has a 0.5 Ω resistance to the earth, the resultant IR voltage drop would be 10,000 V. It is clear that such 10,000-V IR voltage drop could cause serious problems to communication lines in and around the substation in the event that the communication equipment and facilities are not properly insulated and/or neutralized.

The ground grid resistance can be found from

$$R_g = \rho \left[\frac{1}{L_T} + \frac{1}{\sqrt{20A}} \left(1 + \frac{1}{1 + h\sqrt{\frac{20}{A}}} \right) \right] \tag{4.110}$$

where
L_T = total buried length of conductors, m
h = depth of the grid, m
A = area of substation ground surface, m^2

To aid the substation grounding design engineer, IEEE Std. 80-2000 includes a design procedure that has a 12-step process, as shown in Figure 4.30, in terms of substation grounding design procedure block diagram, based on a preliminary of a somewhat arbitrary area; that is, the standard suggests the grid be approximately the size of the distribution substation. However, some references state a common practice that is to extend the grid 3 m beyond the perimeter of the substation fence.

Example 4.8: Let the starting grid be a 10-by-10 ground grid. Design a proper substation grounding to provide safety measures for anyone going near or working on a substation. Hence, use the IEEE

12-step process shown in Figure 4.28, then build a grid large enough to dissipate the ground fault current into the earth. A large grounding grid extending far beyond the substation fence and made of a single copper plate would have the most desirable effect for dispersing fault currents to remote earth and thereby ensure the safety of personnel at the surface. Unfortunately, a copper plate of such size is not an economically viable option. The alternative is to design a grid by using a series of horizontal conductors and vertical ground rods. Of course, the application of conductors and rods depends on the resistivity of the substation ground. Change the variables as necessary in order to meet specifications for grounding of the substation. The variables include the size of the grid, the size of the conductors used, the amount of conductors used, and the spacing of each grounding rod. Use 12,906 A as the maximum value fault current, a maximum clearing time of 0.5 s, and a conductor diameter of 211.6 kcmil, based on the given information. The soil resistivity is 35 $\Omega \cdot$ m and the crushed rock resistivity on the surface of the substation is 2000 $\Omega \cdot$ m. Assume that the substation has no transmission line shield wires and but four distribution neutrals. Design a grid by using a series of horizontal conductors and vertical ground rods based on the resistivity of the soil in ohm meters.

Solution:

STEP 1: FIELD DATA

Assume that the uniform average soil resistivity of the substation ground is measured to be 35 $\Omega \cdot$ m.

STEP 2: CONDUCTOR SIZE

The analysis of the grounding grid should be based on the most conservative fault safety conditions. For example, the fault current 3Ia0 is the assumed maximum value with all current dispersed through the grid, that is, there is no alternative path for ground other than through the grid to remote earth. Since the maximum value of the fault current is given as 12,906 A, the conductor size is selected based on the current-carrying capacity, in addition to the amount of time the fault is going to take place. Thus, use a maximum fault current of 12,906 A, a maximum clearing time of 0.5 s, and a conductor diameter of 211.6 kcmil. The crushed rock resistivity is 2000 $\Omega \cdot$ m. The surface derating factor is 0.714. The diameter of the conductor can be found from Table A1.

STEP 3: TOUCH AND STEP VOLTAGE CRITERIA

According to the federal law, all known hazards must be eliminated where the GPR takes place for the safety of workers at a work site. To remove the hazards associated with GPR, a grounding grid is designed to reduce the hazardous potentials at the surface. First, it is necessary to determine what is not hazardous to the body. For two body types, the potential safe step and touch voltages a human could withstand before the fault is cleared need to be determined from Equations 4.62a and 4.62b for touch voltages, and from Equations 4.62c and 4.62d for step voltages as

$$V_{touch\ 50} = 516\ V \text{ and } V_{touch\ 70} = 698\ V$$

and

$$V_{step\ 50} = 1517\ V \text{ and } V_{step\ 70} = 2126\ V$$

STEP 4: INITIAL DESIGN

The initial design consists of factors obtained from the general knowledge of the substation. The preliminary size of the grounding grid system is largely based on the size of the substation to include

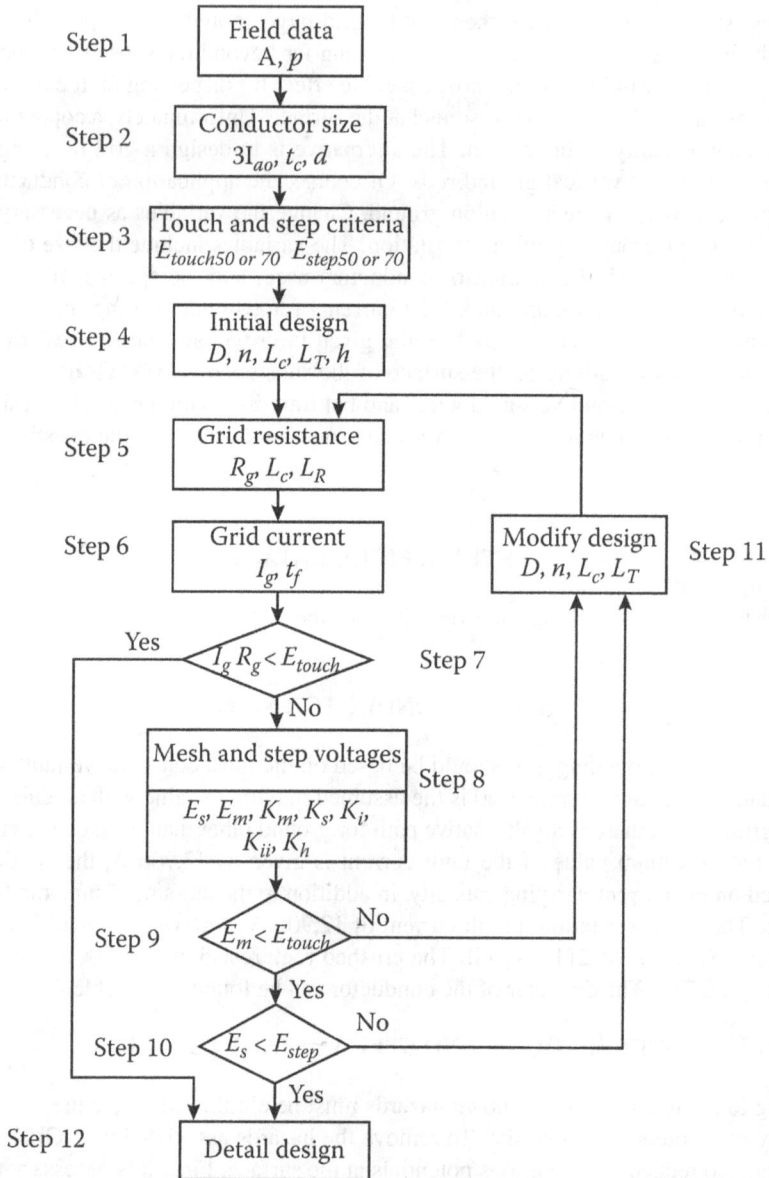

Figure 4.28 Substation grounding design procedure block diagram.

all dimensions within the perimeter of the fence. To establish economic viability, the maximum area is considered and formed in the shape of a square with an area of 100 m². However, the touch voltage has exceeded the limit.

Therefore, an alternative grid size is developed as shown in Figure 4.29, this time in the shape of a rectangle with ground rods. The horizontal distance is called L_X and is measured 24 m, while the vertical distance L_Y is measured 12 m. The area of the grid is 288 m². The horizontal conductors were conservatively spaced at the minimum distance of 3 m apart. The total length of the horizontal conductors L_T is 258 m. A total of 12 grounding rods at a length of 2.5 m each accounted for a total ground rod length L_C of 228 m. The grid burial depth is 0.5 m. From Equation 4.95, the effective number of parallel conductors (n) is found as 6.5.

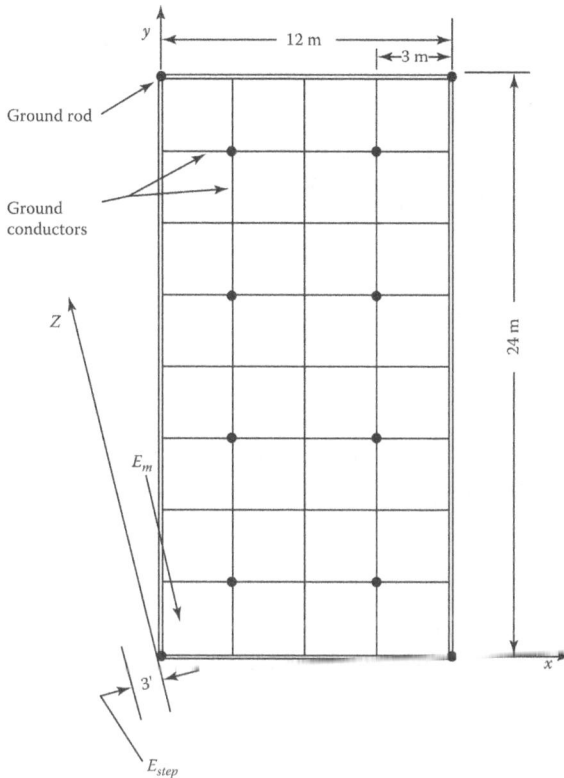

Figure 4.29 Preliminary design.

STEP 5: GRID RESISTANCE

A good grounding system provides a low resistance to remote earth in order to minimize the GPR. The next step is to evaluate the grid resistance by using Equation 4.103. All design parameters can be found in Tables 4.12A and 4.13B, and in the preliminary design outline is shown in Figure 4.29. Table 4.14 gives approximate equivalent impedance of transmission line overhead shield wires and distribution feeder neutrals, according to their numbers. From Equation 4.110 for $L_T = 258$ m, a grid area of $A = 288$ m^2, $\rho = 35$ Ω-m, and $h = 0.5$ m, the grid resistance is

$$R_g = \rho \left[\frac{1}{L_T} + \frac{1}{\sqrt{20A}} \left(1 + \frac{1}{1 + h\sqrt{\frac{20}{A}}} \right) \right]$$

$$= 35 \left[\frac{1}{258} + \frac{1}{\sqrt{20 \times 288}} \left(1 + \frac{1}{1 + 0.5\sqrt{\frac{20}{288}}} \right) \right]$$

$$= 1.0043 \ \Omega$$

STEP 6: GRID CURRENT

From Equation 4.109, the GPR is determined as

$$V_{\text{GPR}} = I_G \times R_g$$

Table 4.12

Initial Design Parameters

ρ	A	L_T	L_C	L_R	L_T	h	L_X	L_Y	D
35 Ω	288 m	258 Ω	228 Ω	30 m	2.5 Ω	0.5 m	24 m	12 m	3 m

Table 4.13

Initial Design Parameters

t_c	h_s	D	$3I_{a0}$	ρ_s	D_f	L_p	n_c	nd	t_f
0.05 s	0.11 m	0.01 m	12,906 A	2000 Ω	1.03	75 m	1	1	0.5 s

This is important because to determine the GPR and compare it to the tolerable touch voltage is the first step to find out whether the grid design is a safe design for the people in and around the substation. The next step is to find the grid current I_G. However, the split factor should first be determined from the following equation

$$S_f = \left| \frac{Z_{eq}}{Z_{eq} + R_g} \right|$$

Since the substation has no impedance line shield wires and four distribution neutrals, from Table 4.14, the equivalent impedance can be found as Zeq = 0.322 + j0.242 Ω. Thus, R_g = 1.0043 Ω and a total fault current of $3I_{a0}$ = 12,906 A, a decrement factor of D_f = 1.026. Thus, the current division factor (or the split factor) can be found as

$$S_f = \left| \frac{Z_{eq}}{Z_{eq} + R_g} \right|$$

$$= \left| \frac{(0.322 + j0.242)}{(0.322 + j0.242) + 1.0043} \right|$$

$$\cong 0.2548$$

Table 4.14

Approximate Equivalent Impedance of Transmission Line Overhead Shield Wires and Distribution Feeder Neutrals

No. of Transmission Lines	No. of Distribution Neutrals	$R_{tg} = 15$ and $R_{d\varphi} = 25$ $R + jX\Omega$	$R_{tg} = 15$ and $R_{d\varphi} = 25$ $R + jX\Omega$
1	1	0.91 + j 0.485 Ω	3.27 + j 0.652 Ω
1	2	0.54 + j 0.33 Ω	2.18 + j 0.412 Ω
1	4	0.295 + j 0.20 Ω	1.32 + j 0.244 Ω
4	4	0.23 + j 0.12 Ω	0.817 + j 0.16 Ω
0	4	0.322 + j 0.242 Ω	1.65 + j 0.291 Ω

since

$$I_g = S_f \times 3I_{a0}$$
$$= 0.2548 \times 12{,}906$$
$$= 3288 \text{ A}$$

thus,

$$I_G = D_f \times I_g$$
$$= 1.026 \times 3288$$
$$= 3375 \text{ A}$$

STEP 7: DETERMINATION OF GPR

As said before, the product of I_G and R_g is the GPR. It is necessary to compare the GPR to the tolerable touch voltage, $V_{\text{touch } 70}$. If the GPR is larger than the $V_{\text{touch } 70}$, further design evaluations are necessary and the tolerable touch and step voltages should be compared with the maximum mesh and step voltages. Hence, first determine the GPR as

$$GPR = I_f \times R_g$$
$$= 3375 \times 1.0043$$
$$= 3390 \text{ V}$$

Check to see whether

$$GPR > V_{\text{touch } 70}$$

Indeed,

$$3390 \text{ V} > 698 \text{ V}$$

As it can be observed from the results, the GPR is much larger than the step voltage. Therefore, further design considerations are necessary and thus the step and mesh voltages must be calculated and compared with the tolerable touch and step voltage as follows.

STEP 8: MESH AND STEP VOLTAGE CALCULATIONS

(a) **Determination of the Mesh Voltage**
    To calculate the mesh equation by using Equation 4.90, it is necessary first to calculate the variables K_m and K_i. Here, K_m can be determined from Equation 4.91. However, again letting $D = 3$ m, $h = 0.5$ m, $d = 0.01$ m, find the following equations as

$$n = n_a \times n_b \times n_c \times n_d$$

where

$$n = \frac{2 \times L_C}{L_P}$$
$$= \frac{2 \times 228}{72}$$
$$= 6.33$$

and

$$n_b = \sqrt{\frac{L_P}{4 \times \sqrt{A}}}$$

$$= \sqrt{\frac{72}{4\sqrt{288}}}$$

$$= 1.03$$

$n_c = 1$, for rectangular grid
$n_d = 1$, for rectangular grid
Thus,

$$n = 6.33 \times 1.03 \times 1 \times 1 = 6.52$$

Since
$K_{ii} = 1$, for rectangular and square grids

$$K_h = \sqrt{1 + \frac{h}{h_0}}$$

$$= \sqrt{1 + \frac{0.5}{1}}$$

$$= 1.22$$

So that from Equation 4.91,

$$K_M = \frac{1}{2\pi}\left[\ln\left\{\frac{D^2}{16h \times d} + \frac{(D+2h)2}{8D \times d} - \frac{h}{4d}\right\} + \frac{K_{ii}}{K_h} \times \ln\left\{\frac{8}{\pi(2n-1)}\right\}\right]$$

$$= \frac{1}{2\pi}\left[\ln\left\{\frac{3^2}{16 \times 0.5 \times 0.01} + \frac{(3+2 \times 0.5)2}{8 \times 3 \times 0.01} - \frac{0.5}{4 \times 0.01}\right\} + \frac{1}{1.22} \times \ln\left\{\frac{8}{\pi(26.52-1)}\right\}\right]$$

$$= 0.61$$

Also, since

$$K_i = 0.644 + 0.148 \times n$$

$$= 0.644 + 0.148 \times 6.52$$

$$= 1.61$$

Thus, the mesh voltage is determined from Equation 4.90 as

$$E_{mesh} = \frac{\rho \times K_m \times K_i \times I_G}{L_C + \left[1.55 + 1.22\left(\frac{L_R}{\sqrt{L_X^2 + L_Y^2}}\right)\right]}$$

$$= \frac{35 \times 0.61 \times 1.61 \times 3375}{288 + \left[1.55 + 1.22\left(\frac{12}{\sqrt{24^2 + 12^2}}\right)\right] 12}$$

$$= 470 \text{ V}$$

(b) **Determination of the Step Voltage**

For the ground to be safe, the step voltage has to be less than the tolerable step voltage. Also, step voltages within a grid system designed for safe mesh voltages will be well within the tolerable limits. This is because both feet and legs are in series rather than in parallel, and the current takes the path from one leg to the other rather than through vital organs. The step voltages are calculated from Equation 4.102

$$E_{\text{step}} = \frac{\rho \times K_s \times K_i \times I_G}{L_S}$$

where
K_s = step coefficient
L_S = buried conductor length, m

For grids with or without ground rods, L_S is determined from Equation 4.103 as

$$
\begin{aligned}
L_S &= 0.75L_C + 0.85L_R \\
&= 0.75 \times 228 + 0.85 \times 30 \\
&= 196.5 \text{ m}
\end{aligned}
$$

The step coefficient is found from Equation 4.104

$$
\begin{aligned}
K_s &= \frac{1}{2\pi}\left[\frac{1}{2h} + \frac{2}{D+h} + \frac{1}{D}(1 - 0.5^{n-2})\right] \\
&= \frac{1}{2\pi}\left[\frac{1}{2 \times 0.5} + \frac{2}{3+0.5} + \frac{1}{3}(1 - 0.5^{6.52-2})\right] \\
&= 0.511 \text{ V}
\end{aligned}
$$

where h = depth of ground grid conductors in meters, usually between 0.25 m ¡ h ¡ 2.5 m. All other variables are as defined before.
Thus, the step voltage can be found as

$$
\begin{aligned}
K_{\text{step}} &= \frac{\rho \times K_s \times K_i \times I_G}{L_s} \\
&= \frac{35 \times 0.511 \times 1.61 \times 3375}{181.2} \\
&= 536.11 \text{ V}
\end{aligned}
$$

STEP 9: COMPARISON OF E_{MESH} VS. V_{TOUCH}

Here, the mesh voltage that is calculated in Step 8 is compared with the tolerable touch voltages calculated in Step 4. If the calculated mesh voltage E_{mesh} is greater than the tolerable $V_{\text{touch 70}}$, further design evaluations are necessary. If the mesh voltage E_{mesh} is smaller than the $V_{\text{touch 70}}$, then it can be moved to the next step and compare E_{step} with $V_{\text{step 70}}$. Accordingly,

$$470 \text{ V} < 700 \text{ V}$$

$$E_{\text{mesh}} < V_{\text{touch 70}}$$

Here, the present grid design passes the second critical criteria. Hence, it can be moved to Step 10 to find out whether the final criterion is met.

STEP 10: COMPARISON OF F_{STEP} VS. $V_{\text{STEP 70}}$

At this step, E_{step} is compared with the calculated tolerable step voltage $V_{step\ 70}$. If

$$E_{step} > V_{step\ 70}$$

A refinement is of the preliminary design is necessary and can be accomplished by decreasing the total grid resistance, closer grid spacing, adding more ground grid rods, if possible, and/or limiting the total fault current. On the other hand, if

$$E_{step} < V_{step\ 70}$$

then the designed grounding grid is considerably safe. Since

$$536.11\ V < 2126\ V$$

then for the design

$$E_{step} < V_{step\ 70}$$

In summary, according to the calculations, the calculated mesh and step voltages are smaller than the tolerable touch and step voltages; therefore, in a typical shock situation, humans that become part of the circuit during a fault will have only what is considered a safe amount of current through their bodies.

There are many variables that can be changed in order to meet specifications for grounding a substation. Some variables include the size of the grid, the size of the conductors used, the amount of conductors used, and the spacing of each ground rod. After many processes an engineer has to go through, the project would then be put into construction if it is approved. Designing safe substation grounding is obviously not an easy task; however, there are certain procedures that an engineer can follow to make the designing substation grounding easier.

4.16   TRANSMISSION LINE GROUNDS

High-voltage transmission lines are designed and built to withstand the effects of lightning with minimum damage and interruption of operation. If the lightning strikes an overhead ground wire (also called *static wire*) on a transmission line, the lightning current is conducted to ground through the ground wire installed along the pole or through the metal tower. The top of the line structure is raised in potential to a value determined by the magnitude of the lightning current and the surge impedance of the ground connection.

In the event that the impulse resistance of the ground connection is large, this potential can be in the magnitude of thousands of volts. If the potential is greater than the insulation level of the apparatus, a flashover will take place, causing an arc. The arc, in turn, will start the operation of protective relays, causing the line to be taken out of service. In the event that the transmission structure is well grounded and there is a sufficient coordination between the conductor insulation and the ground resistance, a flashover can generally be avoided.

The transmission line grounds can be designed in various ways to achieve a low ground resistance. For example, a pole butt grounding plate or butt coil can be used on wood poles. A butt coil is a spiral coil of bare copper wire installed at the bottom of a pole. The wire of the coil is extended up the pole as the ground wire lead. In practice, usually one or more ground rods are used instead to achieve the required low ground resistance.

The sizes of the rods used are usually 5/8 or 3/4 in. in diameter and 10 ft in length. The thickness of the rod does not play a major role in reducing the ground resistance as does the length of the rod. Multiple rods are usually used to provide the low ground resistance require by the high-capacity structures. However, if the rods are moderately close to each other, the overall resistance will be

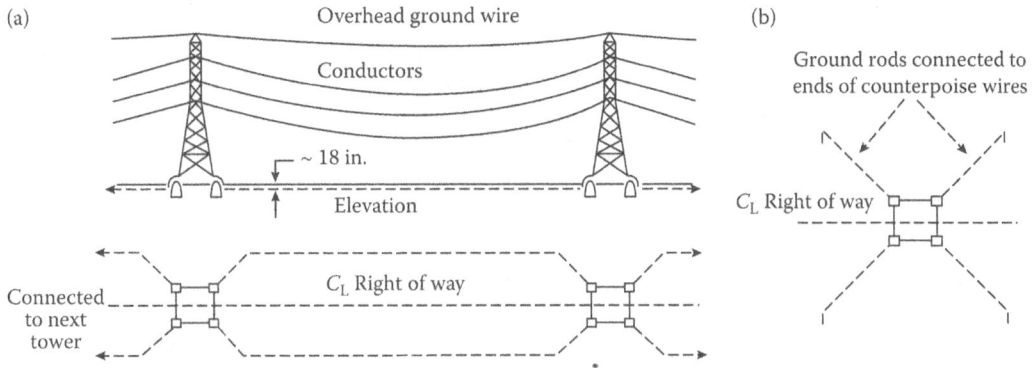

Figure 4.30 Two basic types of counterpoises: (a) continuous (parallel); (b) radial. (From Gönen, T., *Electric Power Transmission System Engineering*, 2nd ed., CRC Press, Boca Raton, FL, 2009.)

more than if the same number of rods were spaced far apart. In other words, adding a second rod does not provide a total resistance of half that of a single rod unless the two are several rod lengths apart (actually infinite distance). Lewis [171] has shown that at 2 ft apart, the resistance of two pipes (used as ground rods) in parallel is about 61% of the resistance of one of them, and at 6 ft apart it is about 55% of the resistance of one pipe.

Where there is bedrock near the surface or where sand is encountered, the soil is usually very dry and therefore has high resistivity. Such situations may require a grounding system known as the counterpoise, made of buried metal (usually galvanized steel wire) strips, wires, or cables. The counterpoise for an overhead transmission line consists of a special grounding terminal that reduces the surge impedance of the ground connection and increases the coupling between the ground wire and the conductors.

The basic types of counterpoises used for transmission lines located in areas with sandy soil or rock close to the surface are the continuous type (also called the *parallel type*) and the radial type (also called the *crowfoot type*), as shown in Figure 4.30. The continuous counterpoise is made of one or more conductors buried under the transmission line for its entire length.

The counterpoise wires are connected to the overhead ground (or *static*) wire at all towers or poles. However, the radial-type counterpoise is made of a number of wires, and extends radially (in some fashion) from the tower legs. The number and length of the wires are determined by the tower location and the soil conditions. The counterpoise wires are usually installed with a cable plow at a length of 18 in. or more so that they will not be disturbed by cultivation of the land.

A multigrounded, common neutral conductor used for a primary distribution line is always connected to the substation grounding system where the circuit originates and to all grounds along the length of the circuit. If separate primary and secondary neutral conductors are used, the conductors have to be connected together provided that the primary neutral conductor is effectively grounded. The resistance of a single buried horizontal wire, when it is used as radial counterpoise, can be expressed [151] as

$$R = \frac{\rho}{\pi \cdot l}\left(\ln\frac{2l}{2\sqrt{a \cdot d}} - 1\right) \quad \text{when } d \ll l \tag{4.111}$$

where
ρ = ground resistivity in ohm meters
l = length of wire in meters
a = radius of wire in meters
d = burial depth in meters

It is assumed that the potential is uniform over the entire length of the wire. This is only true when the wire has ideal conductivity. If the wire is very long, such as with the radial counterpoise, the potential is not uniform over the entire length of the wire. Hence, Equation 4.83 cannot be used. Instead, the resistance of such a continuous counterpoise when $l(r/\rho)^{1/2}$ is large can be expressed as

$$R = \sqrt{(r\rho)}\coth\sqrt{\frac{r}{\rho}} \qquad (4.112)$$

where r = resistance of wire in ohm meters. If the lightning current flows through a counterpoise, the effective resistance is equal to the surge impedance of the wire. The wire resistance decreases as the surge propagates along the wire. For a given length counterpoise, the transient resistance will diminish to the steady-state resistance if the same wire is used in several shorter radial counterpoises rather than as a continuous counterpoise. Thus, the first 250 ft of counterpoise is most effective when it comes to grounding of lightning currents.

4.17   TYPES OF GROUNDING

In general, transmission and subtransmission systems are solidly grounded. Transmission systems are usually connected grounded wye, but subtransmission systems are often connected in delta. Delta systems may also be grounded through grounding transformers. In most high-voltage systems, the neutrals are solidly grounded, that is, connected directly to the ground. The advantages of such grounding are

1. Voltages to ground are limited to the phase voltage
2. Intermittent ground faults and high voltages due to arcing faults are eliminated
3. Sensitive protective relays operated by ground fault currents clear these faults at an early stage

The grounding transformers used are normally either small distribution transformers (that are connected normally in wye–delta, having their secondaries in delta), or small grounding autotransformers with interconnected wye or "zig-zag" windings, as shown in Figure 4.31. The three-phase autotransformer has a single winding. If there is a ground fault on any line, the ground current flows equally in the three legs of the autotransformer. The interconnection offers the minimum impedance to the flow of the single-phase fault current.

The transformers are only used for grounding and carry little current except during a ground fault. Because of that, they can be fairly small. Their ratings are based on the stipulation that they carry current for no more than 5 min since the relays normally operate long before that. The grounding transformers are connected to the substation ground.

All substation equipment and structures are connected to the ground grid with large conductors to minimize the grounding resistance and limit the potential between equipment and the ground surface to a safe value under all conditions. As shown in Figure 4.31, all substation fences are built inside the ground grid and attached to the grid at short intervals to protect the public and personnel. Furthermore, the surface of the substation is usually covered with crushed rock or concrete to reduce the potential gradient when large currents are discharged to ground and to increase the contact resistance to the feet of the personnel in the substation.

As said before, the substation grounding system is connected to every individual equipment, structure, and installation in order to provide the means by which grounding currents are conducted to remote areas. Thus, it is extremely important that the substation ground has a low ground resistance, adequate current-carrying capacity, and safety features for the personnel.

It is crucial to have the substation ground resistance very low so that the total rise of the grounding system potential will not reach values that are unsafe for human contact. Therefore, the substation

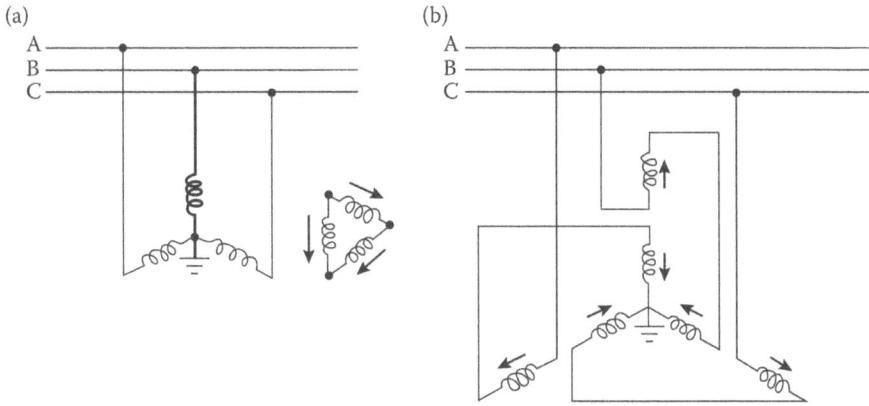

Figure 4.31 Grounding transformers used in delta-connected systems: (a) using wye–delta-connected small distribution transformers, or (b) using grounding autotransformers with interconnected wye or "zig-zag" windings. (From Gönen, T., *Electric Power Transmission System Engineering*, 2nd ed., CRC Press, Boca Raton, FL, 2009.)

grounding system normally is made up of buried horizontal conductors and driven ground rods interconnected (by clamping, welding, or brazing) to form a continuous grid (also called mat) network, as shown in Figure 4.32.

Notice that, a continuous cable (usually it is 4/0 bare stranded copper cable buried 12–18 in. below the surface) surrounds the grid perimeter to enclose as much ground as possible and to prevent current concentration and thus high gradients at the ground cable terminals. Inside the grid, cables are buried in parallel lines and with uniform spacing (e.g., about 10 *times* 20 ft).

All substation equipment and structures are connected to the ground grid with large conductors to minimize the grounding resistance and limit the potential between equipment and the ground surface to a safe value under all conditions. As shown in Figure 4.31, all substation fences are built inside the ground grid and attached to the grid at short intervals to protect the public and personnel.

Furthermore, the surface of the substation is usually covered with crushed rock or concrete to reduce the potential gradient when large currents are discharged to the ground and to increase the contact resistance to the feet of the personnel in the substation.

Today, many utilities have computer programs for performing grounding grid studies. Thus, the number of tedious calculations that must be performed to develop an accurate and sophisticated model of a system is no longer a problem. For example, Figure 4.33a shows a typical computerized grounding grid design with all relevant soil and system data. Figure 4.33b shows the meshes with hazardous potentials as determined by the computer. Figure 4.33c shows the results of the first refinement in the grid design indicating the hazardous touch potentials. Finally, Figure 4.33d shows the final refinement with no hazardous touch potentials.

PROBLEMS

1. Verify Equation 4.5 by using the
    (a) Classical calculus approach
    (b) Laplace transformation approach
2. Use Equation 4.6 and verify that if a fault occurs at $t = 0$ when the angle $\alpha - \theta = 90°$, then the value of the maximum transient current becomes twice the steady-state maximum value.
3. Repeat Example 4.1 assuming that the fault is located on bus 2.
4. Repeat Example 4.1 assuming that the fault is located on bus 3.
5. Use the results of Problem 3 and repeat Example 4.2.

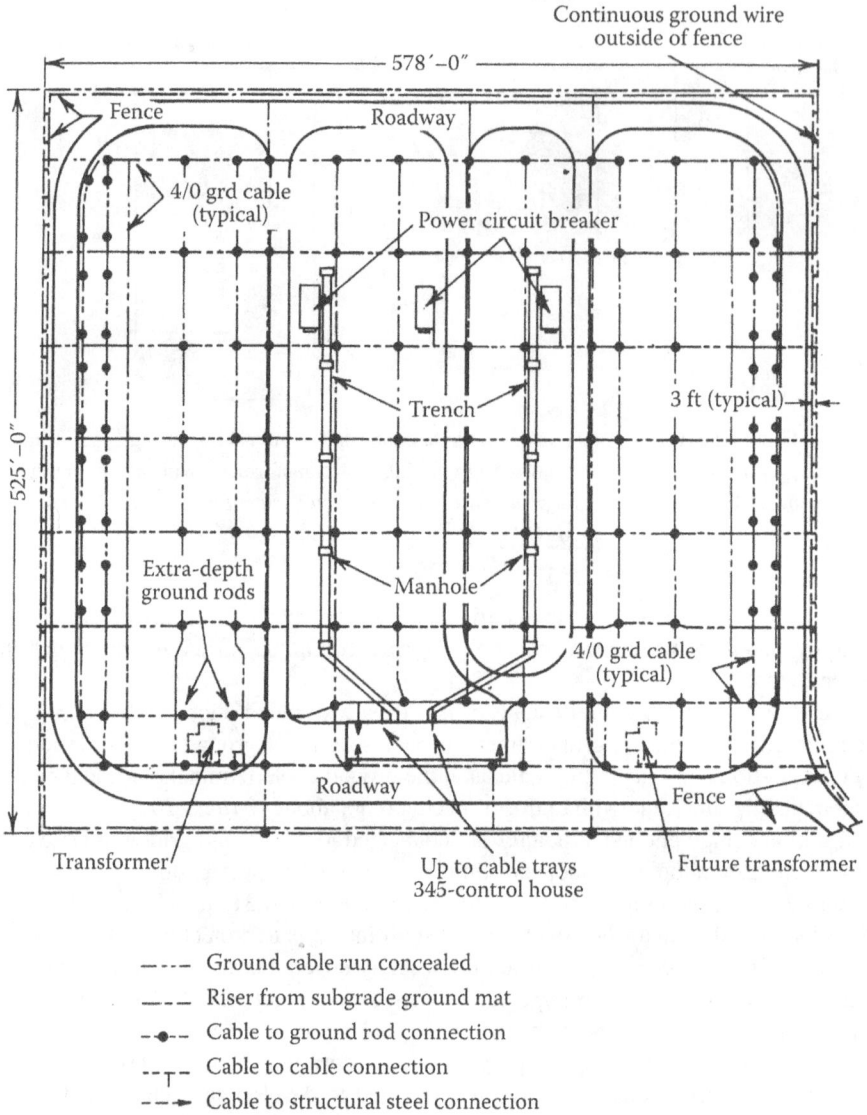

Figure 4.32 Typical grounding (grid) system design for 345-kV substation. (From Fink, D. G., and Beaty, H. W., eds., *Standard Handbook for Electrical Engineers*, 11th ed. Used with permission. 1978 McGraw-Hill.) [5]

6. Use the results of Problem 4 and repeat Example 4.2.
7. Consider the system shown in Figure 4.34 and the following data:
   Generator G_1: 15 kV, 50 MVA, $X_1 = X_2 = 0.10$ pu and $X_0 = 0.05$ pu based on its own ratings
   Synchronous motor: 15 kV, 20 MVA, $X_1 = X_2 = 0.20$ pu and $X_0 = 0.07$ pu based on its own ratings
   Transformer T_1: 15/115 kV, 30 MVA, $X_1 = X_2 = X_0 = 0.06$ pu based on its own ratings
   Transformer T_2: 115/15 kV, 25 MVA, $X_1 = X_2 = X_0 = 0.07$ pu based on its own ratings
   Transmission line TL23: $X_1 = X_2 = 0.03$ pu and $X_0 = 0.10$ pu based on its own ratings
   Assume a three-phase fault at bus 1 and determine the fault current. Use 50 MVA as the megavolt-ampere base.
8. Repeat Problem 7 assuming that bus 2 is faulted.
9. Repeat Problem 7 assuming that bus 3 is faulted.

$\rho_{soil} = 1316\ \Omega-m$
$\rho_{surface} = 3000\ \Omega-m$
$I_{fault} = 1560\ A$
Clearing time = 0.5
Depth of burial = 0.305 m
$E_{touch/tolerable} = 885\ V$
$E_{step/tolerable} = 3134\ V$
$K_m = 0.568$
$K_s = 0.814$
$K_i = 2.0\ (touch),\ 2.5\ (step)$
$E_{touch/worse\ case} = 1121\ V$
$E_{step/worse\ case} = 2010\ V$

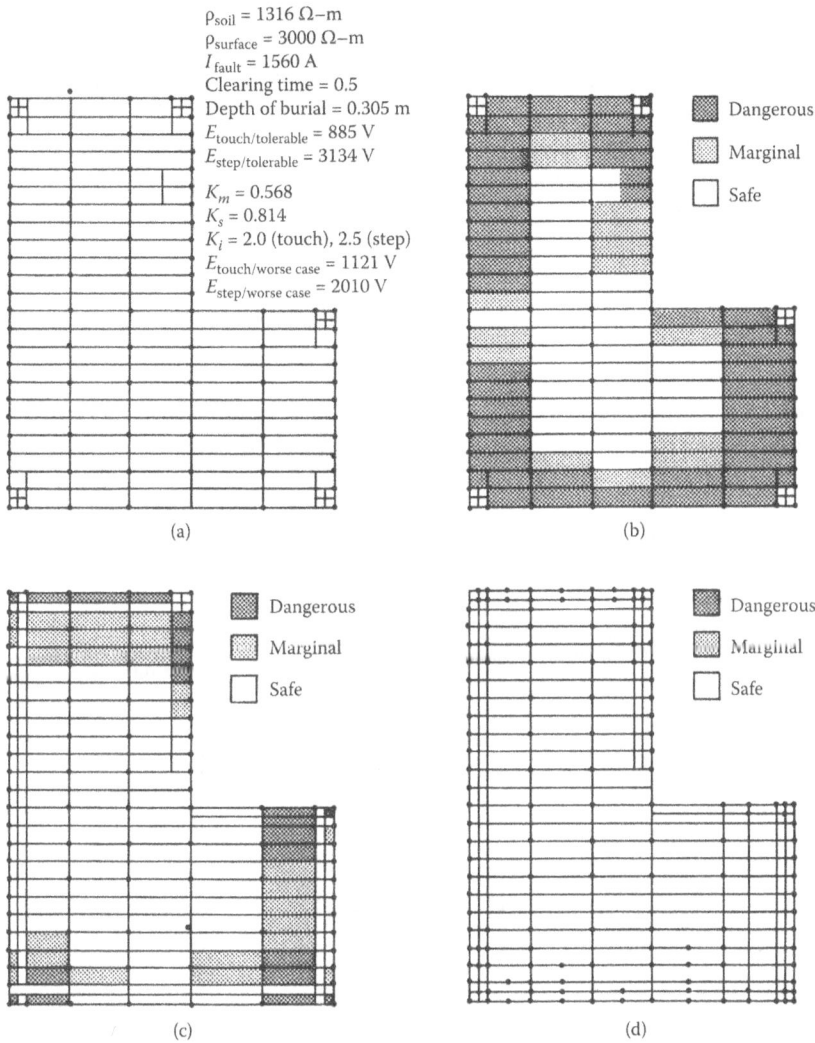

Figure 4.33 Computerized grounding grid design: (a) typical grounding grid design with its data; (b) meshes with hazardous potentials as identified by computer; (c) first refinement of design; (d) final refinement of design with no hazardous touch potentials. (From Institute of Electrical and Electronics Engineers, *IEEE Recommended Practice for Industrial and Commercial Power System Analysis*, IEEE Stand. 399-1980 1980 IEEE.)

10. Repeat Problem 7 assuming that bus 4 is faulted.
11. A generator is connected through an eight-cycle circuit breaker to an unloaded transformer. Its subtransient, transient, and steady-state reactances are given as 8%, 16%, and 100%, respectively. It is operating at no load and rated voltage 1.0 ∠ 0° pu when a three-phase short circuit occurs between the breaker and the transformer. Determine the following:

    (a) Sustained (i.e., steady-state) short-circuit current in breaker in per units
    (b) Initial symmetrical rms current in breaker in per units
    (c) Maximum possible dc component of short circuit in breaker in per units

12. Consider Example 4.5 and assume that the tie–bus arrangement is replaced by an equivalent ring arrangement (as shown in Figure 4.7e), so that the resulting short-circuit megavolt-ampere base is equal to the one found in Example 4.5. Determine the following:

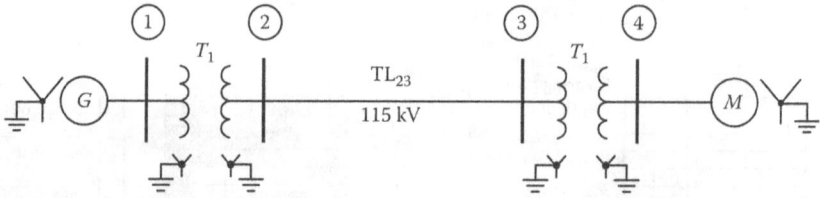

Figure 4.34 Two-substation system for Problem 7.

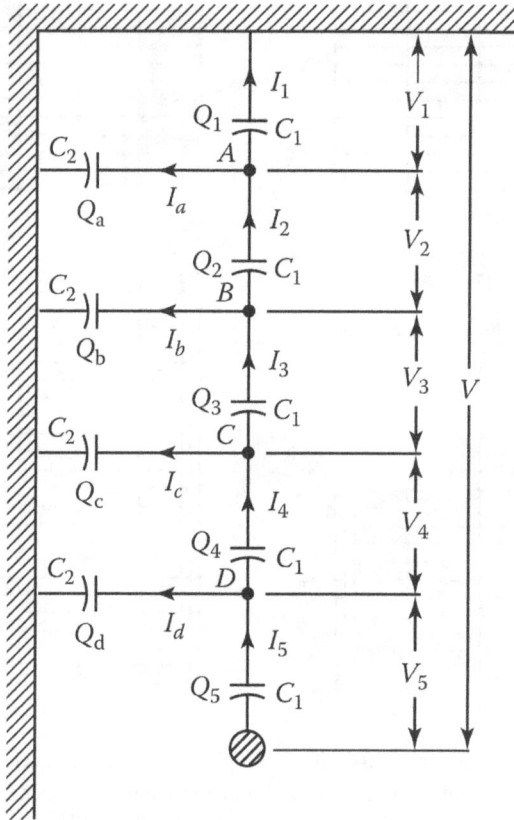

Figure 4.35 Five-insulator string for Problem 13.

    (a) Necessary per-unit reactance of each reactor

    (b) Fault current distribution

13. Assume that a string of five insulators is used to suspend one line conductor, as shown in Figure 4.35. Consider only the series capacitance C_1 and the shunt capacitance C_2 and define the ratio C_1/C_2 as k. Using the charging currents and Kirchhoff's current and voltage laws, derive an expression to determine the voltage distribution across various units of the suspension insulator in terms of the ratio k. [Do not use Equations 4.48 or 4.50]

14. Use the results of Problem 13 and assume that the string of five insulators is used to suspend one conductor of a 6–9 kV, three-phase, overhead line. If the ratio of k is 0.1, determine voltages V_1, V_2, V_3, V_4, and V_5.

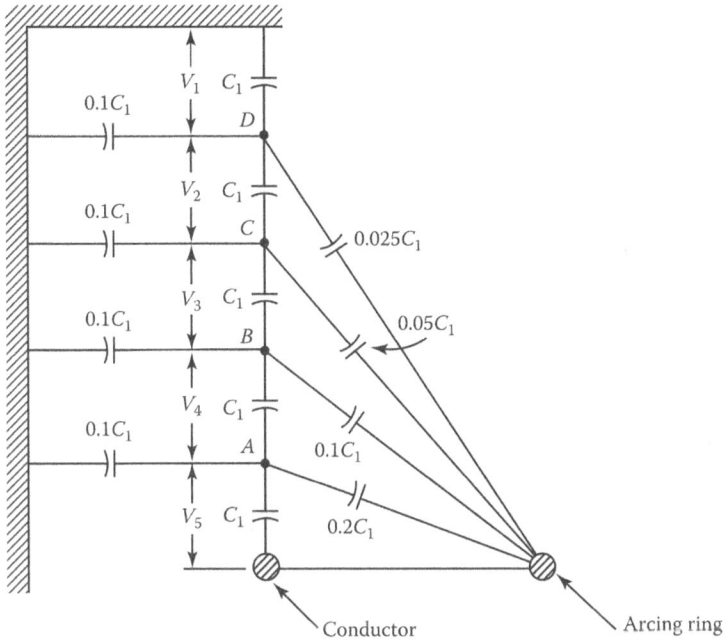

Figure 4.36 Five-insulator string with an arcing ring for Problem 16.

15. Repeat Problem 13 using the Ω charges on the various capacitors. Assume that the potential differences across the capacitors at any instant are as shown in Figure 4.35.
16. Consider Figure 4.36 and assume that an arcing ring has been installed at the line end of the insulator string.
    (a) Repeat Problem 14
    (b) Determine the sting efficiency
17. Consider Figure 4.36, and in order to achieve uniform voltage distribution over the five insulators, determine the following required values:
    (a) Air capacitance C_a
    (b) Air capacitance C_b
    (c) Air capacitance C_c
    (d) Air capacitance C_d
18. Assume that a post-type insulator consists of three-pin insulators fixed one above another and used to support a bus of one 115 kV three-phase system, as shown in Figure 4.38. Use the results of Problem 13. If the voltage across the top pin insulator is twice that of the voltage across the bottom insulator, determine the voltage across the middle insulator.
19. Assume that a string of eight insulators is used to suspend a line conductor of a 138-kV three-phase system, as shown in Figure 4.3. Assume that $C_2 = 0.1 \times C_1$ and $C_3 = 0.02 \times C_1$ and use Equation 4.48. Determine the following:
    (a) Voltage across three units from the ground end
    (b) Voltage across five units from the ground end
20. Repeat Problem 19 using Equation 4.50.
21. Consider Problem 19 and assume that $C_3 = 0$. Use the results of Problem 13 and repeat Problem 19.
22. Repeat Example 4.6 assuming wet organic soil.

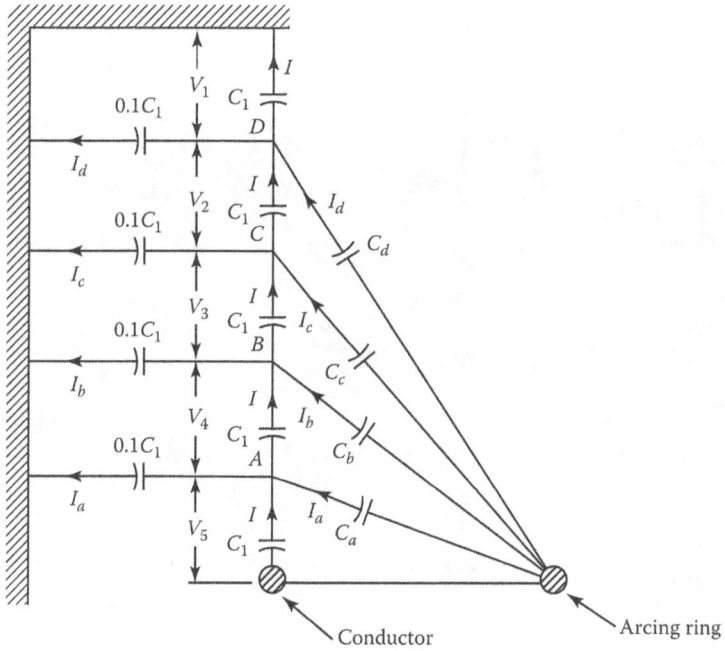

Figure 4.37 Five-insulator string for Problem 17.

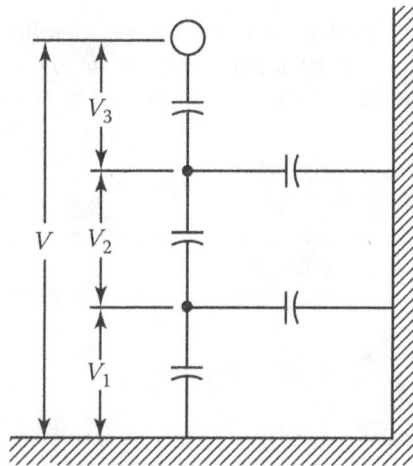

Figure 4.38 Post-type insulator for Problem 18.

23. Repeat Example 4.6 assuming dry soil.
24. Consider Example 4.6 and Equations 4.62 and 4.63. Assume that $K_s = 1.4$, $K_m = 7.0$, and $K_i = 2$ and 2.5 for the touch and step potentials, respectively. Determine the required minimum total length of corresponding grid conductors.

5 Symmetrical Components and Sequence Impedances

5.1 INTRODUCTION

In many ways of how engineers approach a problem, it often begins with simpler establishment of modeling with assumptions that would help to characterize a power system analytically and enable validation of the experimental models. This chapter instills that cultural value by initiating a balanced three-phase power system. This theoretical ground would allow further elaboration of unbalanced fault in next chapter as the system complexity increases. In reality, many systems are very nearly balanced and for practical purposes can be analyzed as if they were truly balanced systems. However, there are also emergency conditions (e.g., unsymmetrical faults, unbalanced loads, open conductors, or unsymmetrical conditions arising in rotating machines) where the degree of unbalance cannot be neglected. To protect the system against such contingencies, it is necessary to size protective devices, such as fuses and circuit breakers, and set the protective relays. Therefore, to achieve this, currents and voltages in the system under such unbalanced operating conditions have to be evaluated in advance.

In 1918, Fortescue [172] proposed a method for resolving an unbalanced set of *n* related phasors into *n* sets of balanced phasors called the *symmetrical components* of the original unbalanced set. The phasors of each set are of equal magnitude and spaced 120° or 0° apart. The method is applicable to systems with any number of phases; however, in this book, only balanced and mesh three-phase power system networks will be discussed.

Today, the symmetrical component theory is widely used in studying unbalanced systems. Furthermore, many electrical devices have been developed and are operating based on the concept of symmetrical components. The examples include (1) the negative-sequence relay to detect system faults, (2) the positive-sequence filter to make generator voltage regulators respond to voltage changes in all three phases rather than in one phase alone, and (3) the Westinghouse-type HCB pilot wire relay using positive- and zero-sequence filters to detect faults.

5.2 SYMMETRICAL COMPONENTS

Any unbalanced three-phase system of phasors can be resolved into three balanced systems of phasors: (1) positive-sequence system, (2) negative-sequence system, and (3) zero-sequence system, as illustrated in Figure 5.1.

The *positive-sequence system* is represented by a balanced system of phasors having the same phase sequence (and therefore positive phase rotation) as the original unbalanced system. The phasors of the positive-sequence system are equal in magnitude and displaced from each other by 120°, as shown in Figure 5.1b.

The *negative-sequence system* is represented by a balanced system of phasors having the opposite phase sequence (and therefore negative phase rotation) to the original system. The phasors of the negative-sequence system are also equal in magnitude and displaced from each other by 120°, as shown in Figure 5.1c.

The zero-sequence system is represented by three single phasors that are equal in magnitude and angular displacements, as shown in Figure 5.1d. Note that, in the hook, the subscripts 0, 1, and 2 denote the zero sequence, positive sequence, and negative sequence, respectively. Therefore, three voltage phasors V_a, V_b, and V_c of an unbalanced set, as shown in Figure 5.1a can be expressed in terms of their symmetrical components as

DOI: 10.1201/9781003129769-5

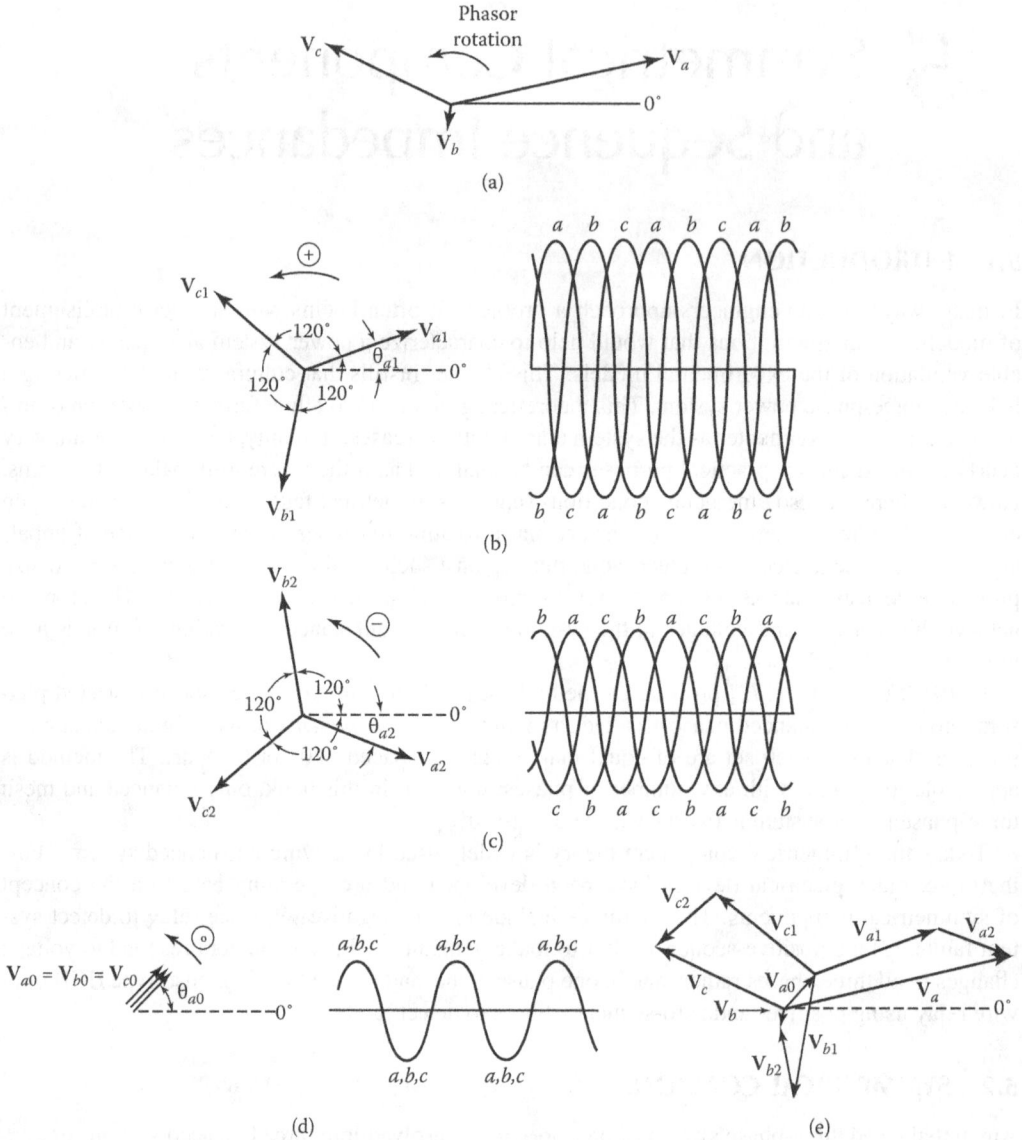

Figure 5.1 Analysis and synthesis of set of three unbalanced voltage phasors: (a) original system of un-balanced phasors; (b) positive-sequence components; (c) negative-sequence components; (d) zero-sequence components; (e) graphical addition of phasors to get original unbalanced phasors.

$$\mathbf{V}_a = \mathbf{V}_{a1} + \mathbf{V}_{a2} + \mathbf{V}_{a0} \tag{5.1}$$

$$\mathbf{V}_b = \mathbf{V}_{b1} + \mathbf{V}_{b2} + \mathbf{V}_{b0} \tag{5.2}$$

$$\mathbf{V}_c = \mathbf{V}_{c1} + \mathbf{V}_{c2} + \mathbf{V}_{c0} \tag{5.3}$$

Figure 5.1e shows the graphical additions of the symmetrical components of Figures 5.1b–5.1d to obtain the original three unbalanced phasors shown in Figure 5.1a.

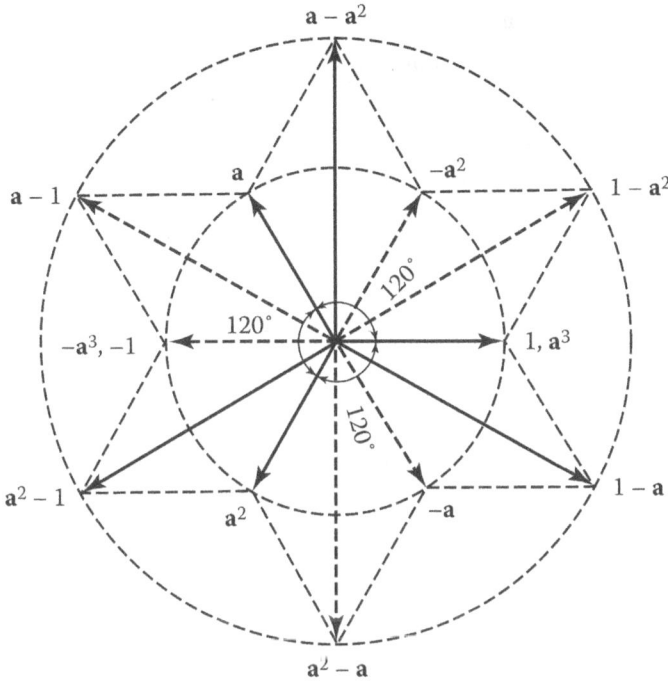

Figure 5.2 Phasor diagram of various powers and functions of operator a.

5.3 OPERATOR "A"

Because of the application of the symmetrical components theory to three-phase systems, there is a need for a *unit phasor* (or *operator*) that will rotate another phasor by 120° in the counterclockwise direction (i.e., it will add 120° to the phase angle of the phasor) but leave its magnitude unchanged when it is multiplied by the phasor (see Figure 5.2). Such an operator is a complex number of unit magnitude with an angle of 120° and is defined by

$$\mathbf{a} = 1\angle 120°$$
$$= 1e^{j\left(\frac{2\pi}{3}\right)}$$
$$= 1(\cos 120° + j\sin 120°)$$
$$= -0.5 + j0.866$$

where

$$j = \sqrt{-1}$$

It is clear that if operator **a** is designated as

$$\mathbf{a} = 1\angle 120°$$

than

$$\mathbf{a}^2 = \mathbf{a} \times \mathbf{a}$$
$$= (1\angle 120°)(1\angle 120°) = 1\angle 240° = 1\angle -120°$$

$$\mathbf{a}^3 = \mathbf{a}^2 \times \mathbf{a}$$
$$= (1\angle 240°)(1\angle 120°) = 1\angle 360° = 1\angle 0°$$

$$\mathbf{a}^4 = \mathbf{a}^3 \times \mathbf{a}$$
$$= (1\angle 0°)(1\angle 120°) = 1\angle 120° = \mathbf{a}$$

$$\mathbf{a}^5 = \mathbf{a}^3 \times \mathbf{a}^2$$
$$= (1\angle 0°)(1\angle 240°) = 1\angle 240° = \mathbf{a}^2$$

$$\mathbf{a}^6 = \mathbf{a}^3 \times \mathbf{a}^3$$
$$= (1\angle 0°)(1\angle 0°) = 1\angle 0° = \mathbf{a}^3$$

$$\vdots$$

$$\mathbf{a}^{n+3} = \mathbf{a}^n \times \mathbf{a}^3 = \mathbf{a}^n$$

Figure 5.2 shows a phasor diagram of the various powers and functions of operator **a**. Various combinations of operator **a** are given in Table 5.1. In manipulating quantities involving operator **a**, it is useful to remember that

$$1 + \mathbf{a} + \mathbf{a}^2 = 0 \tag{5.4}$$

5.4 RESOLUTION OF THREE-PHASE UNBALANCED SYSTEM OF PHASORS INTO ITS SYMMETRICAL COMPONENTS

In the application of the symmetrical component, it is customary to let phase **a** be the reference phase. Therefore, using operator **a**, the symmetrical components of the positive-, negative-, and zero-sequence components can be expressed as

$$\mathbf{V}_{b1} = \mathbf{a}^2 \mathbf{V}_{a1} \tag{5.5}$$

$$\mathbf{V}_{c1} = \mathbf{a} \mathbf{V}_{a1} \tag{5.6}$$

$$\mathbf{V}_{b2} = \mathbf{a} \mathbf{V}_{a2} \tag{5.7}$$

$$\mathbf{V}_{c2} = \mathbf{a}^2 \mathbf{V}_{a2} \tag{5.8}$$

$$\mathbf{V}_{b0} = \mathbf{V}_{c0} \mathbf{V}_{a0} \tag{5.9}$$

Substituting the above equations into Equations 5.2 and 5.3, as appropriate, the phase voltages can be expressed in terms of the sequence voltages as

$$\mathbf{V}_a = \mathbf{V}_{a1} + \mathbf{V}_{a2} + \mathbf{V}_{a0} \tag{5.10}$$

$$\mathbf{V}_b = \mathbf{a}^2 \mathbf{V}_{a1} + \mathbf{a} \mathbf{V}_{a2} + \mathbf{V}_{a0} \tag{5.11}$$

$$\mathbf{V}_c = \mathbf{a} \mathbf{V}_{a1} + \mathbf{a}^2 \mathbf{V}_{a2} + \mathbf{V}_{a0} \tag{5.12}$$

Table 5.1

Powers and Functions of Operator a

Power or Function	In Polar Form	In Rectangular Form
a	$1 \angle 120°$	$-0.5 + j\,0.866$
a^2	$1 \angle 240° = 1 \angle -120°$	$-0.5 - j\,0.866$
a^3	$1 \angle 360° = 1 \angle 0°$	$1.0 + j\,0.0$
a^4	$1 \angle 120°$	$-0.5 + j\,0.866$
$1 + a = -a^2$	$1 \angle 60°$	$0.5 + j\,0.866$
$1 - a$	$\sqrt{3} \angle 30°$	$1.5 - j\,0.866$
$1 + a^2 = -a$	$1 \angle -60°$	$0.5 - j\,0.866$
$1 - a^2$	$\sqrt{3} \angle 30°$	$1.5 + j\,0.866$
$a - 1$	$\sqrt{3} \angle 150°$	$-1.5 + j\,0.866$
$a + a^2$	$1 \angle 180°$	$-1.0 + j\,0.0$
$a - a^2$	$\sqrt{3} \angle 90°$	$0.0 + j\,1.732$
$a^2 - a$	$\sqrt{3} \angle -90°$	$0.0 - j\,1.732$
$a^2 - 1$	$\sqrt{3} \angle -150°$	$-1.5 - j\,0.866$
$1 + a + a^2$	$0 \angle 0°$	$0.0 + j\,0.0$

Equations 5.10 through 5.12 are known as the *synthesis equations*. Therefore, it can be shown that the sequence voltages can be expressed in terms of phase voltages as

$$\mathbf{V}_{a0} = \frac{1}{3}(\mathbf{V}_a + \mathbf{V}_b + \mathbf{V}_c) \tag{5.13}$$

$$\mathbf{V}_{a1} = \frac{1}{3}(\mathbf{V}_a + a\mathbf{V}_b + a^2\mathbf{V}_c) \tag{5.14}$$

$$\mathbf{V}_{a2} = \frac{1}{3}(\mathbf{V}_a + a^2\mathbf{V}_b + a\mathbf{V}_c) \tag{5.15}$$

which are known as the . Alternatively, the synthesis and can be written, respectively, in matrix form as

$$\begin{bmatrix} \mathbf{V}_a \\ \mathbf{V}_b \\ \mathbf{V}_c \end{bmatrix} = \begin{bmatrix} 1 & 1 & 1 \\ 1 & a^2 & a \\ 1 & a & a^2 \end{bmatrix} \begin{bmatrix} \mathbf{V}_{a0} \\ \mathbf{V}_{b1} \\ \mathbf{V}_{c2} \end{bmatrix} \tag{5.16}$$

and

$$\begin{bmatrix} \mathbf{V}_{a0} \\ \mathbf{V}_{a1} \\ \mathbf{V}_{a2} \end{bmatrix} = \frac{1}{3}\begin{bmatrix} 1 & 1 & 1 \\ 1 & a & a^2 \\ 1 & a^2 & a \end{bmatrix} \begin{bmatrix} \mathbf{V}_a \\ \mathbf{V}_b \\ \mathbf{V}_c \end{bmatrix} \tag{5.17}$$

or

$$\begin{bmatrix} \mathbf{V}_{abc} \end{bmatrix} = \begin{bmatrix} \mathbf{A} \end{bmatrix}\begin{bmatrix} \mathbf{V}_{012} \end{bmatrix} \tag{5.18}$$

and

$$\begin{bmatrix} \mathbf{V}_{012} \end{bmatrix} = \begin{bmatrix} \mathbf{A} \end{bmatrix}^{-1}\begin{bmatrix} \mathbf{V}_{abc} \end{bmatrix} \tag{5.19}$$

where

$$[\mathbf{A}] = \begin{bmatrix} 1 & 1 & 1 \\ 1 & \mathbf{a}^2 & \mathbf{a} \\ 1 & \mathbf{a} & \mathbf{a}^2 \end{bmatrix} \qquad (5.20)$$

$$[\mathbf{A}]^{-1} = \frac{1}{3}\begin{bmatrix} 1 & 1 & 1 \\ 1 & \mathbf{a} & \mathbf{a}^2 \\ 1 & \mathbf{a}^2 & \mathbf{a} \end{bmatrix} \qquad (5.21)$$

$$[\mathbf{V}_{abc}] = \begin{bmatrix} \mathbf{V}_a \\ \mathbf{V}_b \\ \mathbf{V}_c \end{bmatrix} \qquad (5.22)$$

$$[\mathbf{V}_{012}] = \begin{bmatrix} \mathbf{V}_{a0} \\ \mathbf{V}_{a1} \\ \mathbf{V}_{a2} \end{bmatrix} \qquad (5.23)$$

The synthesis and in terms of phase and sequence currents can be expressed as

$$\begin{bmatrix} \mathbf{I}_a \\ \mathbf{I}_b \\ \mathbf{I}_c \end{bmatrix} = \begin{bmatrix} 1 & 1 & 1 \\ 1 & \mathbf{a}^2 & \mathbf{a} \\ 1 & \mathbf{a} & \mathbf{a}^2 \end{bmatrix}\begin{bmatrix} \mathbf{I}_{a0} \\ \mathbf{I}_{a1} \\ \mathbf{I}_{a2} \end{bmatrix} \qquad (5.24)$$

and

$$\begin{bmatrix} \mathbf{I}_{a0} \\ \mathbf{I}_{a1} \\ \mathbf{I}_{a2} \end{bmatrix} = \frac{1}{3}\begin{bmatrix} 1 & 1 & 1 \\ 1 & \mathbf{a} & \mathbf{a}^2 \\ 1 & \mathbf{a}^2 & \mathbf{a} \end{bmatrix}\begin{bmatrix} \mathbf{I}_a \\ \mathbf{I}_b \\ \mathbf{I}_c \end{bmatrix} \qquad (5.25)$$

or

$$[\mathbf{I}_{abc}] = [\mathbf{A}][\mathbf{I}_{012}] \qquad (5.26)$$

and

$$[\mathbf{I}_{012}] = [\mathbf{A}]^{-1}[\mathbf{I}_{abc}] \qquad (5.27)$$

Example 5.1: Determine the symmetrical components for the phase voltages of $\mathbf{V}_a = 7.3 \angle 12.5°$, $\mathbf{V}_b = 0.4 \angle -100°$, and $\mathbf{V}_c = 4.4 \angle 154°$ V

Solution:

$$\begin{aligned} \mathbf{V}_{a0} &= \frac{1}{3}(\mathbf{V}_a + \mathbf{V}_b + \mathbf{V}_c) \\ &= \frac{1}{3}(7.3\angle 12.5° + 0.4\angle -100° + 4.4\angle 154°) \\ &= 1.47\angle 45.1° \text{ V} \end{aligned}$$

$$\begin{aligned} \mathbf{V}_{a1} &= \frac{1}{3}(\mathbf{V}_a + \mathbf{a}\mathbf{V}_b + \mathbf{a}^2\mathbf{V}_c) \\ &= \frac{1}{3}[7.3\angle 12.5° + (1\angle 120°)(0.4\angle -100°) + (1\angle 240°)(4.4\angle 154°)] \\ &= 3.97\angle 20.5° \text{ V} \end{aligned}$$

$$\mathbf{V}_{a2} = \frac{1}{3}(\mathbf{V}_a + \mathbf{a}^2\mathbf{V}_b + \mathbf{a}\mathbf{V}_c)$$
$$= \frac{1}{3}[7.3\angle 12.5° + (1\angle 240°)(0.4\angle -100°) + (1\angle 120°)(4.4\angle 154°)]$$
$$= 2.52\angle -19.7°\ \mathrm{V}$$

$$\mathbf{V}_{b0} = \mathbf{V}_{a0}$$
$$= 1.47\angle 45.1°\ \mathrm{V}$$

$$\mathbf{V}_{b1} = \mathbf{a}^2\mathbf{V}_{a1}$$
$$= (1\angle 240°)(3.97\angle 20.5°)$$
$$= 3.97\angle 260.5°\ \mathrm{V}$$

$$\mathbf{V}_{b2} = \mathbf{a}\mathbf{V}_{a2}$$
$$= (1\angle 120°)(2.52\angle -19.7°)$$
$$= 2.52\angle 100.3°\ \mathrm{V}$$

$$\mathbf{V}_{c0} = \mathbf{V}_{a0}$$
$$= 1.47\angle 45.1°\ \mathrm{V}$$

$$\mathbf{V}_{c1} = \mathbf{a}\mathbf{V}_{a1}$$
$$= (1\angle 120°)(3.97\angle 20.5°)$$
$$= 3.97\angle 140.5°\ \mathrm{V}$$

$$\mathbf{V}_{c2} = \mathbf{a}^2\mathbf{V}_{a2}$$
$$= (1\angle 120°)(2.52\angle -19.7°)$$
$$= 2.52\angle 220.3°\ \mathrm{V}$$

Note that, the resulting values for the symmetrical components can be checked numerically (e.g., using Equation 5.11) or graphically, as shown in Figure 5.1e.

5.5　POWER IN SYMMETRICAL COMPONENTS

The three-phase complex power at any point of a three-phase system can be expressed as the sum of the individual complex powers of each phase so that

$$
\begin{aligned}
S_{3\phi} &= P_{3\phi} + jQ_{3\phi} \\
&= S_a + S_b + S_c \\
&= \mathbf{V}_a\mathbf{I}_a^* + \mathbf{V}_b\mathbf{I}_b^* + \mathbf{V}_c\mathbf{I}_c^*
\end{aligned}
\tag{5.28}
$$

or, in matrix notation, where the current and voltage column vectors can be represented as follows:

$$
\mathbf{S}_{3\phi} = \begin{bmatrix} \mathbf{V}_a & \mathbf{V}_b & \mathbf{V}_c \end{bmatrix}^\top \begin{bmatrix} \mathbf{I}_a \\ \mathbf{I}_b \\ \mathbf{I}_c \end{bmatrix}^*
\tag{5.29}
$$

or

$$\mathbf{S}_{3\phi} = \left[\mathbf{V}_{abc}\right]^{\top}\left[\mathbf{I}_{abc}\right]^{*} \qquad (5.30)$$

where

$$\left[\mathbf{V}_{abc}\right] = \left[\mathbf{A}\right]\left[\mathbf{V}_{012}\right]$$

$$\left[\mathbf{I}_{abc}\right] = \left[\mathbf{A}\right]\left[\mathbf{I}_{012}\right]$$

and therefore,

$$\left[\mathbf{V}_{abc}\right]^{\top} = \left[\mathbf{V}_{012}\right]^{\top}\left[\mathbf{A}\right]^{\top} \qquad (5.31)$$

$$\left[\mathbf{I}_{abc}\right]^{*} = \left[\mathbf{A}\right]^{*}\left[\mathbf{V}_{012}\right]^{*} \qquad (5.32)$$

Substituting Equations 5.31 and 5.32 into Equation 5.30,

$$\mathbf{S}_{3\phi} = \left[\mathbf{V}_{012}\right]^{\top}\left[\mathbf{A}\right]^{\top}\left[\mathbf{A}\right]^{*}\left[\mathbf{I}_{012}\right]^{*} \qquad (5.33)$$

where

$$[\mathbf{A}]^{\top}[\mathbf{A}]^{*} = \begin{bmatrix} 1 & 1 & 1 \\ 1 & a^2 & a \\ 1 & a & a^2 \end{bmatrix} \begin{bmatrix} 1 & 1 & 1 \\ 1 & a & a^2 \\ 1 & a^2 & a \end{bmatrix}$$

$$= \begin{bmatrix} 3 & 0 & 0 \\ 0 & 3 & 0 \\ 0 & 0 & 3 \end{bmatrix}$$

$$= 3 \begin{bmatrix} 1 & 0 & 0 \\ 0 & 1 & 0 \\ 0 & 0 & 1 \end{bmatrix}$$

Therefore,

$$\mathbf{S}_{3\phi} = 3\left[\mathbf{V}_{012}\right]^{\top}\left[\mathbf{I}_{012}\right]^{*} \qquad (5.34a)$$

$$= 3\begin{bmatrix} \mathbf{V}_{a0} & \mathbf{V}_{a1} & \mathbf{V}_{a2} \end{bmatrix} \begin{bmatrix} \mathbf{I}_{a0} \\ \mathbf{I}_{a1} \\ \mathbf{I}_{a2} \end{bmatrix}^{*} \qquad (5.34b)$$

or

$$\mathbf{S}_{3\phi} = 3\left[\mathbf{V}_{a0}\mathbf{I}_{a0}^{*} + \mathbf{V}_{a1}\mathbf{I}_{a1}^{*} + \mathbf{V}_{a2}\mathbf{I}_{a2}^{*}\right] \qquad (5.34c)$$

Note that, there are no cross terms (e.g., $\mathbf{V}_{a0}\mathbf{I}_{a1}^{*}$ or $\mathbf{V}_{a1}\mathbf{I}_{a0}^{*}$) in this equation, which indicates that there is no coupling of power among the three sequences. Also, note that the symmetrical components of voltage and current belong to the same phase.

Example 5.2: Assume that the phase voltages and currents of a three-phase system are given as

$$\left[\mathbf{V}_{abc}\right] = \begin{bmatrix} 0 \\ 50 \\ -50 \end{bmatrix} \text{ and } \left[\mathbf{I}_{abc}\right] = \begin{bmatrix} -5 \\ j5 \\ -5 \end{bmatrix}$$

and determine the following:

(a) Three-phase complex power using Equation 5.30
(b) Sequence voltage and current matrices, that is, $[V_{012}]$ and $[I_{012}]$
(c) Three-phase complex power using Equation 5.34

Solution:

(a)

$$S_{3\phi} = [V_{abc}]^T [I_{abc}]^*$$

$$= [0 \quad 50 \quad -50] \begin{bmatrix} 5 \\ -j5 \\ 5 \end{bmatrix}$$

$$= -250 - j250$$

$$= 353.5534\angle 45° \text{ VA}$$

(b)

$$V_{012} = [A]^{-1} [V_{abc}]$$

$$= \frac{1}{3} \begin{bmatrix} 1 & 1 & 1 \\ 1 & a & a^2 \\ 1 & a^2 & a \end{bmatrix} \begin{bmatrix} 0 \\ 50 \\ -50 \end{bmatrix}$$

$$= \begin{bmatrix} 0.0\angle 0° \\ 28.8675\angle 90° \\ 28.8675\angle -90° \end{bmatrix}$$

$$I_{012} = [A]^{-1} [I_{abc}]$$

$$= \frac{1}{3} \begin{bmatrix} 1 & 1 & 1 \\ 1 & a & a^2 \\ 1 & a^2 & a \end{bmatrix} \begin{bmatrix} -5 \\ j5 \\ -5 \end{bmatrix}$$

$$= \begin{bmatrix} 3.7268\angle 153.4° \\ 2.3570\angle 165° \\ 2.3570\angle -75° \end{bmatrix}$$

(c)

$$S_{3\phi} = 3 [V_{a0}I_{a0}^* + V_{a1}I_{a1}^* + V_{a2}I_{a2}^*]$$

$$= 353.5534\angle -45° \text{ VA}$$

5.6   SEQUENCE IMPEDANCES OF TRANSMISSION LINES

5.6.1   SEQUENCE IMPEDANCES OF UNTRANSPOSED LINES

Figure 5.3a shows a circuit representation of an untransposed transmission line with unequal self-impedances and unequal mutual impedances. Here,

$$[V_{abc} =] [Z_{abc}] [I_{abc}] \tag{5.35}$$

where

$$[Z_{abc} =] \begin{bmatrix} Z_{aa} & Z_{ab} & Z_{ac} \\ Z_{ba} & Z_{bb} & Z_{bc} \\ Z_{ca} & Z_{cb} & Z_{cc} \end{bmatrix} \tag{5.36}$$

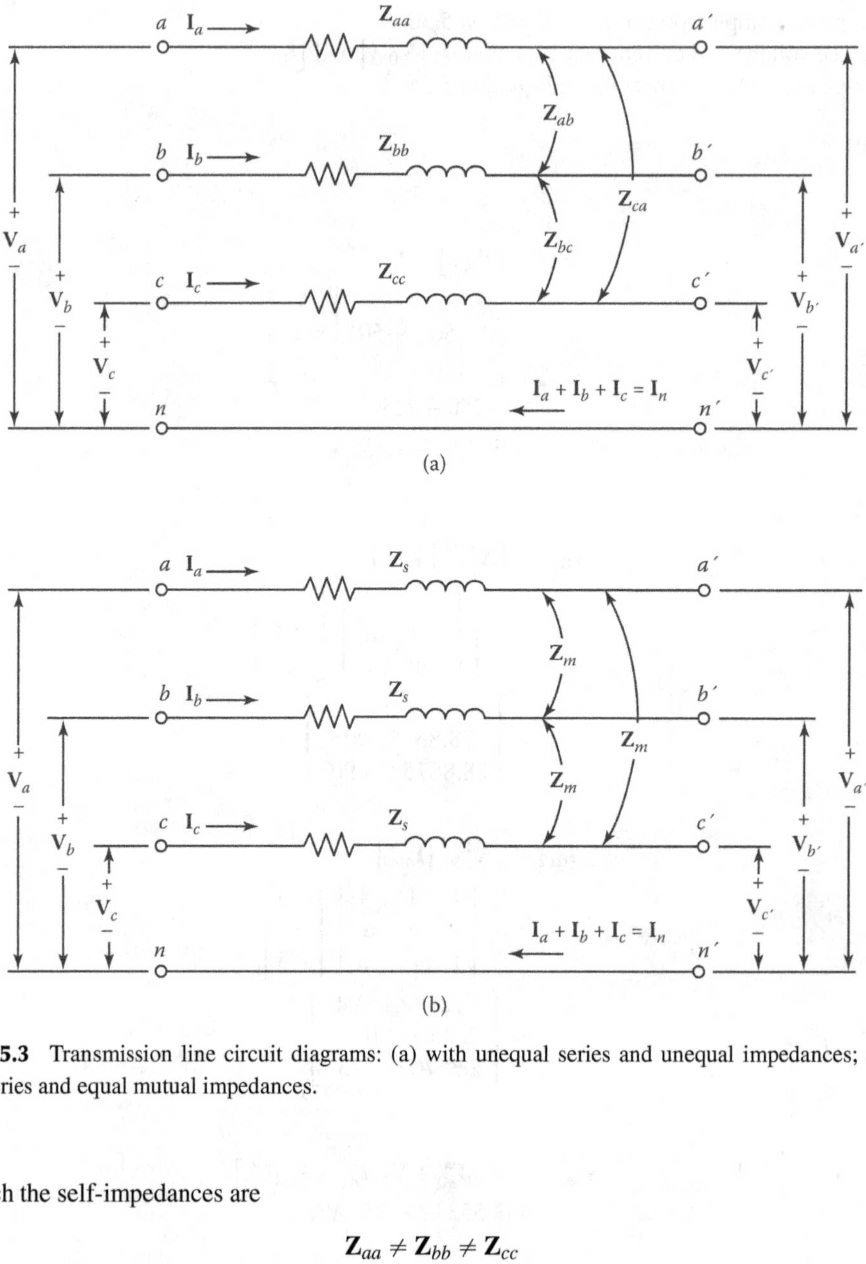

Figure 5.3 Transmission line circuit diagrams: (a) with unequal series and unequal impedances; (b) with equal series and equal mutual impedances.

in which the self-impedances are

$$\mathbf{Z}_{aa} \neq \mathbf{Z}_{bb} \neq \mathbf{Z}_{cc}$$

and the mutual impedances are

$$\mathbf{Z}_{ab} \neq \mathbf{Z}_{bc} \neq \mathbf{Z}_{ca}$$

Multiplying both sides of Equation 5.35 by $[\mathbf{A}]^{-1}$ and also substituting Equation 5.26 into Equation 5.35,

$$[\mathbf{A}]^{-1}[\mathbf{V}_{abc}] = [\mathbf{A}]^{-1}[\mathbf{Z}_{abc}][\mathbf{A}][\mathbf{I}_{012}] \tag{5.37}$$

where the similarity transformation is defined as

$$[\mathbf{Z}_{012}] \triangleq [\mathbf{A}]^{-1}[\mathbf{Z}_{abc}][\mathbf{A}] \tag{5.38}$$

Therefore, the sequence impedance matrix of an untransposed transmission line can be calculated using Equation 5.38 and can be expressed as

$$[\mathbf{Z}_{012}] = \begin{bmatrix} \mathbf{Z}_{00} & \mathbf{Z}_{01} & \mathbf{Z}_{02} \\ \mathbf{Z}_{10} & \mathbf{Z}_{11} & \mathbf{Z}_{12} \\ \mathbf{Z}_{20} & \mathbf{Z}_{21} & \mathbf{Z}_{22} \end{bmatrix} \tag{5.39}$$

or

$$[\mathbf{Z}_{012}] = \begin{bmatrix} (\mathbf{Z}_{s0} + 2\mathbf{Z}_{m0}) & (\mathbf{Z}_{s2} + \mathbf{Z}_{m2}) & (\mathbf{Z}_{s1} + \mathbf{Z}_{m1}) \\ (\mathbf{Z}_{s1} + \mathbf{Z}_{m1}) & (\mathbf{Z}_{s0} + \mathbf{Z}_{m0}) & (\mathbf{Z}_{s2} + 2\mathbf{Z}_{m2}) \\ (\mathbf{Z}_{s2} + \mathbf{Z}_{m2}) & (\mathbf{Z}_{s1} + 2\mathbf{Z}_{m1}) & (\mathbf{Z}_{s0} + \mathbf{Z}_{m0}) \end{bmatrix} \tag{5.40}$$

where, by definition,

$$\mathbf{Z}_{s0} = \text{zero-sequence self-impedance}$$
$$\triangleq \frac{1}{3}(\mathbf{Z}_{aa} + \mathbf{Z}_{bb} + \mathbf{Z}_{cc}) \tag{5.41}$$

$$\mathbf{Z}_{s1} = \text{positive-sequence self-impedance}$$
$$\triangleq \frac{1}{3}(\mathbf{Z}_{aa} + \mathbf{a}\mathbf{Z}_{bb} + \mathbf{a}^2\mathbf{Z}_{cc}) \tag{5.42}$$

$$\mathbf{Z}_{s2} = \text{negative-sequence self-impedance}$$
$$\triangleq \frac{1}{3}(\mathbf{Z}_{aa} + \mathbf{a}^2\mathbf{Z}_{bb} + \mathbf{a}\mathbf{Z}_{cc}) \tag{5.43}$$

$$\mathbf{Z}_{m0} = \text{zero-sequence mutual impedance}$$
$$\triangleq \frac{1}{3}(\mathbf{Z}_{bc} + \mathbf{Z}_{ca} + \mathbf{Z}_{ab}) \tag{5.44}$$

$$\mathbf{Z}_{m1} = \text{positive-sequence mutual impedance}$$
$$\triangleq \frac{1}{3}(\mathbf{Z}_{bc} + \mathbf{a}\mathbf{Z}_{ca} + \mathbf{a}^2\mathbf{Z}_{ab}) \tag{5.45}$$

$$\mathbf{Z}_{m2} = \text{negative-sequence mutual impedance}$$
$$\triangleq \frac{1}{3}(\mathbf{Z}_{bc} + \mathbf{a}^2\mathbf{Z}_{ca} + \mathbf{a}\mathbf{Z}_{ab}) \tag{5.46}$$

Therefore,

$$[\mathbf{V}_{012}] = [\mathbf{I}_{012}][\mathbf{Z}_{012}] \tag{5.47}$$

Note that the matrix in Equation 5.40 is not a symmetrical matrix, and therefore, the application of Equation 5.47 will show that there is a mutual coupling among the three sequences, which is not a desirable result.

5.6.2 SEQUENCE IMPEDANCES OF TRANSPOSED LINES

The remedy is either to completely transpose the line or to place the conductors with equilateral spacing among them so that the resulting mutual impedances* are equal to each other, that is, $\mathbf{Z}_{ab} = \mathbf{Z}_{bc} = \mathbf{Z}_{ca} = \mathbf{Z}_m$, as shown in Figure 5.3b. Furthermore, if the self-impedances of conductors are equal to each other, that is, $\mathbf{Z}_{aa} = \mathbf{Z}_{bb} = \mathbf{Z}_{cc} = \mathbf{Z}_s$, Equation 5.36 can be expressed as

$$[\mathbf{Z}_{abc} =] \begin{bmatrix} \mathbf{Z}_s & \mathbf{Z}_m & \mathbf{Z}_m \\ \mathbf{Z}_m & \mathbf{Z}_s & \mathbf{Z}_m \\ \mathbf{Z}_m & \mathbf{Z}_m & \mathbf{Z}_s \end{bmatrix} \tag{5.48}$$

Table 5.2
Resistivity of Different Soils

Ground Type (Ω/m)	Resistivity ρ
Seawater	0.01–1.0
Wet organic soil	10
Moist soil (average earth)	100
Dry soil	1000
Bedrock	10^4
Pure slate	10^7
Sandstone	10^9
Crushed rock	1.5×10^8

where

$$\mathbf{Z}_s = \left[(\mathbf{r}_a + \mathbf{r}_e) + j0.1213\ln\frac{D_e}{D_s} \right] l \ \Omega \tag{5.49}$$

$$\mathbf{Z}_m = \left[\mathbf{r}_e + j0.1213\ln\frac{D_e}{D_{eq}} \right] l \ \Omega \tag{5.50}$$

$$D_{eq} \triangleq D_m = (D_{ab} \times D_{bc} \times D_{ca})^{1/3}$$

r_a = resistance of a single conductor a.
r_e is the resistance of Carson's [173] equivalent (and fictitious) earth return conductor. It is a function of frequency and can be expressed as

$$r_e = 1.588 \times 10^{-3} f \ \Omega/\text{mi} \tag{5.51}$$

or

$$r_e = 9.868 \times 10^{-4} f \ \Omega/\text{mi} \tag{5.52}$$

At 60 Hz, $r_e = 0.09528$ Ω/mi. The quantity D_e is a function of both the earth resistivity ρ and the frequency f and can be expressed as

$$D_e = 2160 \left(\frac{\rho}{f} \right)^{1/2} \ \text{ft} \tag{5.53}$$

where ρ is the earth resistivity and is given in Table 5.2 for various earth types. If the actual earth resistivity is unknown, it is customary to use an average value of 100 Ω/m for ρ. Therefore, at 60 Hz, $D_e = 2788.55$ ft. D_s is the GMR of the phase conductor as before. Therefore, by applying Equation 5.38,

$$[\mathbf{Z}_{012}] = \begin{bmatrix} (\mathbf{Z}_s + 2\mathbf{Z}_m) & 0 & 0 \\ 0 & (\mathbf{Z}_s - \mathbf{Z}_m) & 0 \\ 0 & 0 & (\mathbf{Z}_s - \mathbf{Z}_m) \end{bmatrix} \tag{5.54}$$

where, by definition,

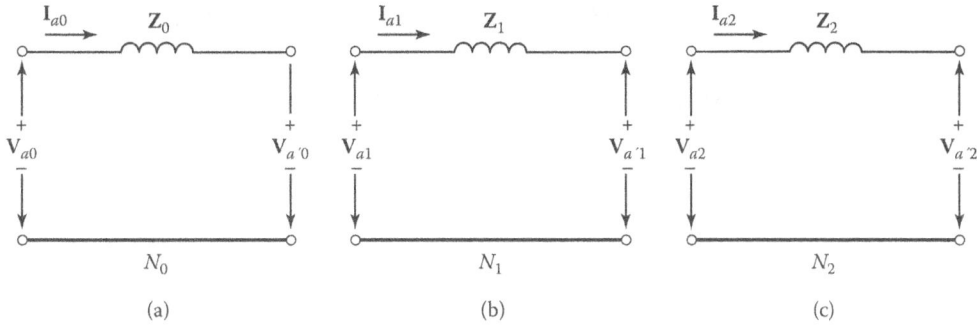

Figure 5.4 Sequence networks of a transmission line: (a) zero-sequence network; (b) positive-sequence network; (c) negative-sequence network.

$$\mathbf{Z}_0 = \text{zero-sequence impedance at 60 Hz}$$
$$\triangleq \mathbf{Z}_{00} = \mathbf{Z}_s + 2\mathbf{Z}_m \tag{5.55a}$$

$$= \left[(r_a + 3r_e) + j0.1213\ln\frac{D_e^3}{D_s \times D_{eq}^2} \right] l \ \Omega \tag{5.55b}$$

$$\mathbf{Z}_1 = \text{positive-sequence impedance at 60 Hz}$$
$$\triangleq \mathbf{Z}_{11} = \mathbf{Z}_s - \mathbf{Z}_m \tag{5.56a}$$

$$= \left[r_a + j0.1213\ln\frac{D_{eq}}{D_s} \right] l \ \Omega \tag{5.56b}$$

$$\mathbf{Z}_2 = \text{negative-sequence impedance at 60 Hz}$$
$$\triangleq \mathbf{Z}_{22} = \mathbf{Z}_s - \mathbf{Z}_m \tag{5.57a}$$

$$= \left[r_a + j0.1213\ln\frac{D_{eq}}{D_s} \right] l \ \Omega \tag{5.57b}$$

Thus, Equation 5.54 can be expressed* as

$$[\mathbf{Z}_{012}] = \begin{bmatrix} \mathbf{Z}_0 & 0 & 0 \\ 0 & \mathbf{Z}_1 & 0 \\ 0 & 0 & \mathbf{Z}_2 \end{bmatrix} \tag{5.58}$$

Both Equations 5.54 and 5.58 indicate that there is no mutual coupling among the three sequences, which is the desirable result. Therefore, the zero-, positive-, and negative-sequence currents cause voltage drops only in the zero-, positive-, and negative-sequence networks, respectively, of the transmission line. Also note, in Equation 5.54, that the positive- and negative-sequence impedances of the transmission line are equal to each other but they are far less than the zero-sequence impedance of the line. Figure 5.4 shows the sequence networks of a transmission line.

5.6.3 ELECTROMAGNETIC UNBALANCES DUE TO UNTRANSPOSED LINES

If the line is neither transposed nor its conductors equilaterally spaced, Equation 5.48 cannot be used. Instead, use the following equation:

$$[\mathbf{Z}_{abc}] = \begin{bmatrix} \mathbf{Z}_{aa} & \mathbf{Z}_{ab} & \mathbf{Z}_{ac} \\ \mathbf{Z}_{ba} & \mathbf{Z}_{bb} & \mathbf{Z}_{bc} \\ \mathbf{Z}_{ca} & \mathbf{Z}_{cb} & \mathbf{Z}_{cc} \end{bmatrix} \tag{5.59}$$

where

$$\mathbf{Z}_{aa} + \mathbf{Z}_{bb} + \mathbf{Z}_{cc} = \left[(\mathbf{r}_a + \mathbf{r}_e) + j0.1213 \ln \frac{D_e}{D_s} \right] l \tag{5.60}$$

$\mathbf{Z}_0$ = zero-sequence impedance at 50 Hz

$$\triangleq \mathbf{Z}_{00} = \mathbf{Z}_s + 2\mathbf{Z}_m$$

$$= \left[(r_a + 3r_e) + j0.1213 \left(\frac{50\text{Hz}}{60\text{ Hz}} \right) \ln \frac{D_e^3}{D_s \times D_{eq}^2} \right] l \ \Omega$$

$$= \left[(r_a + 3r_e) + j0.10108 \ln \frac{D_e^3}{D_s \times D_{eq}^2} \right] l \ \Omega$$

$\mathbf{Z}_1$ = positive-sequence impedance at 50 Hz

$$\triangleq \mathbf{Z}_{11} = \mathbf{Z}_s - \mathbf{Z}_m$$

$$= \left[r_a + j0.10108 \ln \frac{D_{eq}}{D_s} \right] l \ \Omega$$

$\mathbf{Z}_2$ = negative-sequence impedance at 50 Hz

$$\triangleq \mathbf{Z}_{22} = \mathbf{Z}_s - \mathbf{Z}_m$$

$$= \left[r_a + j0.10108 \ln \frac{D_{eq}}{D_s} \right] l \ \Omega$$

$$\mathbf{Z}_{ab} = \mathbf{Z}_{ba} = \left[r_e + j0.1213 \ln \frac{D_e}{D_{ab}} \right] l \tag{5.61}$$

$$\mathbf{Z}_{ac} = \mathbf{Z}_{ca} = \left[r_e + j0.1213 \ln \frac{D_e}{D_{ac}} \right] l \tag{5.62}$$

$$\mathbf{Z}_{bc} = \mathbf{Z}_{cb} = \left[r_e + j0.1213 \ln \frac{D_e}{D_{bc}} \right] l \tag{5.63}$$

The corresponding sequence impedance matrix can be found from Equation 5.38 as before. Therefore, the associated sequence admittance matrix can be found as

$$[\mathbf{Y}_{012}] = [\mathbf{Z}_{012}]^{-1} \tag{5.64a}$$

$$= \begin{bmatrix} \mathbf{Y}_{00} & \mathbf{Y}_{01} & \mathbf{Y}_{02} \\ \mathbf{Y}_{10} & \mathbf{Y}_{11} & \mathbf{Y}_{12} \\ \mathbf{Y}_{20} & \mathbf{Y}_{21} & \mathbf{Y}_{22} \end{bmatrix} \tag{5.64b}$$

Therefore,

$$[\mathbf{I}_{012}] = [\mathbf{Y}_{012}][\mathbf{V}_{012}] \tag{5.65}$$

Since the line is neither transposed nor its conductors equilaterally spaced, there is an electromagnetic unbalance in the system. Such unbalance is determined from Equation 5.65 with only positive-sequence voltage applied. Therefore,

$$\begin{bmatrix} \mathbf{I}_{a0} \\ \mathbf{I}_{b1} \\ \mathbf{I}_{c2} \end{bmatrix} = \begin{bmatrix} \mathbf{Y}_{00} & \mathbf{Y}_{01} & \mathbf{Y}_{02} \\ \mathbf{Y}_{10} & \mathbf{Y}_{11} & \mathbf{Y}_{12} \\ \mathbf{Y}_{20} & \mathbf{Y}_{21} & \mathbf{Y}_{22} \end{bmatrix} \begin{bmatrix} 0 \\ \mathbf{V}_{a1} \\ 0 \end{bmatrix} \tag{5.66a}$$

$$= \begin{bmatrix} \mathbf{Y}_{01} \\ \mathbf{Y}_{11} \\ \mathbf{Y}_{21} \end{bmatrix} \mathbf{V}_{a1} \tag{5.66b}$$

According to Gross and Hesse [174], the per-unit unbalances for zero sequence and negative sequence can be expressed, respectively, as

$$\mathbf{m}_0 \triangleq \frac{\mathbf{I}_{a0}}{\mathbf{I}_{a1}} \text{ pu} \tag{5.67a}$$

$$= \frac{\mathbf{Y}_{01}}{\mathbf{Y}_{11}} \text{ pu} \tag{5.67b}$$

and

$$\mathbf{m}_2 \triangleq \frac{\mathbf{I}_{a2}}{\mathbf{I}_{a1}} \text{ pu} \tag{5.68a}$$

$$= \frac{\mathbf{Y}_{01}}{\mathbf{Y}_{11}} \text{ pu} \tag{5.68b}$$

Since, in physical systems [174],

$$\mathbf{Z}_{22} \gg \mathbf{Z}_{02} \text{ or } \mathbf{Z}_{21}$$

and

$$\mathbf{Z}_{00} \gg \mathbf{Z}_{20} \text{ or } \mathbf{Z}_{01}$$

the approximate values of the per-unit unbalances for zero and negative sequences can be expressed, respectively, as

$$\mathbf{m}_0 \cong \frac{\mathbf{Z}_{01}}{\mathbf{Z}_{00}} \text{ pu} \tag{5.69a}$$

$$\mathbf{m}_2 \cong \frac{\mathbf{Z}_{21}}{\mathbf{Z}_{22}} \text{ pu} \tag{5.69b}$$

Example 5.3: Consider the compact-line configuration shown in Figure 5.5. The phase conductors used are made up of 500-kcmil, 30/7-strand ACSR (aluminum conductor steel reinforced conductor) conductor. The line length is 40 mi and the line is not transposed. Use 50 ° C and 60 Hz. Ignore the overhead ground wire. If the earth has an average resistivity, determine the following:

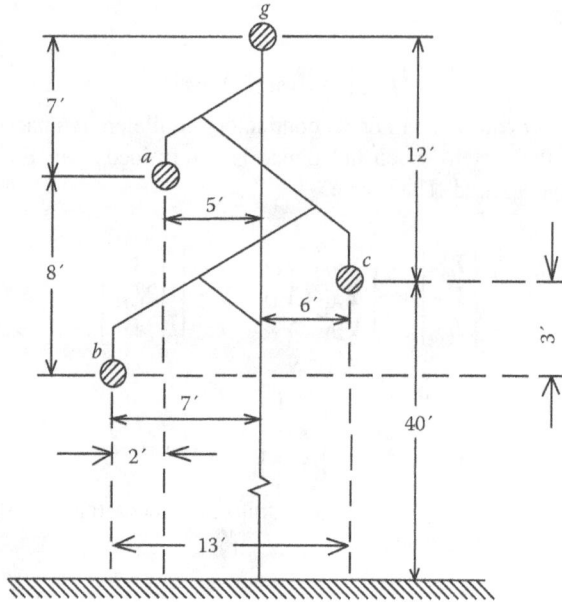

Figure 5.5 Compact-line configuration for Example 5.3.

(a) Line impedance matrix
(b) Sequence impedance matrix of line

Solution:

(a) The conductor parameters can be found from Table A.3 (Appendix A) as

$$r_a = r_b = r_c = 0.206 \ \Omega/\text{mi}$$

$$D_s = D_{sa} = D_{sb} = D_{sc} = 0.0311 \ \text{ft}$$

$$D_{ab} = (2^2 + 8^2)^{1/2} = 8.2462 \ \text{ft}$$
$$D_{bc} = (3^2 + 13^2)^{1/2} = 13.3417 \ \text{ft}$$
$$D_{ac} = (5^2 + 11^2)^{1/2} = 12.0830 \ \text{ft}$$

Since the earth has an average resistivity, $D_e = 2788.5$ ft. At 60 Hz, $r_e = 0.09528 \ \Omega/\text{mi}$. From Equation 5.60, the self-impedances of the line conductors are

$$
\begin{aligned}
\mathbf{Z}_{aa} + \mathbf{Z}_{bb} + \mathbf{Z}_{cc} &= \left[(r_a + r_e) + j0.1213 \ln\frac{D_e}{D_s} \right] l \\
&= \left[(0.206 + 0.09528) + j0.1213 \ln\frac{2788.5}{0.0311} \right] \times 40 \\
&= 12.0512 + j55.3495 \ \Omega
\end{aligned}
$$

The mutual impedances calculated from Equations 5.61 through 5.63 are

$$\mathbf{Z}_{ab} = \mathbf{Z}_{ba} = \left[r_e + j0.1213 \ln \frac{D_e}{D_{ab}} \right] l$$

$$= \left[0.09528 + j0.1213 \ln \frac{2788.5}{0.2462} \right] \cdot \times 40$$

$$= 3.8112 + j28.2650 \ \Omega$$

$$\mathbf{Z}_{bc} = \mathbf{Z}_{cb} = \left[r_e + j0.1213 \ln \frac{D_e}{D_{bc}} \right] l$$

$$= \left[0.09528 + j0.1213 \ln \frac{2788.5}{13.3417} \right] \times 40$$

$$= 3.8112 + j25.9297 \ \Omega$$

$$\mathbf{Z}_{ac} = \mathbf{Z}_{ca} = \left[r_e + j0.1213 \ln \frac{D_e}{D_{bc}} \right] l$$

$$= \left[0.09528 + j0.1213 \ln \frac{2788.5}{12.0830} \right] \times 40$$

$$= 3.8112 + j26.4107 \ \Omega$$

Therefore,

$$[\mathbf{Z}_{abc}] = \begin{bmatrix} (12.0512 + j55.3495) & (3.8112 + j26.4107) & (3.8112 + j26.4107) \\ (3.8112 + j28.2650) & (12.0512 + j55.3495) & (3.8112 + j25.9297) \\ (3.8112 + j26.4107) & (3.8112 + j25.9297) & (12.0512 + j55.3495) \end{bmatrix}$$

(b) Thus, the sequence impedance matrix of the line can be found from Equation 5.38 as

$$[\mathbf{Z}_{012}] [\mathbf{A}]^{-1} = [\mathbf{Z}_{abc}] [\mathbf{A}]$$

$$= \begin{bmatrix} (19.67 + j109.09) & (0.54 + j0.47) & (-0.54 + j0.47) \\ (-0.54 + j0.47) & (8.24 + j28.48) & (-1.07 + j0.94) \\ (0.54 + j0.47) & (1.07 + j0.94) & (8.24 + j28.48) \end{bmatrix}$$

Example 5.4: Repeat Example 5.3 assuming that the line is *completely transposed.*
Solution:

(a) From Equation 5.49,

$$\mathbf{Z}_s = \left[(r_a + r_e + j0.1213 \ln \frac{D_e}{D_s} \right] l \ \Omega$$

$$= 12.0512 + j53.3495 \ \Omega$$

From Equation 5.50,

$$\mathbf{Z}_m = \left[r_e + j0.1213 \ln \frac{D_e}{D_{eq}} \right] l \ Omega$$

$$= 12.0512 + j53.3495 \ \Omega$$

where

$$D_{eq} = (8.2462 \times 13.3417 \times 12.0830)^{1/3} = 11 \text{ ft}$$

Thus,

$$\mathbf{Z}_m = \left[0.09528 + j0.1213 \ln\frac{2788.5}{11}\right] \times 40$$
$$= 3.8112 + j26.8684 \ \Omega$$

Therefore,

$$[\mathbf{Z}_{abc}] = \begin{bmatrix} \mathbf{Z}_s & \mathbf{Z}_m & \mathbf{Z}_m \\ \mathbf{Z}_m & \mathbf{Z}_s & \mathbf{Z}_m \\ \mathbf{Z}_m & \mathbf{Z}_m & \mathbf{Z}_s \end{bmatrix}$$

$$= \begin{bmatrix} (12.0512 + j55.3495) & (3.8112 + j26.4107) & (3.8112 + j26.4107) \\ (3.8112 + j28.2650) & (12.0512 + j55.3495) & (3.8112 + j25.9297) \\ (3.8112 + j26.4107) & (3.8112 + j25.9297) & (12.0512 + j55.3495) \end{bmatrix}$$

(b) From Equation 5.54

$$[\mathbf{Z}_{012}] = \begin{bmatrix} (\mathbf{Z}_s + 0) & 0 & 0 \\ 0 & (\mathbf{Z}_s - \mathbf{Z}_m) & 0 \\ 0 & 0 & (\mathbf{Z}_s - \mathbf{Z}_m) \end{bmatrix}$$

$$= \begin{bmatrix} 19.6736 + j109.086 & 0 & 0 \\ 0 & 8.2400 + j28.4811 & 0 \\ 0 & 0 & 8.2400 + j28.4811 \end{bmatrix}$$

or, by substituting Equations 5.55b and 5.56b into Equation 5.58,

$$[\mathbf{Z}_{012}] = \begin{bmatrix} 19.6736 + j109.086 & 0 & 0 \\ 0 & 8.2400 + j28.4811 & 0 \\ 0 & 0 & 8.2400 + j28.4811 \end{bmatrix}$$

Example 5.5: Consider the results of Example 5.3 and determine the following:

(a) Per-unit electromagnetic unbalance for zero sequence
(b) Approximate value of per-unit electromagnetic unbalance for zero sequence
(c) Per-unit electromagnetic unbalance for negative sequence
(d) Approximate value of per-unit electromagnetic unbalance for negative sequence

Solution:
The sequence admittance of the line can be found as

$$[\mathbf{Y}_{012}] = [\mathbf{Z}_{012}]^{-1}$$
$$= \begin{bmatrix} (1.60 \times 10^{-3} - j8.88 \times 10^{-3}) & (7.57 \times 10^{-5} + 1.93 \times 10^{-4}) & (-2.01 \times 10^{-4} + j6.15 \times 10^{-5}) \\ (-2.01 \times 10^{-4} + j6.15 \times 10^{-5}) & (9.44 \times 10^{-3} - 3.25 \times 10^{-2}) & (-4.55 \times 10^{-4} - j1.55 \times 10^{-3}) \\ (7.57 \times 10^{-5} - j1.93 \times 10^{-4}) & (1.60 \times 10^{-3} - j2.54 \times 10^{-4}) & (9.44 \times 10^{-3} - j3.25 \times 10^{-2}) \end{bmatrix}$$

(a) From Equation 5.67b,

$$\mathbf{m}_0 = \frac{\mathbf{Y}_{01}}{\mathbf{Y}_{11}}$$

$$= \frac{7.57 \times 10^{-5} + j1.93 \times 10^{-4}}{9.44 \times 10^{-3} - j3.25 \times 10^{-2}}$$

$$= 0.61 \angle 142.4° \ \%$$

(b) From Equation 5.69a,

$$\mathbf{m}_0 \cong \frac{\mathbf{Z}_{01}}{\mathbf{Z}_{11}}$$

$$= -\frac{0.54 + j0.47}{19.67 + j0.47}$$

$$= 0.64 \angle 141.3° \ \%$$

(c) From Equation 5.68b,

$$\mathbf{m}_2 = \frac{\mathbf{Y}_{21}}{\mathbf{Y}_{11}}$$

$$= \frac{160 \times 10^{-3} \quad j2.54 \times 10^{-4}}{9.44 \times 10^{-3} - j3.25 \times 10^{-2}}$$

$$= 4.79 \angle 64.8° \ \%$$

(d) From Equation 5.69b,

$$\mathbf{m}_2 = \frac{\mathbf{Z}_{21}}{\mathbf{Z}_{22}}$$

$$= \frac{1.07 \times - j0.94}{8.24 + j28.48}$$

$$= 4.8 \angle 64.8° \ \%$$

5.6.4 SEQUENCE IMPEDANCE OF UNTRANSPOSED LINE WITH OVERHEAD GROUND WIRE

Assume that the untransposed line shown in Figure 5.5 is *shielded* against direct lightning strikes by the overhead ground wire u (used instead of g).

Therefore,

$$[\mathbf{Y}_{abcu}] = [\mathbf{Z}_{abcu}][\mathbf{I}_{abcu}] \tag{5.70}$$

but since for the ground wire $\mathbf{V}_a = 0$,

$$\begin{bmatrix} \mathbf{V}_a \\ \mathbf{V}_b \\ \mathbf{V}_c \\ -- \\ 0 \end{bmatrix} = \begin{bmatrix} \mathbf{Z}_{aa} & \mathbf{Z}_{ab} & \mathbf{Z}_{ac} & | & \mathbf{Z}_{au} \\ \mathbf{Z}_{ba} & \mathbf{Z}_{bb} & \mathbf{Z}_{bc} & | & \mathbf{Z}_{bu} \\ \mathbf{Z}_{ca} & \mathbf{Z}_{cb} & \mathbf{Z}_{cc} & | & \mathbf{Z}_{cu} \\ -- & -- & -- & | & -- \\ \mathbf{Z}_{ua} & \mathbf{Z}_{ub} & \mathbf{Z}_{uc} & | & \mathbf{Z}_{uu} \end{bmatrix} \begin{bmatrix} \mathbf{I}_a \\ \mathbf{I}_b \\ \mathbf{I}_c \\ -- \\ \mathbf{I}_u \end{bmatrix} \tag{5.71}$$

The matrix $[\mathbf{Z}_{abcu}]$ can be determined using Equations 5.59 through 5.63, as before, and also using the following equations:

$$\mathbf{Z}_{au} = \mathbf{Z}_{ua} = \left[r_e + j0.1213 \ln \frac{D_e}{D_{au}} \right] l \tag{5.72}$$

$$\mathbf{Z}_{bu} = \mathbf{Z}_{ub} = \left[r_e + j0.1213 \ln \frac{D_e}{D_{bu}} \right] l \tag{5.73}$$

$$\mathbf{Z}_{cu} = \mathbf{Z}_{uc} = \left[r_e + j0.1213 \ln \frac{D_e}{D_{cu}} \right] l \tag{5.74}$$

$$\mathbf{Z}_{uu} = \mathbf{Z}_{uu} = \left[r_e + j0.1213 \ln \frac{D_e}{D_{uu}} \right] l \tag{5.75}$$

where r_u and D_{uu} are the resistance and GMR of the overhead ground wire, respectively.

The matrix $\left[\mathbf{Z}_{abcu} \right]$ given in Equation 5.71 can be reduced to $\left[\mathbf{Z}_{abc} \right]$ by using the *Kron reduction technique*. Therefore, Equation 5.71 can be reexpressed as

$$\left[\begin{array}{c} \mathbf{V}_{abc} \\ - - - \\ 0 \end{array} \right] = \left[\begin{array}{c|c} \mathbf{Z}_1 & \mathbf{Z}_2 \\ \hline \mathbf{Z}_3 & \mathbf{Z}_4 \end{array} \right] \left[\begin{array}{c} \mathbf{I}_{abc} \\ - - - \\ 0 \end{array} \right] \tag{5.76}$$

where the submatrices $\left[\mathbf{Z}_1 \right], \left[\mathbf{Z}_2 \right], \left[\mathbf{Z}_3 \right]$, and $\left[\mathbf{Z}_4 \right]$ are specified in the partitioned matrix $\left[\mathbf{Z}_{abcu} \right]$ in Equation 5.71. Therefore, after the reduction,

$$\left[\mathbf{Y}_{abc} \right] = \left[\mathbf{Z}_{abc} \right] \left[\mathbf{I}_{abc} \right] \tag{5.77}$$

where

$$\left[\mathbf{Z}_{abc} \right] \triangleq \left[\mathbf{Z}_1 \right] - \left[\mathbf{Z}_2 \right] \left[\mathbf{Z}_4 \right]^{-1} \left[\mathbf{Z}_3 \right] \tag{5.78}$$

Therefore, the sequence impedance matrix can be found from

$$\left[\mathbf{Z}_{012} \right] = \left[\mathbf{A} \right]^{-1} \left[\mathbf{Z}_{abc} \right] \left[\mathbf{A} \right] \tag{5.79}$$

Thus, the sequence admittance matrix becomes

$$\left[\mathbf{Y}_{012} \right] = \left[\mathbf{Z}_{012} \right]^{-1} \tag{5.80}$$

5.7 SEQUENCE CAPACITANCES OF TRANSMISSION LINE

5.7.1 THREE-PHASE TRANSMISSION LINE WITHOUT OVERHEAD GROUND WIRE

Consider Figure 3.47 and assume that the three-phase conductors are charged. Therefore, for sinusoidal steady-state analysis, both voltage and charge density can be represented by phasors. Thus,

$$\left[\mathbf{V}_{abc} \right] = \left[P_{abc} \right] \left[Q_{abc} \right] \tag{5.81}$$

or

$$\begin{bmatrix} \mathbf{V}_a \\ \mathbf{V}_b \\ \mathbf{V}_c \end{bmatrix} = \begin{bmatrix} P_{aa} & P_{ab} & P_{ac} \\ P_{ba} & P_{bb} & P_{bc} \\ P_{ca} & P_{cb} & P_{cc} \end{bmatrix} \begin{bmatrix} q_a \\ q_b \\ q_c \end{bmatrix} \tag{5.82}$$

where

$\left[P_{abc} \right]$ = matrix of potential coefficients, based on Figure 3.47

where

$$p_{aa} = \frac{1}{2\pi\varepsilon}\ln\frac{h_{11}}{r_a}\mathrm{F}^{-1}\ \mathrm{m} \tag{5.83}$$

$$p_{bb} = \frac{1}{2\pi\varepsilon}\ln\frac{h_{22}}{r_b}\mathrm{F}^{-1}\ \mathrm{m} \tag{5.84}$$

$$p_{cc} = \frac{1}{2\pi\varepsilon}\ln\frac{h_{33}}{r_c}\mathrm{F}^{-1}\ \mathrm{m} \tag{5.85}$$

$$p_{ab} = p_{ba} = \frac{1}{2\pi\varepsilon}\ln\frac{l_{12}}{D_{12}}\mathrm{F}^{-1}\ \mathrm{m} \tag{5.86}$$

$$p_{bc} = p_{cb} = \frac{1}{2\pi\varepsilon}\ln\frac{l_{23}}{D_{23}}\mathrm{F}^{-1}\ \mathrm{m} \tag{5.87}$$

$$p_{ac} = p_{ca} = \frac{1}{2\pi\varepsilon}\ln\frac{l_{31}}{D_{31}}\mathrm{F}^{-1}\ \mathrm{m} \tag{5.88}$$

Therefore, from Equation 5.81,

$$[Q_{abc}] = [P_{abc}]^{-1}[V_{abc}]\,\mathrm{C/m} \tag{5.89a}$$

$$= [C_{abc}][V_{abc}]\,\mathrm{C/m} \tag{5.89b}$$

since

$$[C_{abc}] = [P_{abc}]^{-1}\,\mathrm{F/m} \tag{5.90}$$

or

$$[C_{abc}] = \begin{bmatrix} C_{aa} & -C_{ab} & C_{ac} \\ -C_{ba} & C_{bb} & -C_{bc} \\ -C_{ca} & -C_{cb} & C_{cc} \end{bmatrix}\mathrm{F/m} \tag{5.91}$$

where $[C_{abc}]$ is the *matrix of Maxwell's coefficients*, the diagonal terms are *Maxwell's (or capacitance) coefficients*, and the off-diagonal terms are *electrostatic induction coefficients*.

Therefore, the sequence capacitances can be found by using the similarity transformation as

$$[C_{012}] \triangleq [\mathbf{A}]^{-1}[C_{abc}][\mathbf{A}]\,\mathrm{F/m} \tag{5.92a}$$

$$= \begin{bmatrix} C_{00} & C_{01} & C_{02} \\ C_{10} & C_{11} & C_{12} \\ C_{20} & C_{21} & C_{22} \end{bmatrix}\ \mathrm{F/m} \tag{5.92b}$$

Note that, if the line is *transposed*, the matrix of potential coefficients can be expressed in terms of self-and mutual-potential coefficients as

$$[P_{abc}] = \begin{bmatrix} p_s & p_m & p_m \\ p_m & p_s & p_m \\ p_m & p_m & p_s \end{bmatrix}\mathrm{F/m} \tag{5.93}$$

Therefore, using the similarity transformation,

$$[P_{012}] \triangleq [A]^{-1} [P_{abc}] [A] \tag{5.94a}$$

$$= \begin{bmatrix} p_0 & 0 & 0 \\ 0 & p_1 & 0 \\ 0 & 0 & p_2 \end{bmatrix} \tag{5.94b}$$

Thus,

$$[C_{012}] \triangleq [P_{012}]^{-1} \tag{5.95a}$$

$$= \begin{bmatrix} 1/p_0 & 0 & 0 \\ 0 & 1/p_1 & 0 \\ 0 & 0 & 1/p_2 \end{bmatrix} \tag{5.95b}$$

$$= \begin{bmatrix} C_0 & 0 & 0 \\ 0 & C_1 & 0 \\ 0 & 0 & C_2 \end{bmatrix} \tag{5.95c}$$

Alternatively, the sequence capacitances can approximately be calculated without using matrix algebra. For example, the zero-sequence capacitance can be calculated [63] from

$$C_0 = \frac{29.842}{\ln\left(\frac{H_{aa}}{D_{aa}}\right)} \text{ nF/mi} \tag{5.96}$$

where

H_{aa} = GMD between three conductors and their images

$$= [h_{11} \times h_{22} \times h_{33}(l_{12} \times l_{13})^2]^{1/9} \tag{5.97}$$

D_{aa} = self-GMD of overhead conductors as composite group but with D_s of each conductor taken as its radius

$$= [r_a \times r_b \times r_c(D_{12} \times D_{22} \times D_{31})^2]^{1/9} \tag{5.98}$$

Note that, D_s has been replaced by the conductor radius since all charge on a conductor resides on its surface. The positive- and negative-sequence capacitances of a line are the same owing to the fact that the physical parameters do not vary with a change in sequence of the applied voltage. Therefore, they are the same as the line-to-neutral capacitance C_n and can be calculated from Equation 3.298 or 3.299.

Note that, the mutual capacitances of the line can be found from Equation 5.91. The capacitances to ground can be expressed as

$$\begin{bmatrix} C_{ag} \\ C_{bg} \\ C_{cg} \end{bmatrix} = \begin{bmatrix} C_{aa} & C_{ab} & C_{ac} \\ -C_{aa} & -C_{bb} & C_{bc} \\ -C_{ac} & C_{ba} & -C_{cc} \end{bmatrix} \begin{bmatrix} 1 \\ -1 \\ -1 \end{bmatrix} \tag{5.99}$$

If the line is transposed, the capacitance to ground is an average value that can be determined from

$$C_{g,\text{avg}} = \frac{1}{3}(C_{ag} + C_{bg} + C_{cg}) \tag{5.100}$$

Also, note that the shunt admittance matrix of the line is

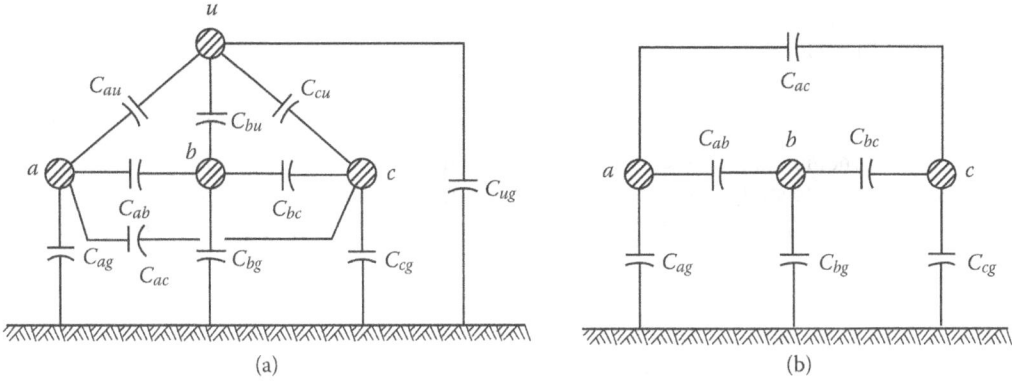

Figure 5.6 Three-phase line with one overhead ground wire u: (a) equivalent circuit showing ground wire; (b) equivalent circuit without showing ground wire.

$$[\mathbf{Y}_{abc}] = j\omega [C_{abc}] \tag{5.101}$$

Therefore,

$$[\mathbf{Y}_{012}] = [\mathbf{A}]^{-1} [\mathbf{Y}_{abc}] [\mathbf{A}] \tag{5.102}$$

Thus,

$$[\mathbf{C}_{012}] = \frac{[\mathbf{Y}_{012}]}{j\omega} \tag{5.103}$$

Hence,

$$[\mathbf{I}_{012}] = j\omega [C_{012}] [\mathbf{V}_{012}] = j [B_{012}] [\mathbf{V}_{012}] \tag{5.104}$$

and

$$[\mathbf{I}_{abc}] = [\mathbf{A}] [\mathbf{I}_{012}] \tag{5.105}$$

or

$$\begin{aligned}[\mathbf{I}_{abc}] &= j\omega [C_{abc}] [\mathbf{V}_{abc}] \\ &= j [B_{abc}] [\mathbf{V}_{abc}] \end{aligned} \tag{5.106}$$

5.7.2   THREE-PHASE TRANSMISSION LINE WITH OVERHEAD GROUND WIRE

Assume that the line is transposed and that the overhead ground wire is denoted by u and that there are nine capacitances involved. The voltages and charge densities involved can be represented by phasors. Therefore,
but since, for the ground wire, $\mathbf{V}_u = 0$,

$$[\mathbf{V}_{abcu}] = [P_{abcu}][Q_{abcu}] \tag{5.107}$$

$$\begin{bmatrix} V_a \\ V_b \\ V_c \\ 0 \end{bmatrix} = \begin{bmatrix} p_{aa} & p_{ab} & p_{ac} & p_{au} \\ p_{ba} & p_{bb} & p_{bc} & p_{bu} \\ p_{ca} & p_{cb} & p_{cc} & p_{cu} \\ p_{ua} & p_{ub} & p_{uc} & p_{uu} \end{bmatrix} \begin{bmatrix} q_a \\ q_b \\ q_c \\ q_u \end{bmatrix} \tag{5.108}$$

The matrix $[P_{abcu}]$ can be calculated as before. The corresponding matrix of the Maxwell coefficients can be found as

$$[C_{abcu}] = [P_{abcu}]^{-1} \tag{5.109}$$

The corresponding equivalent circuit is shown in Figure 5.6a. Such equivalent circuit representation is convenient to study switching transients, traveling waves, overvoltages, etc.

The matrix $[P_{abcu}]$ given in Equation 5.108 can be reduced to $[P_{abc}]$ by using the Kron reduction technique. Therefore, Equation 5.97 can be reexpressed as

$$\left[\begin{array}{c} V_{abc} \\ \hline 0 \end{array} \right] = \left[\begin{array}{cc} P_1 & P_2 \\ \hline P_3 & P_4 \end{array} \right] \left[\begin{array}{c} Q_{abc} \\ \hline Q_u \end{array} \right] \tag{5.110}$$

Where the submatrices $[P_1]$, $[P_2]$, $[P_3]$, and $[P_4]$ are specified in the partitioned matrix $[P_{abcu}]$ in Equation 5.108. Thus, after the reduction,

$$[\mathbf{V}_{abc}] = [P_{abc}][Q_{abc}] \tag{5.111}$$

where

$$[P_{abc}] \triangleq [P_1] - [P_2][P_4]^{-1}[P_3] \tag{5.112}$$

Thus, the corresponding matrix of the Maxwell coefficients can be found as

$$[C_{abc}] = [P_{abc}]^{-1} \tag{5.113}$$

as before. The corresponding equivalent circuit is shown in Figure 5.6b, and such representation is convenient to study a load-flow problem. Of course, the average capacitances to ground can be found as before.

Alternatively, the sequence capacitances can approximately be calculated without using the matrix algebra. For example, the zero-sequence capacitance can be calculated [63] from

$$C_0 = \frac{29.842 \ln \left(\frac{h_{gg}}{D_{gg}} \right)}{\ln \left(\frac{H_{aa}}{D_{aa}} \right) \times \ln \left(\frac{H_{gg}}{D_{gg}} \right) \left[\ln \left(\frac{H_{ag}}{D_{ag}} \right) \right]^2} \quad \text{nF/mi} \tag{5.114}$$

where
H_{aa} = given by Equation 5.97
D_{aa} = given by Equation 5.98
h_{gg} = GMD between ground wires and their images
D_{gg} = self-GMD of ground wires with $D_s = r_g$
H_{ag} = GMD between phase conductors and images of ground wires
D_{ag} = GMD between phase conductors and ground wires

If the transmission line is untransposed, both electrostatic and electromagnetic unbalances exist in the system. If the system neutral is (*solidly*) grounded, in the event of an electrostatic unbalance, there will be a neutral residual current flow in the system due to the unbalance in the charging currents of the line. Such residual current flow is continuous and independent of the load. Since the neutral is grounded, $\mathbf{V}_n = \mathbf{V}_{a0} = 0$, and the *zero-sequence displacement* or *unbalance* is

$$\mathbf{d}_0 \triangleq \frac{\mathbf{C}_{01}}{\mathbf{C}_{11}} \tag{5.115}$$

and the *negative-sequence unbalance* is

$$\mathbf{d_2} \triangleq \frac{\mathbf{C}_{21}}{\mathbf{C}_{11}} \tag{5.116}$$

If the system neutral is not grounded, there will be the neutral voltage $\mathbf{V}_n \neq 0$, and therefore, the neutral point will be shifted. Such zero-sequence neutral displacement or unbalance is defined as

$$\mathbf{d_0} \triangleq \frac{\mathbf{C}_{01}}{\mathbf{C}_{00}} \tag{5.117}$$

Example 5.6: Consider the line configuration shown in Figure 5.5. Assume that the 115-kV line is not transposed and its conductors are made up of 500-kcmil, 30/7-strand ACSR conductors. Ignore the overhead ground wire and determine the following:

(a) Matrix of potential coefficients
(b) Matrix of Maxwell's coefficients
(c) Matrix of sequence capacitances

Zero- and negative-sequence electrostatic unbalances, assuming that the system neutral is solidly grounded
Solution:

(a) The corresponding potential coefficients are calculated using Equations 5.83 through 5.88. For example,

$$P_{aa} = \frac{1}{2\pi\varepsilon} \ln \frac{h_{11}}{r_a}$$
$$= 11.185 \ln \left(\frac{90}{0.037667} \right)$$
$$= 87.0058 \ \text{F}^{-1}\text{m}$$

$$P_{ab} = \frac{1}{2\pi\varepsilon} \ln \frac{l_{11}}{D_{12}}$$
$$= 11.185 \ln \left(\frac{82.0244}{8.2462} \right)$$
$$= 53.6949 \ \text{F}^{-1}\text{m}$$

where

$$l_{12} = \sqrt{22^2 + (45 + 37)^2}$$
$$= 84.9 \ \text{ft}$$

$$D_{12} = \sqrt{2^2 + 8^2}$$
$$= 8.2462 \ \text{ft}$$

The others can also be found similarly. Therefore,

$$[P_{abc}] = \begin{bmatrix} 87.0058 & 25.6949 & 21.9131 \\ 25.6949 & 84.8164 & 19.7635 \\ 21.9131 & 19.7635 & 85.6884 \end{bmatrix}$$

(b)

$$[C_{abc}] = [P_{abc}]^{-1}$$

$$= \begin{bmatrix} 1.31 \times 10^{-2} & -3.38 \times 10^{-3} & -2.58 \times 10^{-3} \\ -3.38 \times 10^{-3} & 1.33 \times 10^{-2} & -2.21 \times 10^{-3} \\ -2.58 \times 10^{-3} & -2.21 \times 10^{-3} & 1.28 \times 10^{-2} \end{bmatrix}$$

(c)

$$[C_{012}] = [A]^{-1} [C_{abc}] [A]$$

$$= \begin{bmatrix} 7.666 \times 10^{-3} + j3.1 \times 10^{-2} & -2.38 \times 10^{-4} - j8.94 \times 10^{-5} & -2.38 \times 10^{-4} - j8.94 \times 10^{-5} \\ -2.38 \times 10^{-4} - j8.94 \times 10^{-5} & 1.58 \times 10^{-2} - j4.37 \times 10^{-19} & 5.33 \times 10^{-4} - j6.02 \times 10^{-4} \\ -2.38 \times 10^{-4} + j8.94 \times 10^{-5} & 5.33 \times 10^{-4} - j6.02 \times 10^{-4} & -1.58 \times 10^{-2} - j1.30 \times 10^{-18} \end{bmatrix}$$

(d) From Equation 5.115,

$$\mathbf{d}_0 = \frac{\mathbf{C}_{01}}{\mathbf{C}_{11}}$$

$$= \frac{-2.38 \times 10^{-4} + j8.94 \times 10^{-5}}{1.58 \times 10^{-2}}$$

$$= 0.0160\angle 159° \text{ or } 1.60\%$$

and from Equation 5.116,

$$\mathbf{d}_2 = \frac{\mathbf{C}_{21}}{\mathbf{C}_{11}}$$

$$= -\frac{5.32 \times 10^{-4} + j6.02 \times 10^{-4}}{1.58 \times 10^{-2}}$$

$$= 0.0508\angle 228.5° \text{ or } 5.08\%$$

5.8   SEQUENCE IMPEDANCES OF SYNCHRONOUS MACHINES

In general, the impedances to positive-, negative-, and zero-sequence currents in synchronous machines (as well as other rotating machines) have different values. The positive-sequence impedance of the synchronous machine can be selected to be its *subtransient* (X_d'') *transient* (X_d'), or synchronous* (X_d) reactance depending on the time assumed to elapse from the instant of fault initiation to the instant at which values are desired (e.g., for relay response, breaker opening, or sustained fault conditions). Usually, however, in fault studies, the subtransient reactance is taken as the positive-sequence reactance of the synchronous machine.

The negative-sequence impedance of a synchronous machine is usually determined from

$$\mathbf{Z}_2 = jX_2 = j\left(\frac{X_d'' + X_q''}{2}\right) \tag{5.118}$$

In a cylindrical-rotor synchronous machine, the subtransient and negative-sequence reactances are the same, as shown in Table 5.3.

The zero-sequence impedance of a synchronous machine varies widely and depends on the pitch of the armature coils. It is much smaller than the corresponding positive- and negative-sequence reactances. It can be measured by connecting the three armature windings in series and applying

Table 5.3

Typical Reactances of Three-Phase Synchronous Machines

	Turbine Generators						Salient-Pole Generators						Synchronous Condensers					
	Two Pole			Four Pole			With Dampers			With Dampers			Air Cooled			Hydrogen Cooled		
	Low	Avg.	High	Low	Avg.	High	Low	Avg.	High	Low	Avg.	High	Low	Avg.	High	Low	Avg.	High
X_d	0.95	1.2	1.45	1	1.2	1.45	0.6	1.25	1.5	0.6	1.25	1.5	1.25	1.85	2.2	1.5	2.2	2.65
X'_d	0.12	0.15	0.21	0.2	0.23	0.28	0.2	0.3	0.5	0.2	0.3	0.5	0.3	0.4	0.5	0.36	0.48	0.6
X''_d	0.07	0.09	0.14	0.12	0.14	0.17	0.13	0.2	0.32	0.2	0.3	0.5	0.19	0.27	0.3	0.23	0.32	0.36
X_q	0.92	1.16	1.42	0.92	1.16	1.42	0.4	0.7	0.8	0.4	0.7	0.8	0.95	1.15	1.3	1.1	1.35	1.55
X_2	0.07	0.09	0.14	0.12	0.14	0.17	0.13	0.2	0.32	0.35	0.48	0.65	0.18	0.26	0.4	0.22	0.31	0.48
X_0	0.01	0.03	0.08	0.015	0.08	0.14	0.03	0.18	0.23	0.03	0.19	0.24	0.025	0.12	0.15	0.03	0.14	0.18

a single-phase voltage. The ratio of the terminal voltage of one phase winding to the current is the zero-sequence reactance. It is approximately equal to the zero-sequence reactance. Table 5.3 [138] gives typical reactance values of three-phase synchronous machines. Note that, in the above discussion, the resistance values are ignored because they are much smaller than the corresponding reactance values.

Figure 5.7 shows the equivalent circuit of a cylindrical-rotor synchronous machine with constant field current. Since the coil groups of the three-phase stator armature windings are displaced from each other by 120 electrical degrees, balanced three-phase sinusoidal voltages are induced in the stator windings. Furthermore, each of the three self-impedances and mutual impedances are equal to each other, respectively, owing to the machine symmetry. Therefore, taking into account the neutral impedance $\mathbf{Z}_n$ and applying Kirchhoff's voltage law it can be shown that

$$\mathbf{E}_a = (R_\phi + jX_s + \mathbf{Z}_n)\mathbf{I}_a + (jX_m + \mathbf{Z}_n)\mathbf{I}_b + (jX_m + \mathbf{Z}_n)\mathbf{I}_c + \mathbf{V}_a \qquad (5.119)$$

$$\mathbf{E}_b = (jX_m + \mathbf{Z}_n)\mathbf{I}_a + (R_\phi + jX_m + \mathbf{Z}_n)\mathbf{I}_b + (jX_m + \mathbf{Z}_n)\mathbf{I}_c + \mathbf{V}_b \qquad (5.120)$$

$$\mathbf{E}_c = (jX_m + \mathbf{Z}_n)\mathbf{I}_a + (jX_m + \mathbf{Z}_n)\mathbf{I}_b + (R_\phi + jX_m + \mathbf{Z}_n)\mathbf{I}_c + \mathbf{V}_c \qquad (5.121)$$

or, in matrix form,

$$\begin{bmatrix} \mathbf{E}_a \\ \mathbf{E}_b \\ \mathbf{E}_c \end{bmatrix} = \begin{bmatrix} \mathbf{Z}_s & \mathbf{Z}_m & \mathbf{Z}_s \\ \mathbf{Z}_m & \mathbf{Z}_s & \mathbf{Z}_m \\ \mathbf{Z}_m & \mathbf{Z}_m & \mathbf{Z}_s \end{bmatrix} \begin{bmatrix} \mathbf{I}_a \\ \mathbf{I}_b \\ \mathbf{I}_c \end{bmatrix} + \begin{bmatrix} \mathbf{V}_a \\ \mathbf{V}_b \\ \mathbf{V}_c \end{bmatrix} \qquad (5.122)$$

where

$$\mathbf{Z}_s = R_\phi + jX_s + \mathbf{Z}_n \qquad (5.123)$$

$$\mathbf{Z}_m = jX_m + \mathbf{Z}_n \qquad (5.124)$$

$$\mathbf{E}_a = \mathbf{E}_a \qquad (5.125)$$

$$\mathbf{E}_b = \mathbf{a}^2 \mathbf{E}_a \qquad (5.126)$$

$$\mathbf{E}_c = \mathbf{a}\mathbf{E}_a \qquad (5.127)$$

Alternatively, Equation 5.122 can be written in shorthand matrix notation as

$$\begin{bmatrix} \mathbf{E}_{abc} \end{bmatrix} = \begin{bmatrix} \mathbf{Z}_{abc} \end{bmatrix} \begin{bmatrix} \mathbf{I}_{abc} \end{bmatrix} + \begin{bmatrix} \mathbf{V}_{abc} \end{bmatrix} \qquad (5.128)$$

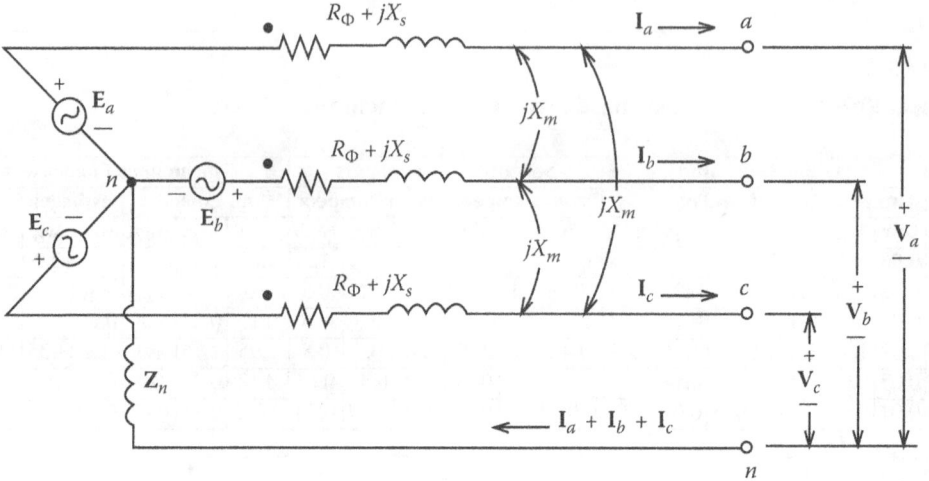

Figure 5.7 Equivalent circuit of cylindrical-rotor synchronous machine.

Multiplying both sides of this equation by $[\mathbf{A}]^{-1}$ and also substituting Equation 5.26 into it,

$$[\mathbf{A}]^{-1}[\mathbf{E}_{abc}] = [\mathbf{A}]^{-1}[\mathbf{Z}_{abc}][\mathbf{A}][\mathbf{I}_{012}] + [\mathbf{A}]^{-1}[\mathbf{V}_{abc}] \tag{5.129}$$

where

$$[\mathbf{A}]^{-1}[\mathbf{E}_{abc}] = \frac{1}{3}\begin{bmatrix} 1 & 1 & 1 \\ 1 & \mathbf{a} & \mathbf{a}^2 \\ 1 & \mathbf{a}^2 & \mathbf{a} \end{bmatrix}\begin{bmatrix} \mathbf{E} \\ \mathbf{a}^2\mathbf{E} \\ \mathbf{a}\mathbf{E} \end{bmatrix}$$

$$= \begin{bmatrix} 0 \\ \mathbf{E} \\ 0 \end{bmatrix} \tag{5.130}$$

$$[\mathbf{Z}_{012}] \triangleq [\mathbf{A}]^{-1}[\mathbf{Z}_{abc}][\mathbf{A}]$$

$$[\mathbf{V}_{012}] \triangleq [\mathbf{A}]^{-1}[\mathbf{V}_{abc}]$$

Also, due to the symmetry of the machine,

$$[\mathbf{Z}_{012}]\begin{bmatrix} \mathbf{Z}_s + 2\mathbf{Z}_m & 0 & 0 \\ 0 & \mathbf{Z}_s - \mathbf{Z}_m & 0 \\ 0 & 0 & \mathbf{Z}_s - \mathbf{Z}_m \end{bmatrix} \tag{5.131}$$

or

$$[\mathbf{Z}_{012}]\begin{bmatrix} \mathbf{Z}_{00} & 0 & 0 \\ 0 & \mathbf{Z}_{11} & 0 \\ 0 & 0 & \mathbf{Z}_{22} \end{bmatrix} \tag{5.132}$$

where

$$\mathbf{Z}_{00} = \mathbf{Z}_s + 2\mathbf{Z}_m = R_\phi + j(X_s + 2X_m) + 3\mathbf{Z}_n \tag{5.133}$$

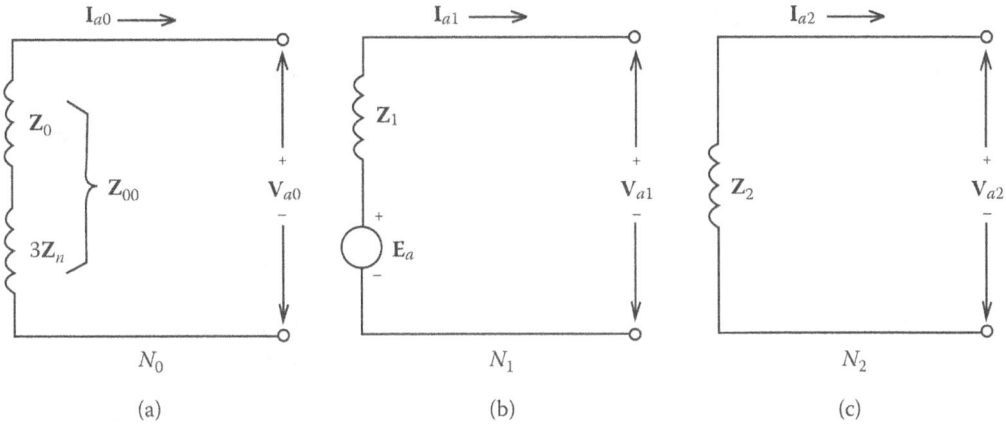

Figure 5.8 Sequence networks of synchronous machine: (a) zero-sequence network; (b) positive-sequence network; (c) negative-sequence network.

$$\mathbf{Z}_{11} = \mathbf{Z}_s - \mathbf{Z}_m = R_\phi + j(X_s + X_m) \tag{5.134}$$

$$\mathbf{Z}_{22} = \mathbf{Z}_s - \mathbf{Z}_m = R_\phi + j(X_s + X_m) \tag{5.135}$$

Therefore, Equation 5.128 in terms of the symmetrical components can be expressed as

$$\begin{bmatrix} 0 \\ \mathbf{E}_a \\ 0 \end{bmatrix} = \begin{bmatrix} \mathbf{Z}_{00} & 0 & 0 \\ 0 & \mathbf{Z}_{11} & 0 \\ 0 & 0 & \mathbf{Z}_{22} \end{bmatrix} \begin{bmatrix} \mathbf{I}_{a0} \\ \mathbf{I}_{a1} \\ \mathbf{I}_{a2} \end{bmatrix} + \begin{bmatrix} \mathbf{V}_{a0} \\ \mathbf{V}_{a1} \\ \mathbf{V}_{a2} \end{bmatrix} \tag{5.136}$$

or, in shorthand matrix notation,

$$\begin{bmatrix} \mathbf{E} \end{bmatrix} = \begin{bmatrix} \mathbf{Z}_{012} \end{bmatrix} \begin{bmatrix} \mathbf{I}_{012} \end{bmatrix} + \begin{bmatrix} \mathbf{V}_{012} \end{bmatrix} \tag{5.137}$$

Similarly,

$$\begin{bmatrix} \mathbf{V}_{a0} \\ \mathbf{V}_{a1} \\ \mathbf{V}_{a2} \end{bmatrix} = \begin{bmatrix} 0 \\ \mathbf{E}_a \\ 0 \end{bmatrix} - \begin{bmatrix} \mathbf{Z}_{00} & 0 & 0 \\ 0 & \mathbf{Z}_{11} & 0 \\ 0 & 0 & \mathbf{Z}_{22} \end{bmatrix} \begin{bmatrix} \mathbf{I}_{a0} \\ \mathbf{I}_{a1} \\ \mathbf{I}_{a2} \end{bmatrix} \tag{5.138}$$

or

$$\begin{bmatrix} \mathbf{V}_{012} \end{bmatrix} = \begin{bmatrix} \mathbf{E} \end{bmatrix} - \begin{bmatrix} \mathbf{Z}_{012} \end{bmatrix} \begin{bmatrix} \mathbf{I}_{012} \end{bmatrix} \tag{5.139}$$

Note that, the machine sequence impedances in the above equations are

$$\mathbf{Z}_0 \triangleq \mathbf{Z}_{00} - 3\mathbf{Z}_n \tag{5.140}$$

$$\mathbf{Z}_1 \triangleq \mathbf{Z}_{11} \tag{5.141}$$

$$\mathbf{Z}_2 \triangleq \mathbf{Z}_{22} \tag{5.142}$$

The expression given in Equation 5.140 is due to the fact that the impedance $\mathbf{Z}_n$ is external to the machine. Figure 5.8 shows the sequence networks of a synchronous machine.

5.9 ZERO-SEQUENCE NETWORKS

It is important to note that the zero-sequence system is not a three-phase system but a single-phase system. This is because the zero-sequence currents and voltages are equal in magnitude and in phase at any point in all the phases of the system. However, the zero-sequence currents can only exist in a circuit if there is a complete path for their flow. Therefore, if there is no complete path for zero-sequence currents in a circuit, the zero-sequence impedance is infinite. In a zero-sequence network drawing, this infinite impedance is indicated by an open circuit.

Figure 5.9 shows zero-sequence networks for wye- and delta-connected three-phase loads. Note that, a wye-connected load with an ungrounded neutral has infinite impedance to zero-sequence currents since there is no return path through the ground or a neutral conductor, as shown in Figure 5.9a. On the other hand, a wye-connected load with solidly grounded neutral, as shown in Figure 5.9b, provides a return path for the zero-sequence currents flowing through the three phases and their sum, $3I_{a0}$, flowing through the ground. If the neutral is grounded through some impedance Z_n as shown in Figure 5.9c, an impedance of $3Z_n$ should be inserted between the neutral point n and the zero-potential bus N_0 in the zero-sequence network. The reason for this is that a current of $3I_{a0}$ produces a zero-sequence voltage drop of $3I_{a0}Z_n$ between the neutral point n and the ground. Therefore, to reflect this voltage drop in the zero-sequence network, where the zero-sequence current $3I_{a0}$ flows, the neutral impedance should be $3Z_n$. A delta-connected load, as shown in Figure 5.9d, provides no path for zero-sequence currents flowing in the line. Therefore, its zero-sequence impedance, as seen from its terminals, is infinite. Yet, it is possible to have zero-sequence currents circulating within the delta circuit. However, they have to be produced in the delta by zero-sequence voltages or by induction from an outside source.

5.10 SEQUENCE IMPEDANCES OF TRANSFORMERS

A three-phase transformer may be made up of three identical single-phase transformers. If this is the case, it is called a three-phase *transformer bank*. Alternatively, it may be built as a three-phase transformer having a single common core (either with shell-type or core-type design) and a tank. For the sake of simplicity, here only the three-phase transformer banks will be reviewed. The impedance of a transformer to both positive- and negative-sequence currents is the same. Even though the zero-sequence series impedances of three-phase units are little different than the positive- and negative-sequence series impedances, it is often assumed in practice that series impedances of all sequences are the same without paying attention to the transformer type

$$\mathbf{Z}_0 = \mathbf{Z}_1 = \mathbf{Z}_2 = \mathbf{Z}_{trf} \qquad (5.143)$$

If the flow of zero-sequence current is prevented by the transformer connection, $\mathbf{Z}_0$ is infinite. Figure 5.10 shows zero-sequence network equivalents of three-phase transformer banks made up of three identical single-phase transformers having two windings with excitation currents neglected. The possible paths for the flow of zero-sequence current are indicated on the connection diagrams, as shown in Figure 5.10a, 5.10c, and 5.10e. If there is no path shown on the connection diagram, this means that the transformer connection prevents the flow of the zero-sequence current by not providing a path for it, as indicated in Figure 5.10b, 5.10d, and 5.10f.

Note that even though the delta-delta bank can have zero-sequence currents circulating within its delta windings, it also prevents the flow of the zero-sequence current outside the delta windings by not providing a return path for it, as shown in Figure 5.10.

Also, note that if the neutral point n of the wye winding (shown in Figure 5.10a or c) is grounded through $\mathbf{Z}_n$, the corresponding zero-sequence impedance $\mathbf{Z}_0$ should be replaced by $\mathbf{Z}_0 + 3\mathbf{Z}_n$.

If the wye winding is solidly grounded, the $\mathbf{Z}_n$ is zero, and therefore $3\mathbf{Z}_n$ should be replaced by a short circuit. On the other hand, if the connection is ungrounded, the $\mathbf{Z}_n$ is infinite, and therefore

Load connection diagram   Zero-sequence network equivalent

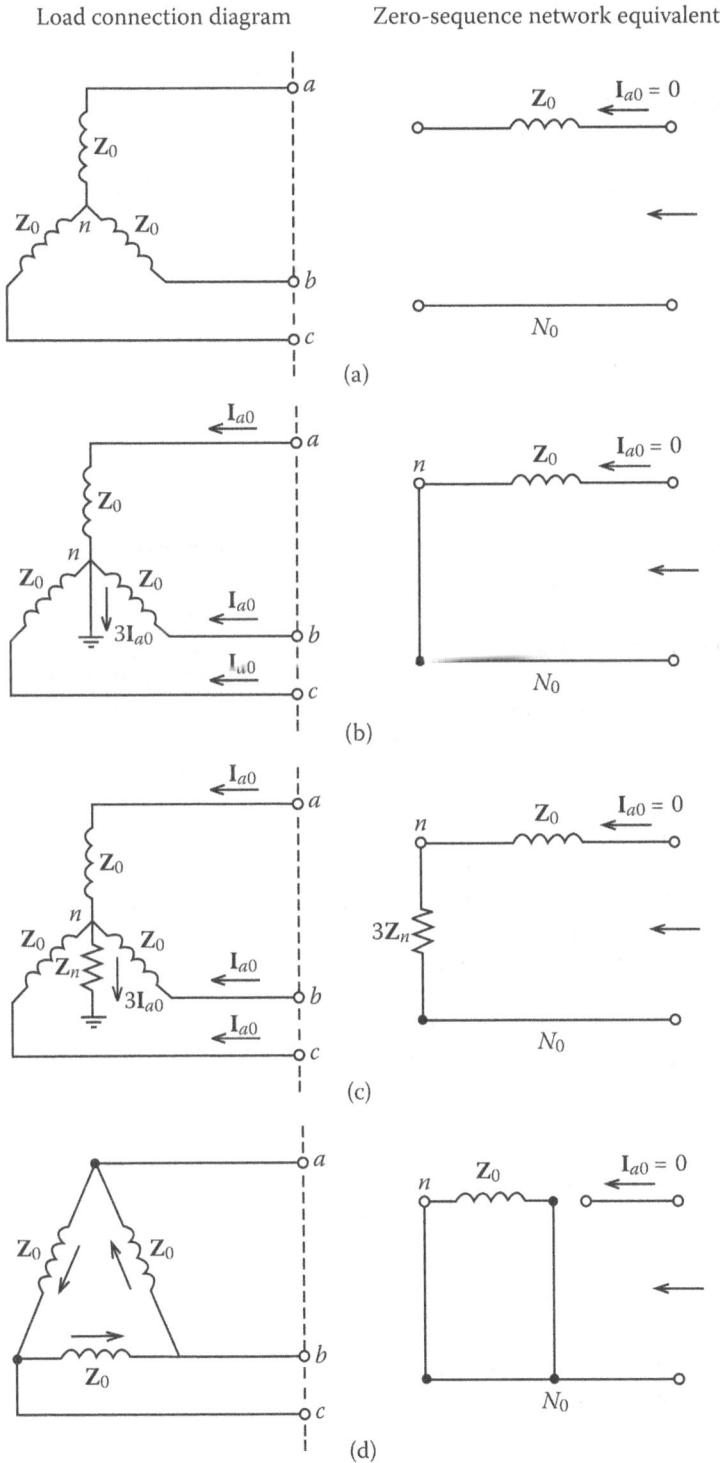

Figure 5.9 Zero-sequence network for wye- and delta-connected three-phase loads: (a) wye-connected load with undergrounded neutral; (b) wye-connected load with grounded neutral; (c) wye-connected load grounded through neutral impedance; (d) delta-connected load.

Table 5.4
System Data for Example 5.7

Network component	MVA Rating	Voltage Rating (kV)	X_1 (pu)	X_2 (pu)	X_0 (pu)
G_1	200	20	0.2	0.14	0.06
G_2	200	13.2	0.2	0.14	0.06
T_1	200	20/230	0.2	0.2	0.2
T_2	200	13.2/230	0.3	0.3	0.3
T_3	200	20/230	0.25	0.25	0.25
T_4	200	13.2/230	0.35	0.35	0.35
TL_{23}	200	230	0.15	0.15	0.3
TL_{56}	200	230	0.22	0.22	0.5

$3\mathbf{Z}_n$ should be replaced with an open circuit. It is interesting to observe that the type of grounding only affects the zero-sequence network, not the positive- and negative-sequence networks.

It is interesting to note that there is no path for the flow of zero-sequence current in a wye-grounded-wye-connected three-phase transformer bank, as shown in Figure 5.10b. This is because there is no zero-sequence current in any given winding on the wye side of the transformer bank since it has an ungrounded wye connection. Therefore, because of the lack of equal and opposite ampere turns in the wye side of the transformer bank, there cannot be any zero-sequence current in the corresponding winding on the wye-grounded side of the transformer, with the exception of a negligible small magnetizing current.

Figure 5.11 shows zero-sequence network equivalents of three-phase transformer banks made of three identical single-phase transformers with three windings. The impedances of the three-winding transformer between primary, secondary, and tertiary terminals, indicated by P, S, and T, respectively, taken two at a time with the other winding open, are $\mathbf{Z}_{PS}$, $\mathbf{Z}_{PT}$, and $\mathbf{Z}_{ST}$, the subscripts indicating the terminals between which the impedances are measured. Note that, only the wye-wye connection with delta tertiary, shown in 5.11a, permits zero-sequence current to flow in from either wye line (as long as the neutrals are grounded).

Example 5.7: Consider the power system shown in Figure 5.12 and the associated data given in Table 5.4. Assume that each three-phase transformer bank is made of three single-phase transformers. Do the following:

(a) Draw the corresponding positive-sequence network
(b) Draw the corresponding negative-sequence network
(c) Draw the corresponding zero-sequence network

Solution:

(a) The positive-sequence network is shown in Figure 5.13a
(b) The negative-sequence network is shown in Figure 5.13b
(c) The zero-sequence network is shown in Figure 5.13c

Example 5.8: Consider the power system given in Example 5.7 and assume that there is a fault on bus 3. Reduce the sequence networks drawn in Example 5.7 to their Thévenin equivalents "looking in" at bus 3.

(a) Show the steps of the positive-sequence network reduction

Symbols	Transformer connection diagram	Zero-sequence network equivalent

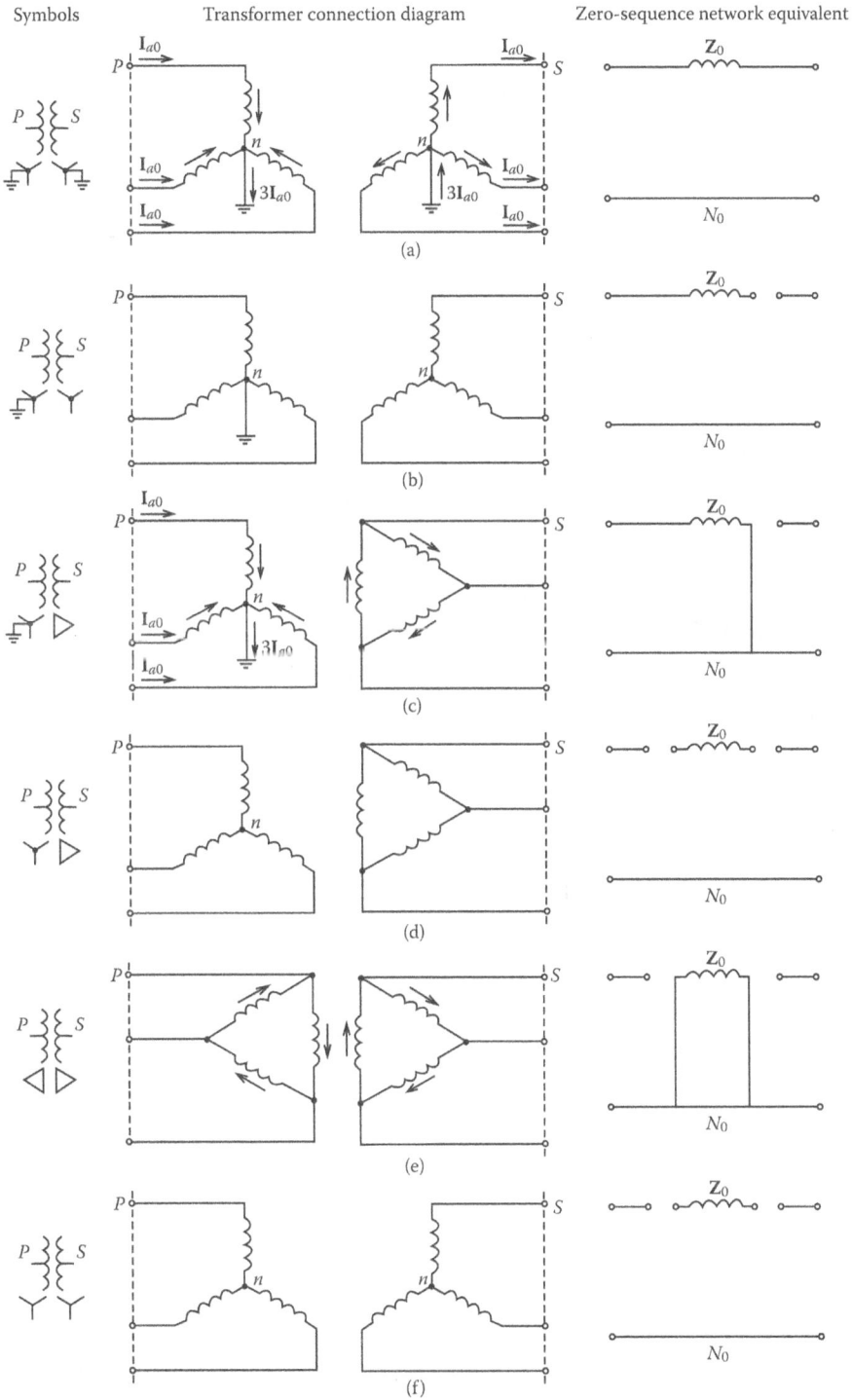

Figure 5.10 Zero-sequence network equivalents of three-phase transformer banks made of three identical single-phase transformers with two windings.

Transformer connection diagram   Zero-sequence network equivalent

(a)

(b)

(c)

(d)

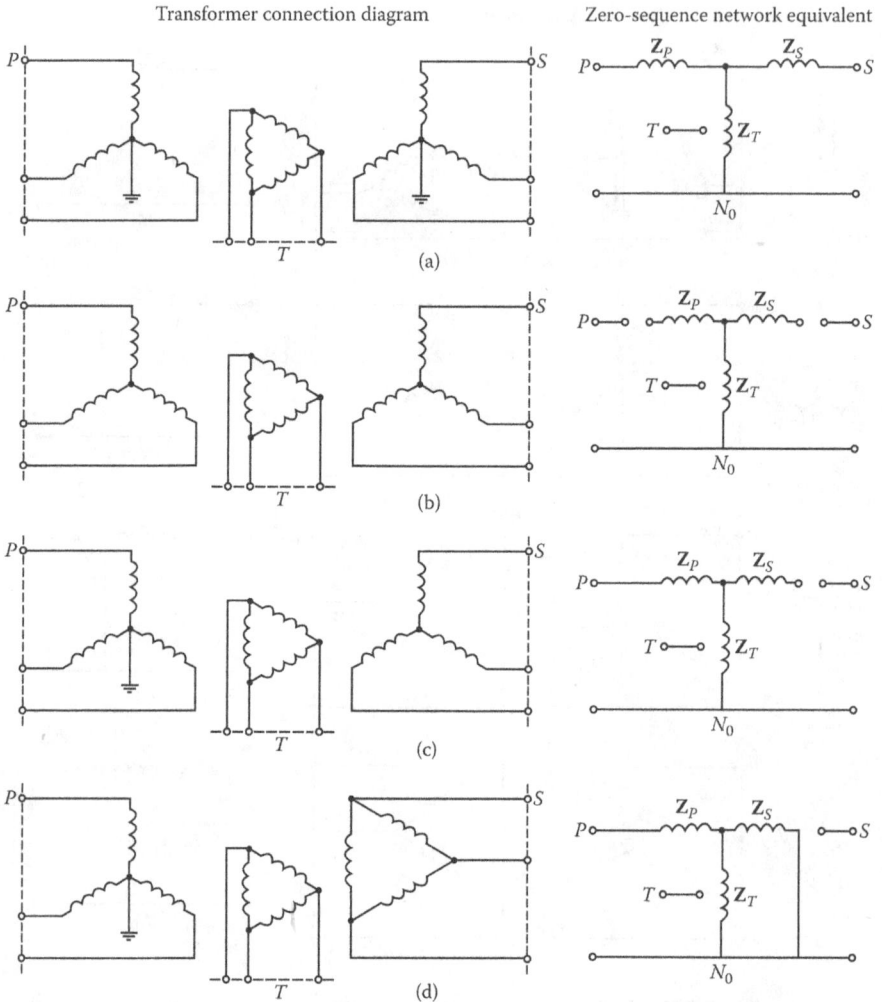

Figure 5.11 Zero-sequence network equivalents of three-phase transformer banks made of three identical single-phase transformers with three windings.

(b) Show the steps of the negative-sequence network reduction
(c) Show the steps of the zero-sequence network reduction

Solution:

(a) Figure 5.14 shows the steps of the positive-sequence network reduction. Note that the delta that exits between nodes 1, 3, and 4, as shown in Figure 5.14a, must be replaced by its equivalent wye configuration, as shown in Figure 5.14b, by performing the following calculations:

$$\mathbf{Z}_1 = j\frac{0.35 \times 0.82}{0.35 + 0.82 + 0.3}$$
$$= j0.1952 \text{ pu}$$

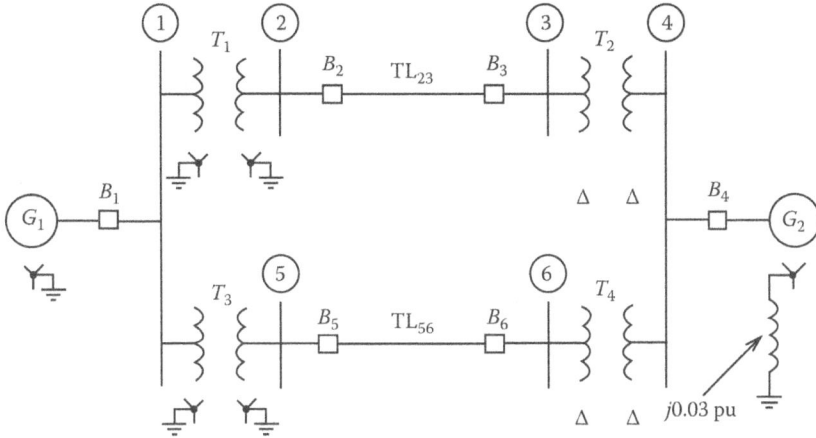

Figure 5.12   Power system for Example 5.7.

$$Z_2 = j\frac{0.3 \times 0.82}{0.35 + 0.82 + 0.3}$$
$$= j0.1673 \text{ pu}$$

$$Z_3 = j\frac{0.35 \times 0.3}{0.35 + 0.82 + 0.3}$$
$$= j0.0714 \text{ pu}$$

(b) Figure 5.15 shows the steps of the negative-sequence network reduction. Note that, the delta that exits between nodes 1, 3, and 4, as shown in Figure 5.14a, must be replaced by its equivalent wye configuration, as shown in Figure 5.14b, by performing the calculations as in part (a) above.

(c) Figure 5.16 shows the steps of the zero-sequence network reduction.

PROBLEMS

1. Determine the symmetrical components for the phase currents of $I_a = 125\angle 20°$, $I_b = 175\angle -100°$, and, $I_c = 95\angle 155°$ A.

2. Assume that the unbalanced phase currents are $I_a = 100\angle 180°$, $I_b = 100\angle 0°$, and, $I_c = 10\angle 20°$ A.

   (a) Determine the symmetrical components.
   (b) Draw a phasor diagram showing I_{a0}, I_{a1}, I_{a2}, I_{b0}, I_{b1}, I_{b2}, I_{c0}, I_{c1}, I_{c2} (i.e., the positive-, negative-, and zero-sequence currents for each phase).

3. Assume that $V_{a1} = 180\angle 0°$, $V_{a2} = 100\angle 100°$, and, $V_{a0} = 250\angle -40°$ V.

   (a) Draw a phasor diagram showing all the nine symmetrical components.
   (b) Find the phase voltages $[V_{abc}]$ using the equation

$$[V_{abc}] = [A][V_{012}]$$

   (c) Find the phase voltages $[V_{abc}]$ graphically and check the results against the ones found in part b.

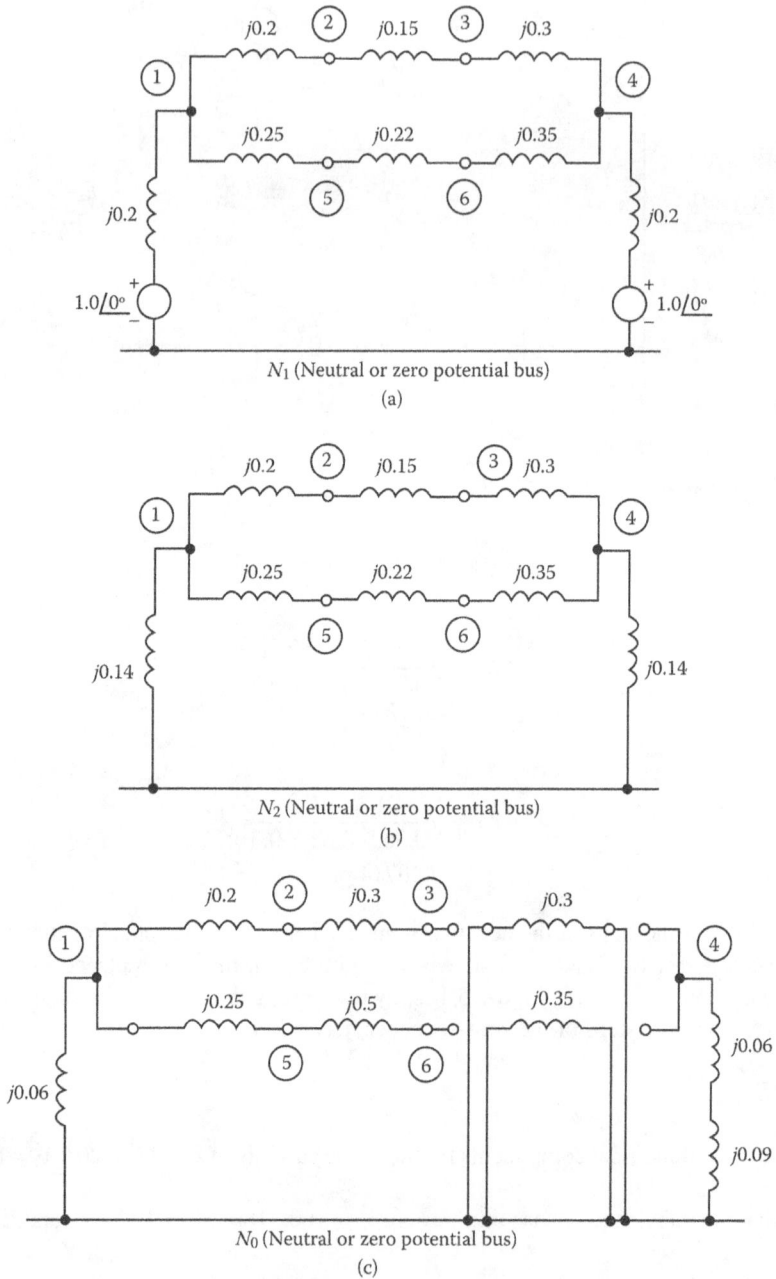

Figure 5.13 Sequence networks for Example 5.7.

4. Repeat Example 5.2 assuming that the phase voltages and currents are given as

$$\left[\mathbf{V}_{abc}\right] = \begin{bmatrix} 100\angle 0° \\ 100\angle 60° \\ 100\angle -60° \end{bmatrix} \text{ and } \left[\mathbf{I}_{abc}\right] = \begin{bmatrix} 10\angle -30° \\ 10\angle 30° \\ 10\angle -90° \end{bmatrix}$$

5. Determine the symmetrical components for the phase currents of $\mathbf{I}_a = 100\angle 20°$, $\mathbf{I}_b = 50\angle -20°$,

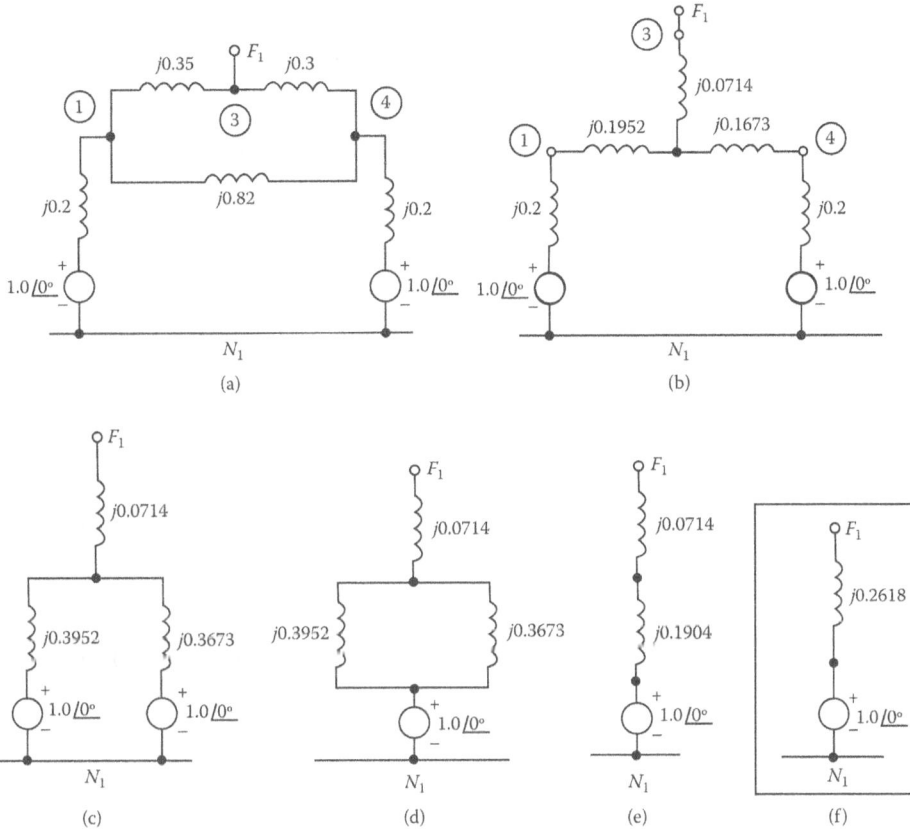

Figure 5.14 Reduction steps for positive-sequence network of Example 5.8.

and, $I_c = 150\angle180°$ A. Draw a phasor diagram showing all the nine symmetrical components.

6. Assume that $I_{a0} = 50 - j86.6$, $I_{a1} = 200\angle0°$, and, $I_{a2} = 400\angle0°$ A. Determine the following:

   (a) The negative sequence current I_{a2}
   (b) The faulted phase b current I_b
   (c) The faulted phase c current I_c

7. Determine the symmetrical components for the phase currents of $I_a = 200\angle0°$, $I_b = 175\angle -90°$, and, $I_c = 100\angle90°$ A.

8. Use the symmetrical components for the phase voltages and verify the following line-to-line voltage equations:

   (a) $V_{ab} = \sqrt{3}(V_{a1}\angle30° + V_{a2}\angle -30°)$
   (b) $V_{bc} = \sqrt{3}(V_{a1}\angle -90° + V_{a2}\angle90°)$
   (c) $V_{ca} = \sqrt{3}(V_{a1}\angle150° + V_{a2}\angle -150°)$

9. Consider Example 5.3 and assume that the voltage applied at the sending end of the line is $69\angle 0°$ kV Determine the phase current matrix from Equation 5.35.

10. Consider a three-phase horizontal line configuration and assume that the phase spacings are $D_{ab} = 30$ ft, $D_{bc} = 30$ ft, and $D_{ca} = 60$ ft. The line conductors are made of 500 kcmil, 37-strand copper conductors. Assume that the 100-mi-long untransposed transmission line operates at 50°C, 60 Hz. If the earth has an average resistivity, determine the following:

   (a) Self-impedances of line conductors in ohms per mile
   (b) Mutual impedances of line conductors in ohms per mile

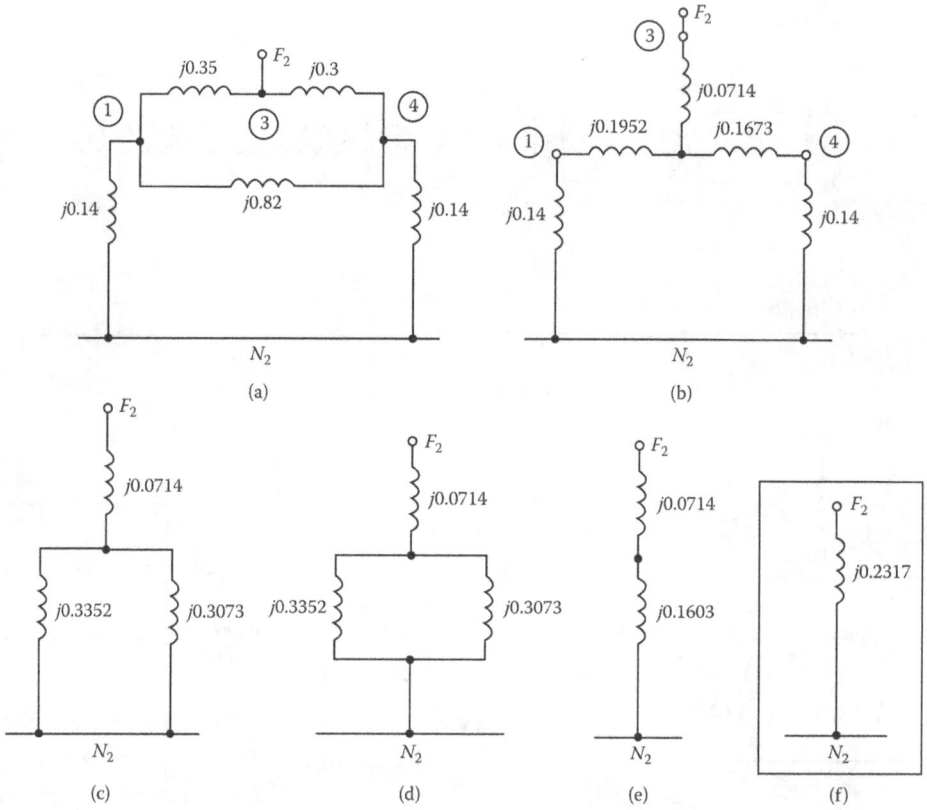

Figure 5.15 Reduction steps for negative-sequence network of Example 5.8.

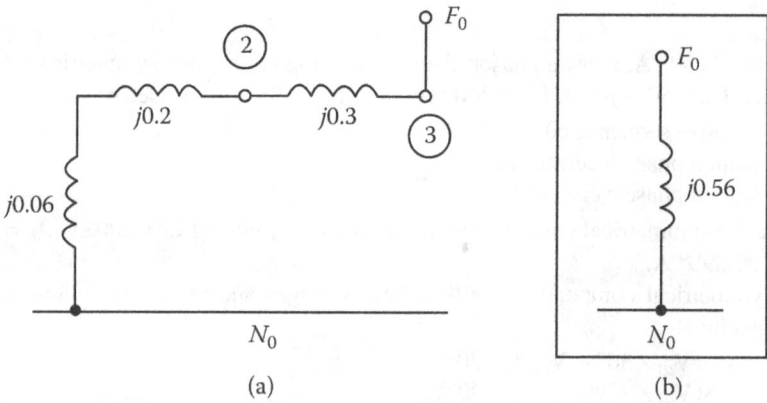

Figure 5.16 Reduction steps for zero-sequence network of Example 5.8.

   (c) Phase impedance matrix of line in ohms

11. Consider a 50-mi-long completely transposed transmission line operating at 25°C, 50 Hz, and
    having 500-kcmil ACSR conductors. The three-phase conductors have a triangular configuration
    with spacings of $D_{ab} = 6$ ft, $D_{bc} = 10$ ft, and $D_{ca} = 8$ ft. If the earth is considered to be dry earth,

determine the following:

    (a) Zero-sequence impedance of line

    (b) Positive-sequence impedance of line

    (c) Negative-sequence impedance of line

12. Consider a three-phase, vertical pole-top conductor configuration. Use 50°C, 60 Hz and assume that the phase spacings are D_{ab} = 72 in., D_{bc} = 72 in., and D_{ca} = 144 in. The line conductors are made of 795-kcmil, 30/19-strand ACSR. If the line is 100 mi long and not transposed, determine the following:

    (a) Phase impedance matrix of line

    (b) Phase admittance matrix of line

    (c) Sequence impedance matrix of line

    (d) Sequence admittance matrix of line

13. Repeat Problem 12 assuming that the phase spacings are D_{ab} = 144 in., D_{bc} = 144 in., and D_{ca} = 288 in.

14. Repeat Problem 12 assuming that the conductor is 795-kcmil, 61% conductivity, 37-strand, hard-drawn aluminum.

15. Repeat Problem 12 assuming that the conductor is 750-kcmil, 97.3% conductivity, 37-strand, hard-drawn copper conductor.

16. Consider the line configuration shown in Figure 5.5. Assume that the 115-kV line is transposed and its conductors are made up of 500-kcmil, 30/7-strand ACSR conductors. Ignore the overhead ground wire but consider the heights of the conductors and determine the zero-sequence capacitance of the line in nanofarads per mile and nanofarads per kilometer.

17. Solve Problem 16 taking into account the overhead ground wire. Assume that the overhead ground wire is made of 3/8-in. E.B.B. steel conductor.

18. Repeat Example 5.6 without ignoring the overhead ground wire. Assume that the overhead ground wire is made of 3/8-in. E.B.B. steel conductor.

19. Consider the line configuration shown in Figure 5.5. Assume that the 115-kV line is transposed and its conductors are made of 500-kcmil, 30/7-strand ACSR conductors. Ignore the effects of conductor heights and overhead ground wire and determine the following:

    (a) Positive- and negative-sequence capacitances to ground of line in nanofarads per mile

    (b) The 60-Hz susceptance of line in microsiemens per mile

    (c) Charging kilovolt-ampers per phase per mile of line

    (d) Three-phase charging kilovolt-amperes per mile of line

20. Repeat Problem 19 without ignoring the effects of conductor heights.

21. Consider the untransposed line shown in Figure 5.17. Assume that the 50-mi-long line has an overhead ground wire of 3/0 ACSR and that the phase conductors are of 556.5-kcmil, 30/7 strand, ACSR. Use a frequency of 60 Hz, an ambient temperature of 50° C, and average earth resistivity and determine the following:

    (a) Phase impedance matrix of line

    (b) Sequence impedance matrix of line

    (c) Sequence admittance matrix of line

    (d) Electrostatic zero- and negative-sequence unbalance factors of line

22. Repeat Problem 21 assuming that there arc two overhead ground wires, as shown in Figure 5.18.

23. Consider the power system given in Example 5.7 and assume that transformers T_1 and T_3 are connected as delta-wye grounded, and T_2 and T_4 are connected as wye grounded-delta, respectively. Assume that there is a fault on bus 3 and do the following:

    (a) Draw the corresponding zero-sequence network.

    (b) Reduce the zero-sequence network to its Thévenin equivalent looking in at bus 3.

24. Consider the power system given in Example 5.7 and assume that all four transformers are connected as wye grounded-wye grounded. Assume there is a fault on bus 3 and do the following:

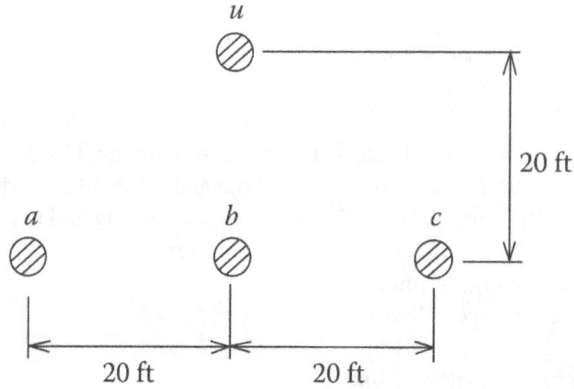

Figure 5.17 System for Problem 27.

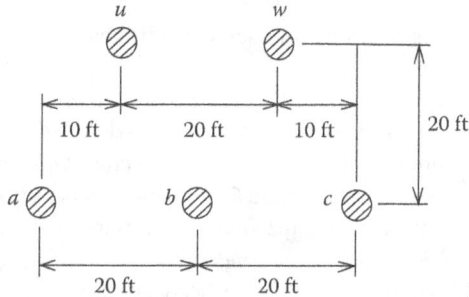

Figure 5.18 System for Problem 22.

(a) Draw the corresponding zero-sequence network.
(b) Reduce the zero-sequence network to its Thévenin equivalent looking in at bus 3.

25. Consider the power system given in Problem 10. Use 25 MVA as the megavolt-ampere base and draw the positive-, negative-, and zero-sequence networks (but do not reduce them). Assume that the two three-phase transformer bank connections are:
(a) Both wye-grounded
(b) Delta-wye grounded for transformer T_1 and wye grounded-delta for transformer T_2
(c) Wye grounded-wye for transformer T_1 and delta-wye for transformer T_2

26. Assume that a three-phase, 45-MVA, 34.5/115-kV transformer bank of three single-phase transformers, with nameplate impedances of 7.5%, is connected wye-delta with the high- voltage side delta. Determine the zero-sequence equivalent circuit (in per-unit values) under the following conditions:
(a) If neutral is ungrounded
(b) If neutral is solidly grounded
(c) If neutral is grounded through 10-Ω resistor
(d) If neutral is grounded through 4000-μF capacitor

27. Consider the system shown in Figure 5.19. Assume that the following data are given based on 20

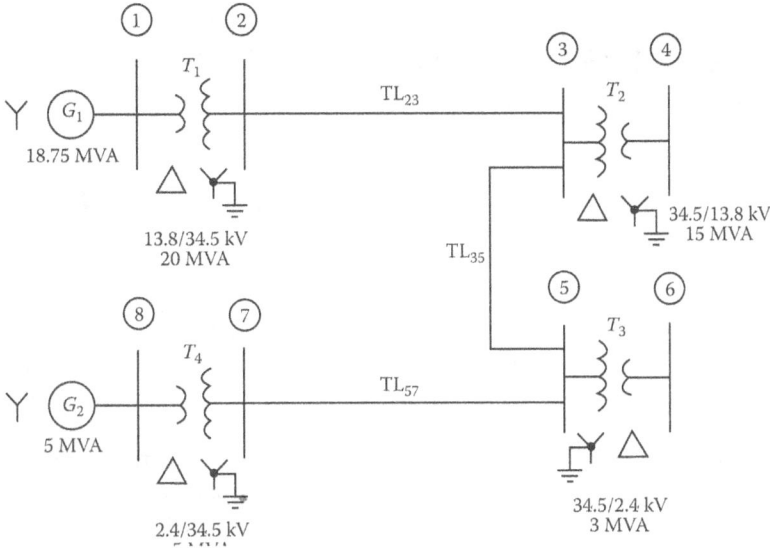

Figure 5.19 System for Problem 27.

MVA and the line-to-line base voltages as shown in Figure 5.19.

$$\text{Generator } G_1 : X_1 = 0.25 \text{ pu}, X_2 = 0.15 \text{ pu}, X_0 = 0.05 \text{ pu}$$
$$\text{Generator } G_2 : X_2 = 0.90 \text{ pu}, X_2 = 0.60 \text{ pu}, X_0 = 0.05 \text{ pu}$$
$$\text{Transformer } T_1 : X_1 = X_2 = X_0 = 0.10 \text{ pu}$$
$$\text{Transformer } T_2 : X_1 = X_2 = 0.10 \text{ pu}, X_0 = \infty$$
$$\text{Transformer } T_3 : X_1 = X_2 = X_0 = 0.50 \text{ pu}$$
$$\text{Transformer } T_4 : X_1 = X_2 = 0.30 \text{pu}, X_0 = \infty$$
$$\text{Transmission line TL}_{23} : X, = X_2 = 0.15 \text{ pu}, X_0 = 0.50 \text{ pu}$$
$$\text{Transmission line TL}_{35} : X_1, = X_2 = 0.30 \text{ pu}, X_0 = 1.00 \text{ pu}$$
$$\text{Transmission line TL}_{57} : X_1, = X_2 = 0.30 \text{ pu}, X_0 = 1.00 \text{ pu}$$

(a) Draw the corresponding positive-sequence network
(b) Draw the corresponding negative-sequence network
(c) Draw the corresponding zero-sequence network

[69, 175–182]

6 Analysis of Unbalanced Faults

6.1 INTRODUCTION

The analytical discussion in this chapter is expanded from balanced faults to unbalanced faults in a balanced three-phase transmission network. In actual power systems, most of the faults are not balanced, but rather unbalanced faults, particularly the single line-to-ground (SLG) faults. According to Reference [183], the frequency of occurrence for three-phase, SLG, line-to-line, and double line-to-ground (DLG) faults is 5%, 70%, 15%, and 10%, respectively.

Although the three-phase fault is generally considered the most severe type of fault, the SLG fault may be more severe than the three-phase fault under two specific circumstances: (1) when the generators involved in the fault have solidly grounded neutrals or low-impedance neutral impedances and (2) when the fault occurs on the wye-grounded side of delta-wye-grounded transformer banks. The current of a line-to-line fault is about 86.6% of the three-phase fault current.

Faults can be classified as shunt faults (short circuits), series faults (open conductor), and simultaneous faults (multiple faults happening at the same time). Unbalanced faults can be solved easily by using the symmetrical components of an unbalanced system of currents or voltages. Therefore, an unbalanced system can be transformed into three fictitious networks: the positive-sequence (which is the only one that has a driving voltage), the negative-sequence, and the zero-sequence networks, which are interconnected in a specific manner depending on the fault type involved. This book only considers and examines shunt faults.

6.2 SHUNT FAULTS

The voltage to ground of phase a at the fault point F before the fault occurred is $\mathbf{V}_F$, and it is usually selected as $1.0\angle 0°$ pu. However, it is possible to have a $\mathbf{V}_F$ value that is not $1.0\angle 0°$ pu. If so, Table 6.1 [184] gives formulas to calculate the fault currents and voltages at the fault point F and their corresponding symmetrical components for various types of faults. Note that, the positive-, negative-, and zero-sequence impedances are viewed from the fault point as $\mathbf{Z}_1$, $\mathbf{Z}_2$, and $\mathbf{Z}_0$, respectively. In the table, $\mathbf{Z}_f$ is the fault impedance and $\mathbf{Z}_{eq}$ is the equivalent impedance to replace the fault in the positive-sequence network. Also, note that the value of the impedance $\mathbf{Z}_g$ is zero in Table 6.1.

6.2.1 SLG FAULT

In general, a single line-to-ground (SLG) fault can occur on a transmission system when a conductor falls to ground or contacts the neutral wire. Figure 6.1a provides a general representation of an SLG fault at a fault point F, with a fault impedance of $\mathbf{Z}f^*$. The fault impedance $\mathbf{Z}f$ is usually ignored in fault studies. Figure 6.1b shows the resulting sequence networks' interconnection. For simplicity in fault calculations, the faulted phase is usually assumed to be phase a, as shown in Figure 6.1b.

If a phase other than phase a is faulted in reality (e.g., phase b), the phases of the system can be relabeled (i.e., a, b, c becomes c, a, b) [175]. Another method to deal with this situation is to use the "generalized fault diagram" introduced by Atabekov [175] and further developed by Anderson [63]. It can be observed from Figure 6.1b that the zero-, positive-, and negative-sequence currents are equal to each other. Therefore,

$$\mathbf{I}_{a0} = \mathbf{I}_{a1} = \mathbf{I}_{a2} = \frac{1.0\angle 0°}{\mathbf{Z}_0 + \mathbf{Z}_1 + \mathbf{Z}_2 + 3\mathbf{Z}_f} \tag{6.1}$$

DOI: 10.1201/9781003129769-6

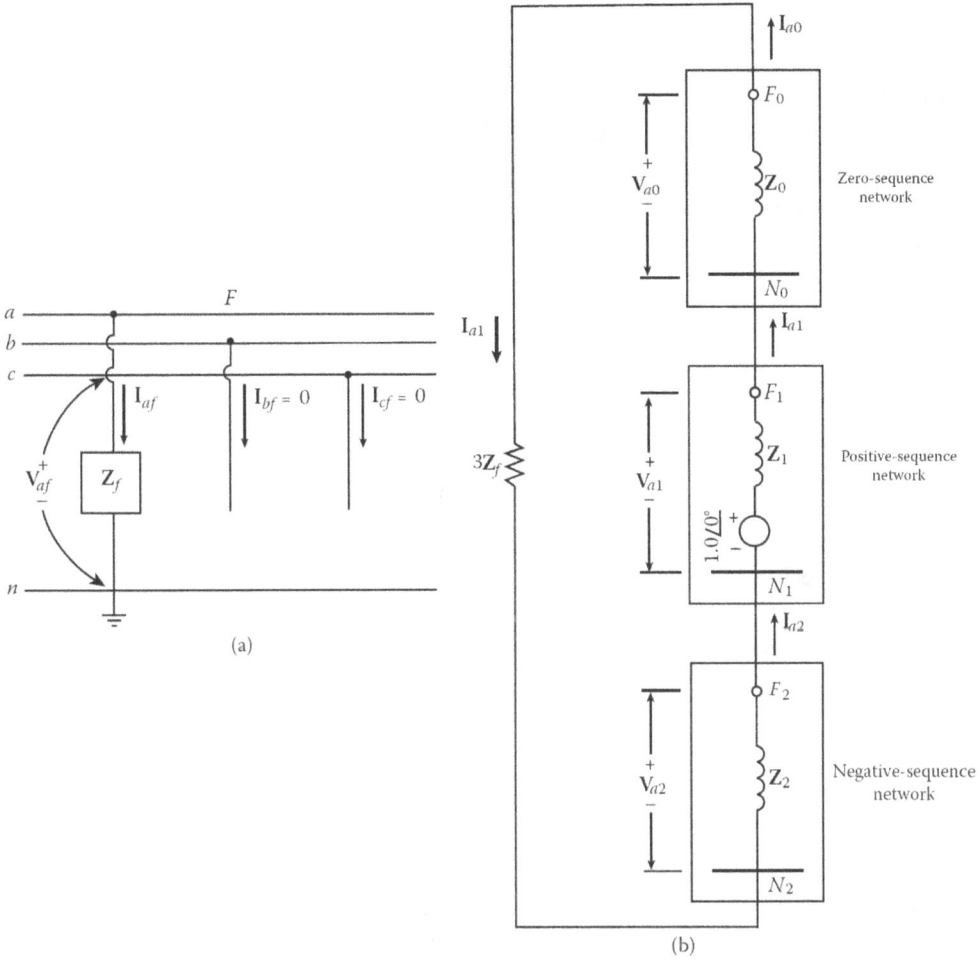

Figure 6.1 SLG fault: (a) general representation; (b) interconnection of sequence networks.

Since

$$\begin{bmatrix} \mathbf{I}_{af} \\ \mathbf{I}_{bf} \\ \mathbf{I}_{cf} \end{bmatrix} = \begin{bmatrix} 1 & 1 & 1 \\ 1 & \mathbf{a}^2 & \mathbf{a} \\ 1 & \mathbf{a} & \mathbf{a}^2 \end{bmatrix} \begin{bmatrix} \mathbf{I}_{a0} \\ \mathbf{I}_{a1} \\ \mathbf{I}_{a2} \end{bmatrix}$$ (6.2)

the fault current for phase a can be found as

$$\mathbf{I}_{af} = \mathbf{I}_{a0} + \mathbf{I}_{a1} + \mathbf{I}_{a2}$$

or

$$\mathbf{I}_{af} = 3\mathbf{I}_{a0} = 3\mathbf{I}_{a1} = 3\mathbf{I}_{a2}$$ (6.3)

From Figure 6.1a,

$$\mathbf{V}_{af} = \mathbf{Z}_f \mathbf{I}_{af}$$ (6.4)

Substituting Equation 6.3 into Equation 6.4, the voltage at faulted phase a can be expressed as

$$\mathbf{V}_{af} = 3\mathbf{Z}_f \mathbf{I}_{a1}$$ (6.5)

But,

$$\mathbf{V}_{af} = \mathbf{V}_{a0} + \mathbf{V}_{a1} + \mathbf{V}_{a2} \tag{6.6}$$

Therefore,

$$\mathbf{V}_{a0} + \mathbf{V}_{a1} + \mathbf{V}_{a2} = 3\mathbf{Z}_f\mathbf{I}_{a1} \tag{6.7}$$

which justifies the interconnection of sequence networks in series, as shown in Figure 6.1b.

Once the sequence currents are found, the zero-, positive-, and negative-sequence voltages can be found from

$$\begin{bmatrix} \mathbf{V}_{a0} \\ \mathbf{V}_{a1} \\ \mathbf{V}_{a2} \end{bmatrix} = \begin{bmatrix} 0 \\ 1.0\angle 0° \\ 0 \end{bmatrix} - \begin{bmatrix} \mathbf{Z}_0 & 0 & 0 \\ 0 & \mathbf{Z}_1 & 0 \\ 0 & 0 & \mathbf{Z}_2 \end{bmatrix} \begin{bmatrix} \mathbf{I}_{a0} \\ \mathbf{I}_{a1} \\ \mathbf{I}_{a2} \end{bmatrix} \tag{6.8}$$

as

$$\mathbf{V}_{a0} = -\mathbf{Z}_0\mathbf{I}_{a0} \tag{6.9}$$

$$\mathbf{V}_{a1} = 1.0 - \mathbf{Z}_1\mathbf{I}_{a1} \tag{6.10}$$

$$\mathbf{V}_{a2} = -\mathbf{Z}_2\mathbf{I}_{a2} \tag{6.11}$$

In the event of having an SLG fault on phase b or c, the voltages related to the known phase a voltage components can be found from

$$\begin{bmatrix} \mathbf{V}_{af} \\ \mathbf{V}_{bf} \\ \mathbf{V}_{cf} \end{bmatrix} = \begin{bmatrix} 1 & 1 & 1 \\ 1 & \mathbf{a}^2 & \mathbf{a} \\ 1 & \mathbf{a} & \mathbf{a}^2 \end{bmatrix} \begin{bmatrix} \mathbf{V}_{a0} \\ \mathbf{V}_{a1} \\ \mathbf{V}_{a2} \end{bmatrix} \tag{6.12}$$

as

$$\mathbf{V}_{bf} = \mathbf{V}_{a0} + \mathbf{a}^2\mathbf{V}_{a1} + \mathbf{a}\mathbf{V}_{a2} \tag{6.13}$$

and

$$\mathbf{V}_{cf} = \mathbf{V}_{a0} + \mathbf{a}\mathbf{V}_{a1} + \mathbf{a}^2\mathbf{V}_{a2} \tag{6.14}$$

Example 6.1: Consider the system described in Examples 5.7 and 5.8 and assume that there is an SLG fault, involving phase a, and that the fault impedance is $5 + j0\ \Omega$. Also, assume that $\mathbf{Z}_0$ and $\mathbf{Z}_2$ are $j0.56$ and $j0.3619\ \Omega$, respectively.

(a) Show the interconnection of the corresponding equivalent sequence networks.
(b) Determine the sequence and phase currents.
(c) Determine the sequence and phase voltages.
(d) Determine the line-to-line voltages.

Solution:

(a) Figure 6.2 shows the interconnection of the resulting equivalent sequence networks.
(b) The impedance base on the 230-kV line is

$$\mathbf{Z}_B = \frac{230^2}{200}$$
$$= 264.5\ \Omega$$

Therefore,

$$\mathbf{Z}_f = \frac{5\ \Omega}{264.5\ \Omega}$$
$$= 0.0189\ \text{pu}\ \Omega$$

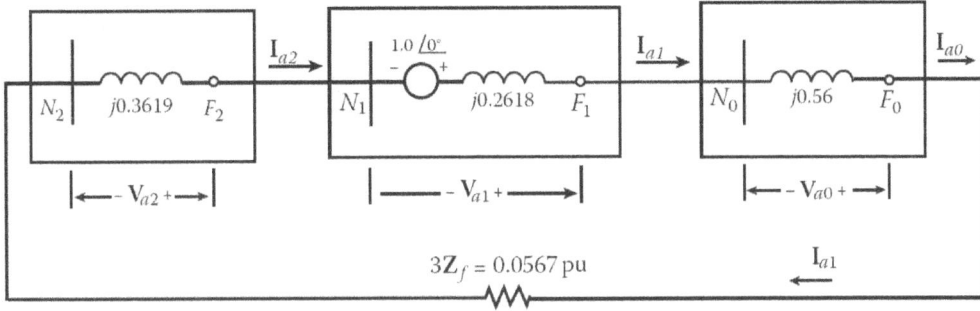

Figure 6.2 Interconnection of resultant equivalent sequence networks of Example 6.2.

Thus, the sequence currents and the phase currents are

$$\mathbf{I}_{a0} = \mathbf{I}_{a1} = \mathbf{I}_{a2} = \frac{1.0\angle 0°}{\mathbf{Z}_0 + \mathbf{Z}_1 + \mathbf{Z}_2 + 3\mathbf{Z}_f}$$

$$= \frac{1.0\angle 0°}{j0.56 + j0.2618 + j0.3619 + 0.0567}$$

$$- 0.8438\angle \quad 87.3° \text{ pu A}$$

and

$$\begin{bmatrix} \mathbf{I}_{af} \\ \mathbf{I}_{bf} \\ \mathbf{I}_{cf} \end{bmatrix} = \begin{bmatrix} 1 & 1 & 1 \\ 1 & a^2 & a \\ 1 & a & a^2 \end{bmatrix} \begin{bmatrix} 0.8438\angle -87.3° \\ 0.8438\angle -87.3° \\ 0.8438\angle -87.3° \end{bmatrix}$$

$$= \begin{bmatrix} 2.5314\angle -87.3° \\ 0 \\ 0 \end{bmatrix} \text{ pu A}$$

(c) The sequence and phase voltages are

$$\begin{bmatrix} \mathbf{V}_{a0} \\ \mathbf{V}_{a1} \\ \mathbf{V}_{a2} \end{bmatrix} = \begin{bmatrix} 0 \\ 1.0\angle 0° \\ 0 \end{bmatrix} \begin{bmatrix} j0.56 & 0 & 0 \\ 0 & j0.2618 & 0 \\ 0 & 0 & j0.3619 \end{bmatrix} \begin{bmatrix} 0.8438\angle -87.3° \\ 0.8438\angle -87.3° \\ 0.8438\angle -87.3° \end{bmatrix}$$

$$= \begin{bmatrix} 0.4725\angle -177.7° \\ 0.7794\angle -0.8° \\ 0.3054\angle -177.7° \end{bmatrix} \text{ pu V}$$

and

$$\begin{bmatrix} \mathbf{V}_{af} \\ \mathbf{V}_{bf} \\ \mathbf{V}_{cf} \end{bmatrix} = \begin{bmatrix} 1 & 1 & 1 \\ 1 & a^2 & a \\ 1 & a & a^2 \end{bmatrix} \begin{bmatrix} 0.4725\angle -177.7° \\ 0.7794\angle -0.8° \\ 0.3054\angle -177.7° \end{bmatrix}$$

$$= \begin{bmatrix} 0.0479\angle 86.26° \\ 1.823\angle -127° \\ 1.1709\angle 127.5° \end{bmatrix} \text{ pu V}$$

(d) The fine-to-line voltages at the fault point are

$$\mathbf{V}_{abf} = \mathbf{V}_{af} - \mathbf{V}_{bf}$$

$$= 0.0479\angle 87.26° - 1.1823\angle 207.07°$$

$$= 1.146\angle 51.85° \text{ pu V}$$

$$\mathbf{V}_{bcf} = \mathbf{V}_{bf} - \mathbf{V}_{cf}$$
$$= 1.823\angle -127° - 1.1709\angle 127.5°$$
$$= 1.878\angle -89.79° \text{ pu V}$$

$$\mathbf{V}_{caf} = \mathbf{V}_{cf} - \mathbf{V}_{af}$$
$$= 1.709\angle 121.5° - 0.0479\angle 86.26°$$
$$= 1.2106\angle 126.2° \text{ pu V}$$

Example 6.2: Consider the system given in Figure 6.3a and assume that the given impedance values are based on the same megavolt-ampere value. The two three-phase transformer banks are made of three single-phase transformers. Assume that there is an SLG fault, involving phase a, at the middle of the transmission line TL_{23}, as shown in the figure.

(a) Draw the corresponding positive-, negative-, and zero-sequence networks, without reducing them, and their corresponding interconnections.
(b) Determine the sequence currents at fault point F.
(c) Determine the sequence currents at the terminals of generator G_1.
(d) Determine the phase currents at the terminals of generator G_1.
(e) Determine the sequence voltages at the terminals of generator G_1.
(f) Determine the phase voltages at the terminals of generator G_1.
(g) Repeat parts (c) through (f) for generator G_2.

Solution:

(a) Figure 6.3b shows the corresponding sequence networks.
(b) The sequence currents at fault point F are

$$\mathbf{I}_{a0} = \mathbf{I}_{a1} = \mathbf{I}_{a2} = \frac{1.0\angle 0°}{\mathbf{Z}_0 + \mathbf{Z}_1 + \mathbf{Z}_2}$$
$$= \frac{1.0\angle 0°}{j0.2619 + j0.25 + j0.25}$$
$$= -j1.3125 \text{ pu A}$$

(c) Therefore, the sequence current contributions of generator G_1 can be found by symmetry as

$$\mathbf{I}_{a1,G_1} = \frac{1}{2} \times \mathbf{I}_{a1}$$
$$= -j0.6563 \text{ pu A}$$

and

$$\mathbf{I}_{a2,G_1} = \frac{1}{2} \times \mathbf{I}_{a2}$$
$$= -j0.6563 \text{ pu A}$$

and by current division,

$$\mathbf{I}_{a0,G_1} = \frac{0.5}{0.55 + 0.5} \times \mathbf{I}_{a0}$$
$$= -j0.6250 \text{ pu A}$$

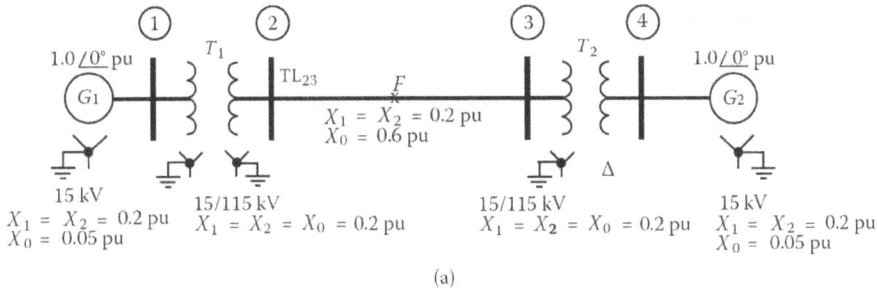

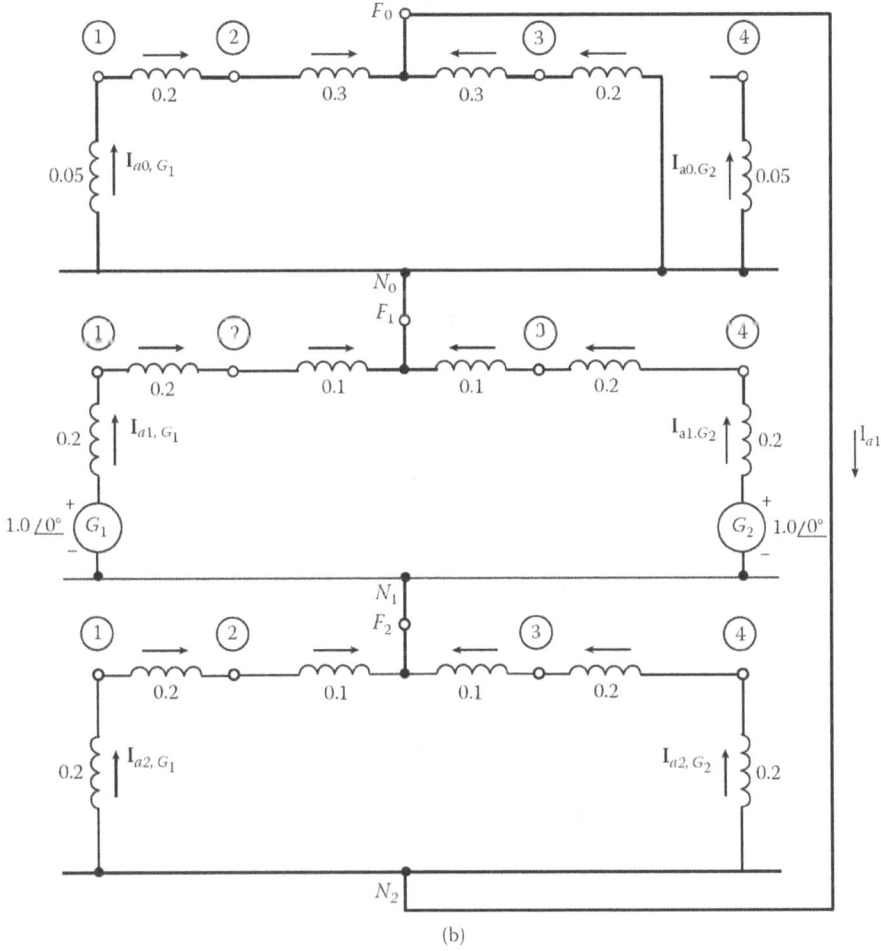

Figure 6.3 The system and the solution for Example 6.2.

(d) The phase currents at the terminals of generator G_1, are

$$\begin{bmatrix} \mathbf{I}_{af} \\ \mathbf{I}_{bf} \\ \mathbf{I}_{cf} \end{bmatrix} = \begin{bmatrix} 1 & 1 & 1 \\ 1 & \mathbf{a}^2 & \mathbf{a} \\ 1 & \mathbf{a} & \mathbf{a}^2 \end{bmatrix} \begin{bmatrix} 0.6250\angle -90° \\ 0.6563\angle -90° \\ 0.6563\angle -90° \end{bmatrix}$$

$$= \begin{bmatrix} 1.9376\angle -90° \\ 0.0313\angle 90° \\ 0.0313\angle 90° \end{bmatrix} \text{ pu A}$$

(e) The sequence voltages at the terminals of generator G_1, are

$$\begin{bmatrix} \mathbf{V}_{a0} \\ \mathbf{V}_{a1} \\ \mathbf{V}_{a2} \end{bmatrix} = \begin{bmatrix} 0 \\ 1.0\angle 0° \\ 0 \end{bmatrix} \begin{bmatrix} j0.2619 & 0 & 0 \\ 0 & j0.25 & 0 \\ 0 & 0 & j0.25 \end{bmatrix} \begin{bmatrix} 0.6250\angle -90° \\ 0.6563\angle -90° \\ 0.6563\angle -90° \end{bmatrix}$$

$$= \begin{bmatrix} 0.1637\angle 180° \\ 0.8360\angle 0° \\ 0.1641\angle 180° \end{bmatrix} \text{ pu V}$$

(f) Therefore, the phase voltages are

$$\begin{bmatrix} \mathbf{V}_{af} \\ \mathbf{V}_{bf} \\ \mathbf{V}_{cf} \end{bmatrix} = \begin{bmatrix} 1 & 1 & 1 \\ 1 & \mathbf{a}^2 & \mathbf{a} \\ 1 & \mathbf{a} & \mathbf{a}^2 \end{bmatrix} \begin{bmatrix} 0.1637\angle 180° \\ 0.8360\angle 0° \\ 0.1641\angle 180° \end{bmatrix}$$

$$= \begin{bmatrix} 0.5082\angle 0° \\ 0.9998\angle 240° \\ 0.9998\angle 120° \end{bmatrix} \text{ pu V}$$

(g) Similarly, for generator G_2, by symmetry,

$$\mathbf{I}_{a1,G_2} = \frac{1}{2} \times \mathbf{I}_{a1}$$
$$= -j0.6563 \text{ pu A}$$

and

$$\mathbf{I}_{a2,G_2} = \frac{1}{2} \times \mathbf{I}_{a2}$$
$$= -j0.6563 \text{ pu A}$$

and by inspection

$$\mathbf{I}_{a0,G_2} = 0$$

However, since transformer T_2 has wye–delta connections and the U.S. Standard terminal markings provide that $\mathbf{V}_{a1(HV)}$ leads $\mathbf{V}_{a1(LV)}$ by 30° and $\mathbf{V}_{a2(HV)}$ lags $\mathbf{V}_{a2(LV)}$ by 30°, regardless of which side has the delta-connected windings, taking into account the 30° phase shifts,

$$\mathbf{I}_{a1,G_2} = 0.6563\angle -90° - 30°$$
$$= 0.6563\angle -120° \text{ pu A}$$

and

$$\mathbf{I}_{a2,G_2} = 0.6563\angle -90° + 30°$$
$$= 0.6563\angle -60° \text{ pu A}$$

This is because generator G2 is on the low-voltage side of the transformer. Therefore,

$$\begin{bmatrix} \mathbf{I}_{af} \\ \mathbf{I}_{bf} \\ \mathbf{I}_{cf} \end{bmatrix} = \begin{bmatrix} 1 & 1 & 1 \\ 1 & \mathbf{a}^2 & \mathbf{a} \\ 1 & \mathbf{a} & \mathbf{a}^2 \end{bmatrix} \begin{bmatrix} 0 \\ 0.6563\angle -120° \\ 0.6563\angle -60° \end{bmatrix}$$

$$= \begin{bmatrix} 1.1368\angle -90° \\ 1.1368\angle 90° \\ 0 \end{bmatrix} \text{ pu A}$$

The positive- and negative-sequence voltages on the G_2 side are the same as on the G_1 side. Thus,

$$\mathbf{V}_{a1} = 0.8434\angle 0° \text{ pu V}$$

$$\mathbf{V}_{a2} = 0.1641\angle 180° \text{ pu V}$$

Again, taking into account the 30° phase shifts,

$$\mathbf{V}_{a1} = 0.8434\angle 0° - 30°$$
$$= 0.8434\angle - 30° \text{ pu V}$$

$$\mathbf{V}_{a2} = 0.1641\angle 180° + 30°$$
$$= 0.1641\angle - 210° \text{ pu V}$$

Obviously

$$\mathbf{V}_{a0} = 0$$

Therefore, the phase voltages at the terminals of generator G_2 are

$$\begin{bmatrix} \mathbf{V}_{af} \\ \mathbf{V}_{bf} \\ \mathbf{V}_{cf} \end{bmatrix} = \begin{bmatrix} 1 & 1 & 1 \\ 1 & \mathbf{a}^2 & \mathbf{a} \\ 1 & \mathbf{a} & \mathbf{a}^2 \end{bmatrix} \begin{bmatrix} 0 \\ 0.8434\angle - 30° \\ 0.1641\angle 210° \end{bmatrix}$$
$$= \begin{bmatrix} 0.7745\angle - 40.6° \\ 0.7745\angle 220.6° \\ 1.0775\angle 90° \end{bmatrix} \text{ pu V}$$

6.2.2 LINE-TO-LINE FAULT

A line-to-line (L–L) fault on a transmission system occurs when two conductors are short-circuited. The general representation of a line-to-line fault at fault point F with a fault impedance $\mathbf{Z}_f$ is shown in Figure 6.4a. The resulting sequence networks are interconnected, as shown in Figure 6.4b. For the sake of symmetry, it is assumed that the line-to-line fault is between phases b and c. From Figure 6.4a, it can be observed that:

$$\mathbf{I}_{af} = 0 \tag{6.15}$$

$$\mathbf{I}_{bf} = -\mathbf{I}_{cf} \tag{6.16}$$

$$\mathbf{V}_{bc} = \mathbf{V}_b - \mathbf{V}_c = \mathbf{Z}_f \mathbf{I}_{bf} \tag{6.17}$$

From Figure 6.4b, the sequence currents can be found as

$$\mathbf{I}_{a0} = 0 \tag{6.18}$$

$$\mathbf{I}_{a1} = -\mathbf{I}_{a2} = \frac{1.0\angle 0°}{\mathbf{Z}_1 + \mathbf{Z}_2 + \mathbf{Z}_f} \tag{6.19}$$

If $\mathbf{Z}_f = 0$

$$\mathbf{I}_{a1} = -\mathbf{I}_{a2} = \frac{1.0\angle 0°}{\mathbf{Z}_1 + \mathbf{Z}_2} \tag{6.20}$$

Substituting Equations 6.18 and 6.19 into Equation 6.2, the fault currents for phases a and b can be found as

$$\mathbf{I}_{cf} = -\mathbf{I}_{cf} = \sqrt{3}\mathbf{I}_{a1}\angle - 90° \tag{6.21}$$

(a)

(b)

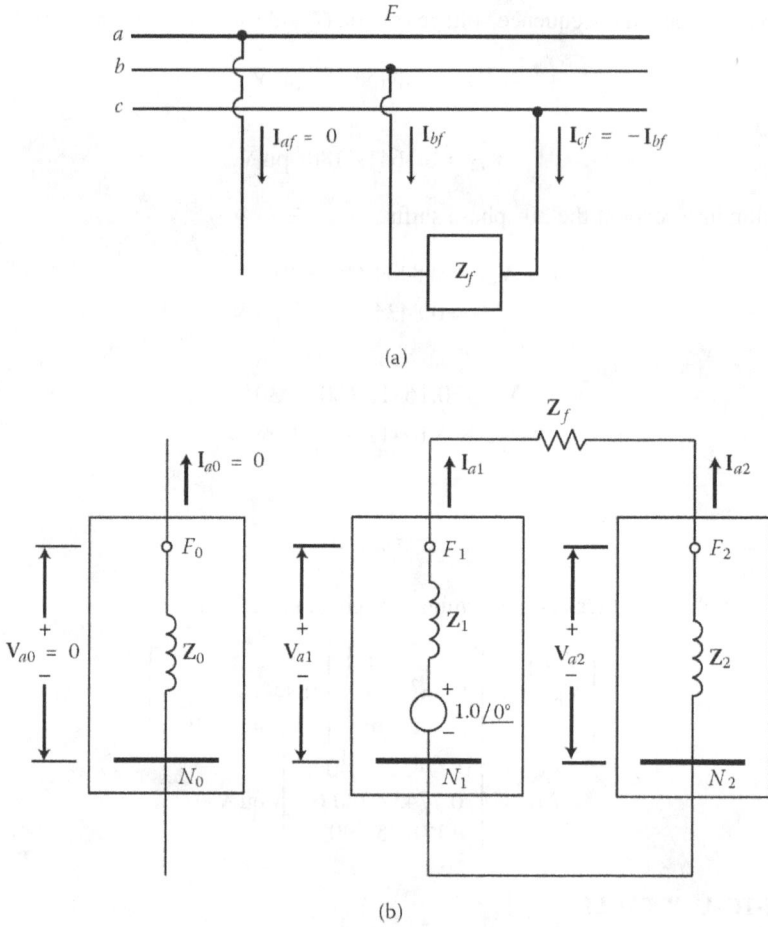

Figure 6.4 Line-to-line fault: (a) general representation; (b) interconnection of sequence networks.

Similarly, substituting Equations 6.18 and 6.19 into Equation 6.8, the sequence voltages can be found as

$$\mathbf{V}_{a0} = 0 \tag{6.22}$$

$$\mathbf{V}_{a1} = 1.0 - \mathbf{Z}_1 \mathbf{I}_{a1} \tag{6.23}$$

$$\mathbf{V}_{a2} = -\mathbf{Z}_2 \mathbf{I}_{a2} = \mathbf{Z}_2 \mathbf{I}_{a1} \tag{6.24}$$

Also, substituting Equations 6.22 through 6.24 into Equation 6.12,

$$\mathbf{V}_{af} = \mathbf{V}_{a1} + \mathbf{V}_{a2} \tag{6.25}$$

or

$$\mathbf{V}_{af} = 1.0 + \mathbf{I}_{a1}(\mathbf{Z}_2 - \mathbf{Z}_1) \tag{6.26}$$

and

$$\mathbf{V}_{bf} = \mathbf{a}^2 \mathbf{V}_{a1} + \mathbf{a} \mathbf{V}_{a2} \tag{6.27}$$

or

$$\mathbf{V}_{bf} = \mathbf{a}^2 + \mathbf{I}_{a1}(\mathbf{a}\mathbf{Z}_2 - \mathbf{a}^2\mathbf{Z}_1) \tag{6.28}$$

and

$$\mathbf{V}_{cf} = \mathbf{a}\mathbf{V}_{a1} + +\mathbf{a}^2\mathbf{V}_{a2} \tag{6.29}$$

or

$$\mathbf{V}_{cf} = \mathbf{a} + \mathbf{I}_{a1}(\mathbf{a}^2\mathbf{Z}_2 - \mathbf{a}\mathbf{Z}_1) \tag{6.30}$$

Thus, the line-to-line voltages can be expressed as

$$\mathbf{V}_{ab} = \mathbf{V}_{af} - \mathbf{V}_{bf} \tag{6.31}$$

or

$$\mathbf{V}_{ab} = \sqrt{3}(\mathbf{V}_{a1}\angle 30° + \mathbf{V}_{a2}\angle - 30°) \tag{6.32}$$

and

$$\mathbf{V}_{bc} = \mathbf{V}_{bf} - \mathbf{V}_{cf} \tag{6.33}$$

or

$$\mathbf{V}_{bc} = \sqrt{3}(\mathbf{V}_{a1}\angle - 90° + \mathbf{V}_{a2}\angle 90°) \tag{6.34}$$

and

$$\mathbf{V}_{ca} = \mathbf{V}_{cf} - \mathbf{V}_{af} \tag{6.35}$$

or

$$\mathbf{V}_{ca} = \sqrt{3}(\mathbf{V}_{a1}\angle 150° + \mathbf{V}_{a2}\angle - 150°) \tag{6.36}$$

Example 6.3: Repeat Example 6.1 assuming that there is a line-to-line fault, involving phases b and c, at bus 3.
Solution:

(a) Figure 6.5 shows the interconnection of the resulting equivalent sequence networks.

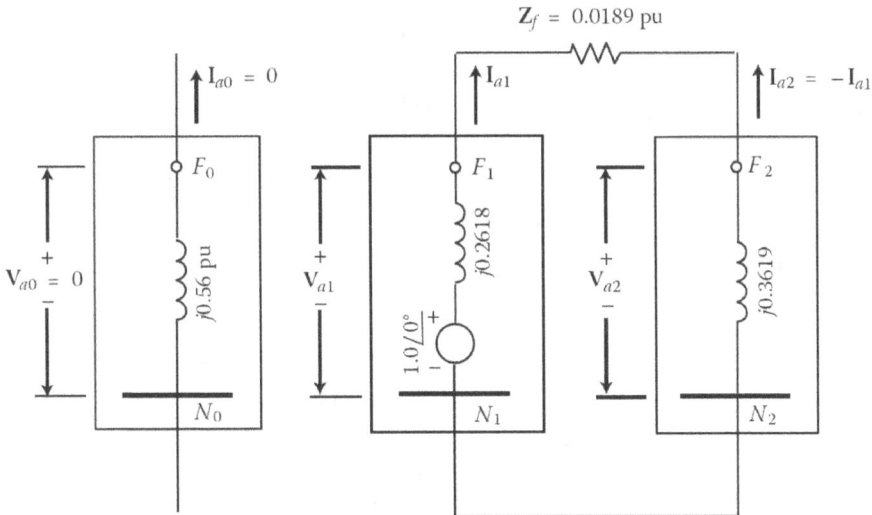

Figure 6.5   Interconnection of resultant equivalent sequence networks of Example 6.3.

(b) The sequence and the phase currents are

$$\mathbf{I}_{a0} = 0$$

$$\mathbf{I}_{a1} = -\mathbf{I}_{a2} = \frac{1.0\angle 0°}{\mathbf{Z}_1 + \mathbf{Z}_2 + \mathbf{Z}_f}$$

$$= \frac{1.0\angle 0°}{j0.2618 + j0.3619 + 0.0189}$$

$$= 1.6026\angle -88.3° \text{ pu A}$$

and

$$\begin{bmatrix} \mathbf{I}_{af} \\ \mathbf{I}_{bf} \\ \mathbf{I}_{cf} \end{bmatrix} = \begin{bmatrix} 1 & 1 & 1 \\ 1 & a^2 & a \\ 1 & a & a^2 \end{bmatrix} \begin{bmatrix} 0 \\ 1.6026\angle -88.3° \\ 1.6026\angle 91.7° \end{bmatrix}$$

$$= \begin{bmatrix} 0 \\ 2.7758\angle -178.3° \\ 2.7758\angle 1.7° \end{bmatrix} \text{ pu A}$$

(c) The sequence and phase voltages are

$$\begin{bmatrix} \mathbf{V}_{a0} \\ \mathbf{V}_{a1} \\ \mathbf{V}_{a2} \end{bmatrix} = \begin{bmatrix} 0 \\ 1.0\angle 0° \\ 0 \end{bmatrix} \begin{bmatrix} j0.56 & 0 & 0 \\ 0 & j0.2618 & 0 \\ 0 & 0 & j0.3619 \end{bmatrix} \begin{bmatrix} 0 \\ 1.6026\angle -88.3° \\ 1.6026\angle 91.7° \end{bmatrix}$$

$$= \begin{bmatrix} 0 \\ 0.5808\angle -1.2° \\ 0.5800\angle 1.7° \end{bmatrix} \text{ pu V}$$

and

$$\begin{bmatrix} \mathbf{V}_{af} \\ \mathbf{V}_{bf} \\ \mathbf{V}_{cf} \end{bmatrix} = \begin{bmatrix} 1 & 1 & 1 \\ 1 & a^2 & a \\ 1 & a & a^2 \end{bmatrix} \begin{bmatrix} 0 \\ 0.5808\angle -1.3° \\ 0.5800\angle 1.7° \end{bmatrix}$$

$$= \begin{bmatrix} 1.1604\angle 0.2° \\ 0.6061\angle -179.7° \\ 0.5540\angle -171.8° \end{bmatrix} \text{ pu V}$$

(d) The line-to-line voltages at the fault point are

$$\begin{aligned} \mathbf{V}_{abf} &= \mathbf{V}_{af} - \mathbf{V}_{bf} \\ &= 1.7667 + j0.0008 \\ &= 1.7668\angle 0.3° \text{ pu V} \end{aligned}$$

$$\begin{aligned} \mathbf{V}_{bcf} &= \mathbf{V}_{bf} - \mathbf{V}_{cf} \\ &= -0.0524 - j0.0016 \\ &= 0.0525\angle -178.3° \text{ pu V} \end{aligned}$$

$$\begin{aligned} \mathbf{V}_{caf} &= \mathbf{V}_{cf} - \mathbf{V}_{af} \\ &= -1.7143 - j0.0065 \\ &= 1.7143\angle -179.8° \text{ pu V} \end{aligned}$$

6.2.3 DLG FAULT

When two conductors come into contact with the ground or when two conductors touch the neutral of a three-phase grounded system, it results in a double-line-ground (DLG) fault on a transmission system. Figure 6.6a shows the general representation of a DLG fault at a fault point F, including a fault impedance $\mathbf{Z}f$ and the impedance from the line to the ground $\mathbf{Z}g$ (which can be zero or infinite). The faulted phase is assumed to be between phases b and c for symmetry purposes. Figure 6.6b shows the resulting sequence networks interconnected due to the fault. It can be observed from Figure 6.6a that:

$$\mathbf{I}_{af} = 0 \tag{6.37}$$

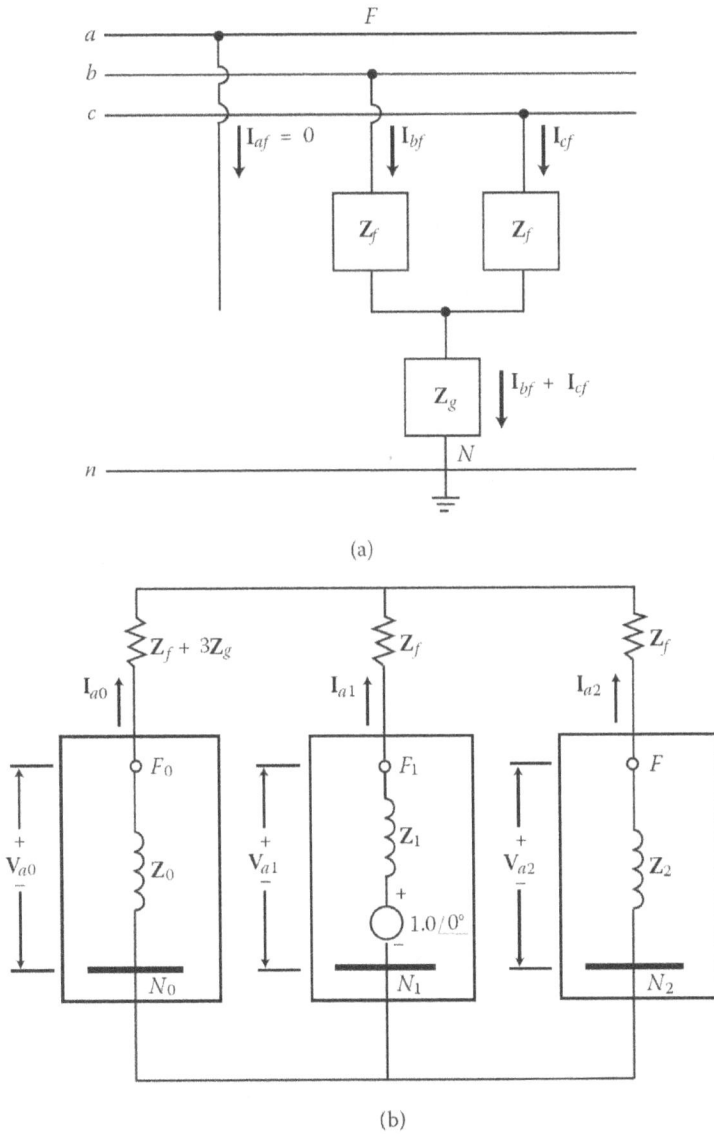

(a)

(b)

Figure 6.6 DLG fault: (a) general representation; (b) interconnection of sequence networks.

$$\mathbf{V}_{bf} = (\mathbf{Z}_f + \mathbf{Z}_g)\mathbf{I}_{bf} + \mathbf{Z}_g\mathbf{I}_{cf} \tag{6.38}$$

$$\mathbf{V}_{cf} = (\mathbf{Z}_f + \mathbf{Z}_g)\mathbf{I}_{cf} + \mathbf{Z}_g\mathbf{I}_{bf} \tag{6.39}$$

From Figure 6.6b, the positive-sequence currents can be found as

$$\mathbf{I}_{a1} = \frac{1.0\angle 0°}{(\mathbf{Z}_1 + \mathbf{Z}_f) + \frac{(\mathbf{Z}_2+\mathbf{Z}_f)(\mathbf{Z}_0+\mathbf{Z}_f+3\mathbf{Z}_g)}{(\mathbf{Z}_2+\mathbf{Z}_f)+(\mathbf{Z}_0+\mathbf{Z}_f+3\mathbf{Z}_g)}} \tag{6.40a}$$

$$= \frac{1.0\angle 0°}{(\mathbf{Z}_1 + \mathbf{Z}_f) + \frac{(\mathbf{Z}_2+\mathbf{Z}_f)(\mathbf{Z}_0+\mathbf{Z}_f+3\mathbf{Z}_g)}{\mathbf{Z}_0+\mathbf{Z}_2+2\mathbf{Z}_f+3\mathbf{Z}_g}} \tag{6.40b}$$

The negative- and zero-sequence currents can be found, by using current division, as

$$\mathbf{I}_{a2} = -\left[\frac{(\mathbf{Z}_0 + \mathbf{Z}_f + 3\mathbf{Z}_g)}{(\mathbf{Z}_2 + \mathbf{Z}_f)(\mathbf{Z}_0 + \mathbf{Z}_f + 3\mathbf{Z}_g)}\right]\mathbf{I}_{a1} \tag{6.41}$$

and

$$\mathbf{I}_{a0} = -\left[\frac{(\mathbf{Z}_0 + \mathbf{Z}_f)}{(\mathbf{Z}_2 + \mathbf{Z}_f)(\mathbf{Z}_0 + \mathbf{Z}_f + 3\mathbf{Z}_g)}\right]\mathbf{I}_{a1} \tag{6.42}$$

or as an alternative method, since

$$\mathbf{I}_{af} = 0 = \mathbf{I}_{a0} + \mathbf{I}_{a1} + \mathbf{I}_{a2}$$

then if $\mathbf{I}_{a1}$ and $\mathbf{I}_{a2}$ are known,

$$\mathbf{I}_{a0} = -(\mathbf{I}_{a1} + \mathbf{I}_{a2}) \tag{6.43}$$

Note that, in the event of having $\mathbf{Z}_f = 0$ and $\mathbf{Z}_g = 0$, the positive-, negative-, and zero-sequences can be expressed as

$$\mathbf{I}_{a1} = \frac{1.0\angle 0°}{\mathbf{Z}_1 + \frac{\mathbf{Z}_0 \times \mathbf{Z}_2}{\mathbf{Z}_0 + \mathbf{Z}_2}} \tag{6.44}$$

and by current division

$$\mathbf{I}_{a2} = -\left[\frac{\mathbf{Z}_0}{\mathbf{Z}_0 + \mathbf{Z}_2}\right]\mathbf{I}_{a1} \tag{6.45}$$

$$\mathbf{I}_{a0} = -\left[\frac{\mathbf{Z}_2}{\mathbf{Z}_0 + \mathbf{Z}_2}\right]\mathbf{I}_{a1} \tag{6.46}$$

Note that, the fault current for phase a is already known to be

$$\mathbf{I}_{af} = 0$$

the fault currents for phases $a0$ and b can be found by substituting Equations 6.40 through 6.42 into Equation 6.2 so that

$$\mathbf{I}_{bf} = \mathbf{I}_{a0} + \mathbf{a}^2\mathbf{I}_{a1} + \mathbf{a}\mathbf{I}_{a2} \tag{6.47}$$

and

$$\mathbf{I}_{cf} = \mathbf{I}_{a0} + \mathbf{a}\mathbf{I}_{a1} + \mathbf{a}^2\mathbf{I}_{a2} \tag{6.48}$$

It can be shown that the total fault current flowing into the neutral is

$$\mathbf{I}_n = \mathbf{I}_{bf} + \mathbf{I}_{cf} + 3\mathbf{I}_{a0} \tag{6.49}$$

The sequence voltages can be found from Equation 6.8 as

$$\mathbf{V}_{a0} = -\mathbf{Z}_0\mathbf{I}_{a0} \tag{6.50}$$

$$\mathbf{V}_{a1} = 1.0 - \mathbf{Z}_1 \mathbf{I}_{a1} \tag{6.51}$$

$$\mathbf{V}_{a2} = -\mathbf{Z}_2 \mathbf{I}_{a2} \tag{6.52}$$

Similarly, the phase voltages can be found from Equation 6.12 as

$$\mathbf{V}_{af} = \mathbf{V}_{a0} + \mathbf{V}_{a1} + \mathbf{V}_{a2} \tag{6.53}$$

$$\mathbf{V}_{bf} = \mathbf{V}_{a0} + \mathbf{a}^2 \mathbf{V}_{a1} + \mathbf{a} \mathbf{V}_{a2} \tag{6.54}$$

$$\mathbf{V}_{cf} = \mathbf{V}_{a0} + \mathbf{a} \mathbf{V}_{a1} + \mathbf{a}^2 \mathbf{V}_{a2} \tag{6.55}$$

or, alternatively, the phase voltages $\mathbf{V}_{bf}$ and $\mathbf{V}_{cf}$ can be determined from Equations 6.38 and 6.39. As before, the line-to-line voltages can be found from

$$\mathbf{V}_{ab} = \mathbf{V}_{af} - \mathbf{V}_{bf} \tag{6.56}$$

$$\mathbf{V}_{bc} = \mathbf{V}_{bf} - \mathbf{V}_{cf} \tag{6.57}$$

$$\mathbf{V}_{ca} = \mathbf{V}_{cf} - \mathbf{V}_{af} \tag{6.58}$$

Note that, in the event of having $\mathbf{Z}_f = 0$ and $\mathbf{Z}_g = 0$, the sequence voltages become

$$\mathbf{V}_{a0} = \mathbf{V}_{a1} = \mathbf{V}_{a2} = 1.0 - \mathbf{Z}_1 \mathbf{I}_{a1} \tag{6.59}$$

where the positive-sequence current is found by using Equation 6.44. Once the sequence voltages are determined from Equation 6.59, the negative- and zero-sequence currents can be determined from

$$\mathbf{I}_{a2} = -\frac{\mathbf{V}_{a2}}{\mathbf{Z}_2} \tag{6.60}$$

and

$$\mathbf{I}_{a0} = -\frac{\mathbf{V}_{a0}}{\mathbf{Z}_0} \tag{6.61}$$

Using the relationship given in Equation 6.59, the resultant phase voltages can be expressed as

$$\mathbf{V}_{af} = \mathbf{V}_{a0} + \mathbf{V}_{a1} + \mathbf{V}_{a2} = 3\mathbf{V}_{a1} \tag{6.62}$$

$$\mathbf{V}_{bf} = \mathbf{V}_{cf} = 0 \tag{6.63}$$

Therefore, the line-to-line voltages become

$$\mathbf{V}_{abf} = \mathbf{V}_{af} - \mathbf{V}_{bf} = \mathbf{V}_{af} \tag{6.64}$$

$$\mathbf{V}_{bcf} = \mathbf{V}_{bf} - \mathbf{V}_{cf} = 0 \tag{6.65}$$

$$\mathbf{V}_{caf} = \mathbf{V}_{cf} - \mathbf{V}_{af} = -\mathbf{V}_{af} \tag{6.66}$$

Example 6.4: Repeat Example 6.1 assuming that there is a DLG fault with $Z_f = 5\ \Omega$ and $Z_g = 5\ \Omega$, involving phases b and c, at bus 3.
Solution:

(a) Figure 6.7 shows the interconnection of the resulting equivalent sequence networks.

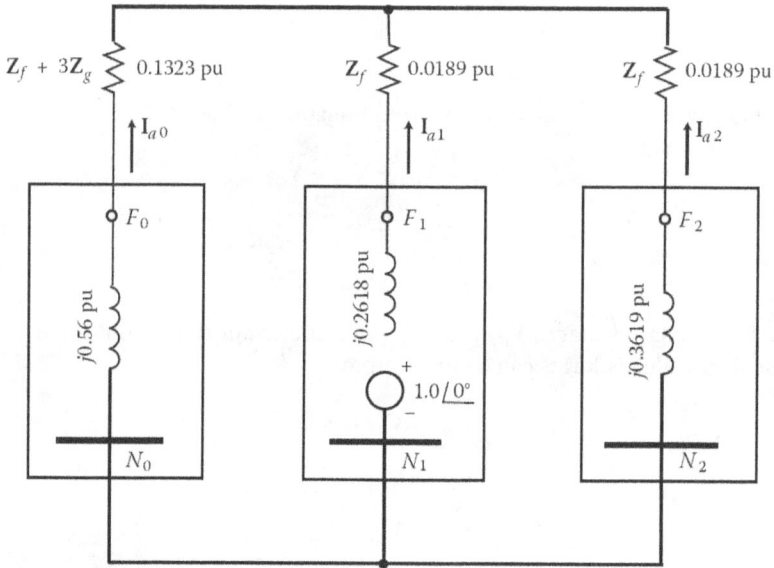

Figure 6.7   Interconnection of resultant equivalent sequence networks of Example 6.4.

(b) Since

$$Z_f + 3Z_g = \frac{5 + 30}{264.5}$$

$$= 0.1323 \text{ pu } \Omega$$

the sequence currents are

$$\mathbf{I}_{a1} = \frac{1.0\angle 0°}{(\mathbf{Z}_1 + \mathbf{Z}_f) + \frac{(\mathbf{Z}_2 + \mathbf{Z}_f)(\mathbf{Z}_0 + \mathbf{Z}_f + 3\mathbf{Z}_g)}{(\mathbf{Z}_2 + \mathbf{Z}_f) + (\mathbf{Z}_0 + \mathbf{Z}_f + 3\mathbf{Z}_g)}}$$

$$= \frac{1.0\angle 0°}{(j0.2618 + 0.0189) + \frac{(j0.3619 + 0.0189)(j0.56 + 0.1323)}{j0.3619 + 0.0189 + j0.56 + 0.1323}}$$

$$= 2.0597\angle -84.5° \text{ pu } \Omega$$

$$\mathbf{I}_{a2} = -\left[\frac{\mathbf{Z}_0 + \mathbf{Z}_f + 3\mathbf{Z}_g}{(\mathbf{Z}_2 + \mathbf{Z}_f) + (\mathbf{Z}_0 + \mathbf{Z}_f + 3\mathbf{Z}_g)} \right] \mathbf{I}_{a1}$$

$$= \left[\frac{0.5754\angle 76.7}{0.9342\angle 80.7} \right] (2.0597\angle -84.5°)$$

$$= -1.2686\angle -88.5° \text{ pu } \Omega$$

$$\mathbf{I}_{a0} = -\left[\frac{(\mathbf{Z}_2 + \mathbf{Z}_f)}{(\mathbf{Z}_2 + \mathbf{Z}_f) + (\mathbf{Z}_0 + \mathbf{Z}_f + 3\mathbf{Z}_g)} \right] \mathbf{I}_{a1}$$

$$= \left[\frac{0.3624\angle 87}{0.9342\angle 80.7} \right] (2.0597\angle -84.5°)$$

$$= -0.799\angle -78.2° \text{ pu } \Omega$$

and the phase currents are

$$
\begin{bmatrix} \mathbf{I}_{af} \\ \mathbf{I}_{bf} \\ \mathbf{I}_{cf} \end{bmatrix} = \begin{bmatrix} 1 & 1 & 1 \\ 1 & a^2 & a \\ 1 & a & a^2 \end{bmatrix} \begin{bmatrix} -0.799\angle-78.2° \\ 2.0597\angle-84.5° \\ -1.2686\angle-88.5° \end{bmatrix}
$$

$$
= \begin{bmatrix} 0 \\ 3.2677\angle162.7° \\ 2.9653\angle27.6° \end{bmatrix} \text{ pu A}
$$

(c) The sequence and phase voltages are

$$
\begin{bmatrix} \mathbf{V}_{a0} \\ \mathbf{V}_{a1} \\ \mathbf{V}_{a2} \end{bmatrix} = \begin{bmatrix} 0 \\ 1.0\angle0° \\ 0 \end{bmatrix} - \begin{bmatrix} j0.56 & 0 & 0 \\ 0 & j0.2618 & 0 \\ 0 & 0 & j0.3619 \end{bmatrix} \begin{bmatrix} -0.799\angle-78.2° \\ 2.0597\angle-84.5° \\ -1.2686\angle-88.5° \end{bmatrix}
$$

$$
= \begin{bmatrix} 0.4474\angle11.8° \\ 0.4662\angle-6.4° \\ 0.4591\angle1.5° \end{bmatrix} \text{ pu V}
$$

and

$$
\begin{bmatrix} \mathbf{V}_{af} \\ \mathbf{V}_{bf} \\ \mathbf{V}_{cf} \end{bmatrix} = \begin{bmatrix} 1 & 1 & 1 \\ 1 & a^2 & a \\ 1 & a & a^2 \end{bmatrix} \begin{bmatrix} 0.4474\angle11.8° \\ 0.4662\angle-6.4° \\ 0.4591\angle1.5° \end{bmatrix}
$$

$$
= \begin{bmatrix} 1.3611\angle2.2° \\ 0.1333\angle126.1° \\ 0.1198\angle74.4° \end{bmatrix} \text{ pu V}
$$

(d) The line-to-line voltages at the fault point are

$$
\mathbf{V}_{abf} = \mathbf{V}_{af} - \mathbf{V}_{bf}
$$
$$
= 1.4386 - j0.0555
$$
$$
= 1.4397\angle-2.2° \text{ pu V}
$$

$$
\mathbf{V}_{bcf} = \mathbf{V}_{bf} - \mathbf{V}_{cf}
$$
$$
= -0.1107 - j0.0077
$$
$$
= 0.111\angle184° \text{ pu V}
$$

$$
\mathbf{V}_{caf} = \mathbf{V}_{cf} - \mathbf{V}_{af}
$$
$$
= -1.3279 + j0.0632
$$
$$
= 1.3294\angle177.3° \text{ pu V}
$$

6.2.4 SYMMETRICAL THREE-PHASE FAULTS

The three-phase (3ϕ) fault is a balanced (i.e., symmetrical) fault that can be analyzed using symmetrical components. In contrast, the unbalanced (i.e., unsymmetrical) fault is not a three-phase fault. The general representation of a balanced three-phase fault at a fault point F with impedances $\mathbf{Z}f$ and $\mathbf{Z}g$ is shown in Figure 6.8a. In Figure 6.8b, the resulting sequence networks are not interconnected but short-circuited over their own fault impedances, and therefore, they are isolated from

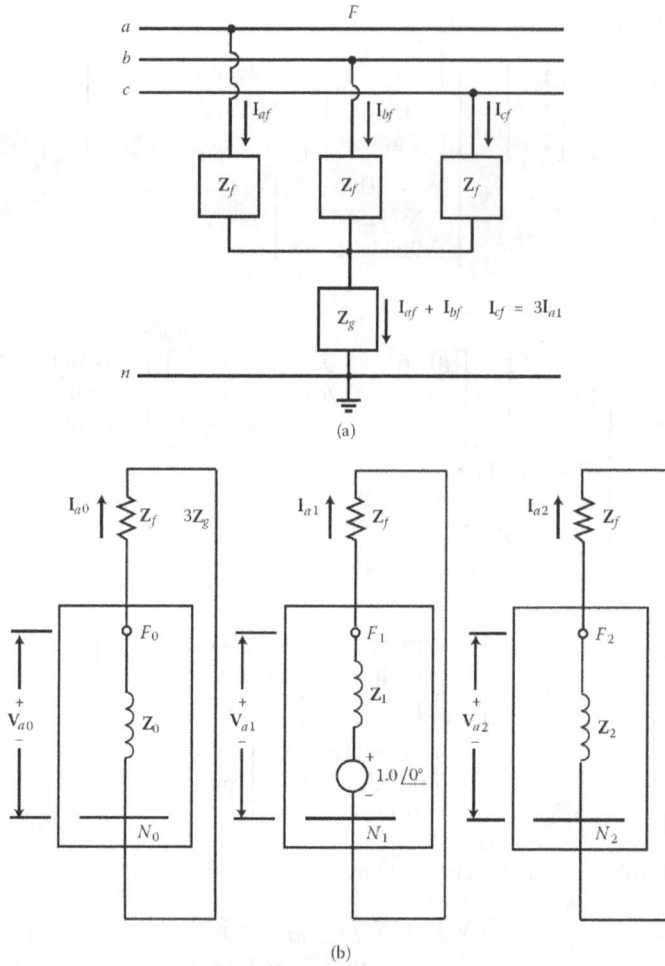

Figure 6.8 Three-phase fault: (a) general representation; (b) interconnection of sequence networks.

each other. As only the positive-sequence network is considered to have an internal voltage source, the positive-, negative-, and zero-sequence currents can be expressed as:

$$\mathbf{I}_{a0} = 0 \tag{6.67}$$

$$\mathbf{I}_{a2} = 0 \tag{6.68}$$

$$\mathbf{I}_{a1} = \frac{1.0\angle 0°}{\mathbf{Z}_1 + \mathbf{Z}_f} \tag{6.69}$$

If the fault impedance $\mathbf{Z}_f$ is zero,

$$\mathbf{I}_{a1} = \frac{1.0\angle 0°}{\mathbf{Z}_1} \tag{6.70}$$

Substituting Equations 6.67 and 6.68 into Equation 6.2,

$$\begin{bmatrix} \mathbf{I}_{af} \\ \mathbf{I}_{bf} \\ \mathbf{I}_{cf} \end{bmatrix} = \begin{bmatrix} 1 & 1 & 1 \\ 1 & \mathbf{a}^2 & \mathbf{a} \\ 1 & \mathbf{a} & \mathbf{a}^2 \end{bmatrix} \begin{bmatrix} 0 \\ \mathbf{I}_{a1} \\ 0 \end{bmatrix} \tag{6.71}$$

from which

$$\mathbf{I}_{af} = \mathbf{I}_{a1} = \frac{1.0\angle 0°}{\mathbf{Z}_1 + \mathbf{Z}_f} \tag{6.72}$$

$$\mathbf{I}_{bf} = \mathbf{a}^2\mathbf{I}_{a1} = \frac{1.0\angle 240°}{\mathbf{Z}_1 + \mathbf{Z}_f} \tag{6.73}$$

$$\mathbf{I}_{cf} = \mathbf{a}\mathbf{I}_{a1} = \frac{1.0\angle 120°}{\mathbf{Z}_1 + \mathbf{Z}_f} \tag{6.74}$$

Since the sequence networks are short-circuited over their own fault impedances,

$$\mathbf{V}_{a0} = 0 \tag{6.75}$$

$$\mathbf{V}_{a1} = \mathbf{Z}_f\mathbf{I}_{a1} \tag{6.76}$$

$$\mathbf{V}_{a2} = 0 \tag{6.77}$$

Therefore, substituting Equations 6.75 through 6.67 into Equation 6.12,

$$\begin{bmatrix} \mathbf{V}_{af} \\ \mathbf{V}_{bf} \\ \mathbf{V}_{cf} \end{bmatrix} = \begin{bmatrix} 1 & 1 & 1 \\ 1 & \mathbf{a}^2 & \mathbf{a} \\ 1 & \mathbf{a} & \mathbf{a}^2 \end{bmatrix} \begin{bmatrix} 0 \\ \mathbf{V}_{a1} \\ 0 \end{bmatrix} \tag{6.78}$$

Thus,

$$\mathbf{V}_{af} = \mathbf{V}_{a1} - \mathbf{Z}_f\mathbf{I}_{a1} \tag{6.79}$$

$$\mathbf{V}_{bf} = \mathbf{a}^2\mathbf{V}_{a1} = \mathbf{Z}_f\mathbf{I}_{a1}\angle 240° \tag{6.80}$$

$$\mathbf{V}_{cf} = \mathbf{a}\mathbf{V}_{a1} = \mathbf{Z}_f\mathbf{I}_{a1}\angle 120° \tag{6.81}$$

Hence, the line-to-line voltages become

$$\mathbf{V}_{ab} = \mathbf{V}_{af} - \mathbf{V}_{bf} = \mathbf{V}_{a1}(1 - \mathbf{a}^2) = \sqrt{3}\mathbf{Z}_f\mathbf{I}_{a1}\angle 30° \tag{6.82}$$

$$\mathbf{V}_{bc} = \mathbf{V}_{bf} - \mathbf{V}_{cf} = \mathbf{V}_{a1}(\mathbf{a}^2 - \mathbf{a}) = \sqrt{3}\mathbf{Z}_f\mathbf{I}_{a1}\angle - 90° \tag{6.83}$$

$$\mathbf{V}_{ca} = \mathbf{V}_{cf} - \mathbf{V}_{af} = \mathbf{V}_{a1}(\mathbf{a} - 1) = \sqrt{3}\mathbf{Z}_f\mathbf{I}_{a1}\angle 150° \tag{6.84}$$

Note that, in the event of having $\mathbf{Z}_f = 0$,

$$\mathbf{I}_{af} = \frac{1.0\angle 0°}{\mathbf{Z}_1} \tag{6.85}$$

$$\mathbf{I}_{bf} = \frac{1.0\angle 240°}{\mathbf{Z}_1} \tag{6.86}$$

$$\mathbf{I}_{cf} = \frac{1.0\angle 120°}{\mathbf{Z}_1} \tag{6.87}$$

and

$$\mathbf{V}_{af} = 0 \tag{6.88}$$

$$\mathbf{V}_{bf} = 0 \tag{6.89}$$

$$\mathbf{V}_{cf} = 0 \tag{6.90}$$

and, of course,

$$\mathbf{V}_{a0} = 0 \tag{6.91}$$

$$\mathbf{V}_{a1} = 0 \tag{6.92}$$

$$\mathbf{V}_{a2} = 0 \tag{6.93}$$

Example 6.5: Repeat Example 6.1 assuming that there is a symmetrical three-phase fault with $\mathbf{Z}_f$ = 5 Ω and $\mathbf{Z}_g$ = 10 Ω at bus 3. Let $\mathbf{Z}_0$ = $j0.5$ pu. Use $\mathbf{Z}_B$ = 264.5 Ω.
Solution:

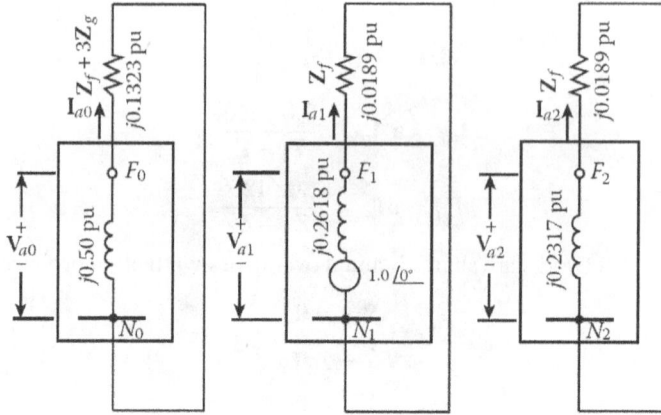

Figure 6.9 Interconnection of sequence networks of Example 6.5.

(a) Figure 6.9 shows the interconnection of the resulting equivalent sequence networks.
(b) The sequence and phase currents are

$$\mathbf{I}_{a0} = \mathbf{I}_{a2} = 0$$

$$
\begin{aligned}
\mathbf{I}_{a1} &= \frac{1.0\angle 0°}{\mathbf{Z}_1 + \mathbf{Z}_f} \\
&= \frac{1.0\angle 0°}{j0.2618 + 0.0189} \\
&= 3.8098\angle -85.9° \text{ pu A}
\end{aligned}
$$

and

$$
\begin{bmatrix} \mathbf{I}_{af} \\ \mathbf{I}_{bf} \\ \mathbf{I}_{cf} \end{bmatrix} =
\begin{bmatrix} 1 & 1 & 1 \\ 1 & a^2 & a \\ 1 & a & a^2 \end{bmatrix}
\begin{bmatrix} 0 \\ 3.8098\angle -85.9° \\ 0 \end{bmatrix}
$$

$$
= \begin{bmatrix} 3.8098\angle -85.9° \\ 3.8098\angle 154.1° \\ 3.8098\angle 34.1° \end{bmatrix} \text{ pu A}
$$

(c) The sequence and phase voltages are

$$
\begin{bmatrix} \mathbf{V}_{a0} \\ \mathbf{V}_{a1} \\ \mathbf{V}_{a2} \end{bmatrix} =
\begin{bmatrix} 0 \\ 1.0\angle 0° \\ 0 \end{bmatrix} -
\begin{bmatrix} j0.5 & 0 & 0 \\ 0 & j0.2618 & 0 \\ 0 & 0 & j0.2317 \end{bmatrix}
\begin{bmatrix} 0 \\ 3.8098\angle -85.9° \\ 0 \end{bmatrix}
$$

$$
= \begin{bmatrix} 0 \\ 0.0720\angle -85.9° \\ 0 \end{bmatrix} \text{ pu V}
$$

and

$$\begin{bmatrix} \mathbf{V}_{af} \\ \mathbf{V}_{bf} \\ \mathbf{V}_{cf} \end{bmatrix} = \begin{bmatrix} 1 & 1 & 1 \\ 1 & \mathbf{a}^2 & \mathbf{a} \\ 1 & \mathbf{a} & \mathbf{a}^2 \end{bmatrix} \begin{bmatrix} 0 \\ 0.0720\angle -85.9° \\ 0 \end{bmatrix}$$

$$= \begin{bmatrix} 0.0720\angle -85.9° \\ 0.0720\angle 154.9° \\ 0.0720\angle 34.1° \end{bmatrix} \text{pu V}$$

(d) The line-to-line voltages at the fault point are

$$\mathbf{V}_{abf} = \mathbf{V}_{af} - \mathbf{V}_{bf}$$
$$= 0.0601 - j0.1023$$
$$= 0.1247\angle -55.9° \text{ pu V}$$

$$\mathbf{V}_{bcf} = \mathbf{V}_{bf} - \mathbf{V}_{cf}$$
$$= -0.1248 - j0.0099$$
$$= 0.1252\angle 184.5° \text{ pu V}$$

$$\mathbf{V}_{caf} = \mathbf{V}_{cf} - \mathbf{V}_{af}$$
$$= 0.0545 + j0.1122$$
$$= 0.1247\angle 64.1° \text{ pu V}$$

6.2.5 UNSYMMETRICAL THREE-PHASE FAULTS

However, it is possible that not all three-phase faults are symmetrical. There are also unsymmetrical faults, as depicted in Figures 6.10–6.14. For instance, Figure 6.10 shows an unsymmetrical three-phase fault comprising a single line-to-ground (SLG) fault and a line-to-line fault, both occurring at the same fault point F. Figure 6.11 illustrates an unsymmetrical three-phase fault with a fault impedance $\mathbf{Z}f$ connected between two lines. Figure 6.12 illustrates an unsymmetrical three-phase fault with a different fault impedance on each phase, where $\mathbf{Z}f1 \neq \mathbf{Z}f2$. Figure 6.13 illustrates an unsymmetrical three-phase fault to ground. Lastly, Figure 6.14 depicts an unsymmetrical three-phase fault with delta-connected fault impedances, where $\mathbf{Z}f1 \neq \mathbf{Z}_{f2}$. The derivation of the equations required to compute fault currents is not included here, but it can be found in the relevant literature.

6.3 GENERALIZED FAULT DIAGRAMS FOR SHUNT FAULTS

If an SLG fault occurs on a phase other than phase a, there are two methods that can be used to calculate the fault current. The first method is analytical and involves using Kron's primitive network concept, which is not discussed in this book. However, interested readers can find more information about this subject in Anderson's review [63]. The second method is based on Atabekov's generalized fault diagram [175], which is developed from Harder [185] and Hobson and Whitehead [186]. This method is briefly reviewed in this section and later in Section 6.6. For more information, interested readers are highly recommended to read Anderson's comprehensive review on this subject [63].

Figure 6.15 illustrates the generalized fault diagram that applies to an SLG fault on any phase. The resulting positive-, negative-, and zero-sequence networks are connected by ideal transformers, or phase shifters, having complex turns ratios of 1, a, or a^2. These transformers provide the output currents and voltages, which are phase-shifted by 0, 120°, or 240°, respectively, compared to the

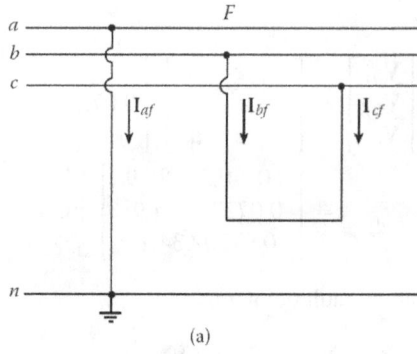

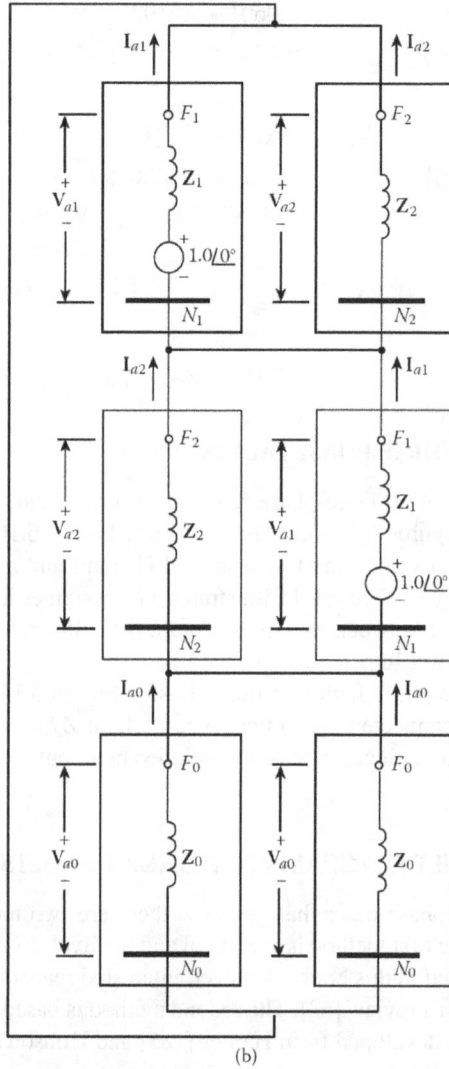

Figure 6.10 (a) Representation of an unsymmetrical three-phase fault and (b) interconnection of resultant equivalent sequence networks of Example 6.5.

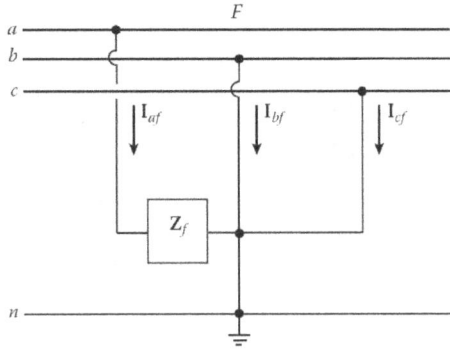

(a)

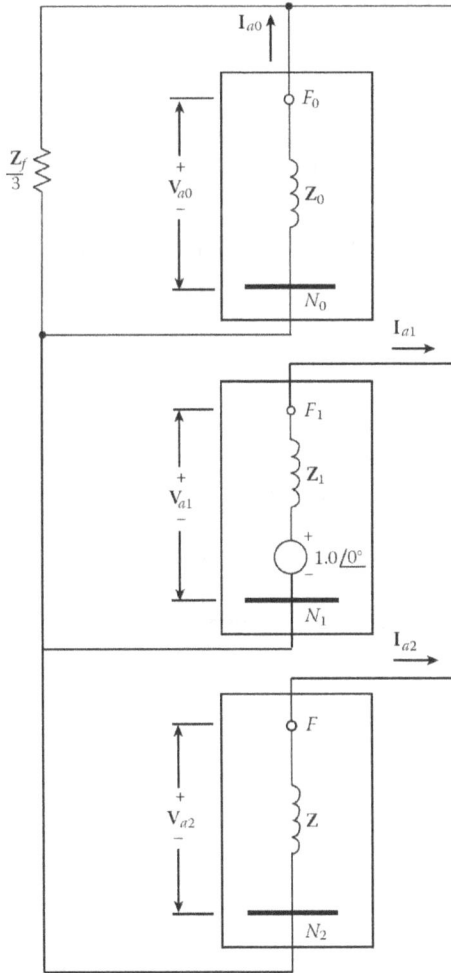

(b)

Figure 6.11   Generalized fault diagram for DLG fault.

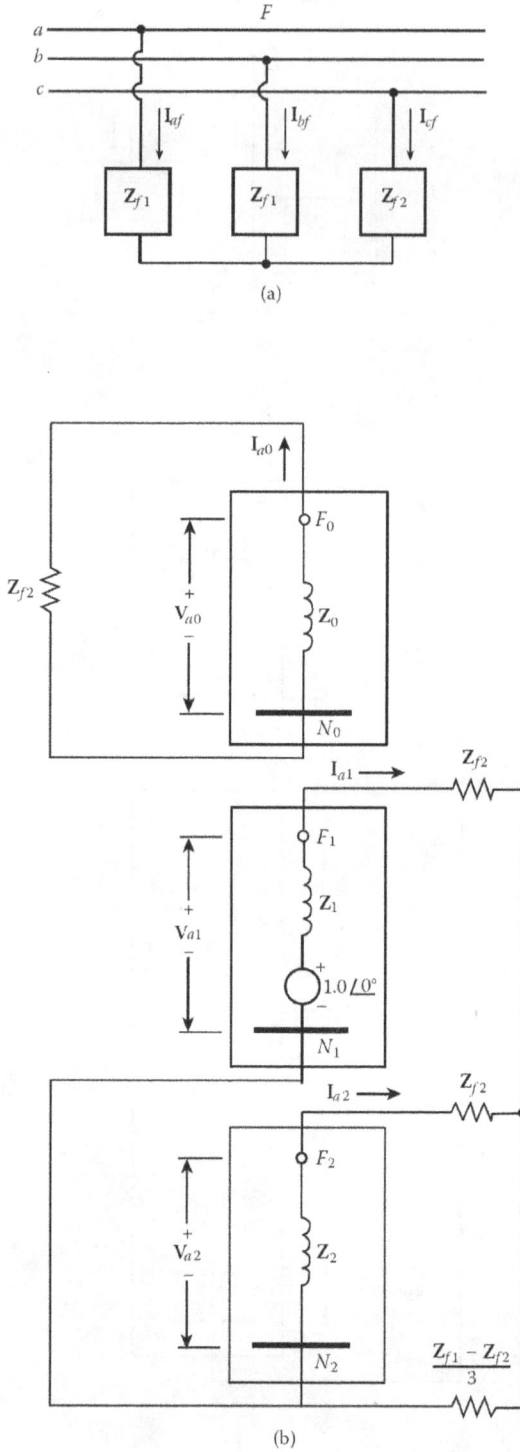

Figure 6.12   Generalized fault diagram for Example 6.6.

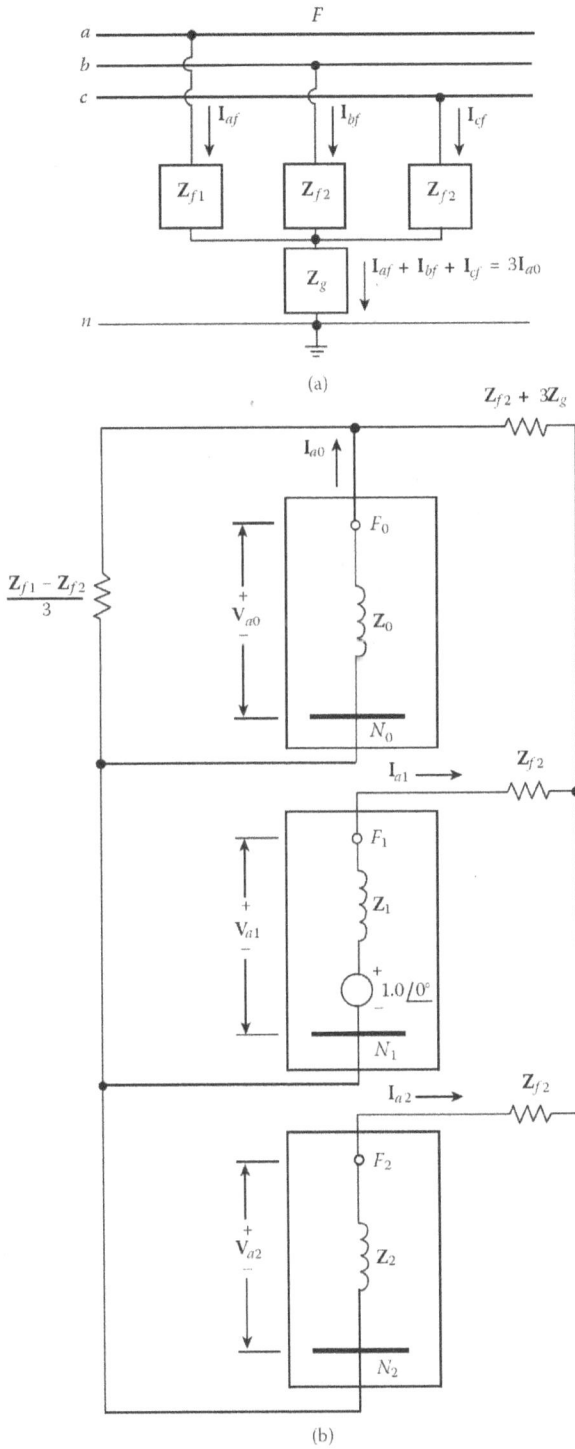

Figure 6.13 Generalized fault diagram for Example 6.7.

(a)

(b)

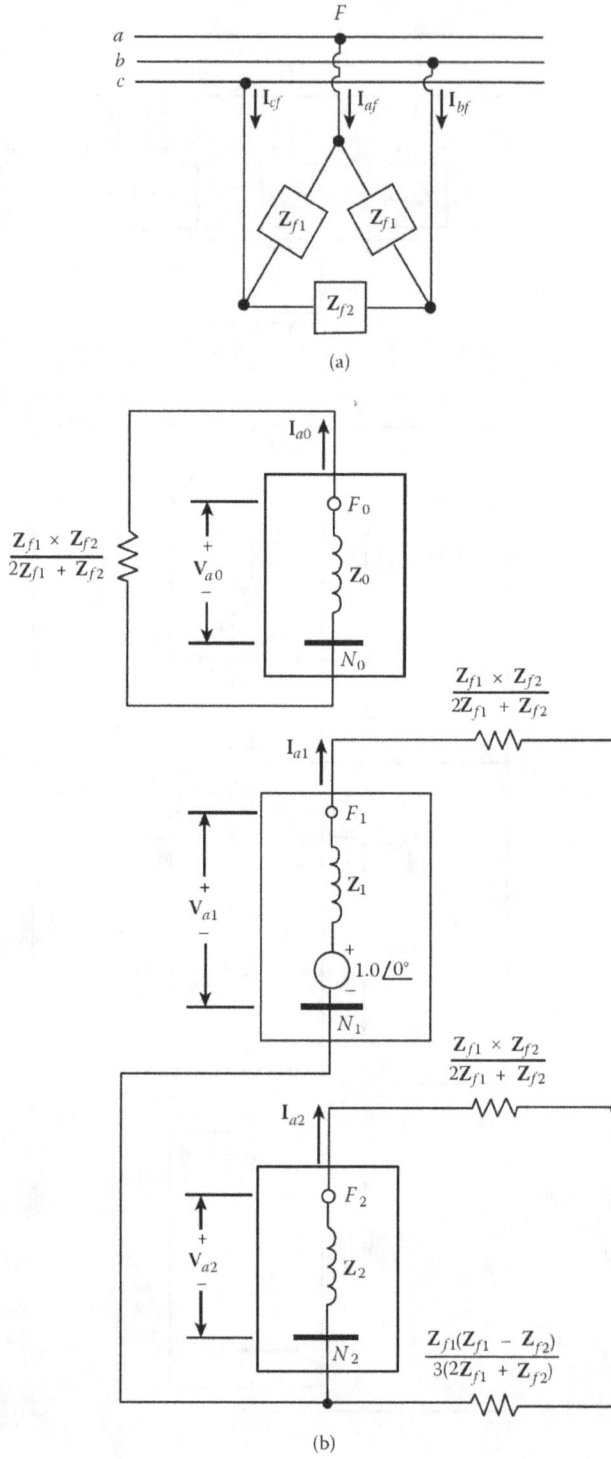

Figure 6.14 One line open: (a) general representation; (b) connection of sequence networks.

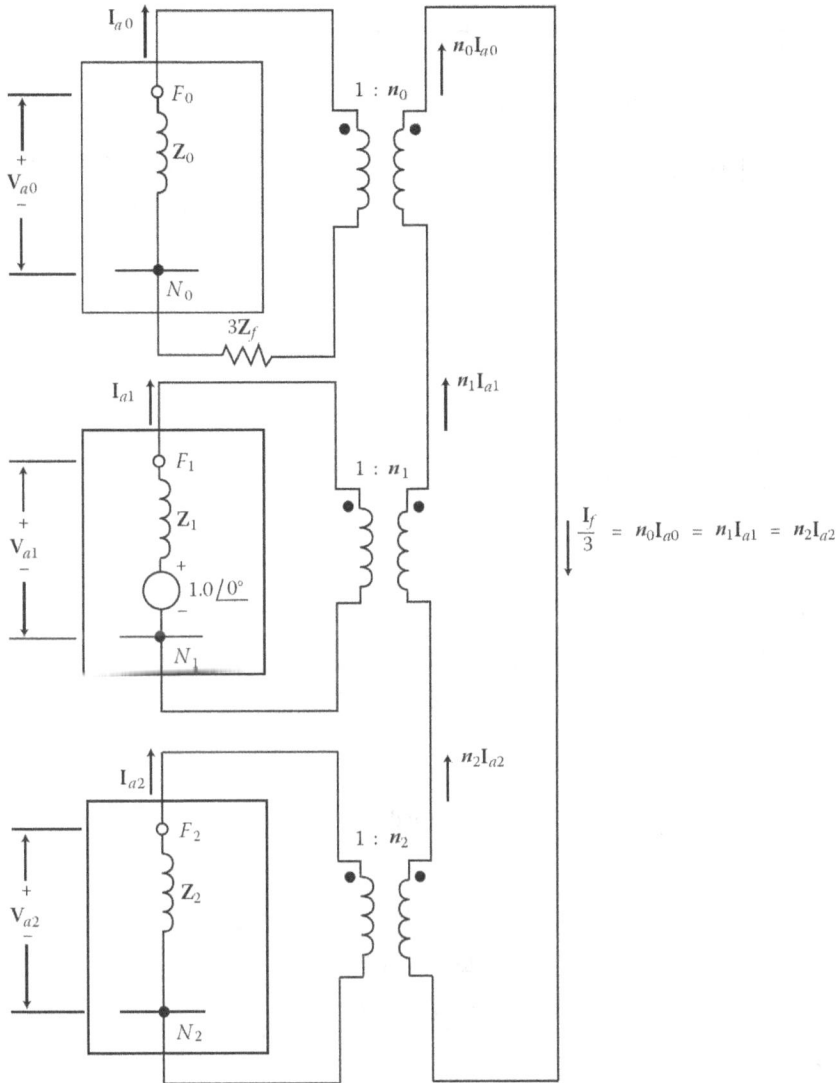

Figure 6.15 Generalized fault diagram for SLG fault.

input currents and voltages, without changing their magnitudes. It is not necessary to consider the ideal transformer as a physical device but rather as a symbolic phase shifter with a phase shift of $\mathbf{n} = je^{j\theta}$. The appropriate phase shifts for a given SLG fault can be obtained from Table 6.2. For example, if the SLG fault occurs on phase b, then the correct phase shifts are $\mathbf{n}_0 = 1$, $\mathbf{n}_1 = \mathbf{a}^2$, and $\mathbf{n}_2 = \mathbf{a}$.

The generalized fault diagram for a DLG fault on any two phases is presented in Figure 6.16. The phase shifts required for a given DLG fault can be determined using Table 6.1. For instance, if the DLG fault is between phases a and b, then the appropriate phase shifts are $\mathbf{n}_0 = 1$, $\mathbf{n}_1 = \mathbf{a}$, and $\mathbf{n}_2 = \mathbf{a}^2$.

Example 6.6: Consider the system described in Example 6.1 and assume that the SLG fault involves phase b.

Table 6.1

Complex Turns Ratios of Phase Shifters for SLG and GLG Faults

SLG Fault on Phase	DLG Fault on Phases	Phase Shift		
		n_0	n_1	n_2
a	$b - c$	1	1	1
b	$c - a$	1	a^2	a
c	$a - b$	1	a	a^2

(a) Draw the generalized fault diagram.
(b) Determine the sequence currents.
(c) Determine the phase currents.

Solution:

(a) Figure 6.17 shows the resulting generalized fault diagram.
(b) Since

$$\mathbf{I}_{a0} = a^2 \mathbf{I}_{a1} = a \mathbf{I}_{a2}$$

then

$$\mathbf{I}_a = 0 \text{ and } \mathbf{I}_c = 0$$

Thus, as before,

$$\mathbf{I}_{a1} = \frac{1.0\angle 0°}{\mathbf{Z}_0 + \mathbf{Z}_1 + \mathbf{Z}_2 + 3\mathbf{Z}_f}$$
$$= 0.8438\angle - 87.3° \text{ pu}$$

but

$$\mathbf{I}_{a0} = \mathbf{a}^2 \mathbf{I}_{a1}$$
$$= 1.0\angle 240° (0.8438\angle - 87.3°)$$
$$= 0.8438\angle 152.7° \text{ pu}$$

and

$$\mathbf{I}_{a2} = \frac{\mathbf{I}_{a0}}{a}$$
$$= \frac{0.8438\angle 152.7°}{1.0\angle 120°}$$
$$= 0.8438\angle 32.7° \text{ pu}$$

(c) Therefore,

$$\mathbf{I}_a = \mathbf{I}_c = 0$$

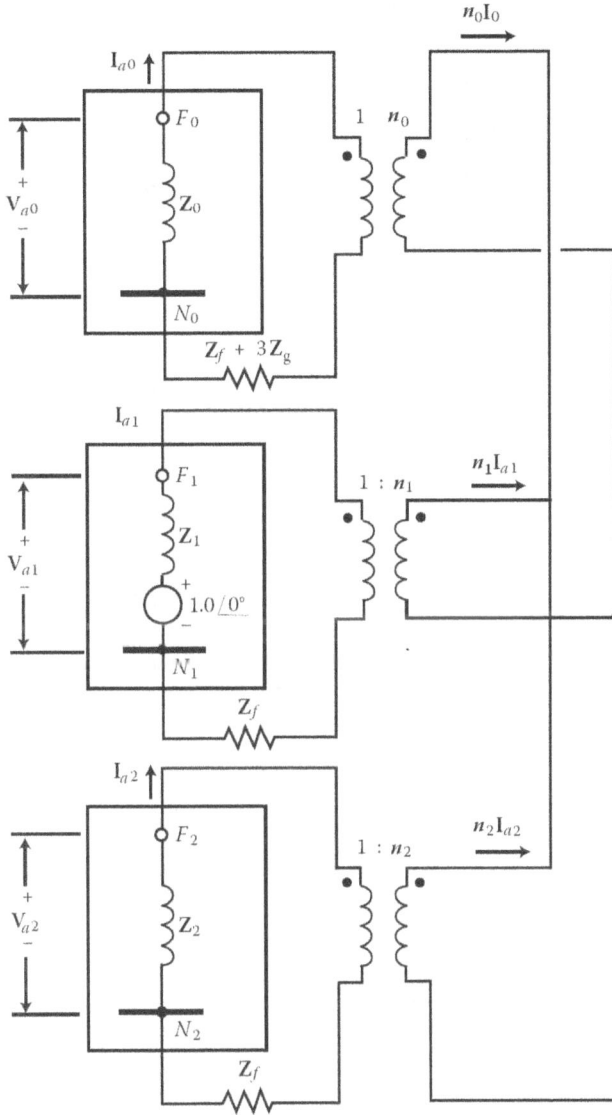

Figure 6.16 Generalized fault diagram for DLG fault.

hence,

$$
\begin{bmatrix} \mathbf{I}_{af} \\ \mathbf{I}_{bf} \\ \mathbf{I}_{cf} \end{bmatrix} = \begin{bmatrix} 1 & 1 & 1 \\ 1 & \mathbf{a}^2 & \mathbf{a} \\ 1 & \mathbf{a} & \mathbf{a}^2 \end{bmatrix} \begin{bmatrix} 0.8438\angle 152.7° \\ 0.8438\angle -87.3° \\ 0.8438\angle 32.7° \end{bmatrix}
$$

$$
= \begin{bmatrix} 0 \\ 2.5314\angle 152.7° \\ 0 \end{bmatrix} \text{ pu}
$$

Example 6.7: Consider the system described in Example 6.4 and assume that the DLG fault involves phases a and b.

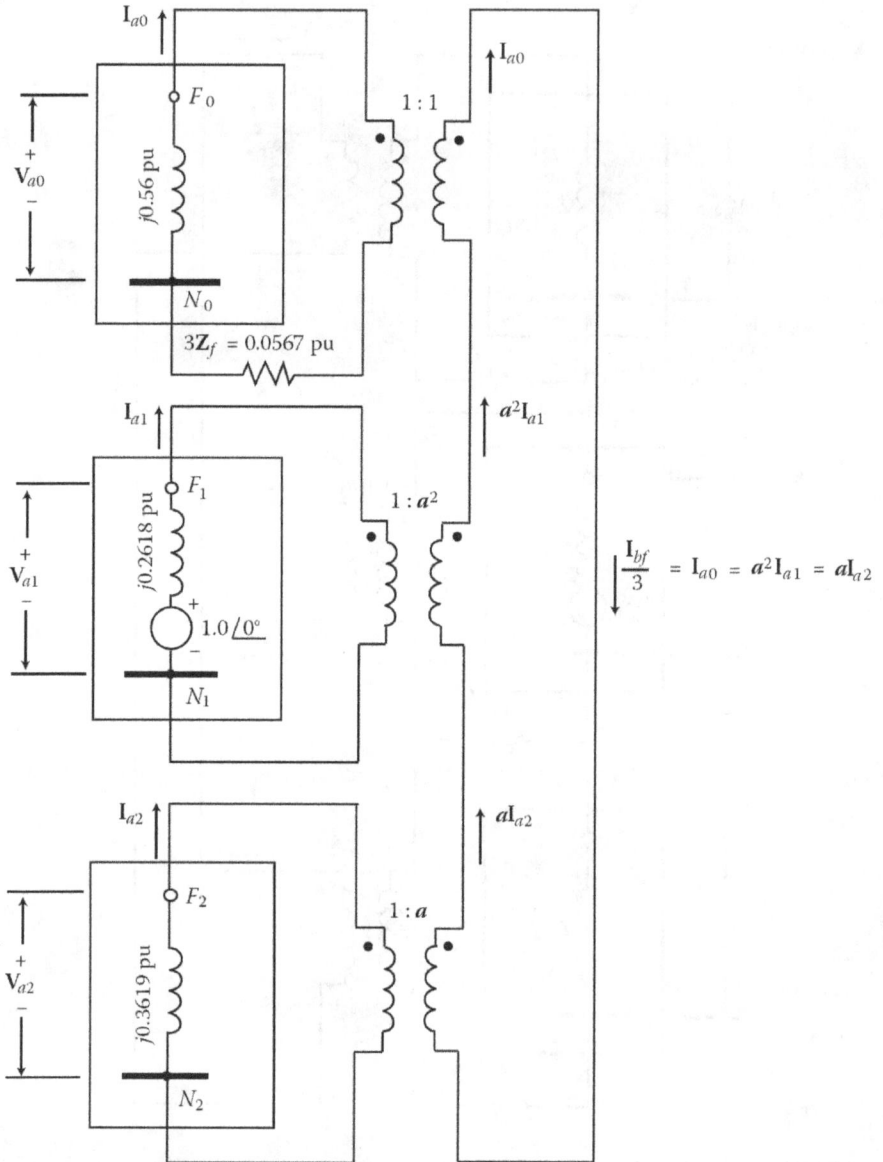

Figure 6.17 Generalized fault diagram for SLG fault of Example 6.6.

(a) Draw the generalized fault diagram.
(b) Determine the sequence currents.
(c) Determine the phase currents.

Solution:

(a) Figure 6.18 shows the resulting generalized fault diagram.
(b) From Example 6.4

$$\mathbf{I}_{a1} = 2.5097\angle -84.5° \text{ pu}$$

hence,

$$\mathbf{I}_{a0} = -\left(\frac{j0.3808}{1.0731}\right)\mathbf{a}\mathbf{I}_{a1}$$
$$= -0.3549(1\angle120°)(2.0597\angle-84.5°)$$
$$= 0.7309\angle215.5° \text{ pu}$$

and

$$\mathbf{I}_{a2} = -\left(\frac{j0.6923}{1.0731}\right)\frac{\mathbf{I}_{a1}}{\mathbf{a}}$$
$$= -0.6451(1\angle120°)(2.0597\angle-84.5°)$$
$$= 1.3288\angle-24.5° \text{ pu}$$

(c)

$$\mathbf{I}_f = \mathbf{I}_a + \mathbf{I}_b = 3\mathbf{I}_{a0}$$
$$= 3(0.7309\angle215.5°)$$
$$= 2.1927\angle215.5° \text{ pu}$$

$$\mathbf{I}_a = \mathbf{I}_{a0} + \mathbf{I}_{a1} + \mathbf{I}_{a2}$$
$$= 0.7309\angle215.5° + 2.0597\angle-84.5° + 1.3288\angle-24.5°$$
$$= 0.8116 - j3.0256$$
$$= 3.1326\angle-75° \text{ pu}$$

$$\mathbf{I}_b = \mathbf{I}_{a0} + \mathbf{a}^2\mathbf{I}_{a1} + \mathbf{a}\mathbf{I}_{a2}$$
$$= 0.7309\angle215.5° + (1\angle240°)(2.0597\angle-84.5°) + (1\angle120°)(1.3288\angle-24.5°)$$
$$= 2.5966 + j1.7524$$
$$= 2.1326\angle146° \text{ pu}$$

$$\mathbf{I}_c \triangleq 0$$

Also, as a check,

$$\mathbf{I}_f = \mathbf{I}_a + \mathbf{I}_b$$
$$= 3.1326\angle-75° + 3.1326\angle146°$$
$$= 2.1927\angle215.5°$$

6.4 SERIES FAULTS

Generally, series (longitudinal) faults occur due to an unbalanced series impedance condition of the lines. A series fault can be caused by one or two broken lines or an impedance inserted in one or two lines. Such faults occur when line (or circuits) are controlled by circuit breakers or fuses, which do not open all three phases, leading to one or two phases of the line being open while the other phase(s) is closed.

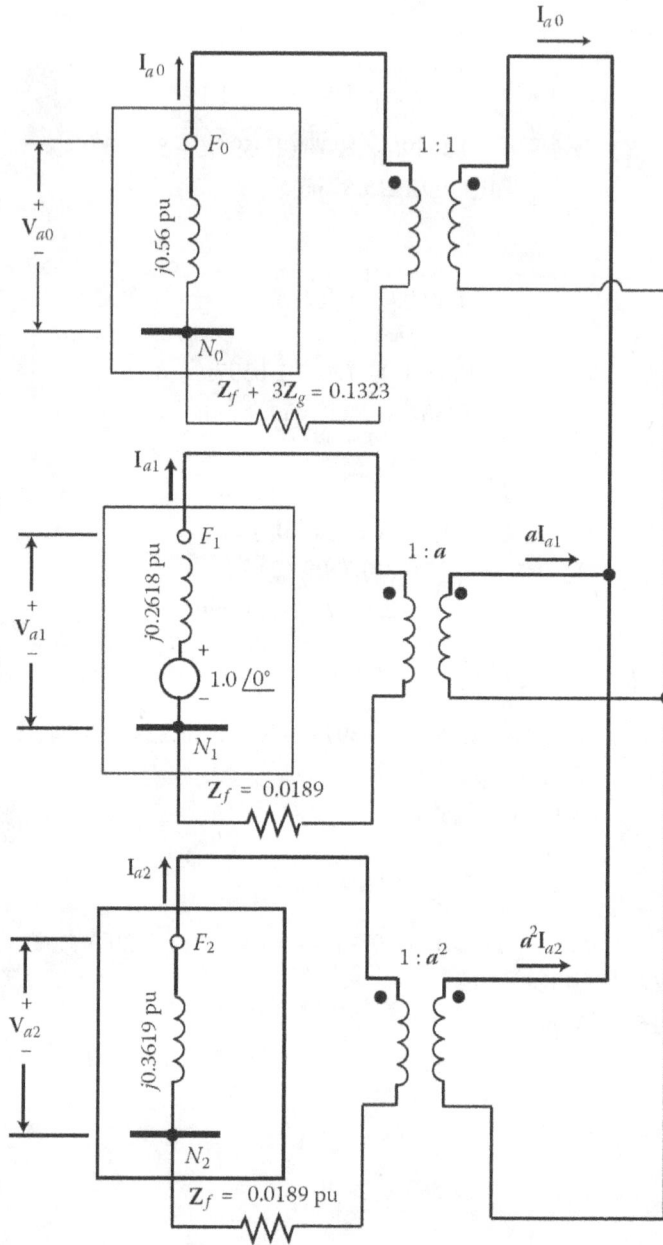

Figure 6.18 Generalized fault diagram for SLG fault of Example 6.7.

Figure 6.19 depicts a series fault resulting from one line (phase a) being open, leading to a series unbalance. In a series fault, unlike a shunt fault, there are two fault points, F and F', located on either side of the unbalance. The series line impedances $\mathbf{Z}'s$ can take any value between zero and infinity. In this particular case, the line impedance between points F and F' of phase a is infinite.

The sequence networks include the symmetrical portions of the system, looking back to the left of F and to the right of F'. Since in series faults, there is no connection between lines or between line(s) and neutral, only the sequence voltages of $\mathbf{V}_{aa'-0}$, $\mathbf{V}_{aa'-1}$, and $\mathbf{V}_{aa'-2}$ are of interest, not the sequence voltages of $\mathbf{V}_{a0}$, $\mathbf{V}_{a1}$, $\mathbf{V}_{a2}$, etc. (as it was the case with the shunt faults).

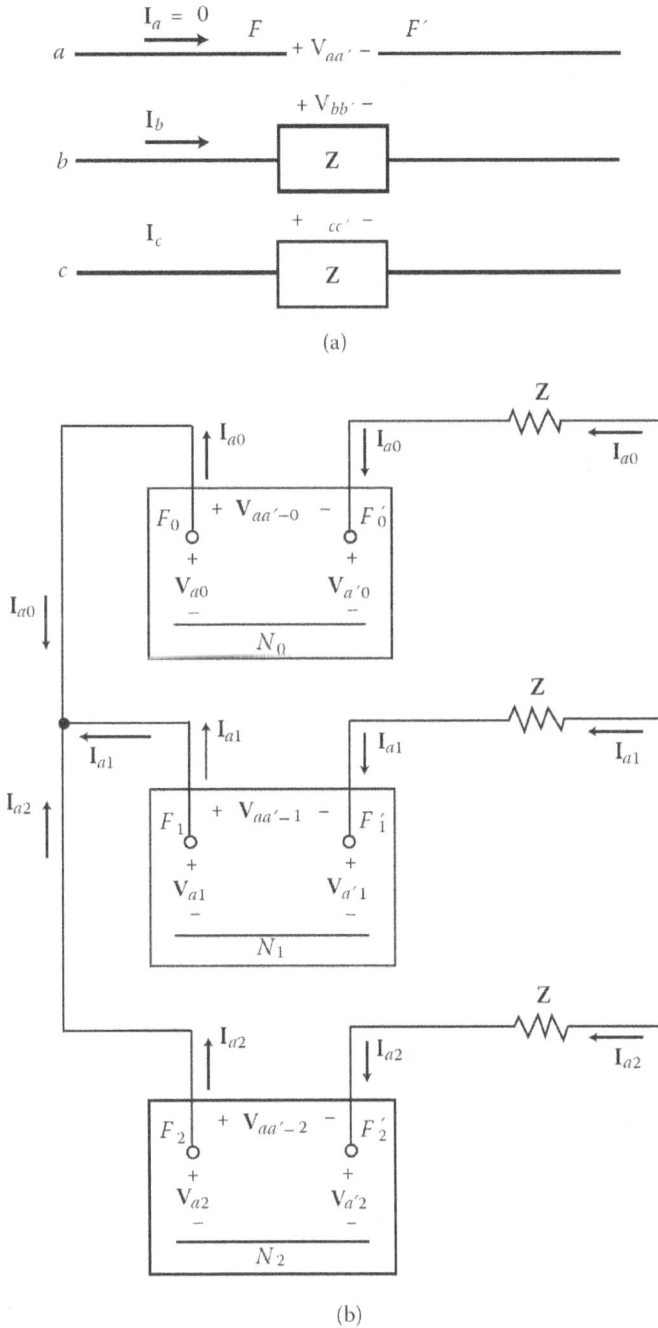

Figure 6.19 One line open: (a) general representation; (b) connection of sequence networks.

6.4.1 ONE LINE OPEN

From Figure 6.19, it can be observed that the line impedance for the open-line conductor in phase a is infinity, whereas the line impedances for the other two phases have some finite values. Hence, the positive-, negative-, and zero-sequence currents can be expressed as

$$\mathbf{I}_{a1} = \frac{\mathbf{V}_F}{\mathbf{Z} + \mathbf{Z}_1 + (\mathbf{Z} + \mathbf{Z}_0)(\mathbf{Z} + \mathbf{Z}_2)/(2\mathbf{Z} + \mathbf{Z}_0 + \mathbf{Z}_2)} \tag{6.94}$$

and by current division,

$$\mathbf{I}_{a2} = \left(-\frac{\mathbf{Z} + \mathbf{Z}_0}{2\mathbf{Z} + \mathbf{Z}_0 + \mathbf{Z}_2} \right) \mathbf{I}_{a1} \tag{6.95}$$

and

$$\mathbf{I}_{a0} = \left(-\frac{\mathbf{Z} + \mathbf{Z}_2}{2\mathbf{Z} + \mathbf{Z}_0 + \mathbf{Z}_2} \right) \mathbf{I}_{a1} \tag{6.96}$$

or simply

$$\mathbf{I}_{a0} = -(\mathbf{I}_{a1} + \mathbf{I}_{a2}) \tag{6.97}$$

6.4.2 TWO LINES OPEN

If two lines are open as shown in Figure 6.20, then the line impedances for one line open (OLO) in phases b and c are infinity, whereas the line impedance of phase a has some finite value. Thus,

$$\mathbf{I}_b = \mathbf{I}_c = 0 \tag{6.98}$$

and

$$\mathbf{V}_{aa'} = \mathbf{Z}\mathbf{I}_a \tag{6.99}$$

By inspection of Figure 6.20, the positive-, negative-, and zero-sequence currents can be expressed as

$$\mathbf{I}_{a1} = \mathbf{I}_{a2} = \mathbf{I}_{a0} = \frac{\mathbf{V}_F}{\mathbf{Z}_0 + \mathbf{Z}_1 + \mathbf{Z}_2 + 3\mathbf{Z}_f} \tag{6.100}$$

6.5 DETERMINATION OF SEQUENCE NETWORK EQUIVALENTS FOR SERIES FAULTS

Thévenin's theorem cannot be directly applied to series faults as they have two fault points (F and F'), unlike shunt faults, which only have one. Therefore, a two-port Thévenin equivalent of the sequence networks is required for series faults, as proposed by Anderson in [63] and [184].

6.5.1 BRIEF REVIEW OF TWO-PORT THEORY

Figure 6.21 shows a general two-port network for which it can be written that

$$\begin{bmatrix} \mathbf{V}_1 \\ \mathbf{V}_2 \end{bmatrix} \begin{bmatrix} \mathbf{Z}_{11} & \mathbf{Z}_{12} \\ \mathbf{Z}_{21} & \mathbf{Z}_{22} \end{bmatrix} \begin{bmatrix} \mathbf{I}_1 \\ \mathbf{I}_2 \end{bmatrix} \tag{6.101}$$

where the *open-circuit impedance parameters* can be determined by leaving the ports open and expressed in terms of voltage and current as

$$\mathbf{Z}_{11} = \left. \frac{\mathbf{V}_1}{\mathbf{I}_1} \right|_{I_2=0} \tag{6.102}$$

$$\mathbf{Z}_{12} = \left. \frac{\mathbf{V}_1}{\mathbf{I}_2} \right|_{I_1=0} \tag{6.103}$$

$$\mathbf{Z}_{21} = \left. \frac{\mathbf{V}_2}{\mathbf{I}_1} \right|_{I_2=0} \tag{6.104}$$

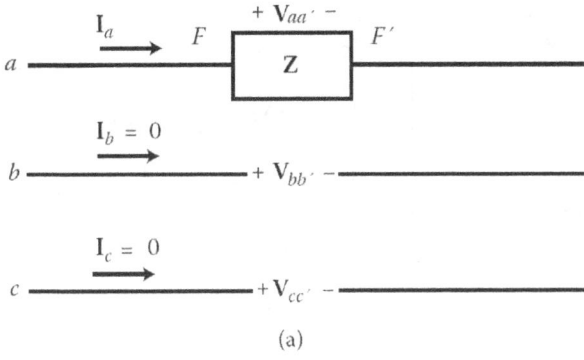

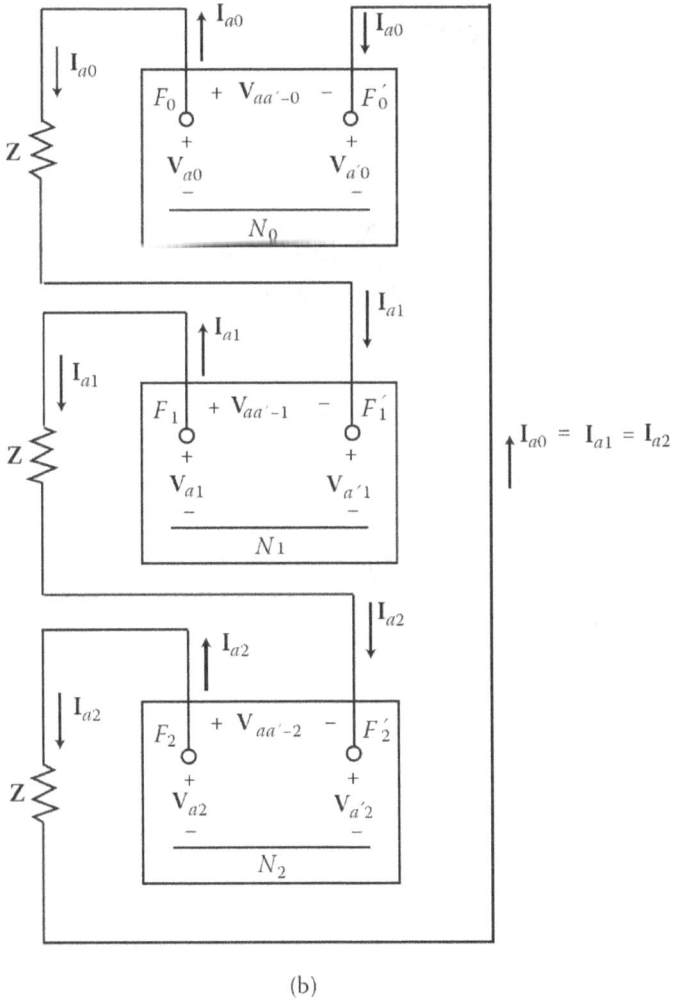

Figure 6.20 Two-lines open: (a) general representation; (b) interconnection of sequence networks.

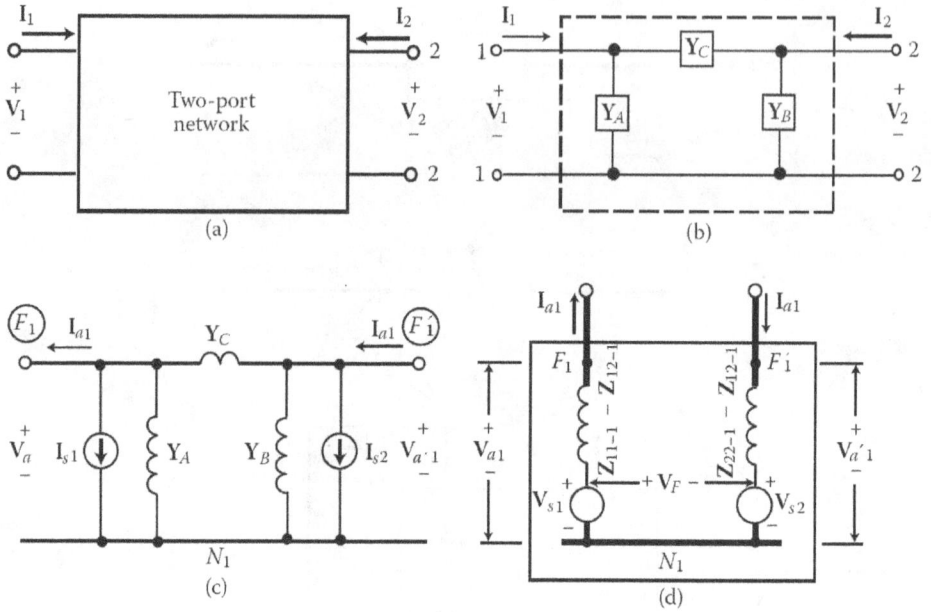

Figure 6.21 Application of two-port network theory for determining equivalent positive-sequence network for series faults: (a) general two-port network; (b) general π-equivalent positive sequence network; (c) equivalent positive-sequence network; (d) uncoupled positive-sequence network.

$$\mathbf{Z}_{22} = \left.\frac{\mathbf{V}_2}{\mathbf{I}_2}\right|_{I_1=0} \tag{6.105}$$

Alternatively, it can be observed that

$$\begin{bmatrix} \mathbf{I}_1 \\ \mathbf{I}_2 \end{bmatrix} \begin{bmatrix} \mathbf{Y}_{11} & \mathbf{Y}_{12} \\ \mathbf{Y}_{21} & \mathbf{Y}_{22} \end{bmatrix} \begin{bmatrix} \mathbf{V}_1 \\ \mathbf{V}_2 \end{bmatrix} \tag{6.106}$$

where the *short-circuit admittance parameters* can be determined (by short-circuiting the ports) from

$$\mathbf{Y}_{11} = \left.\frac{\mathbf{I}_1}{\mathbf{V}_1}\right|_{V_2=0} \tag{6.107}$$

$$\mathbf{Y}_{21} = \left.\frac{\mathbf{I}_2}{\mathbf{V}_1}\right|_{V_2=0} \tag{6.108}$$

$$\mathbf{Y}_{12} = \left.\frac{\mathbf{I}_1}{\mathbf{V}_2}\right|_{V_1=0} \tag{6.109}$$

$$\mathbf{Z}_{22} = \left.\frac{\mathbf{I}_2}{\mathbf{V}_2}\right|_{V_1=0} \tag{6.110}$$

Figure 6.21b shows a general π-equivalent of a two-port network in terms of admittances. The $\mathbf{Y}_A$, $\mathbf{Y}_B$, and $\mathbf{Y}_C$ admittances can be found from

$$\mathbf{Y}_A = \mathbf{Y}_{11} + \mathbf{Y}_{12} \tag{6.111}$$

$$\mathbf{Y}_B = \mathbf{Y}_{22} + \mathbf{Y}_{12} \tag{6.112}$$

$$\mathbf{Y}_C = -\mathbf{Y}_{12} \tag{6.113}$$

6.5.2 EQUIVALENT ZERO-SEQUENCE NETWORKS

By comparing the zero-sequence network shown in Figure 6.19 with the general two-port network shown in Figure 6.21a, it can be observed that

$$I_1 = -I_{a0} \tag{6.114}$$

$$I_2 = I_{a0} \tag{6.115}$$

$$V_1 = V_{a0} \tag{6.116}$$

$$V_2 = V_{a'0} \tag{6.117}$$

Hence, substituting Equations 6.114 through 6.117 into Equation 6.106, it can be expressed for the Thévenin equivalent of the zero-sequence network that

$$\begin{bmatrix} -I_{a0} \\ I_{a0} \end{bmatrix} \begin{bmatrix} Y_{11-0} & Y_{12-0} \\ Y_{21-0} & Y_{22-0} \end{bmatrix} \begin{bmatrix} V_{a0} \\ V_{a'0} \end{bmatrix} \tag{6.118}$$

6.5.3 EQUIVALENT POSITIVE- AND NEGATIVE-SEQUENCE NETWORKS

The active two-port network with internal sources, which is the equivalent positive-sequence network, is shown in Figure 6.16c. Therefore, the two-port Thévenin equivalent of the positive-sequence network can be expressed as:

$$\begin{bmatrix} -I_{a1} \\ I_{a1} \end{bmatrix} \begin{bmatrix} Y_{11-1} & Y_{12-1} \\ Y_{21-1} & Y_{22-1} \end{bmatrix} \begin{bmatrix} V_{a1} \\ V_{a'1} \end{bmatrix} + \begin{bmatrix} I_{s1} \\ I_{s2} \end{bmatrix} \tag{6.119}$$

or, alternatively,

$$\begin{bmatrix} -V_{a1} \\ V_{a'1} \end{bmatrix} \begin{bmatrix} Z_{11-1} & Z_{12-1} \\ Z_{21-1} & Z_{22-1} \end{bmatrix} \begin{bmatrix} -I_{a1} \\ I_{a1} \end{bmatrix} + \begin{bmatrix} V_{s1} \\ V_{s2} \end{bmatrix} \tag{6.120}$$

where I_{s1}, V_{s1} and I_{s2}, V_{s2} represent internal sources 1 and 2, respectively. As before, the admittances Y_A, Y_B, and Y_C can be determined from

$$Y_A = Y_{11-1} + Y_{12-1} \tag{6.121}$$

$$Y_B = Y_{22-1} + Y_{12-1} \tag{6.122}$$

$$Y_C = -Y_{12-1} \tag{6.123}$$

Anderson [63] simplifies Equation 6.120 and shows that the two-port Thévenin equivalent of the negative-sequence network is the same as the one shown in Figure 6.16c, but without the internal sources.

$$\begin{bmatrix} V_{a1} \\ V_{a'1} \end{bmatrix} = \begin{bmatrix} V_{s1} \\ V_{s2} \end{bmatrix} - \begin{bmatrix} (Z_{11-1} - Z_{12-1})I_{a1} \\ -(Z_{22-1} - Z_{12-1})I_{a1} \end{bmatrix} \tag{6.124}$$

due to the fact that I_{a1} leaves the network at fault point F and enters at fault point F' due to external connection. This facilitates the voltage V_{a1} to be expressed in terms of the equivalent impedance Z_{s1} and the current I_{a1} to be expressed as it has been done for the shunt faults, where

$$V_{a1} = V_F - Z_1 I_{a1} \tag{6.125}$$

Therefore, it can be concluded that the port of the positive-sequence network are completely uncoupled and that the resulting uncoupled positive-sequence network can be shown as in Figure 6.16d. The voltages V_{a1} and $V_{a'1}$ can be found from Equation 6.120 as

$$\begin{bmatrix} V_{a1} \\ V_{a'1} \end{bmatrix} = \frac{1}{\Delta_y} \begin{bmatrix} (Y_{12-1}I_{s2} - Y_{22-1}I_{s1}) - (Y_{12-1}Y_{22-1})I_{a1} \\ (Y_{12-1}I_{s1} - Y_{11-1}I_{s2}) - (Y_{12-1}Y_{11-1})I_{a1} \end{bmatrix} \tag{6.126}$$

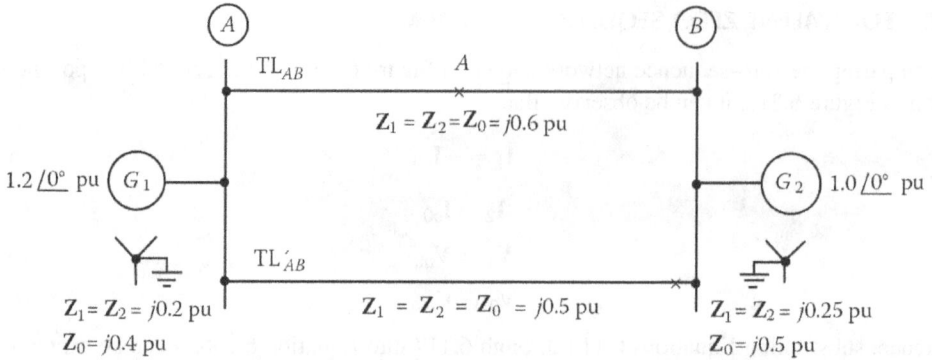

Figure 6.22 System diagram for Example 6.8.

where

$$\Delta_y = \det \begin{bmatrix} \mathbf{Y}_{11-1} & \mathbf{Y}_{12-1} \\ \mathbf{Y}_{21-1} & \mathbf{Y}_{22-1} \end{bmatrix} \tag{6.127}$$

or

$$\Delta_y = \mathbf{Y}_{11-1}\mathbf{Y}_{22-1} - \mathbf{Y}_{12-1}^2 \tag{6.128}$$

Example 6.8: Consider the system shown in Figure 6.22 and assume that there is a series fault at fault point A, which is located at the middle of the transmission line TL_{AB}, as shown in the figure, and determine the following:

(a) Admittance matrix associated with the positive-sequence network
(b) Two-port Thévenin equivalent of the positive-sequence network
(c) Two-port Thévenin equivalent of the negative-sequence network

Solution:

Figure 6.18 shows the steps that are necessary to determine the positive- and negative-sequence network equivalents. Figure 6.18a shows the impedance diagram of the system for the positive sequence. Figure 6.18b shows the resulting two-port equivalent with input and output currents and voltages.

(a) To determine the elements of the admittance matrix $\mathbf{Y}$, it is necessary to remove the internal voltage sources, shown in Figure 6.18b, by short-circuiting them. Then, with $\mathbf{V}_2 = 0$ (i.e., by short circuiting the terminals of the second port), apply $\mathbf{V}_1 = 1.0\angle 0°$ pu and determine the parameter

$$\mathbf{Y}_{11} = \frac{\mathbf{I}_1}{\mathbf{V}_1}\bigg|_{\mathbf{V}_2=0} = \mathbf{I}_1 = \frac{1.0\angle 0°}{0.4522\angle 90°} = -j2.2115 \text{ pu}$$

and

$$\mathbf{Y}_{21} = \frac{\mathbf{I}_2}{\mathbf{V}_1}\bigg|_{\mathbf{V}_2=0} = \mathbf{I}_2 = -0.1087\mathbf{I}_1 = j0.2404 \text{ pu}$$

Now, with $\mathbf{V}_1$ and $\mathbf{V}_2 = 1.0\angle 0°$ pu, determine

$$\mathbf{Y}_{22} = \frac{\mathbf{I}_2}{\mathbf{V}_2}\bigg|_{\mathbf{V}_1=0} = \mathbf{I}_2 = \frac{1.0\angle 0°}{0.4782\angle 90°} = -j2.0912 \text{ pu}$$

and

$$\mathbf{Y}_{22} = \mathbf{Y}_{21} = j0.2404 \text{ pu}$$

Hence,

$$\mathbf{Y} = \begin{bmatrix} \mathbf{Y}_{11-1} & \mathbf{Y}_{12-1} \\ \mathbf{Y}_{21-1} & \mathbf{Y}_{22-1} \end{bmatrix}$$

$$= \begin{bmatrix} -j2.2115 & j0.2404 \\ j0.2404 & -2.0912 \end{bmatrix}$$

(b) To find the source currents $\mathbf{I}_{s1}$ and $\mathbf{I}_{s2}$, short circuit both F and F' to neutral and use the super-position theorem, so that

$$\mathbf{I}_{s1} = \mathbf{I}_{s1(1.2)} + \mathbf{I}_{s1(1.0)}$$
$$= 2.0193\angle 90° + 0.7212\angle 90°$$
$$= 2.7405\angle 90° \text{ pu}$$

and

$$\mathbf{I}_{s2} = \mathbf{I}_{s2(1.2)} + \mathbf{I}_{s2(1.0)}$$
$$= 0.4326\angle 90° + 1.4904\angle 90°$$
$$= 1.9230\angle 90° \text{ pu}$$

Figure 6.23c shows the resulting two-port Thévenin equivalent of the positive-sequence network. Figure 6.23d shows the corresponding coupled positive-sequence network.

(c) Figure 6.23e shows the resulting two-port Thévenin equivalent of the negative-sequence network. Notice that it is the same as the one for the positive-sequence network but with-out its current sources. Figure 6.23f shows the corresponding coupled negative-sequence network.

Example 6.9: Consider the solution of Example 6.8 and determine the following:

(a) Uncoupled positive-sequence network
(b) Uncoupled negative-sequence network

Solution:

(a) From Example 6.8,

$$\mathbf{Y} = \begin{bmatrix} -j2.2115 & j0.2404 \\ j0.2404 & -2.0912 \end{bmatrix}$$

where

$$\Delta_y = -4.6247 - (-0.0578)$$
$$= -4.5669$$

Since

$$\begin{bmatrix} \mathbf{V}_{a1} \\ \mathbf{V}_{a'1} \end{bmatrix} = \frac{1}{\Delta_y} \begin{bmatrix} (\mathbf{Y}_{12-1}\mathbf{I}_{s2} - \mathbf{Y}_{22-1}\mathbf{I}_{s1}) - (\mathbf{Y}_{12-1}\mathbf{Y}_{22-1})\mathbf{I}_{a1} \\ (\mathbf{Y}_{12-1}\mathbf{I}_{s1} - \mathbf{Y}_{11-1}\mathbf{I}_{s2}) - (\mathbf{Y}_{12-1}\mathbf{Y}_{11-1})\mathbf{I}_{a1} \end{bmatrix}$$

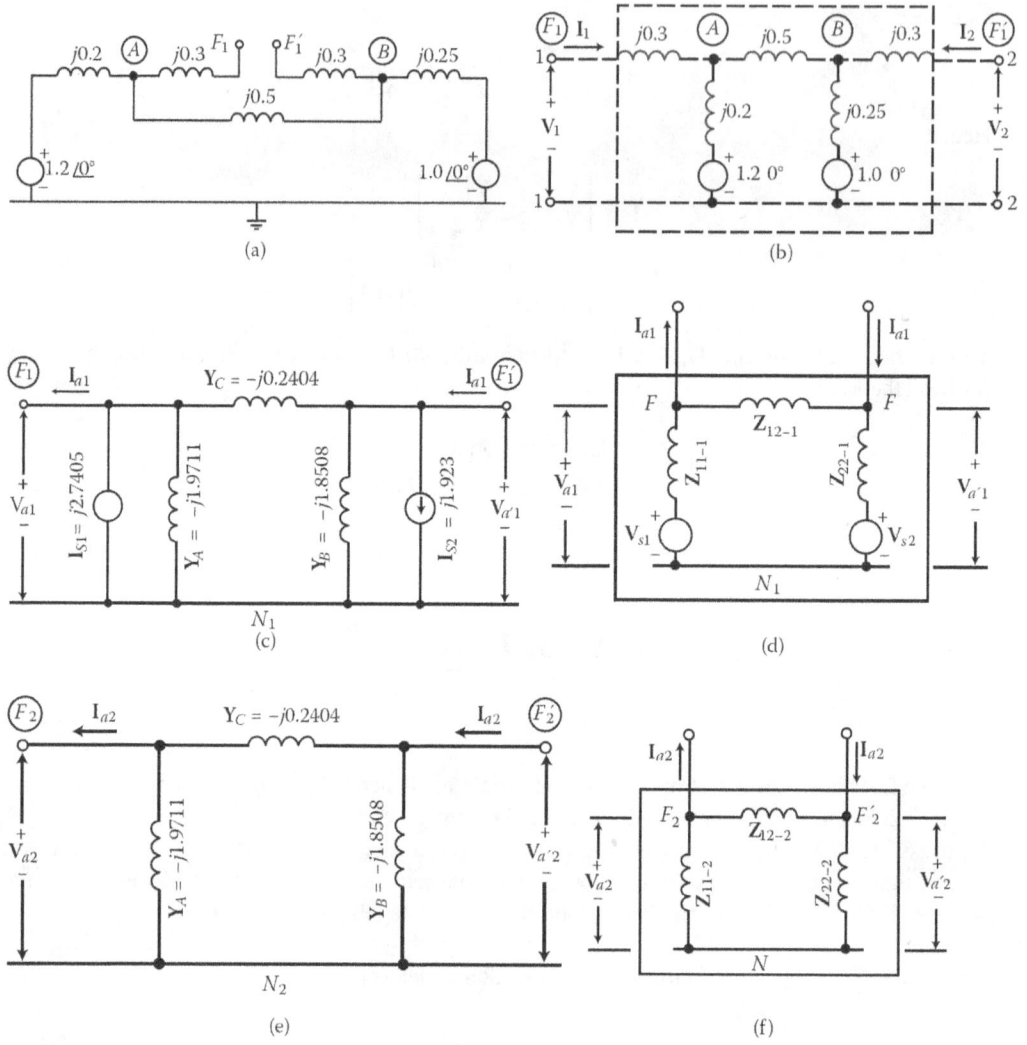

Figure 6.23 The following steps illustrate how to obtain the positive- and negative-sequence network equivalents for Example 6.8: (a) Draw the system diagram; (b) derive the resulting two-port equivalent with input and output currents and voltages; (c) obtain the two-port Thévenin equivalent of the positive-sequence network; (d) generate the resulting coupled positive-sequence network; (e) derive the two-port Thévenin equivalent of the negative-sequence network; (f) obtain the resulting coupled negative-sequence network.

where

$$(\mathbf{Y}_{12-1}\mathbf{I}_{s2} - \mathbf{Y}_{22-1}\mathbf{I}_{s1}) = j0.2404(j1.9230) - (-j2.0912)j2.7405$$
$$= -6.1932$$

$$(\mathbf{Y}_{12-1}\mathbf{Y}_{22-1})\mathbf{I}_{a1}) = (j0.2404 - j2.0912)\mathbf{I}_{a1}$$
$$= -j1.8508\mathbf{I}_{a1}$$

$$(\mathbf{Y}_{12-1}\mathbf{I}_{s1} - \mathbf{Y}_{11-1}\mathbf{I}_{s2}) = j0.2404(j2.7405) - (-j2.2115)j1.923$$
$$= -4.9115$$

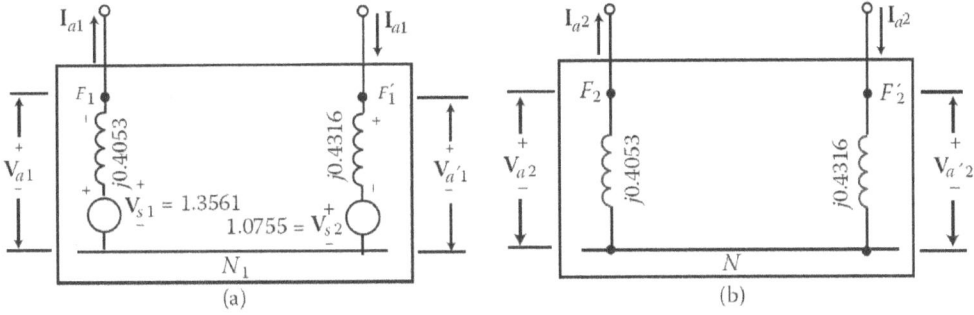

Figure 6.24 Uncoupled sequence networks: (a) positive-sequence network; (b) negative-sequence network.

$$(\mathbf{Y}_{12-1}\mathbf{Y}_{11-1})\mathbf{I}_{a1}) = (j0.2404 - j2.2115)\mathbf{I}_{a1}$$
$$= -j1.9711\mathbf{I}_{a1}$$

Therefore,

$$\begin{bmatrix} \mathbf{V}_{a1} \\ \mathbf{V}_{a'1} \end{bmatrix} = \frac{1}{-4.5669}\begin{bmatrix} (-6.1932) + j1.8508\mathbf{I}_{a1} \\ (-4.9115) - j1.9711\mathbf{I}_{a1} \end{bmatrix}$$
$$= \begin{bmatrix} 1.3561 - j0.4053\mathbf{I}_{a1} \\ 1.0755 + j0.4316\mathbf{I}_{a1} \end{bmatrix}$$

(b) Figure 6.24b shows the corresponding uncoupled negative-sequence network.

6.6 GENERALIZED FAULT DIAGRAM FOR SERIES FAULTS

The generalized fault diagrams for Open Line-to-Ground (OLO) and Two-Line-Open (TLO) faults are presented in Figures 6.25 and 6.26, respectively. The resulting positive-, negative-, and zero-sequence networks in each case are coupled through ideal transformers with appropriate phase shifts, as given in Table 6.2. For instance, if the OLO fault occurs on phase b, then the phase shifts should be $n_0 = 1$, $n_1 = \mathbf{a}^2$, and $n_2 = \mathbf{a}$. On the other hand, if the TLO fault involves phases b and c, then the appropriate phase shifts are $n_0 = 1$, $n_1 = 1$, and $n_2 = 1$

Example 6.10: Assume that a given power system has two generators connected to each other over a transmission line and that there is a TLO series fault on the transmission line at some location involving phases a and b. Figure 6.27 shows the resulting two-port Thévenin equivalents of the sequence networks at fault points F and F'. If the remaining line (i.e., line c) has a fault impedance of 0.1 pu, do the following:

(a) Draw the generalized fault diagram.
(b) Determine the positive-sequence current.
(c) Determine the negative-sequence current.
(d) Determine the negative-sequence current.
(e) Determine the line current for phase a.
(f) Determine the line current for phase b.
(g) Determine the line current for phase c.

Solution:

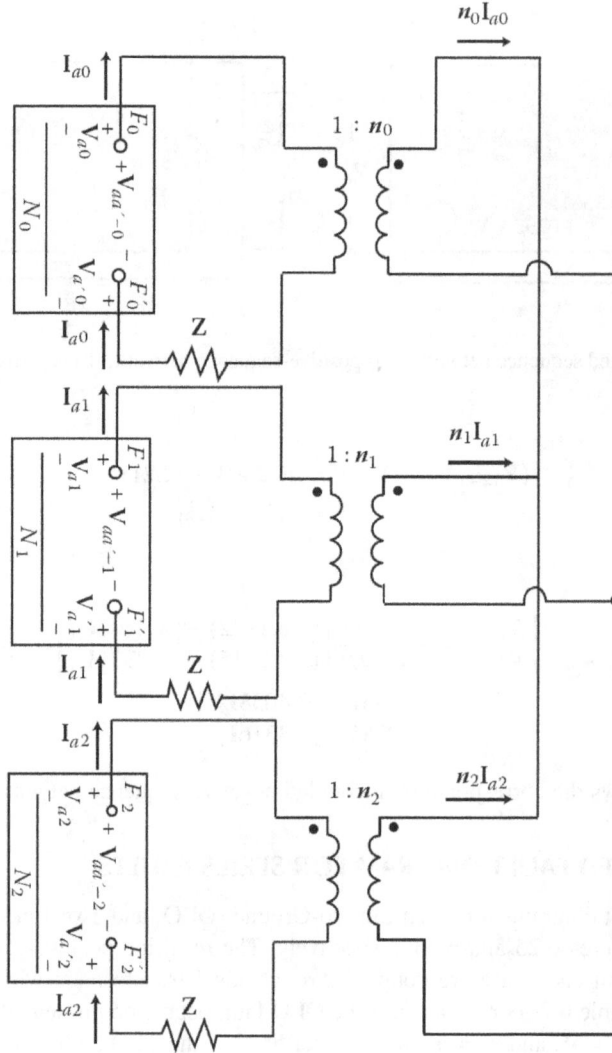

Figure 6.25 Generalized fault diagram for one-line open.

Table 6.2
Complex Turns Ratios of Phase Shifters for OLO and TLO Faults

One Line Open	Two Lines Open	Phase Shift		
		n_0	n_1	n_2
a	$b-c$	1	1	1
b	$c-a$	1	a^2	a
c	$a-b$	1	a	a^2

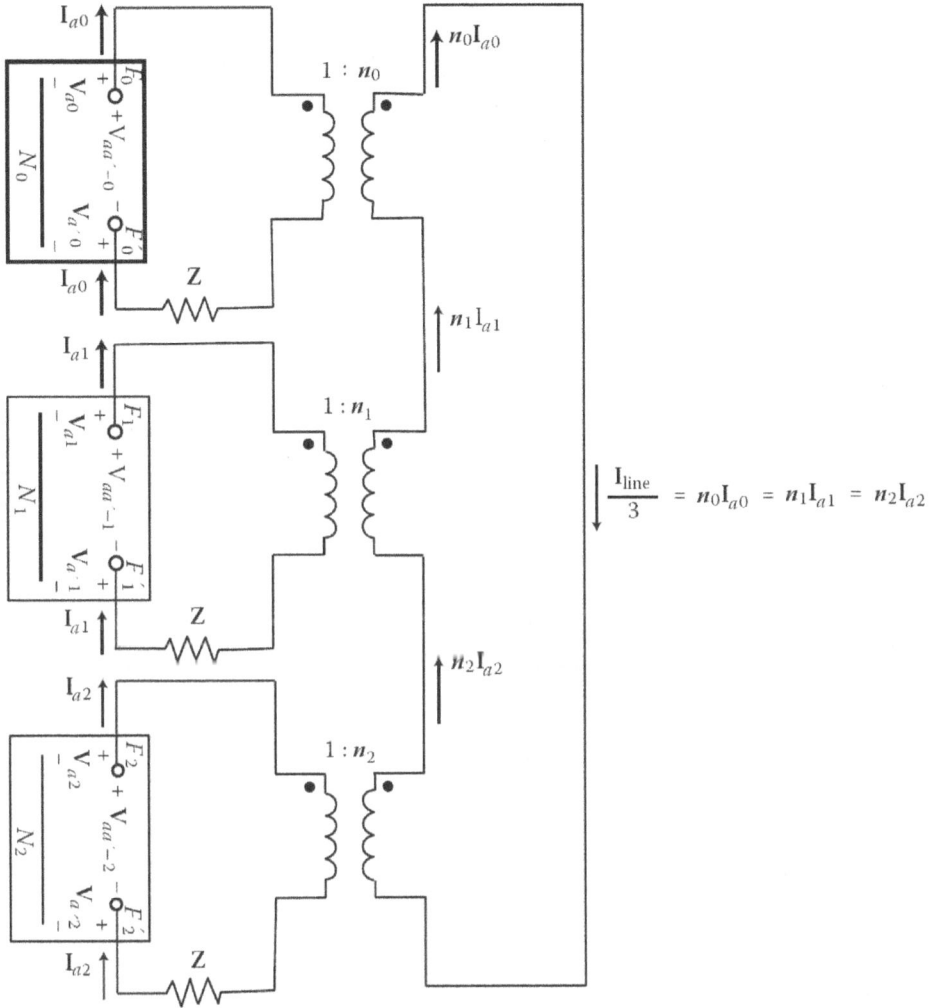

Figure 6.26 Generalized fault diagram for two lines open.

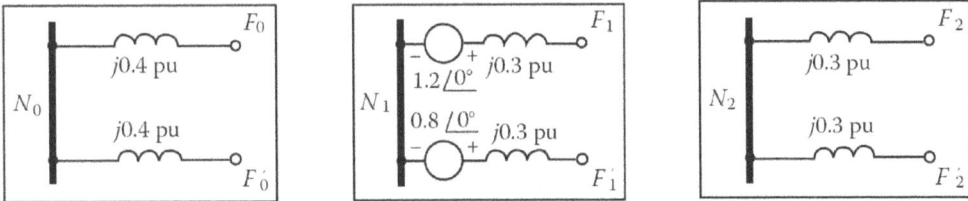

Figure 6.27 Resulting two-port Thévenin equivalents of sequence networks for Example 6.10.

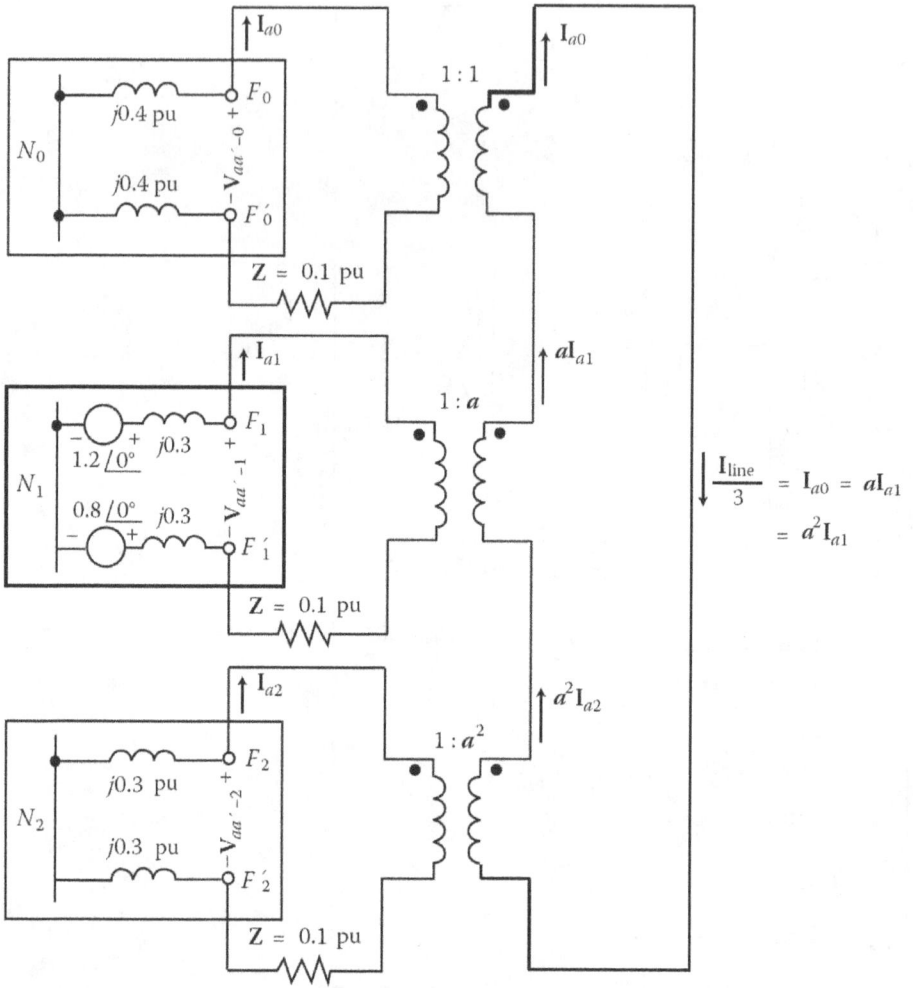

Figure 6.28 Resulting generalized fault diagram for Example 6.10.

(a) Figure 6.28 shows the resulting generalized fault diagram.

(b) From Equation 6.100,

$$I_{a1} = \frac{V_f}{Z_0 + Z_1 + Z_2 + 3Z_f}$$

$$= \frac{1.2\angle 0° - 0.8\angle 0°}{j0.8 + j0.6 + j0.6 + 0.3}$$

$$= \frac{0.4\angle 0°}{0.3 + j2.0}$$

$$= 0.1978\angle - 81.5° \text{ pu}$$

(c) From Figure 6.28,

$$\mathbf{I}_{a0} = \mathbf{a}\mathbf{I}_{a1} = \mathbf{a}^2\mathbf{I}_{a2}$$

Therefore,

$$\mathbf{I}_{a2} = \frac{\mathbf{I}_{a1}}{\mathbf{a}}$$

$$= \frac{0.1978\angle - 81.5°}{1\angle 120°}$$

$$= 0.1978\angle - 201.5°$$

$$= 0.1978\angle 158.5° \text{ pu}$$

(d) From part (c),

$$\mathbf{I}_{a0} = \mathbf{a}\mathbf{I}_{a1}$$

$$= (1\angle 120°)(0.1978\angle - 81.5°)$$

$$= 0.1978\angle 38.5° \text{ pu}$$

(e) By definition,

$$\mathbf{I}_a \triangleq 0$$

but as a check,

$$\mathbf{I}_a = \mathbf{I}_{a0} + \mathbf{I}_{a1} + \mathbf{I}_{a2}$$

$$= 0.1978\angle 38.5° + 0.1978\angle - 81.5° + 0.1978\angle 158.5° = 0$$

(f) Again by definition, $I_b \triangleq 0$; but as a check,

$$\mathbf{I}_b = \mathbf{I}_{a0} + \mathbf{a}^2\mathbf{I}_{a1} + \mathbf{a}\mathbf{I}_{a2}$$

$$= 0.1978\angle 38.5° + (1\angle 240°)(0.1978\angle - 81.5°) + (1\angle 120°)(0.1978\angle 158.5°)$$

$$= 0.1978\angle 38.5° + 0.1978\angle 158.5° + 0.1978\angle 278.5° = 0$$

(g)

$$\mathbf{I}_c = \mathbf{I}_{a0} + \mathbf{a}\mathbf{I}_{a1} + \mathbf{a}^2\mathbf{I}_{a2}$$

$$= 0.1978\angle 38.5° + (1\angle 120°)(0.1978\angle - 81.5°) + (1\angle 240°)(0.1978\angle 158.5°)$$

$$= 0.1978\angle 38.5° + 0.1978\angle 38.5° + 0.1978\angle 38.5.5°$$

$$= 0.5934\angle 38.5°$$

As a check,

$$\mathbf{I}_f = \mathbf{I}_{\text{line}} \triangleq 3\mathbf{I}_{a0} = 3(0.1978\angle 38.5°)$$

$$= 0.5934\angle 38.5° \text{ pu}$$

6.7   SYSTEM GROUNDING

The neutral point or points of a system, rotating machine or transformer can be connected to the ground through a *system neutral ground*. A system that intentionally grounds at least one of its neutral points, either solidly or through a current-limiting device, is called a *grounded system*. For instance, most transformer neutrals in transmission systems are solidly grounded.

However, generator neutrals are commonly grounded through a current-limiting device to restrict the ground fault current. Figure 6.29 illustrates different neutral grounding methods used with generators and the resulting zero-sequence networks. Thus, the methods for grounding the system neutral comprise:

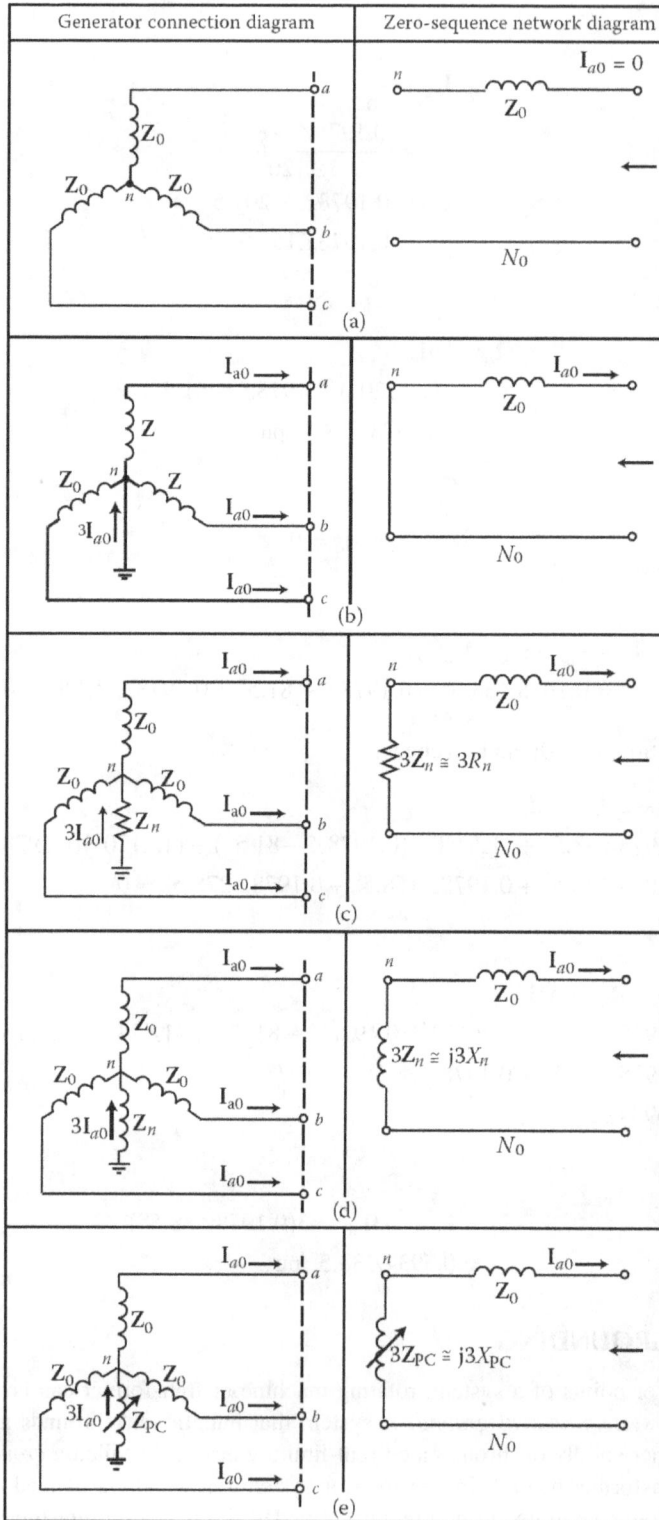

Figure 6.29 Various system (neutral) grounding methods used with generators and resulting zero-sequence network: (a) ungrounded; (b) solidly grounded; (c) resistance grounded; (d) reactance grounded; (e) grounded through Peterson coil.

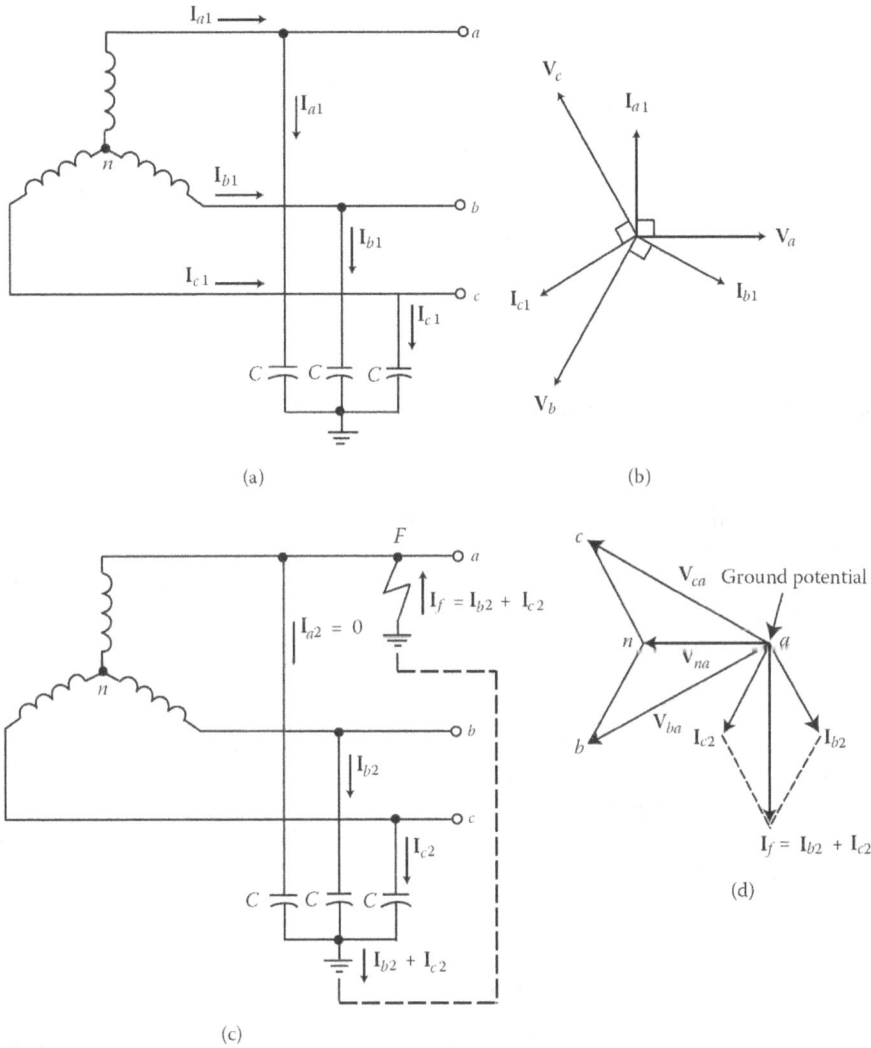

Figure 6.30 Representation of ungrounded system: (a) charging currents under normal condition; (b) phasor diagram under normal condition; (c) charging currents during SLG fault; (d) resulting phasor diagram.

1. Ungrounded
2. Solidly grounded
3. Resistance grounded
4. Reactance grounded
5. Peterson coil grounded

To begin with, the last four methods provide grounded neutrals, while the first method offers an ungrounded, isolated, or free neutral system.

In an ungrounded system, there is no intentional connection between the neutral point or points of the system and the ground, as depicted in Figure 6.30a. The line conductors possess distributed capacitances to each other (not shown in the figure) and to the ground because of capacitive coupling.

Under balanced conditions (assuming a perfectly transposed line), each conductor has the same capacitance to the ground. Hence, the charging current of each phase is the same, as shown in Figure

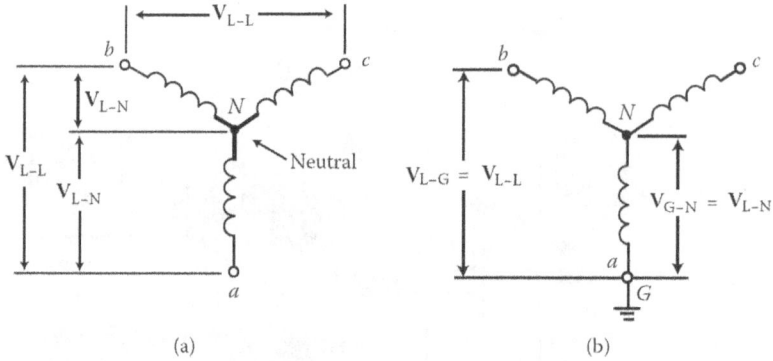

Figure 6.31 Voltage diagrams of ungrounded system: (a) before SLG fault; (b) after SLG fault.

6.30b. Consequently, the potential of the neutral is the same as the ground potential, as illustrated in Figure 6.30a. The charging currents $\mathbf{I}_{a1}$, $\mathbf{I}_{b1}$, and $\mathbf{I}_{c1}$ lead their respective phase voltages by $90°$. Therefore, where X_c is the capacitive reactance of the line to the ground. These phasor currents are in balance, as depicted in Figure 6.30a.

$$|\mathbf{I}_{a1}| = |\mathbf{I}_{b1}| = |\mathbf{I}_{c1}| = \frac{\mathbf{V}_{L-N}}{\mathbf{X}_c} \tag{6.129}$$

Let us consider a scenario where there is a line-to-ground (SLG) fault involving phase a, as depicted in Figure 6.30c. Consequently, the potential of phase a becomes equal to the ground potential, leading to the absence of any charging current in this phase. As a result, the neutral point shifts from its previous position at ground potential to a new position as shown in Figure 6.30d and illustrated in Figure 6.31b. In the faulted phase, a charging current of three times the normal per-phase charging current flows since the phase voltage of each of the two healthy phases increases by three times its normal phase voltage. Thus, the charging current of the faulted phase is given by:

$$\mathbf{I}_f = 3\mathbf{I}_{b1} = 3\mathbf{I}_{c1} \tag{6.130}$$

High voltage surges resulting from SLG faults on ungrounded neutral systems can damage insulation of connected apparatus. If the capacitive current generated by the fault is not large enough to activate protective devices, the insulation may suffer cumulative weakening leading to eventual failure. Additionally, a current that is sufficient to maintain an arc in the ionized part of the fault may exist even after the SLG fault is cleared. This phenomenon, called the *arcing ground*, can produce high-frequency oscillations that lead to surge voltages as high as six times the normal value and damage the insulation at any point in the system. Neutral grounding reduces transient voltage buildup from intermittent ground faults and reduces the destructiveness of high-frequency voltage oscillations. As a result, most modern high-voltage systems ground their neutral systems to avoid these problems with ungrounded neutral systems. It is essential to apply a sufficient current to protective devices to operate them during an SLG fault.

The advantages of neutral grounding include the following:

1. Voltages of phases are restricted to the line-to-ground voltages since the neutral point is not shifted in this system.
2. The ground relays can be used to protect against the ground faults.
3. The high voltages caused by arcing grounds or transient SLG faults are eliminated.
4. The overvoltages caused by lightning are easily eliminated contrary to the case of the isolated neutral systems.

5. The induced static charges do not cause any disturbance since they are conducted to ground immediately.
6. It provides a reliable system.
7. It provides a reduction in operating and maintenance expenses.

When a power system's neutral point is connected directly to the ground, as shown in Figure 6.32a, it is said to be solidly (i.e., directly) grounded. During a single-line-to-ground (SLG) fault on any phase (e.g., phase a), the line-to-ground voltage of that phase becomes zero, but the remaining two phases will still have the same voltages as before since the neutral point remains unshifted, as illustrated in Figure 6.32b. Solid grounding is typically used in circuits where the impedance is high enough to keep the fault current within reasonable limits, as the power source provides the fault current in addition to the charging currents.

To determine how solidly the system should be grounded, the magnitude of the SLG fault current is compared to the system's three-phase current. If the ground fault current is higher than the three-phase current, the system should be more solidly grounded. For insulation cost-saving purposes, equipment rated at 230 kV and above is typically designed to operate only on an effectively grounded system, where the coefficient of grounding does not exceed 0.80, with the coefficient of grounding defined as the ratio of the maximum sustained line-to-ground voltage during faults to the maximum operating line-to-line voltage.

Generators are generally grounded through resistance, reactance, or Peterson coil to limit the fault current to a value that is less than the maximum three-phase fault current the generator can deliver and the short-circuit current for which its windings are rated. A resistance-grounded system connects the system neutral to the ground through one or more resistors to reduce the risk of arcing hazards and permit ground fault protection. In a reactance-grounded system, the system neutral is connected to ground through a reactor, and the system's solidity is determined by the ratio of zero-sequence reactance to positive-sequence reactance, with the system being reactance grounded if the ratio is greater than 3.0 and solidly grounded if the ratio is less than 3.0.

Source transformers or generators with wye-connected windings are the best way to obtain the system neutral for grounding purposes. If the system neutral is not available, a zigzag grounding transformer with no secondary winding or a wye-delta grounding transformer can be used. In this case, the delta side must be closed to provide a path for zero-sequence current, and the wye winding must have the same voltage rating as the circuit to be grounded, while the delta voltage rating can be chosen at any standard voltage level. Table 6.3 summarizes the characteristics of various methods of system neutral grounding [187].

6.8   ELIMINATION OF SLG FAULT CURRENT BY USING PETERSON COILS

When the reactance of a neutral reactor is increased to the point where it is equivalent to the system's capacitance to ground, the system's zero-sequence network becomes parallel resonance for SLG faults. Consequently, a fault current flows through the neutral reactor to the ground. Simultaneously, a current of about the same magnitude and out of phase by 180° with the reactor current passes through the system's capacitance to ground. The two currents cancel each other out, except for a small resistance component, as they pass through the fault. The reactor that achieves this is referred to as a ground fault neutralizer, arc suppression coil, or Peterson coil. It is an adjustable iron-core reactor with winding taps.

Resonant grounding is a practical technique for clearing both transient (such as those caused by lightning, small animals, or tree branches) and sustained SLG faults. Peterson coils have additional benefits, such as extinguishing arcs and reducing voltage dips caused by SLG faults. However, they have some disadvantages, such as the need for re-tuning after network modification or line-switching operations, the need for line transposition, and increased corona and radio interference in DLG fault conditions.

Table 6.3
System Characteristics with Various Grounding Methods

	Ungrounded	Essentially Solid Grounding		Reactance Grounding High Value Reactor	Ground-Fault Neutralizer	Resistance Grounding Resistance
		Low-Value Solid	Low-Value Reactor			
Current for phase-to ground fault in percent of three-phase fault current	Less than 1%	Varies, may be 100% or greater	Usually designed to produce 25–100%	5–25%	Nearly zero fault current	5–20%
Transient overvoltages	Very high	Not excessive	Not excessive	Very high	Not excessive	Not excessive
Automatic segregation of faulty zone	No	Yes	Yes	Yes	No	Yes
Lightning arresters	Ungrounded-neutral type	Grounded-neutral type	Grounded-neutral type	Ungrounded-neutral type	Ungrounded-neutral type	Ungrounded-neutral type
Remarks	Not recommended owing to overvoltages and nonsegregation of fault	Generally used on systems (1) ≤600 V and (2) >15 kV	Generally used on systems (1) ≤600 V and (2) >15 kV	Not used due to excessive overvoltages	Best suited for high voltage overhead may be self-healing	Generally used on industrial systems of 2.4–15 kV

Source: Beeman, D., ed., *Industrial Power System Handbook*. McGraw-Hill, New York, 1955.

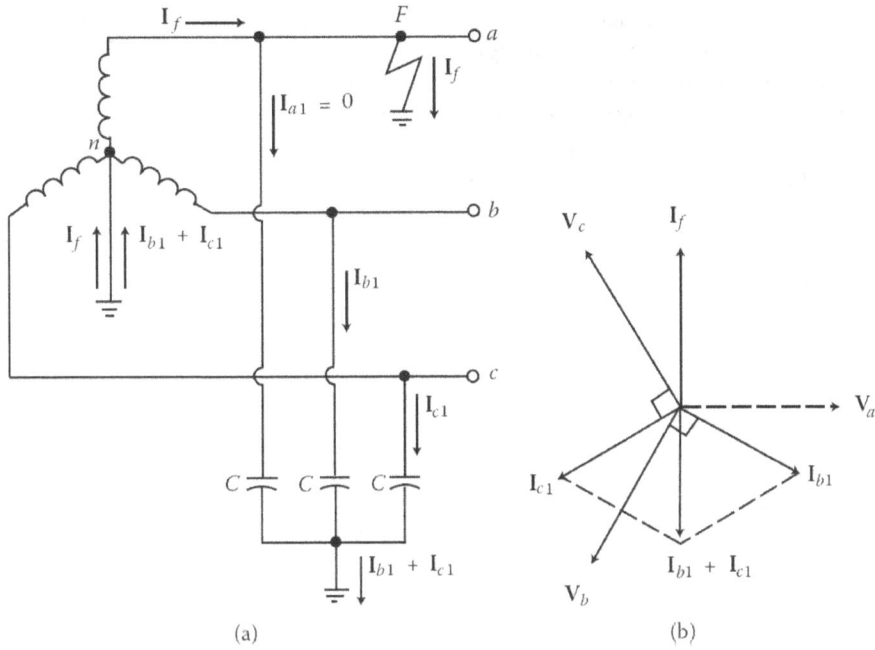

Figure 6.32 Representation of SLG fault on solidly grounded system: (a) solidly grounded system; (b) phasor diagram.

Example 6.11: Consider the subtransmission system shown in Figure 6.33. Assume that loads are connected to buses 2 and 3 and are supplied from bus 1 through 69-kV lines of TL_{12}, TL_{13}, and TL_{23}. The line lengths are 5, 10, and 5 mi for lines TL_{12}, TL_{13}, and TL_{23}, respectively. The lines are transposed and made of three 500-kcmil, 30/7-strand ACSR conductors and there are no ground wires. The geometric mean distance (GMD) between the three conductors and their images (i.e., H_{aa}) is 81.5 ft. The self-GMD of the overhead conductors as a composite group (i.e., D_{aa}) is 1.658 ft. To reduce the SLG faults, a Peterson coil is to be installed between the neutral of the wye-connected secondary of the supply transformer T_1 and ground. The transformer T_1 has a leakage reactance of 5% based on its 25-MVA rating. Do the following:

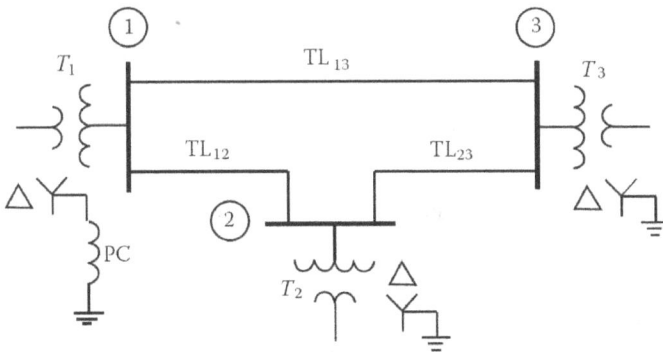

Figure 6.33 Subtransmission system for Example 6.11.

(a) Determine the total zero-sequence capacitance and susceptance per phase of the system at 60 Hz.
(b) Draw the zero-sequence network of the system.
(c) Determine the continues-current rating of the Peterson coil.
(d) Determine the required reactance value for the Peterson coil.
(e) Determine the inductance value of the Peterson coil.
(f) Determine the continuous kVA rating for the Peterson coil.
(g) Determine the continuous-voltage rating for the Peterson coil.

Solution:

(a)

$$C_0 = \frac{29.842}{\ln(H_{aa}/D_{aa})}$$
$$= \frac{29.842}{\ln(81.5/1.658)}$$
$$= 7.6616 \text{ nF/mi}$$

Therefore,

$$b_0 = \omega C_0$$
$$= 2.8884 \mu S/\text{mi}$$

and for the total system,

$$B_0 = b_0 I$$
$$= 2.8884 \times 20$$
$$= 57.7671 \mu S$$

The total zero-sequence reactance is

$$\sum X_{c0} = \frac{1}{B_0} = \frac{10^6}{57.7671}$$
$$= 17,310.8915 \ \Omega$$

and the total zero-sequence capacitance of the system is

$$\sum C_0 = \frac{B_0}{\omega} = \frac{57.7671 \times 10^{-6}}{377}$$
$$= 0.1532 \ \mu F$$

(b) The resulting zero-sequence network is shown in Figure 6.34.
(c) Since the leakage reactance of transformer T_1 is

$$X_1 = X_2 = X_0 = 0.05 \text{ pu}$$

or since

$$Z_B = \frac{kV_B^2}{MVA_B}$$
$$= \frac{69^2}{25}$$
$$= 190.44 \ \Omega$$

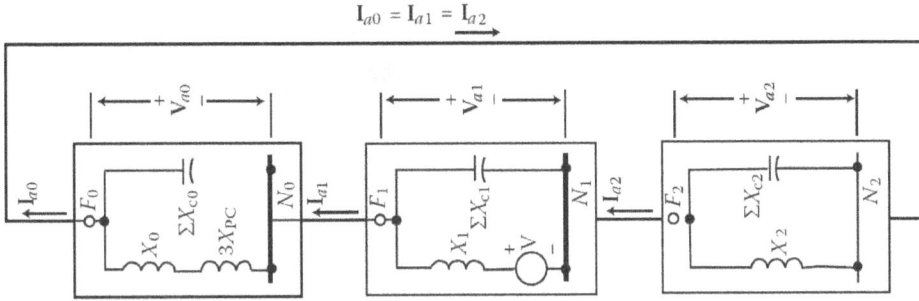

Figure 6.34 Interconnection of sequence networks for Example 6.11.

To have a zero SLG current,

$$\mathbf{I}_{a0} = \mathbf{I}_{a1} = \mathbf{I}_{a2} = 0$$

Thus, it is required that

$$\mathbf{V}_{a0} = -\mathbf{V}_f$$

where

$$\mathbf{V}_F = \frac{69 \times 10^3}{\sqrt{3}}$$
$$= 39,837.17 \text{ V}$$

Since $\sum X_{c1} \gg X_1$ and $\sum X_{c2} \gg 2$, the zero-sequence current component flowing through the Peterson coil (PC) can be expressed as

$$\mathbf{I}_{a0} = \frac{-\mathbf{V}_{a0}}{j(X_0 + 3X_{PC})}$$
$$= \frac{\mathbf{V}_F}{j(X_0 + 3X_{PC})}$$

or

$$\mathbf{I}_{a0(PC)} = \frac{39,837.17\angle 0°}{j17,310.8915}$$
$$= -j2.3013 \text{ A}$$

Therefore, the continuous-current rating for the Peterson coil is

$$\mathbf{I}_{PC} = 3\mathbf{I}_{a0(PC)} = 6.9038 \text{ A}$$

(d) Since

$$3X_{PC} + X_0 = 17,310.8915 \ \Omega$$

where

$$X_0 = 9.522 \ \Omega$$

therefore,

$$3X_{PC} = 17,310.8915 - 9.522$$
$$= 17,301.3695 \ \Omega$$

and thus, the required reactance value for the Peterson coil is

$$X_{PC} = \frac{17,310.8915\Omega}{3}$$
$$= 5767.1232 \ \Omega$$

(e) Hence, its inductance is

$$L_{PC} = \frac{X_{PC}}{\omega}$$
$$= \frac{5767.1232}{377}$$
$$= 15.2928 \ H$$

(f) Its continuous kVA rating is

$$S_{PC} = I_{PC}^2 X_{PC}$$
$$= (6.9030)^2(5767.1232)$$
$$= 274.88 \ kVA$$

(g) The voltage across the Peterson coil is

$$V_{PC} = I_{PC}X_{PC}$$
$$= (6.9030)(5767.1232)$$
$$= 39,815.07 \ V$$

which is approximately equal to the line-to-neutral voltage.

6.9   SIX-PHASE SYSTEMS

The development of six-phase transmission lines has been proposed as a way to increase power transfer capacity while minimizing electrical environmental impacts. This new technology can offer a number of advantages over traditional three-phase systems. For example, six-phase transmission lines have lower voltage gradients of the conductors. This decrease in voltage gradient helps to reduce audible noise and electrostatic effects, which is a significant benefit without requiring additional insulation.

In multiphase transmission lines, if the line-to-ground voltage is fixed, then the line-to-line voltage decreases as the number of phases increases. This decrease in line-to-line voltage enables the insulation distance between the lines to be reduced. This leads to more compact tower designs and lower material costs.

However, the unsymmetrical tower-top configurations can create unbalanced conditions in the system, which can affect the operation and stability of the system. Engineers can use symmetrical components analysis to determine the unbalance factors caused by the unsymmetrical tower-top configurations. By analyzing the system using symmetrical components, engineers can develop solutions to balance the system and ensure its stable operation.

6.9.1   APPLICATION OF SYMMETRICAL COMPONENTS

Symmetrical components refer to the process of decomposing an unbalanced six-phase system into six sets of balanced currents or voltages. These sets of balanced components are called symmetrical components. The balanced voltage sequence sets of a six-phase system are illustrated in Figure 6.35 [188].

The first set, called the first-order positive sequence components, are of equal magnitude and are arranged in the phase sequence *abcdef*. These components have a 60° phase shift. The other five sets are the first-order negative sequence, the second-order positive sequence, the second-order negative sequence, the odd sequence, and finally the zero sequence components.

Expressing sequence components using the phase sequence *abcdef* enables engineers to analyze the system based on its symmetrical components as follows. This analysis approach is helpful in identifying and addressing the impacts of unbalanced conditions on the system.

$$
\left.
\begin{aligned}
V_a &= V_{a0+} + V_{a1+} + V_{a2+} + V_{a0-} + V_{a2-} + V_{a1-} \\
V_b &= V_{b0+} + V_{b1+} + V_{b2+} + V_{b0-} + V_{b2-} + V_{b1-} \\
V_c &= V_{c0+} + V_{c1+} + V_{c2+} + V_{c0-} + V_{c2-} + V_{c1-} \\
V_d &= V_{d0+} + V_{d1+} + V_{d2+} + V_{d0-} + V_{d2-} + V_{d1-} \\
V_e &= V_{e0+} + V_{e1+} + V_{e2+} + V_{e0-} + V_{e2-} + V_{e1-} \\
V_f &= V_{f0+} + V_{f1+} + V_{f2+} + V_{f0-} + V_{f2-} + V_{f1-}
\end{aligned}
\right\}
\tag{6.131}
$$

6.9.2   TRANSFORMATIONS

By taking phase *a* as the reference phase as usual, the set of voltages given in Equation 6.131 can be expressed in matrix form as

$$
\begin{bmatrix} V_a \\ V_b \\ V_c \\ V_d \\ V_e \\ V_f \end{bmatrix}
=
\begin{bmatrix}
1 & 1 & 1 & 1 & 1 & 1 \\
1 & b^5 & b^4 & b^3 & b^2 & b \\
1 & b^4 & b^2 & 1 & b^4 & b^2 \\
1 & b^3 & 1 & b^3 & 1 & b^3 \\
1 & b^2 & b^4 & 1 & b^2 & b^4 \\
1 & b & b^2 & b^3 & b^4 & b^5
\end{bmatrix}
\begin{bmatrix} V_{a0+} \\ V_{a1+} \\ V_{a2+} \\ V_{a0-} \\ V_{a2-} \\ V_{a1-} \end{bmatrix}
\tag{6.132}
$$

or in short-hand matrix notation,

$$
[\mathbf{V}_\phi] = [\mathbf{T}_6][\mathbf{V}_S]
\tag{6.133}
$$

where

$[\mathbf{V}_\phi]$ = the matrix of unbalanced phase voltages
$[\mathbf{V}_s]$ = the matrix of balanced sequence voltages
$[\mathbf{T}_6]$ = the six-phase symmetrical transformation matrix

Similar to the definition of the *a* operator in three-phase systems, it is possible to define a six-phase operator *b* as

$$
b = 1.0\angle 60°
\tag{6.134}
$$

or

$$
\begin{aligned}
b &= \exp(j\pi/3) \\
&= 0.5 + j0.866
\end{aligned}
\tag{6.135}
$$

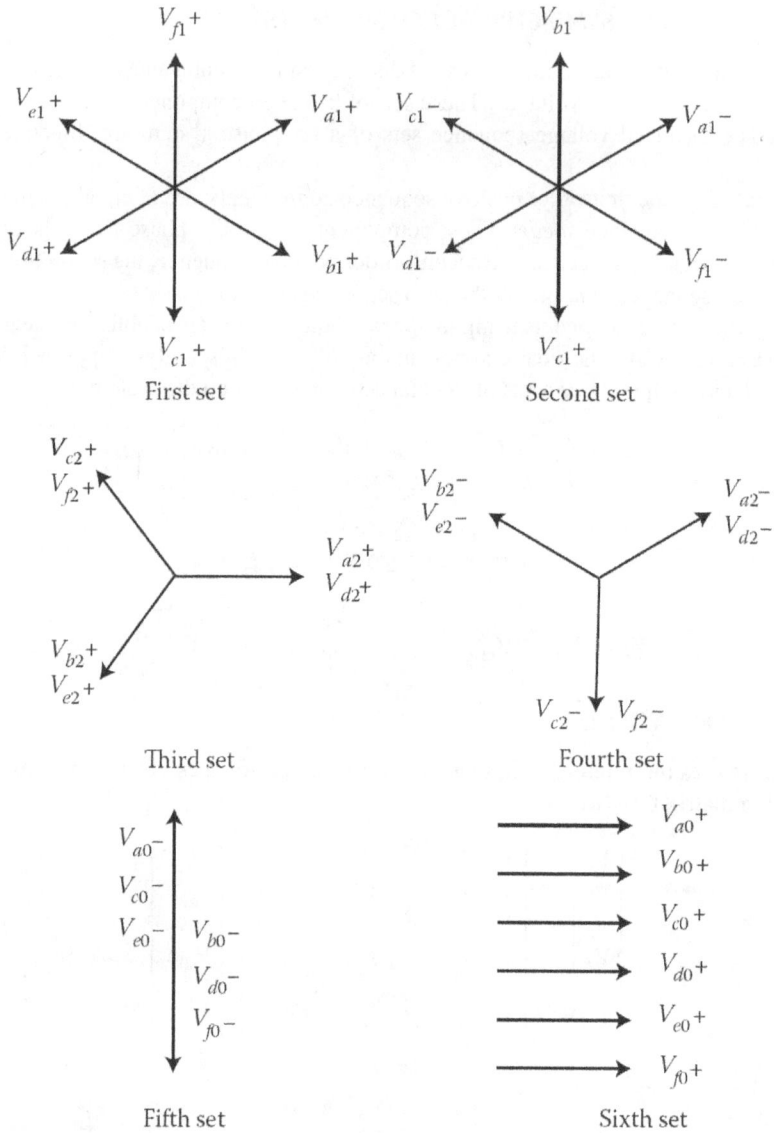

Figure 6.35 Balanced voltage-sequence sets of a six-phase system. (From Gönen, T., *Electric Power Transmission System Engineering*, 2nd ed. CRC Press, Roca Baton, FL, 2009.) [189]

It can be shown that

$$b = -\mathbf{a}^2$$
$$= -1(1.0\angle 120°)^2$$
$$= -(1.0\angle 240°)$$
$$= 1.0\angle 60°$$

(6.136)

The relation between the sequence components and the unbalanced phase voltages can be expressed as

$$[\mathbf{V}_S] = [\mathbf{T}_6]^{-1}[\mathbf{V}_\phi]$$

(6.137)

Similar equations can be written for the phase currents and their sequence components as

$$[\mathbf{I}_\phi] = [\mathbf{T}_6][\mathbf{I}_s]$$

(6.138)

and

$$[\mathbf{I}_s] = [\mathbf{T}_6]^{-1}[\mathbf{I}_\phi] \tag{6.139}$$

The sequence impedance matrix $[\mathbf{Z}_s]$ can be determined from the phase impedance matrix $[\mathbf{Z}_\phi]$ by applying KVL. Hence,
and

$$[\mathbf{V}_\phi] = [\mathbf{Z}_\phi][\mathbf{I}_\phi] \tag{6.140}$$

and

$$[\mathbf{V}_s] = [\mathbf{Z}_s][\mathbf{I}_s] \tag{6.141}$$

where
$[\mathbf{Z}_\phi]$ = phase impedance matrix of the line in 6×6
$[\mathbf{Z}_s]$ = sequence impedance matrix of the line in 6×6

After multiplying both sides of Equation 6.141 by $[\mathbf{Z}_s]^{-1}$,

$$\begin{aligned}
[\mathbf{I}_s] &= [\mathbf{Z}_s]^{-1}[\mathbf{V}_s] \\
&= [\mathbf{Y}_s][\mathbf{V}_s]
\end{aligned} \tag{6.142}$$

where $[\mathbf{Y}_s]$ is the sequence admittance matrix.

Since the unbalanced factors are to be determined after having only the first-order positive sequence voltage applied, the above equation can be reexpressed as

$$\begin{bmatrix} I_{a0+} \\ I_{a1+} \\ I_{a2+} \\ I_{a0-} \\ I_{a2-} \\ I_{a1-} \end{bmatrix} = \begin{bmatrix} Y_{0+0+} & Y_{0+0+} & Y_{0+0+} & Y_{0+0-} & Y_{0+0-} & Y_{0+0-} \\ Y_{0+0+} & Y_{0+0+} & Y_{0+0+} & Y_{0+0-} & Y_{0+0-} & Y_{0+0-} \\ Y_{0+0+} & Y_{0+0+} & Y_{0+0+} & Y_{0+0-} & Y_{0+0-} & Y_{0+0-} \\ Y_{0-0+} & Y_{0-0+} & Y_{0-0+} & Y_{0-0-} & Y_{0-0-} & Y_{0-0-} \\ Y_{0-0+} & Y_{0-0+} & Y_{0-0+} & Y_{0-0-} & Y_{0-0-} & Y_{0-0-} \\ Y_{0-0+} & Y_{0-0+} & Y_{0-0+} & Y_{0-0-} & Y_{0-0-} & Y_{0-0-} \end{bmatrix} \begin{bmatrix} 0 \\ V_{a1+} \\ 0 \\ 0 \\ 0 \\ 0 \end{bmatrix} \tag{6.143}$$

6.9.3 ELECTROMAGNETIC UNBALANCE FACTORS

The unbalanced factors in a six-phase transmission line include five electromagnetic sequence factors, namely zero, second-order positive, odd, second-order negative, and first-order negative sequence factors. These factors are calculated by taking the ratio of the corresponding current to the first-order positive sequence current. The computation of the zero-sequence unbalanced factor can be expressed as follows:

$$\mathbf{m}_{0+} = \frac{I_{a0+}}{I_{a1+}} \tag{6.144a}$$

or

$$\begin{aligned}
\mathbf{m}_{0+} &= \frac{(y_{0+1+})(V_{a1+})}{(y_{1+1+})(V_{a1+})} \\
&= \frac{y_{0+1+}}{y_{1+1+}}
\end{aligned} \tag{6.144b}$$

The second-order positive unbalance factor is

$$\mathbf{m}_{2+} = \frac{I_{a2+}}{I_{a1+}} \tag{6.145a}$$

or

$$m_{2+} = \frac{(y_{2+1+})(V_{a1+})}{(y_{1+1+})(V_{a1+})}$$
$$= \frac{y_{2+1+}}{y_{1+1+}}$$

(6.145b)

The odd unbalance factor is

$$m_{0-} = \frac{I_{a0-}}{I_{a1+}}$$

(6.146a)

or

$$m_{0-} = \frac{(y_{0-1+})(V_{a1+})}{(y_{1+1+})(V_{a1+})}$$
$$= \frac{y_{0-1+}}{y_{1+1+}}$$

(6.146b)

The second-order negative unbalance factor is

$$m_{2-} = \frac{I_{a2-}}{I_{a1+}}$$

(6.147a)

or

$$m_{2-} = \frac{(y_{2-1+})(V_{a1+})}{(y_{1+1+})(V_{a1+})}$$
$$= \frac{y_{2-1+}}{y_{1+1+}}$$

(6.147b)

The first-order negative unbalance factor is

$$m_{1-} = \frac{I_{a1-}}{I_{a1+}}$$

(6.148a)

or

$$m_{1-} = \frac{(y_{1-1+})(V_{a1+})}{(y_{1+1+})(V_{a1+})}$$
$$= \frac{y_{1-1+}}{y_{1+1+}}$$

(6.148b)

Due to the unsymmetrical configuration in a system, there may be circulating residual or zero-sequence currents that flow in the high-voltage system, particularly in systems with solidly grounded neutrals. These circulating currents can interfere with the proper operation of sensitive elements in ground relays. Along with zero-sequence unbalances, capacitive unbalances can also cause negative-sequence charging currents to flow through the system. These charging currents may result in additional power losses in transformers and rotating machines in the system.

6.9.4 TRANSPOSITION OF THE SIX-PHASE LINES

One way to improve the performance of six-phase transmission lines is through transposition. There are several methods of transposition, including complete, cyclic, and reciprocal. In complete transposition, each conductor takes every possible position with respect to every other conductor over an equal length. This ensures that the line is symmetric and that the impedance is the same for all conductors. The impedance matrix for a completely transposed six-phase system can be expressed as follows:

$$[\mathbf{Z}_\phi] = \begin{bmatrix} Z_s & Z_m & Z_m & Z_m & Z_m & Z_m \\ Z_m & Z_s & Z_m & Z_m & Z_m & Z_m \\ Z_m & Z_m & Z_s & Z_m & Z_m & Z_m \\ Z_m & Z_m & Z_m & Z_s & Z_m & Z_m \\ Z_m & Z_m & Z_m & Z_m & Z_s & Z_m \\ Z_m & Z_m & Z_m & Z_m & Z_m & Z_s \end{bmatrix} \tag{6.149}$$

In a six-phase transmission line, it is difficult to achieve a complete transposition. Also, it is not of interest owing to the differences in the line-to-line voltages. Therefore, it is more efficient to implement cyclic transposition or reciprocal cyclic transposition. The impedance matrix for a cyclically transposed line can be expressed as

$$[\mathbf{Z}_\phi] = \begin{bmatrix} Z_s & Z_{m1} & Z_{m2} & Z_{m3} & Z_{m4} & Z_{m5} \\ Z_{m5} & Z_s & Z_{m1} & Z_{m2} & Z_{m3} & Z_{m4} \\ Z_{m4} & Z_{m5} & Z_s & Z_{m1} & Z_{m2} & Z_{m3} \\ Z_{m3} & Z_{m4} & Z_{m5} & Z_s & Z_{m1} & Z_{m2} \\ Z_{m2} & Z_{m3} & Z_{m4} & Z_{m5} & Z_s & Z_{m1} \\ Z_{m1} & Z_{m2} & Z_{m3} & Z_{m4} & Z_{m5} & Z_s \end{bmatrix} \tag{6.150}$$

Similarly, the impedance matrix for a reciprocal cyclically transposed line can be given as

$$[\mathbf{Z}_\phi] = \begin{bmatrix} Z_s & Z_{m1} & Z_{m2} & Z_{m3} & Z_{m2} & Z_{m1} \\ Z_{m1} & Z_s & Z_{m1} & Z_{m2} & Z_{m3} & Z_{m2} \\ Z_{m2} & Z_{m1} & Z_s & Z_{m1} & Z_{m2} & Z_{m3} \\ Z_{m3} & Z_{m2} & Z_{m1} & Z_s & Z_{m1} & Z_{m2} \\ Z_{m2} & Z_{m3} & Z_{m2} & Z_{m1} & Z_s & Z_{m1} \\ Z_{m1} & Z_{m2} & Z_{m3} & Z_{m2} & Z_{m1} & Z_s \end{bmatrix} \tag{6.151}$$

where Z_s is the self-impedance and Z_m is the mutual impedance.

6.9.5 PHASE ARRANGEMENTS

Changing the phase conductors can alter the values of electromagnetic and electrostatic unbalances. Nonetheless, a specific phase configuration exists that results in minimal electromagnetic unbalances. It's important to note that some phasing arrangements may lead to significantly large circulating current unbalances.

6.9.6 OVERHEAD GROUND WIRES

To protect transmission lines from lightning strikes, overhead ground wires are often installed. However, the presence of these wires can have an impact on both self and mutual impedances. Specifically, the resistance of self and mutual impedances increases slightly while their reactance decreases significantly. Depending on the type and size of the configuration, overhead ground wires can either increase or decrease one or more of the unbalances. The equivalent impedance matrix for the transmission lines can be compared using Kron reduction.

6.9.7 DOUBLE-CIRCUIT TRANSMISSION LINES

The voltage equation of a double-circuit line is given by

$$\begin{bmatrix} \sum \mathbf{V}_{ckt1} \\ \hline \sum \mathbf{V}_{ckt2} \end{bmatrix} = \begin{bmatrix} \mathbf{Z}_{ckt1} & \mathbf{Z}_{ckt1\ ckt2} \\ \hline \mathbf{Z}_{ckt2\ ckt1} & \mathbf{Z}_{ckt2} \end{bmatrix} \begin{bmatrix} \sum \mathbf{I}_{ckt1} \\ \sum \mathbf{I}_{ckt2} \end{bmatrix} \tag{6.152}$$

where $\sum \mathbf{V}_{ckt1}$, $\sum \mathbf{V}_{ckt2}$, and $\sum \mathbf{I}_{ckt1}$, $\sum \mathbf{I}_{ckt2}$ are the phase voltages and currents, respectively. Each of them is a column vector of size 6×1. Each $[\mathbf{Z}]$ impedance matrix has dimensions of 6×6. The above matrix is solved for the currents, in order to express the unbalance factors in terms of sequence currents. Hence,

$$\begin{bmatrix} \sum \mathbf{I}_{ckt1} \\ \sum \mathbf{I}_{ckt2} \end{bmatrix} = [\mathbf{Y}_{\text{line}}] \begin{bmatrix} \mathbf{V}_{ckt1} \\ \mathbf{V}_{ckt2} \end{bmatrix} \tag{6.153}$$

where $[\mathbf{Y}_{\text{line}}]$ is the admittance matrix of the line having a size 12×12.

To determine the sequence admittance matrix of the line, the appropriate transformation matrix needs to be defined as

$$[\mathbf{T}_{12}] = \begin{bmatrix} \mathbf{T}_6 & 0 \\ 0 & \mathbf{T}_6 \end{bmatrix} \tag{6.154}$$

Premultiplying Equation 6.153 by the transformation matrix $[\mathbf{T}_{12}]$ and postmultiplying by $[\mathbf{T}_{12}]^{-1}$ and also inserting the unity matrix of $[\mathbf{T}_{12}]^{-1}[\mathbf{T}_{12}] = [\mathbf{U}]$ into the right-hand side of the equation,

$$\begin{bmatrix} \sum \mathbf{I}_{\text{seq1}} \\ \sum \mathbf{I}_{\text{seq2}} \end{bmatrix} = [\mathbf{T}_{12}][\mathbf{Y}_{\text{line}}][\mathbf{T}_{12}]^{-1} \begin{bmatrix} \mathbf{V}_{\text{seq1}} \\ \mathbf{V}_{\text{seq2}} \end{bmatrix} \tag{6.155}$$

where $\sum \mathbf{V}_{seq1}$, $\sum \mathbf{V}_{seq2}$, $\sum \mathbf{I}_{seq1}$, and $\sum \mathbf{I}_{seq2}$ are the sequence voltage drops and currents for the first and second circuits, respectively. The sequence admittance matrix is

$$[\mathbf{Y}_{\text{seq}}] = [\mathbf{T}_{12}][\mathbf{Y}_{\text{line}}][\mathbf{T}_{12}]^{-1} \tag{6.156}$$

To determine the unbalance factors, only the first-order positive sequence voltage is applied, and there are two types of unbalances, namely the net-through and the net-circulating unbalances. The sequence matrix is obtained by expanding the above equation to the complete 12×12 matrix as follows:

$$
\begin{bmatrix} I_{a0+} \\ I_{a1+} \\ I_{a2+} \\ I_{a0-} \\ I_{a2-} \\ I_{a1-} \\ I_{a0+} \\ I_{a1+} \\ I_{a2+} \\ I_{a0-} \\ I_{a2-} \\ I_{a1-} \end{bmatrix}
=
\begin{bmatrix}
y_{0+0+} & y_{0+1+} & y_{0+2+} & y_{0+0-} & y_{0+2-} & y_{0+1-} & y_{0+0+} & y_{0+1+} & y_{0+2+} & y_{0+0-} & y_{0+2-} & y_{0+1-} \\
y_{1+0+} & y_{1+1+} & y_{1+2+} & y_{1+0-} & y_{1+2-} & y_{1+1-} & y_{1+0+} & y_{1+1+} & y_{0+2+} & y_{1+0-} & y_{1+2-} & y_{1+1-} \\
y_{2+0+} & y_{2+1+} & y_{2+2+} & y_{2+0-} & y_{2+2-} & y_{2+1-} & y_{2+0+} & y_{2+1+} & y_{0+2+} & y_{2+0-} & y_{2+2-} & y_{2+1-} \\
y_{0-0+} & y_{0-1+} & y_{0-2+} & y_{0-0-} & y_{0-2-} & y_{0-1-} & y_{0-0+} & y_{0-1+} & y_{0+2+} & y_{0-0-} & y_{0-2-} & y_{0-1-} \\
y_{2-0+} & y_{2-1+} & y_{2-2+} & y_{2-0-} & y_{2-2-} & y_{2-1-} & y_{2-0+} & y_{2-1+} & y_{0-2+} & y_{2-0-} & y_{2-2-} & y_{2-1-} \\
y_{1-0+} & y_{1-1+} & y_{1-2+} & y_{1-0-} & y_{1-2-} & y_{1-1-} & y_{1-0+} & y_{1-1+} & y_{0-2+} & y_{1-0-} & y_{1-2-} & y_{1-1-} \\
y_{0+0+} & y_{0+1+} & y_{0+2+} & y_{0+0-} & y_{0+2-} & y_{0+1-} & y_{0+0+} & y_{0+1+} & y_{0+2+} & y_{0+0-} & y_{0+2-} & y_{0+1-} \\
y_{1+0+} & y_{1+1+} & y_{1+2+} & y_{1+0-} & y_{1+2-} & y_{1+1-} & y_{1+0+} & y_{1+1+} & y_{0+2+} & y_{1+0-} & y_{1+2-} & y_{1+1-} \\
y_{2+0+} & y_{2+1+} & y_{2+2+} & y_{2+0-} & y_{2+2-} & y_{2+1-} & y_{2+0+} & y_{2+1+} & y_{0+2+} & y_{2+0-} & y_{2+2-} & y_{2+1-} \\
y_{0-0+} & y_{0-1+} & y_{0-2+} & y_{0-0-} & y_{0-2-} & y_{0-1-} & y_{0-0+} & y_{0-1+} & y_{0+2+} & y_{0-0-} & y_{0-2-} & y_{0-1-} \\
y_{2-0+} & y_{2-1+} & y_{2-2+} & y_{2-0-} & y_{2-2-} & y_{2-1-} & y_{2-0+} & y_{2-1+} & y_{0-2+} & y_{2-0-} & y_{2-2-} & y_{2-1-} \\
y_{1-0+} & y_{1-1+} & y_{1-2+} & y_{1-0-} & y_{1-2-} & y_{1-1-} & y_{1-0+} & y_{1-1+} & y_{1-2+} & y_{1-0-} & y_{1-2-} & y_{1-1-}
\end{bmatrix}
\begin{bmatrix} 0 \\ \sum \mathbf{V}_{a1+} \\ 0 \\ 0 \\ 0 \\ 0 \\ 0 \\ \sum \mathbf{V}_{a1+} \\ 0 \\ 0 \\ 0 \\ 0 \end{bmatrix}
$$

$$\tag{6.157}$$

The net-through unbalance factors are defined as

$$
\begin{aligned}
\mathbf{m}_{0+_t} &\triangleq \frac{I_{a0+} + I_{a'0+}}{I_{a1+} + I_{a'1+}} \\
&= \frac{(y_{0+1+} + y_{0+1'+} + y_{0'+1+} + y_{0'+1'+})\sum \mathbf{V}_{a1+}}{(y_{1+1+} + y_{1+1'+} + y_{1'+1+} + y_{1'+1'+})\sum \mathbf{V}_{a1+}} \\
&= \frac{y_{0+1+} + y_{0+1'+} + y_{0'+1+} + y_{0'+1+}}{y_k}
\end{aligned}
\tag{6.158}
$$

where

$$
y_k = y_{1+1+} + y_{1+1'+} + y_{1'+1+} + y_{1'+1+}
$$

$$
\mathbf{m}_{2+_t} = \frac{y_{2+1+} + y_{2+1'+} + y_{2'+1+} + y_{2'+1+}}{y_k}
\tag{6.159}
$$

$$
\mathbf{m}_{0-_t} = \frac{y_{0-1+} + y_{0-1'+} + y_{2'-1+} + y_{2'-1+}}{y_k}
\tag{6.160}
$$

$$
\mathbf{m}_{2-_t} = \frac{y_{2-1+} + y_{2-1'+} + y_{2'-1+} + y_{2'-1+}}{y_k}
\tag{6.161}
$$

$$
\mathbf{m}_{1-_t} = \frac{y_{1-1+} + y_{1-1'+} + y_{1'-1+} + y_{1'-1+}}{y_k}
\tag{6.162}
$$

The net-circulating unbalances are defined as

$$
\begin{aligned}
\mathbf{m}_{0+_c} &\triangleq \frac{I_{a0+} + I_{a'0+}}{I_{a1+} + I_{a'1+}} \\
&= \frac{y_{0+1+} + y_{0+1'+} - y_{0'+1+} - y_{0'+1'+}}{y_k}
\end{aligned}
\tag{6.163}
$$

$$
\mathbf{m}_{2+_c} = \frac{y_{2+1+} + y_{2+1'+} - y_{2'+1+} - y_{2'+1+}}{y_k}
\tag{6.164}
$$

$$
\mathbf{m}_{0-_c} = \frac{y_{0-1+} + y_{0-1'+} - y_{0'-1+} - y_{0'-1+}}{y_k}
\tag{6.165}
$$

$$
\mathbf{m}_{2-_c} = \frac{y_{2-1+} + y_{2-1'+} - y_{2'-1+} - y_{2'-1+}}{y_k}
\tag{6.166}
$$

$$
\mathbf{m}_{1-_c} = \frac{y_{1-1+} + y_{1-1'+} - y_{1'-1+} - y_{1'-1+}}{y_k}
\tag{6.167}
$$

PROBLEMS

1. Consider the system shown in Figure 6.36. Assume that the following data are given based on 20 MVA and the line-to-line base voltages as shown in Figure 6.36.
   Generator G_1: $X_1 = 0.25$ pu, $X_2 = 0.15$ pu, $X_0 = 0.05$ pu
   Generator G_2: $X_1 = 0.90$ pu, $X_2 = 0.60$ pu, $X_0 = 0.05$ pu
   Transformer T_1: $X_1 = X_2 = X_0 = 0.10$ pu
   Transformer T_2: $X_1 = X_2 = 0.10$ pu, $X_0 = \infty$
   Transformer T_3: $X_1 = X_2$ $X_0 = 0.50$ pu
   Transformer T_4: $X_1 = X_2 = 0.30$ pu, $X_0 = \infty$
   Transmission line TL_{23}: $X_1 = X_2 = 0.15$ pu, $X_0 = 0.50$ pu
   Transmission line TL_{35}: $X_1 = X_2 = 0.30$ pu, $X_0 = 1.00$ pu
   Transmission line TL_{57}: $X_1 = X_2 = 0.30$ pu, $X_0 = 1.00$ pu

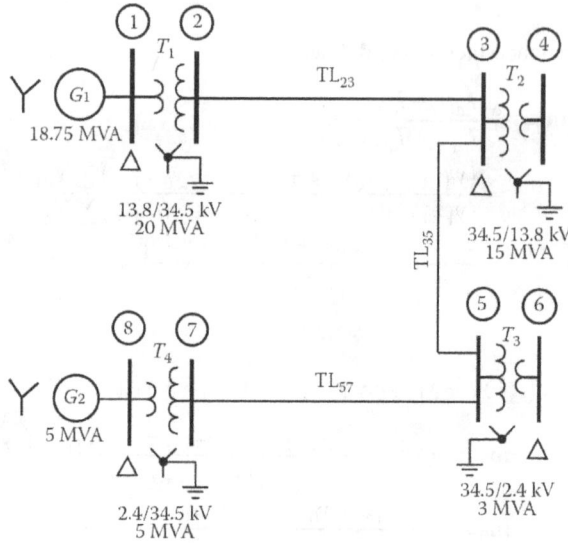

Figure 6.36 SLG fault: (a) general representation; (b) interconnection of sequence networks.

    (a) Draw the corresponding positive-sequence network.
    (b) Draw the corresponding negative-sequence network.
    (c) Draw the corresponding zero-sequence network.

2. Use the system and its data from Problem 7 (Chapter 4) and assume an SLG fault at bus 4. Assume that Z_f is $j0.1$ pu based on 50 MVA. Determine the fault current in per units and amperes.

3. Consider the system given in Problem 7 (Chapter 4) and assume that there is a line-to-line fault at bus 3 involving phases b and c. Determine the fault currents for both phases in per units and amperes.

4. Consider the system given in Problem 7 (Chapter 4) and assume that there is a DLG fault at bus 2, involving phases b and c. Assume that Z_f is $j0.1$ pu and Z_g is $j0.2$ pu (where Z_g is the neutral-to-ground impedance) both based on 50 MVA.

5. Consider the system given in Example 4.1 and determine the following:

    (a) Line-to-ground fault (Also, find the ratio of this line-to-ground fault current to the three-phase fault current found in Example 4.1.)

    (b) Line-to-line fault (Also, find the ratio of this line-to-line fault current to the previously calculated three-phase fault current).

    (c) DLG fault

6. Repeat Problem 5 assuming that the fault is located on bus 2.

7. Repeat Problem 5 assuming that the fault is located on bus 3.

8. Consider the system shown in Figure 6.43a Assume that loads, line capacitance, and transformer-magnetizing currents are neglected and that the following data is given based on 20 MVA and the line-to-line voltages as shown in Figure 6.37a. Do not neglect the resistance of the transmission line TL_{23}. The prefault positive-sequence voltage at bus 3 is $V_{an} = 1.0 \angle 0°$ pu, as shown in Figure 6.37b.

Generator: $X_1 = 0.20$ pu, $X_2 = 0.10$ pu, $X_0 = 0.05$ pu
Transformer T_1: $X_1 = X_2 = 0.05$ pu, $X_0 = X_1$ (looking into high-voltage side)
Transformer T_2: $X_1 = X_2 = 0.05$ pu, $X_0 = \infty$ (looking into high-voltage side)
Transmission line: $Z_1 = Z_2 = 0.2 + j0.2$ pu, $Z_0 = 0.6 + j0.6$ pu

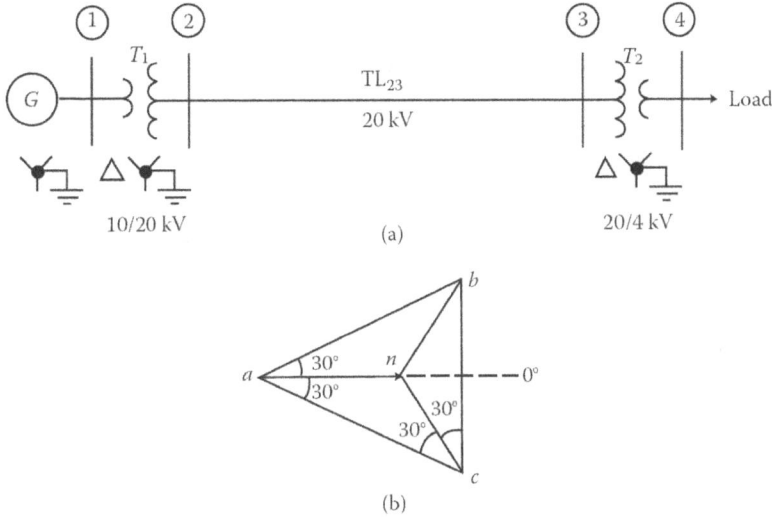

Figure 6.37 System for Problem 8.

Assume that there is a bolted (i.e., with zero fault impedance) line-to-line fault on phases b and c at bus 3 and determine the following:

(a) Fault current $\mathbf{I}_{bf}$ in per units and amperes
(b) Phase voltages $\mathbf{V}_a, \mathbf{V}_b$, and $\mathbf{V}_c$ at bus 2 in per units and kilovolts
(c) Line-to-line voltages $\mathbf{V}_{ab}, \mathbf{V}_{bc}$, and $\mathbf{V}_{ca}$
(d) Generator line currents $\mathbf{I}_a, \mathbf{I}_b$, and $\mathbf{I}_c$

Given: per-unit positive-sequence currents on the low-voltage side of the transformer bank lag positive-sequence currents on the high-voltage side by 30° and similarly for negative-sequence currents excepting that the low-voltage currents lead the high-voltage by 30°.

9. Consider Figure 6.38 and assume that the generator ratings are 2.40/4.16Y kV, 15 MW (3ϕ), 18.75 MVA (3ϕ), 80% power factor, 2 poles, 3600 rpm. The generator reactances are $X_1 = X_2 = 0.10$ pu and $X_0 = 0.05$ pu, all based on generator ratings. Note that, the given value of X_1 is subtransient reactance X'', one of several different positive-sequence reactances of a synchronous machine. The subtransient reactance corresponds to the initial symmetrical fault current (the transient dc component not included) that occurs before demagnetizing armature magnetomotive force begins to weaken the net field excitation. If manufactured in accordance with U.S. standards, the coils of a synchronous generator will withstand the mechanical forces that accompany a three-phase fault current, but not more. Assume that this generator is to supply a four-wire, wye-connected distribution. Therefore, the neutral grounding reactor X_n should have the smallest possible reactance. Consider both SLG and DLG faults. Assume the prefault positive-sequence internal voltage of phase a is 2500∠ 0° or 1.042∠0° pu and determine the following:

(a) Specify X_n in ohms and in per units.
(b) Specify the minimum allowable momentary symmetrical current rating of the reactor in amperes.
(c) Find the initial symmetrical voltage across the reactor, $\mathbf{V}_n$, when a bolted SLG fault occurs on the oil circuit breaker terminal in volts.

10. Consider the system shown in Figure 6.39 and the following data:
Generator G: $X_1, = X_2 = 0.10$ pu and $X_0 = 0.05$ pu based on its ratings
Motor: $X_1 = X_2 = 0.10$ pu and $X_0 = 0.05$ pu based on its ratings

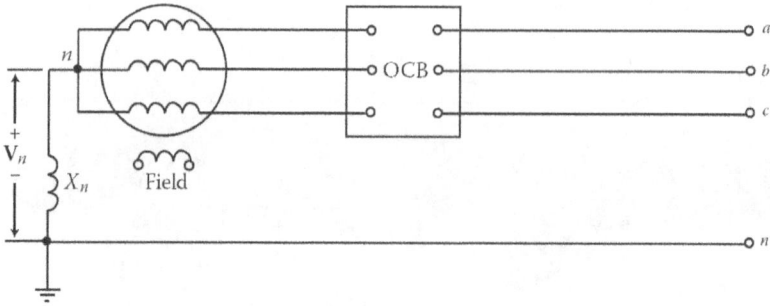

Figure 6.38 Generator system for Problem 9.

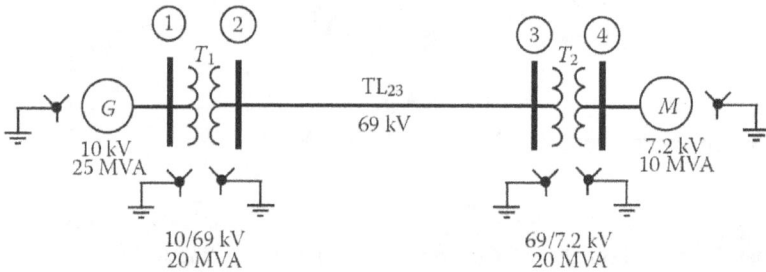

Figure 6.39 Transmission system for Problem 10.

Transformer T_1: $X_1 = X_2 = X_0 = 0.05$ pu based on its ratings
Transformer T_2: $X_1 = X_2 = X_0 = 0.10$ pu based on its ratings
Transmission line TL$_{23}$: $X_1 = X_2 = X_0 = 0.09$ pu based on 25 MVA
Assume that bus 2 is faulted and determine the faulted phase currents.

   (a) Determine the three-phase fault.
   (b) Determine the line-to-ground fault involving phase a.
   (c) Use the results of part (a) and calculate the line-to-neutral phase voltages at the fault point.

11. Consider the system given in Problem 10 and assume a line-to-line fault, involving phases b and c, at bus 2 and determine the faulted phase currents.

12. Consider the system shown in Figure 6.40 and assume that the associated data is given in Table 6.4 and is based on a 100-MVA base and referred to nominal system voltages.
Assume that there is a three-phase fault at bus 6. Ignore the prefault currents and determine the following:

   (a) Fault current in per units at faulted bus 6
   (b) Fault current in per units in transmission line TL$_{25}$

13. Use the results of Problem 12 and calculate the line-to-neutral phase voltages at the faulted bus 6.

14. Repeat Problem 12 assuming a line-to-ground fault, with $Z_f = 0$ pu, at bus 6.

15. Use the results of Problem 14 and calculate the line-to-neutral phase voltages at the following buses:

   (a) Bus 6
   (b) Bus 2

16. Repeat Problem 12 assuming a line-to-line fault at bus 6.

17. Repeat Problem 12 assuming a DLG fault, with $Z_f = 0$ and $Z_g = 0$, at bus 6.

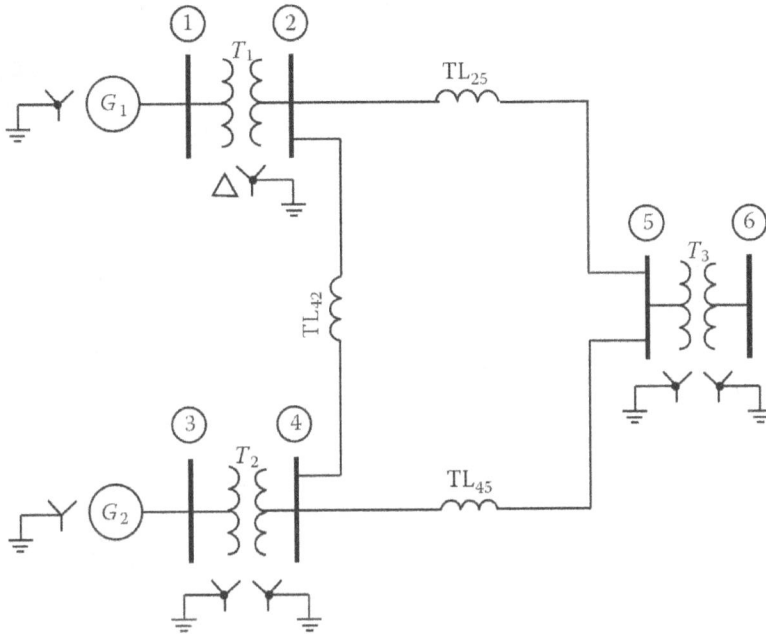

Figure 6.40 Transmission system for Problem 12.

Table 6.4
Data for Problem 6.12

Network Component	X_1 (pu)	X_2 (pu)	X_0 (pu)
G_1	0.35	0.35	0.09
G_2	0.35	0.35	0.09
T_1	0.1	0.1	0.1
T_2	0.1	0.1	0.1
T_3	0.05	0.05	0.05
TL_{42}	0.45	0.45	1.8
TL_{25}	0.35	0.35	1.15
TL_{45}	0.35	0.35	1.15

18. Consider the system shown in Figure 6.41 and data given in Table 6.5. Assume that there is a fault at bus 2. After drawing the corresponding sequence networks, reduce them to their Thévenin equivalents "looking in" at bus 2 for
    (a) Positive-sequence network
    (b) Negative-sequence network
    (c) Zero-sequence network
19. Use the solution of Problem 18 and calculate the fault currents for the following faults and draw the corresponding interconnected sequence networks.
    (a) SLG fault at bus 2 assuming that the faulted phase is phase a
    (b) DLG fault at bus 2 involving phases b and c
    (c) Three-phase fault at bus 2

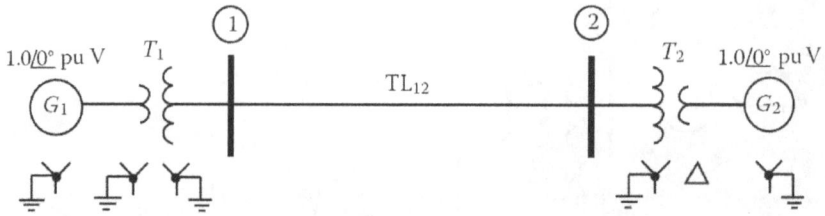

Figure 6.41 Transmission system for Problem 18.

Table 6.5
Table for Problem 18

Network Component	Base MVA	X_1 (pu)	X_2 (pu)	X_0 (pu)
G_1	100	0.2	0.15	0.05
G_2	100	0.3	0.2	0.05
T_1	100	0.2	0.2	0.2
T_2	100	0.15	0.15	0.15
TL$_{12}$	100	0.6	0.6	0.9

20. Use the solution of Problem 18 and calculate the fault currents for the following faults and draw the corresponding interconnected sequence networks and calculate the fault currents, assuming that
    (a) SLG fault at bus 2 involves phase b
    (b) DLG fault at bus 2 involves phases c and a
21. Repeat parts (a) and (b) of Problem 19 assuming that
    (a) SLG fault at bus 2 involves phase c
    (b) DLG fault at bus 2 involves phases a and b
22. Repeat Example 6.6 assuming that the fault impedance is zero.
23. Repeat Example 6.6 assuming that the fault involves phase c.
24. Repeat Example 6.7 assuming that the Z_f and Z_g, are zero.
25. Repeat Example 6.7 assuming that the fault involves phases c and a.
26. Consider the system shown in Figure 6.42 and data given in Table 6.6. Assume that there is an SLG fault at bus 3. Do the following:
    (a) Determine the Thévenin equivalent positive-sequence impedance.
    (b) Determine the Thévenin equivalent negative-sequence impedance.
    (c) Determine the Thévenin equivalent zero-sequence impedance.
    (d) Determine the positive-, negative-, and zero-sequence currents.
    (e) Determine the phase currents in per units and amperes.
    (f) Determine the positive-, negative-, and zero-sequence voltages.
    (g) Determine the phase voltages in per units and kilovolts.
    (h) Determine the line-to-line voltages in per units and kilovolts.
    (i) Draw a voltage phasor diagram using before-the-fault line-to-neutral and line-to-line voltage values.
    (j) Draw a voltage phasor diagram using the resultant after-the-fault line-to-neutral and line-to-line voltage values.

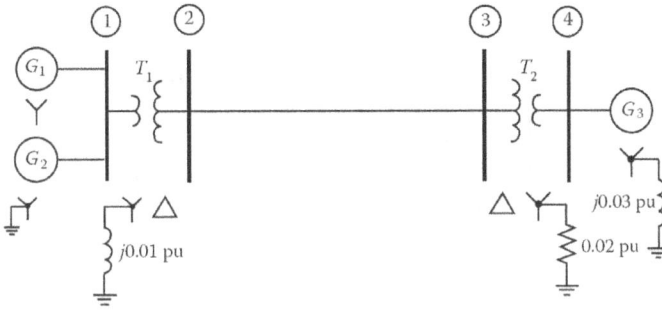

Figure 6.42 Transmission system for Problem 26.

Table 6.6
Table for Problem 26

Network Component	Base MVA	Voltage Rating (kV)	X_1 (pu)	X_2 (pu)	X_0 (pu)
G_1	100	13.8	0.15	0.15	0.05
G_2	100	13.8	0.15	0.15	0.05
G_3	100	13.8	0.15	0.15	0.05
T_1	100	13.8/115	0.2	0.2	0.2
T_2	100	115/13.8	0.18	0.18	0.18
TL_{23}	100	115	0.3	0.3	0.9

27. Consider the system shown in Figure 6.43 and assume that the following data on the same base are given:

Generator G_1: $X_1 = 0.15$ pu, $X_2 = 0.10$ pu, $X_0 = 0.05$ pu
Generator G_2: $X_1 = 0.30$ pu, $X_2 = 0.20$ pu, $X_0 = 0.10$ pu
Transformer T_1: $X_1 = X_2 = X_0 = 0.10$ pu
Transformer T_2: $X_1 = X_2 = X_0 = 0.15$ pu
Transmission line TL_{12}: $X_1 = X_2 = 0.30$ pu, $X_0 = 0.60$ pu
Transmission line TL_{23}: $X_1 = X_2 = 0.30$ pu, $X_0 = 0.60$ pu
Assume that fault point A is located at the middle of the top transmission line, as shown in the figure, and determine the fault current(s) in per units for the following faults:

(a) SLG fault (involving phase a)

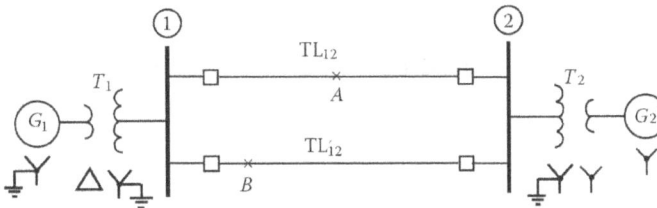

Figure 6.43 Transmission system for Problem 27.

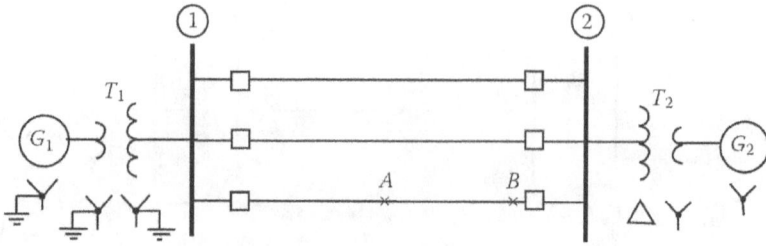

Figure 6.44 Transmission system for Problem 29.

  (b) DLG fault (involving phases b and c)
  (c) Three-phase fault
28. Repeat Problem 27 assuming that the fault point is n and is located at the beginning of the bottom line.
29. Consider the system shown in Figure 6.44 and assume that the following data on the same base are given:
    Generator G_1: $X_1 = 0.15$ pu, $X_2 = 0.10$ pu, $X_0 = 0.05$ pu
    Generator G_2: $X_1 = 0.15$ pu, $X_2 = 0.10$ pu, $X_0 = 0.05$ pu
    Transformer T_1: $X_1 = X_2 = X_0 = 0.10$ pu
    Transformer T_2: $X_1 = X_2 = X_0 = 0.15$ pu
    Transmission lines: $X_1 = X_2 = 0.30$ pu, $X_0 = 0.60$ (all three are identical)
    Assume that the fault point A is located at the middle of the bottom line, as shown in the figure, and determine the fault current(s) in per units for the following faults:
    (a) SLG fault (involving phase a)
    (b) SLG fault (involving phases b and c)
    (c) DLG fault (involving phases b and c)
    (d) Three-phase fault
30. Repeat Problem 29 assuming that the faulted point is B and is located at the end of the bottom line.
31. Consider the system shown in Figure 6.45 and its data given in Table 6.7. Assume that there is an SLG fault involving phase a at fault point F.
    (a) Draw the corresponding equivalent positive-sequence network.
    (b) Draw the corresponding equivalent negative-sequence network.
    (c) Draw the corresponding equivalent zero-sequence network.
32. Use the results of Problem 31 and determine the interior sequence currents flowing in each of the four transmission lines.
    (a) Positive-sequence currents
    (b) Negative-sequence currents
    (c) Zero-sequence currents
33. Use the results of Problem 32 and determine the interior phase currents in each of the four transmission lines.
    (a) Phase a currents
    (b) Phase b currents
    (c) Phase c currents
34. Use the results of Problems 32 and 33 and draw a three-line diagram of the given system. Show the phase and sequence currents on it.
    (a) Determine the SLG fault current.

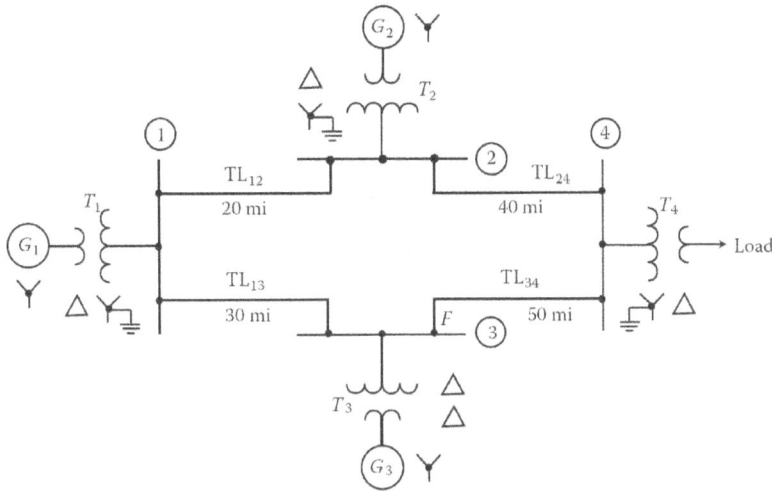

Figure 6.45   Four-bus system for Problem 31.

Table 6.7
Table for Problem 31

Network Component	Base MVA	Base kV$_{(L-L)}$	X_1 (pu)	X_2 (pu)	X_0 (pu)
G_1	100	230	0.15	0.15	
G_2	100	230	0.2	0.2	
G_3	100	230	0.25	0.25	
T_1	100	230	0.1	0.1	0.1
T_2	100	230	0.09	0.09	0.09
T_3	100	230	0.08	0.08	0.08
T_4	100	230	0.11	0.11	0.11
TL$_{12}$	100	230	0.1	0.1	0.36
TL$_{13}$	100	230	0.2	0.2	0.6
TL$_{24}$	100	230	0.35	0.35	1.05
TL$_{34}$	· 100	230	0.4	0.4	1.2

   (b) Is the fault current equal to the sum of the zero-sequence currents (i.e., $\mathbf{I}_{f(\text{SLG})} = \sum 3\mathbf{I}_{a0}$?
   (c) Phase c currents
35. Repeat Example 6.9 assuming that there is a series fault at the fault point B of the bottom line (i.e., TL$'_{AB}$).
36. Repeat Example 6.10 using the results of Problem 35.
37. Consider the system given in Example 6.8 and determine the following:
   (a) Admittance matrix associated with zero-sequence network
   (b) Two-port Thévenin equivalent of zero-sequence network
38. Use the results of Problem 37 and determine the uncoupled zero-sequence network.
39. Consider the system shown in Figure 6.46 and assume that the equivalent of a large system is shown with two buses of interest and two interconnecting lines. Assume that one conductor of the top line becomes open.

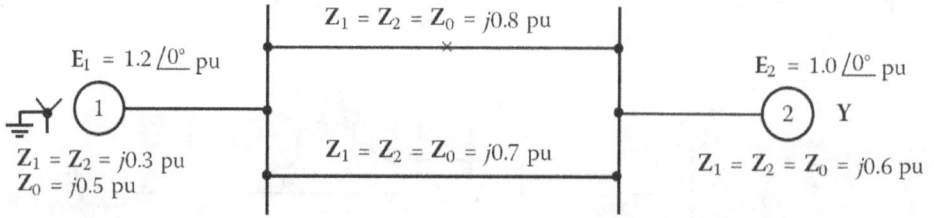

$$E_1 = 1.2\,\underline{/0°}\ \text{pu}$$

$$Z_1 = Z_2 = Z_0 = j0.8\ \text{pu}$$

$$E_2 = 1.0\,\underline{/0°}\ \text{pu}$$

$$Z_1 = Z_2 = j0.3\ \text{pu}$$
$$Z_0 = j0.5\ \text{pu}$$

$$Z_1 = Z_2 = Z_0 = j0.7\ \text{pu}$$

$$Z_1 = Z_2 = Z_0 = j0.6\ \text{pu}$$

Figure 6.46 Transmission system for Problem 39.

   (a) Draw the corresponding positive-, negative-, and zero-sequence networks, without reducing them, and their interconnections.
   (b) Determine the uncoupled positive-sequence Thévenin equivalent.
   (c) Determine the uncoupled negative-sequence Thévenin equivalent.
   (d) Determine the uncoupled zero-sequence Thévenin equivalent.
   (e) Using the uncoupled sequence equivalents found in parts (b) through (d), repeat part (a).
40. Repeat Example 6.10 assuming an OLO fault involving phase a and $Z = j0.1$ pu.
41. Use the solutions of Example 6.9 and Problem 38, and assume that the series fault is an OLO fault on phase b, $Z = j0.1$ pu.
   (a) Draw the generalized fault diagram.
   (b) Determine the positive-sequence current.
   (c) Determine the negative-sequence current.
   (d) Determine the zero-sequence current.
   (e) Determine the line current for phase a.
   (f) Determine the line current for phase b.
   (g) Determine the line current for phase c.
42. Repeat Problem 41 assuming that the open line is phase c.
43. Repeat Problem 41 assuming that there is a TLO series fault involving phases c and a.
44. Consider the OLO series fault representation given in Figure 6.19. Assume that there is a Z_f impedance between the fault points F and F' instead of being open and that the impedances $(Z's)$ shown on lines b and c are zero. Redraw Figure 6.19 to reflect these changes and mathematically verify the resulting interconnection of the sequence networks.
45. Consider 6.19 and assume that there is no OLO series fault on phase a but the previous fault points F and F' are short-circuited with a line impedance Z of zero. However, the line impedances $(Z's)$ on phases b and c are not zero. Redraw Figure 6.19 to reflect these changes and mathematically verify the resulting interconnection of the sequence networks.
46. Consider 6.19 and assume that, between the points F and F', phases a, b, and c have the impedances Z_f, Z, and Z, respectively. Redraw Figure 6.19 to reflect these changes and mathematically verify the resulting interconnection of the sequence networks.
47. Use the result of Problem 6.19c and assume that the prefault load currents in line TL_{12} where $I_{a0} = I_{a2} = 0$ and $I_{a1} = 0.6Z{-}30°$ pu.
48. Assume that there is an OLO fault and an SLG fault with fault impedance Z_f both on phase a at a given fault point.
   (a) Determine the general representation diagram of the simultaneous fault.
   (b) Show the interconnection of sequence networks.
   (c) Mathematically verify the interconnection drawn in part (b).
49. Consider Figure 6.35a and c and verify Equation 6.130.

50. Repeat Example 6.11 assuming that all three lines of the transmission system have ground wires. The self-GMD of ground wires, D_{gg}, is 0.03125 ft. The GMD between phase conductors and ground wires, D_{ag}, is 13.0628 ft. The GMD between ground wires and their images, H_{gg}, is 104 ft. The GMD between phase conductors and images of ground wires, H_{ag}, is 92.8102 ft.

7 Power Flow Analysis

7.1 INTRODUCTION

Historically, load flow was referred to as "power flow" in the technical literature. However, it is often said that "load does not flow, power flows," and the technical community has consistently used the term "power flow" to describe the classical analysis of electron flow through power transmission or distribution circuits. In general, power flow is the solution to the non-linear system for normal, balanced, three-phase steady-state operating conditions of a transmission system. Power flow calculations are typically performed for operational planning and system control. The data obtained from power flow studies are used to analyze normal operating modes, conduct contingency analysis, assess outage security, optimize dispatching, and ensure stability.

A "contingency" refers to the loss of a major transmission element or a large generating unit. When a system is able to withstand any single major contingency, it is now referred to as "N-1 secure" according to the terminology suggested by the National Electric Reliability Council (NERC). A "double contingency" occurs when two transmission lines, two generators, or a line and a generator are lost [176]. In this case, the system can be considered N-2 secure. This higher order of contingency does not necessarily provide an exhaustive list of scenarios, as the hypothesized scenarios should align with the disruptive forces of the abnormal conditions. For example, adverse weather conditions may cause electrical short circuits and result in tripped circuit breakers.

The difficulties and importance of the power flow problem have been succinctly articulated by Shipley [190]. Mathematicians and engineers worldwide have been fascinated by this problem for many years, with many dedicating a significant portion of their professional lives to finding a solution. Power flow has received more attention than all other power system problems combined, resulting in an enormous number of technical publications. The nature of the problem makes it unlikely to develop a perfect procedure, thus progress will continue to be made on improving solutions for a long time.

Prior to 1929, all power flow calculations were performed manually. In 1929, network calculators (developed by Westinghouse) or network analyzers (developed by General Electric) were introduced to perform power flow calculations. The first paper describing a digital method to solve the power flow problem was published in 1954 [191]. However, the first truly successful digital method was developed by Ward and Hale in 1956 [192]. Many early iterative methods were based on the Y-matrix of the Gauss-Seidel method [192–198]. This method required minimal computer storage and a small number of iterations for small networks. However, as the network size increased, the number of iterations required grew dramatically for large systems, and in some cases, the method failed to provide a solution at all.

The slowly converging behavior of the Gauss-Seidel method and its frequent failure to converge in ill-conditioned situations led to the development of Z-matrix methods [199–215]. While these methods have considerably better convergence characteristics, they require significantly more computer storage memory since the Z-matrix is full, unlike the sparse Y-matrix.

The difficulties encountered in power flow studies led to the development of the Newton-Raphson method. Originally developed by Van Ness and Griffin [204, 206], and later improved by others [207–212], this method utilizes the Newton-Raphson algorithm to solve the simultaneous quadratic equations of the power network. Unlike the Gauss-Seidel algorithm, it requires more time per iteration but only a few iterations, and its performance is largely independent of the network size. As a result, the Newton-Raphson method can easily solve power flow problems that the Gauss-Seidel method struggles with, such as systems with negative impedances. However, it was

DOI: 10.1201/9781003129769-7

not computationally competitive on large systems due to the rapid growth of computing platforms and storage requirements, despite efficient algorithm manipulation.

Nevertheless, the development of a highly efficient sparsity-programmed ordered elimination technique by Tinney and others [213, 214] has significantly enhanced the efficiency of the Newton-Raphson method in terms of speed and storage requirements, making it the most widely used power flow method. Furthermore, the method has been further improved through the incorporation of automatic controls and adjustments, such as program-controlled in-phase tap changes, phase-angle regulators, and area interchange control. Given that system planning studies and operations often require multiple-case power flow solutions, recent research efforts have focused on the development of decoupled Newton-Raphson methods [215–219].

These methods are based on the fact that in any power transmission network operating in steady state, the coupling (i.e., interdependence) between $P–\theta$ (i.e., active powers and bus voltage angles) and $Q–V$ (i.e., reactive powers and bus voltage magnitudes) is relatively weak, unlike the strong coupling between P and θ, and between Q and V. Therefore, these methods solve the power flow problem by "decoupling" (i.e., solving separately) the $P–\theta$ and $Q–V$ problems. As a result, solutions are obtained by applying approximations to the Newton-Raphson method. These methods offer adequate accuracy and very fast speed, making them suitable for online applications and contingency security assessments.

A reliable power flow algorithm should provide fast convergence. However, no method currently exists that is both adequately fast and capable of providing a feasible solution for every case study. As a result, significant research efforts have been dedicated to the development of numerical methods and programming techniques for solving power flow problems optimally [213, 214, 220–252]. Additionally, power systems and the size of problems to be solved are continuously increasing at a faster rate than the development of computers with greater capacity. Therefore, the theory of network tearing, or *diakoptics*, as proposed by Gabriel Kron and later developed by Happ and others [240, 253–258], has recently gained attention in power flow studies. According to diakoptics, a large network system can be divided into smaller pieces using a technique known as "tearing." Each piece can then be analyzed separately, and the obtained solutions can be transformed to provide the overall solution.

Another interesting development in the power flow problem has been the introduction of probabilistic power flow, where the power inputs and outputs of a network are treated as random variables [238, 239, 259, 260]. This approach recognizes that the power flow problem has inherent stochastic characteristics, as some of its input data is subject to uncertainty. Consequently, the output data, such as branch flows, can be represented as a range of possible values along with corresponding frequencies of occurrence or a probability distribution function.

The uncertainties associated with the input data arise from various factors, including (1) uncertainties in load forecasts due to economic factors and conservation measures, (2) load variability influenced by weather conditions and load management practices, (3) unavailability of installed generator and transmission facilities due to forced outages or maintenance, (4) delays in the installation of new generation and transmission infrastructure, (5) sudden changes in fuel prices and fuel availability, and (6) advancements in power generation and transmission techniques.

Present-day trends are focused on the development of interactive power flow programs [234], enabling planning engineers to modify data on a computer either through a dialogue mode with the program or via graphic displays. The engineer can then direct the program to solve the problem. Such programs often allow the planner to choose specific portions of the output data for display after obtaining a solution. It is anticipated that future power flow programs will offer improved interactive capabilities and faster algorithms, aiming to minimize the difficulties encountered in finding solutions within appropriate time and effort constraints.

7.2 STATE OF THE ART

In recent decades, power flow analysis has become a commonly implemented computational tool in software systems and is widely used today to validate and verify specific conditions of a power system. The analysis provides binary indicators of "converged" or "diverged" statuses in the solutions. Power flow solutions have been utilized to check various aspects of operations and planning scenarios, ensuring that the power flowing across branches adds up to a zero value.

One of the earliest research conclusions in power flow analysis is the continuation power flow (CPF), which provides steady-state voltage stability analysis for critical points of system "collapse" at each node of a power system (typically referring to a substation in the power grid context) [261]. The lead author of [261] later expanded these studies by identifying the point of collapse through direct tracing of the equilibrium, as elaborated in [262]. This approach avoids the need to rebuild the system Jacobian matrix and continually check for system singularity. A more robust interpretation of CPF has been studied in a realistic South Korean system to determine flow limits using a modified CPF approach, considering high-order contingency analyses [263].

Applications of power flow analysis include optimal power flow for enhancing voltage stability margins with reactive reserve [264], reduction of active power losses using continuation methods [265], and optimal power flow using expert systems [266]. The use of cyber-based contingencies, such as disruptive switching attacks through substation control networks, has been evaluated through combinatorial analysis validated using a power flow module [267]. Furthermore, the cyber-based contingency analysis, focusing on cascading effects by removing overloads on circuits, has been used to validate hypothesized scenarios [268]. This extreme contingency analysis is then employed to anticipate risks for insurance businesses in estimating premiums for IP-based power substations [269]. The research also extends to the determination of feasible spaces in optimal power flow problems [270]. In recent innovations, the application of power flow analysis has been extended to quantum computing platforms [271].

Due to the non-linear nature of power systems, achieving convergence in power flow analysis can be challenging for engineers, especially when encountering new phenomena or case studies where an "error" occurs, such as diverged power flow without a solution. To address this, the feasibility of power flow has been re-analyzed using circuit integration techniques that involve modeling loads and circuits with dependent sources [272]. This represents a fundamental reconstruction of the power flow formulation, improving the equivalent circuit formulation to ensure successful solution convergence for both transmission and distribution systems [273, 274]. Additionally, an enhanced multi-period optimal power flow approach has been developed, incorporating differential dynamic programming specifically for energy storage systems [275].

Despite the extensive literature on power flow formulation in recent decades, it remains an ongoing subject where the challenge lies in balancing the robustness of solution convergence with the potential problematic scenarios that can arise from diverged solutions.

7.3 CAUSES OF NONCONVERGENT POWER FLOW CASES

The topic of power flow divergence is relevant and applicable because it can occur when initiating a new base case with a diverged power solution. Troubleshooting the robustness of power flow convergence is often an ongoing challenge. The following scenarios outline potential causes of divergence, which can be attributed to numeric instability in the initial state setup for base case studies.

1. A line with significantly lower impedance compared to another line connected in series with it. (Solution: If possible, combine the impedance of short and long series-connected lines into a single line.)
2. Missing tie lines due to oversight during model creation or incorrect connection of tie lines.

3. Excessively large magnitude of impedance or susceptance value. This may be caused by a misplaced decimal point or incorrect specification of large cutoff impedance during equivalization.
4. The regulating (slack) bus of a system is assigned to a different system. This is likely due to incorrect data entry during model changes.
5. Inadvertently creating an isolated system or island. Voltage phase divergence will be immediately flagged, and the program will stop calculating after the first iteration.
6. Unrealistic tap limits for transformer controlled under load (TCUL) transformers.
7. Very large radial system.
8. Issues with voltage regulation, such as: (a) Unequal voltage schedules at generating units connected by a low impedance line. (b) Attempting to regulate a radial line at both ends with unequal voltages. (Solution: Avoid regulating a radial bus and maintain the Mvar output of a radial bus at the value obtained in the last iteration.) (c) Conflicting voltage regulation. (d) Unreasonably small voltage range for switched shunts. (e) Remote regulation of more than one bus away.
9. Over-equivalization of outside regions in regional base case models.
10. Not solvable from a flat start.
11. Fictitious regulation of buses.
12. Negative or extremely low tap values.
13. Negative or extremely low voltage schedules.
14. Failure to follow the approved MMWG (Model Management Working Group) sign convention for phase shifters or not adhering to the minimum MW tolerance for phase-shifting-under load transformers (refer to page 3 of this appendix).
15. Zero or very low reactance branches. Minimum reactance should be 0.0001 per unit.
16. Inconsistent representation of delta-wye transformers, typically involving two companies interconnected at both voltage levels.

Convergence in power flow analysis is vital for accurate and reliable results. It involves correct modeling, accurate tie line connections, and proper management of impedance, susceptance, and system configuration to avoid divergence. Additional considerations include preventing isolated systems, setting realistic transformer tap limits, and ensuring proper voltage regulation. Adhering to conventions, reactance values, and consistent transformer representation also contribute to achieving convergence and obtaining meaningful outcomes in power system studies.

7.4 FORMULATING A POWER FLOW PROBLEM

The power flow problem involves calculating the real and reactive powers flowing in each line and determining the voltage magnitude and phase angle at each bus in a transmission system, considering specified generation and load conditions. Power flow studies provide valuable information about the system's ability to transfer energy without overloading lines and to ensure adequate voltage regulation using devices such as capacitors, reactors, transformers with tap-changing capabilities, and rotating machines that supply reactive power.

The bus power at a specific bus can be defined in terms of generated power, load power, and transmitted power. For instance, the net power (bus power) at the i-th bus of an n-bus power system can be represented as:

$$\begin{aligned} \mathbf{S}_i &= P_i + jQ_i \\ &= (P_{Gi} - P_{Li} - P_{Ti}) + j(Q_{Gi} - Q_{Li} - Q_{Ti}) \end{aligned} \tag{7.1}$$

where
  $\mathbf{S}_i$ = three-phase complex bus power at i-th bus
  P_i = three-phase real bus power at i-th bus
  Q_i = three-phase reactive bus power at i-th bus

Table 7.1
Summary of Bus Classification

Bus Type	Known Quantities	Unknown Quantities
Slack	$|V| = 1.0$	P, Q
	$\theta = 0$	
Generator (PV bus)	$P, |V|$	Q, θ
Load (PQ bus)	P, Q	$|V|, \theta$

P_{Gi} = three-phase real generated power flowing into i-th bus
P_{Li} = three-phase real load power flowing out of i-th bus
P_{Ti} = three-phase real transmitted power flowing out of i-th bus
Q_{Gi} = three-phase reactive generated power flowing into i-th bus
Q_{Li} = three-phase reactive load power flowing out of i-th bus
Q_{Ti} = three-phase reactive transmitted power flowing out of i-th bus

In power flow studies, the basic assumption is that the given power system is a balanced three-phase system operating in its steady state with a constant 60-Hz frequency.* Therefore, the system can be represented by its single-phase positive-sequence network with a lumped series and shunt branches. The power flow problem can be solved either by using the bus admittance matrix ($\mathbf{Y}_{bus}$) or the bus impedance matrix ($\mathbf{Z}_{bus}$) representation of the given network. It is customary to use the nodal analysis approach. Thus, if the bus voltages are known, the bus currents can be expressed as

$$[\mathbf{I}_{bus}] = [\mathbf{Y}_{bus}][\mathbf{V}_{bus}] \tag{7.2}$$

or in its inverse form,

$$[\mathbf{V}_{bus}] = [\mathbf{Y}_{bus}]^{-1}[\mathbf{I}_{bus}]$$
$$= [\mathbf{Z}_{bus}][\mathbf{I}_{bus}] \tag{7.3}$$

If the bus voltages are known, then the bus currents can be calculated from Equation 7.2. Otherwise, if the bus currents are known, then the bus voltages can be found from Equation 7.3.

However, in a given power flow problem, nodal active and reactive powers are the independent variables and nodal voltages are the dependent variables. Therefore, the determination of the nodal voltages, which initially seems to be simple, given that the nodal currents are known, becomes a nonlinear problem dictating an iterative solution method since it is nodal power rather than current that is known.

Each bus in a network is characterized by four variable quantities: real power, reactive power, voltage magnitude (line-to-ground), and voltage phase angle. Typically, two of these variables are specified as independent variables, while the other two are unknown and need to be determined. Due to the nature of generation and load, the electrical conditions at each bus are defined in terms of active and reactive power rather than bus current. The complex power flowing into the ith bus can be mathematically represented as follows:

$$\mathbf{V}_i \mathbf{I}_i^* = P_i + jQ_i \tag{7.4}$$

the bus current is related to these variables as

$$\mathbf{I}_i = \frac{P_i - jQ_i}{\mathbf{V}_i^*} \tag{7.5}$$

In general, there are three types of buses in a power flow problem, each with its own specified variables: (1) slack/swing/reference bus, (2) generator buses, and (3) load buses. Since transmission losses in a given system are associated with the bus voltage profile, the total power generation requirement cannot be determined until a solution is obtained. Therefore, the generator at the slack bus is used to supply the additional active and reactive power necessary to compensate for system losses.

In graph theory, a node is the juncture of connectivity where the connectedness of the elements is established. The power engineering literature defines "buses" as the equivalent of nodes. The slack bus is defined with a known magnitude and phase angle of the voltage, which is a magnitude of 1 and a zero angle. The real and reactive power generated are the quantities to be determined. It is only after a solution is converged, meaning all bus voltages are known, that the real and reactive power generation requirements at the slack bus can be determined. In other words, the losses are not known in advance, and consequently, the power at the slack bus cannot be specified. It is imperative to ensure that each power system is defined with only one slack bus. The chosen buses as part of the "slacks" are those nodes that have the highest degree of connectivity with branches associated with the node in a subset of electrical "islands."

To define the power flow problem to be solved, it is necessary to specify the real power and the voltage magnitude at each generator bus. This is because these quantities are controllable through the governor and excitation controls, respectively. The generator bus is also known as the PV bus. Since an overexcited synchronous generator supplies current at a lagging power factor, the reactive power Q of a generator is not required to be specified.

The load bus is also known as the PQ bus. This is because the real and reactive powers are specified at a given load bus. Table 7.1 gives the bus types with corresponding known and unknown variables. Note that the slack bus voltage is set to 1.0 pu with a phase angle of $0°$ (i.e., the slack bus voltage is used as a reference voltage).

It is possible that some load buses may have transformers capable of tap-changing and phase-shifting operations. These types of load buses are known as voltage-controlled load buses. At the voltage-controlled load buses, the known quantities are the voltage magnitude, in addition to the real and reactive powers, and the unknown quantities are the voltage phase angle and the turns ratio.

7.5 SIGN OF REAL AND REACTIVE POWERS

When performing a power flow study, it is important to remember that lagging reactive power is considered positive due to the inductive current, while leading reactive power is negative due to the capacitive current. Additionally, the positive bus current is in the direction that flows toward the bus. As the generator current flows toward the bus and the load current flows away from the bus, the sign of power is positive for the generator bus and negative for the load bus. Therefore, the following observations can be made:

1. The real and reactive powers associated with the inductive load bus (i.e., the lagging power factor load bus) are both negative.
2. The real and reactive powers associated with the capacitive load bus (i.e., the leading power factor load bus) are negative and positive, respectively.
3. The real and reactive powers associated with the inductive generator bus (i.e., the bus with a generator operating in lagging power factor mode) are both positive.
4. The real and reactive powers associated with the capacitive generator bus (i.e., the bus with a generator operating in leading power factor mode) are positive and negative, respectively.
5. The reactive power of a shunt capacitive compensation apparatus located at a bus is positive.

For instance, in the case where a load bus is connected to a load that consumes 5 pu of real power (W) and 3 pu of inductive reactive power (vars), the bus current, as indicated by Equation 7.5, can be formulated as follows:

$$I_i = \frac{P_i - jQ_i}{V_i^*}$$

$$= \frac{-5 - j(-3)}{V_i^*}$$

$$= \frac{-5 + j3}{V_i^*} \text{pu A}$$

However, if a generator bus is connected to a generator that operates in a lagging power factor mode and supplies 5 pu W and 3 pu vars, then the bus current becomes

$$I_i = \frac{P_i - jQ_i}{V_i^*}$$

$$= \frac{5 - j(+3)}{V_i^*}$$

$$= \frac{5 - j3}{V_i^*} \text{pu A}$$

7.6 GAUSS ITERATIVE METHOD

Assume that a set of simultaneous linear equations with n unknowns ($x_1, x_2, x_3, \cdots, x_n$ independent variables) is given as

$$a_{11}x_1 + a_{12}x_2 + a_{13}x_3 + \cdots + a_{1n}x_n = b_1$$
$$a_{21}x_1 + a_{22}x_2 + a_{23}x_3 + \cdots + a_{1n}x_n = b_2$$
$$\cdot$$
$$\cdot \qquad\qquad\qquad\qquad\qquad\qquad (7.6)$$
$$\cdot$$
$$a_{n1}x_1 + a_{n2}x_2 + a_{n3}x_3 + \cdots + a_{nn}x_n = b_n$$

where the a coefficients and the b dependent variables are known. The given Equation set 7.6 can be reexpressed as

$$x_1 = \frac{1}{a_{11}}(b_1 + a_{12}x_2 - a_{23}x_3 - \cdots - a_{1n}x_n)$$

$$x_2 = \frac{1}{a_{22}}(b_2 + a_{21}x_1 - a_{23}x_3 - \cdots - a_{2n}x_n$$

$$\cdot$$
$$\cdot \qquad\qquad\qquad\qquad\qquad\qquad (7.7)$$
$$\cdot$$

$$x_n = \frac{1}{a_{nn}}(b_n + a_{n1}x_1 - a_{n2}x_2 - \cdots - a_{n,n-1}x_{n-1})$$

Assume that the initial approximation values of the independent variables are $x_1^{(0)}$, $x_2^{(0)}$, $x_1^{(0)}$,

$\cdots, x_n^{(0)}$. Thus, after the substitution, the Equation set 7.7 can be written as

$$x_1^{(1)} = \frac{1}{a_{11}}(b_1 + a_{12}x_2^{(0)} - a_{23}x_3^{(0)} - \cdots - a_{1n}x_n^{(0)})$$

$$x_2^{(1)} = \frac{1}{a_{22}}(b_2 + a_{21}x_1^{(0)} - a_{23}x_3^{(0)} - \cdots - a_{2n}x_n^{(0)})$$

$$\cdot$$
$$\cdot$$
$$\cdot$$
(7.8)

$$x_n^{(1)} = \frac{1}{a_{nn}}(b_n + a_{n1}x_1^{(0)} - a_{n2}x_2^{(0)} - \cdots - a_{n,n-1}x_{n-1}^{(0)})$$

where the initial values* are usually selected as

$$x_1^{(0)} = \frac{b_1}{a_{11}}$$

$$x_1^{(0)} = \frac{b_1}{a_{11}}$$

$$\cdot$$
$$\cdot$$
$$\cdot$$
(7.9)

$$x_1^{(0)} = \frac{b_1}{a_{11}}$$

If the results obtained from Equation 7.8 match the initial values within a predetermined toler-ance, a convergence (i.e., solution) has been achieved. Otherwise, the new corrected values of the independent variables (i.e., $x_1^{(1)}, x_2^{(1)}, x_3^{(1)}, \cdots, x_n^{(1)}$) are substituted into the next iteration. Therefore, after the $k + 1$ iteration,

$$x_1^{(k+1)} = \frac{1}{a_{11}}(b_1 - a_{12}x_2^{(k)} - a_{13}x_3^{(k)} - \cdots - a_{1n}x_n^{(k)})$$

$$x_2^{(k+1)} = \frac{1}{a_{22}}(b_2 - a_{21}x_1^{(k)} - a_{23}x_3^{(k)} - \cdots - a_{2n}x_n^{(k)})$$

$$\cdot$$
$$\cdot$$
$$\cdot$$
(7.10)

$$x_n^{(k+1)} = \frac{1}{a_{nn}}(b_n - a_{n1}x_1^{(k)} - a_{n3}x_3^{(k)} - \cdots - a_{nn,n-1}x_{n-1}^{(k)})$$

7.7 GAUSS-SEIDAL ITERATIVE METHOD

The Gauss-Seidel iterative method builds upon the Gauss iterative method but introduces a more ef-ficient substitution technique. In the Gauss-Seidel approach, the newly computed values of variable x are immediately utilized in the right sides of subsequent equations as soon as they are obtained during the iterations. As a result,

$$x_1^{(k+1)} = \frac{1}{a_{11}}(b_1 - a_{12}x_2^{(k)} - a_{13}x_3^{(k)} - \cdots - a_{1n}x_n^{(k)})$$

$$x_2^{(k+1)} = \frac{1}{a_{22}}(b_2 - a_{21}\boxed{x_1^{(k+1)}} - a_{23}x_3^{(k)} - \cdots - a_{2n}x_n^{(k)})$$

$$x_3^{(k+1)} = \frac{1}{a_{22}}(b_3 - a_{31}\boxed{x_1^{(k+1)}} - a_{32}\boxed{x_2^{(k+1)}} - \cdots - a_{3n}x_n^{(k)})$$

.

.

.

$$x_n^{(k+1)} = \frac{1}{a_{nn}}(b_n - a_{n1}\boxed{x_1^{(k+1)}} - a_{n3}\boxed{x_3^{(k+1)}} - \cdots - a_{nn}\boxed{x_{n-1}^{(k+1)}}) \tag{7.11}$$

Please note that the superscript (i) in this context represents the ith approximation for variable x. It signifies the iteration cycle number rather than indicating a power.

In Equation 7.11, the circled values correspond to the immediately substituted values, which are determined in the preceding steps of the $(k + 1)$th iteration.

7.8 APPLICATION OF GAUSS-SEIDAL METHOD USING Y_{BUS}

Assuming that the neutral of an n-bus network system is taken as the reference point, the n current equations can be expressed in terms of the n unknown voltages.

$$[\mathbf{I}_{bus}] = [\mathbf{Y}_{bus}][\mathbf{V}_{bus}]$$

or

$$I_1 = \mathbf{Y}_{11}\mathbf{V}_1 + \mathbf{Y}_{12}\mathbf{V}_2 + \mathbf{Y}_{13}\mathbf{V}_3 + \cdots + \mathbf{Y}_{1n}\mathbf{V}_n$$
$$I_2 = \mathbf{Y}_{21}\mathbf{V}_1 + \mathbf{Y}_{22}\mathbf{V}_2 + \mathbf{Y}_{23}\mathbf{V}_3 + \cdots + \mathbf{Y}_{2n}\mathbf{V}_n$$
$$I_3 = \mathbf{Y}_{31}\mathbf{V}_1 + \mathbf{Y}_{32}\mathbf{V}_2 + \mathbf{Y}_{33}\mathbf{V}_3 + \cdots + \mathbf{Y}_{3n}\mathbf{V}_n$$

.

.   (7.12)

.

$$I_{3n} = \mathbf{Y}_{n1}\mathbf{V}_1 + \mathbf{Y}_{n2}\mathbf{V}_2 + \mathbf{Y}_{n3}\mathbf{V}_3 + \cdots + \mathbf{Y}_{nn}\mathbf{V}_n$$

or, in matrix form,

$$
\begin{bmatrix} \mathbf{I}_1 \\ \mathbf{I}_2 \\ \mathbf{I}_2 \\ \cdot \\ \cdot \\ \cdot \\ \mathbf{I}_n \end{bmatrix} =
\begin{bmatrix}
\mathbf{Y}_{11} & \mathbf{Y}_{12} & \mathbf{Y}_{13} & \cdots & \mathbf{Y}_{1n} \\
\mathbf{Y}_{21} & \mathbf{Y}_{22} & \mathbf{Y}_{23} & \cdots & \mathbf{Y}_{2n} \\
\mathbf{Y}_{31} & \mathbf{Y}_{32} & \mathbf{Y}_{33} & \cdots & \mathbf{Y}_{3n} \\
\cdot & \cdot & \cdot & \cdots & \cdot \\
\cdot & \cdot & \cdot & \cdots & \cdot \\
\cdot & \cdot & \cdot & \cdots & \cdot \\
\mathbf{Y}_{n1} & \mathbf{Y}_{n2} & \mathbf{Y}_{n3} & \cdots & \mathbf{Y}_{nn}
\end{bmatrix}
\begin{bmatrix} \mathbf{V}_1 \\ \mathbf{V}_2 \\ \mathbf{V}_3 \\ \cdot \\ \cdot \\ \cdot \\ \mathbf{V}_n \end{bmatrix} \tag{7.13}
$$

Note that, the diagonal element $\mathbf{Y}_{ii}$ is obtained as the algebraic sum of all primitive admittances incident to node i and that the off-diagonal elements $\mathbf{Y}_{jj} = \mathbf{Y}_{ji}$ are obtained as the negative of the primitive (branch) admittance connection nodes i and j, that is, $\mathbf{Y}_{ij} = -\mathbf{y}_{ij}$. In the case of a π representation of transmission circuits, the shunt susceptance would be included in the nodal admittance term, as would the admittance of shunt reactors or the shunt elements of equivalent circuits as those used for representing tapped transformers.

Therefore, the bus voltages for the $(k + 1)$-th iteration can be determined from Equation 7.12 when $\mathbf{V}_i^{(k)}$ and $\mathbf{I}_i^{(k)}$ are found after the kth iteration. Thus,

$$
\begin{aligned}
\mathbf{V}_1^{(k+1)} &= \frac{1}{\mathbf{Y}_{11}}(\mathbf{I}_1^{(k)} - \mathbf{Y}_{12}\mathbf{V}_2^{(k)} - \mathbf{Y}_{13}\mathbf{V}_3^{(k)} - \cdots - \mathbf{Y}_{1n}\mathbf{V}_n^{(k)}) \\
\mathbf{V}_2^{(k+1)} &= \frac{1}{\mathbf{Y}_{22}}(\mathbf{I}_2^{(k)} - \mathbf{Y}_{21}\mathbf{V}_2^{(k+1)} - \mathbf{Y}_{23}\mathbf{V}_3^{(k)} - \cdots - \mathbf{Y}_{2n}\mathbf{V}_n^{(k)}) \\
\mathbf{V}_3^{(k+1)} &= \frac{1}{\mathbf{Y}_{33}}(\mathbf{I}_3^{(k)} - \mathbf{Y}_{31}\mathbf{V}_2^{(k+1)} - \mathbf{Y}_{32}\mathbf{V}_3^{(k+1)} - \cdots - \mathbf{Y}_{3n}\mathbf{V}_n^{(k)}) \\
\mathbf{V}_n^{(k+1)} &= \frac{1}{\mathbf{Y}_{nn}}(\mathbf{I}_n^{(k)} - \mathbf{Y}_{n1}\mathbf{V}_2^{(k+1)} - \mathbf{Y}_{n2}\mathbf{V}_2^{(k+1)} - \cdots - \mathbf{Y}_{nn}\mathbf{V}_{n-1}^{(k+1)})
\end{aligned}
\tag{7.14}
$$

Even though currents in Equation 7.14 are unknown, they can be expressed in terms of P, Q, and V as

$$
\mathbf{I}_i = \left(\frac{P_i + jQ_i}{\mathbf{V}_i}\right)^* = \frac{P_i - jQ_i}{\mathbf{V}_i^*}
$$

A general formula to determine the bus voltage at the i-th (PQ) bus can be developed by substituting Equation 7.5 into Equation 7.14 so that

$$
\mathbf{V}_i^{(k+1)} = \frac{1}{\mathbf{Y}_{ii}}\left(\frac{P_i - jQ_i}{\mathbf{V}_i^{(k)*}} - \sum_{j=1}^{n}\mathbf{Y}_{ij}\mathbf{V}_{ij}^{(k)}\right) \quad \text{for } i = 2,\cdots,n
\tag{7.15}
$$

Note that, bus 1 is designated as the slack bus with known voltage magnitude and phase angle. Therefore, the bus voltage calculations start with bus 2.

If the i-th bus is a PV bus where real power and voltage magnitude, rather than reactive power, are given, then the unknown reactive power has to be determined first before each iteration. Thus, for the generator bus i,

$$
\mathbf{I}_{\text{gen}} = \frac{P_i - jQ_i}{\mathbf{V}_i^*} = \mathbf{Y}_{i1}\mathbf{V}_1 + \mathbf{Y}_{i2}\mathbf{V}_2 + \mathbf{Y}_{i3}\mathbf{V}_3 + \cdots + \mathbf{Y}_{in}\mathbf{V}_n
\tag{7.16}
$$

$$
P_i - jQ_i = \mathbf{V}_i^{(k)}\left[\sum_{j=1}^{n}\mathbf{Y}_{ij}\mathbf{V}_j^{(k)}\right]
\tag{7.17}
$$

Taking the imaginary part of Equation 7.11,

$$
jQ_i = -\text{Im}\left[\mathbf{V}_i^{(k)}\left(\sum_{j=1}^{n}\mathbf{Y}_{ij}\mathbf{V}_j^{(k)}\right)\right]
\tag{7.18}
$$

Note that the best values of voltages are used in calculating the reactive power Q_i. Once Q_i is found, it is used in Equation 7.15 to determine the new $\mathbf{V}_i$ at the generator bus. Usually, a limit on maximum and/or minimum $Q_{i,\text{spec}}$ may be specified. If the calculated reactive power, $Q_{i,\text{calc}}$, should

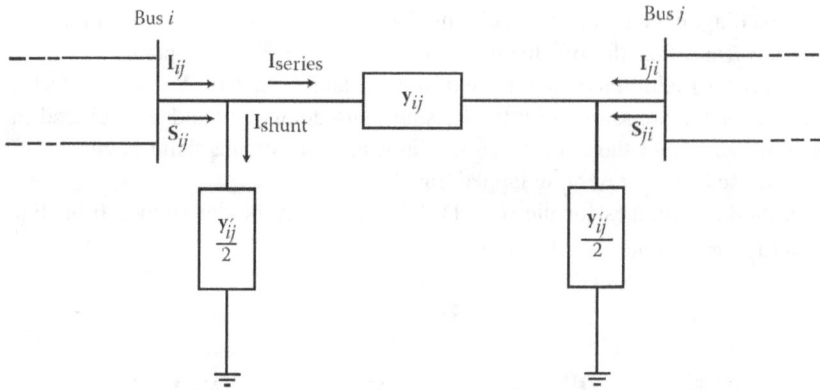

Figure 7.1 Line model used in power flow calculation.

exceed the limits $Q_{i,spec}$, then the limiting value of $Q_{i,spec}$ should be used in place of $Q_t exti, calc$ in Equation 7.15.

If the magnitude of the new calculated voltage is larger than the magnitude of the specified (original) voltage, the new voltage is corrected by multiplying by the ratio of the specified voltage magnitude to the calculated voltage magnitude, keeping the new phase angle of the calculated voltage. In other words, it is only the voltage magnitude that is being corrected.

In summary, the iteration process in the Gauss–Seidel method starts by assuming initial phasor values for the unknown bus voltages (except for the slack bus) and computing their new values, that is, the corrected voltages. At each bus, the corrected voltage is substituted back into Equation 7.15 as the estimated value for V_i^* to calculate the new value V_i.

This process is continued for a predetermined number of iterations, usually twice. As the corrected voltage value is determined at each bus, it is employed in finding the corrected voltage at the next. The process is repeated at each bus successively for the rest of the buses to finish the first iteration. This iteration process for the network system is repeated until the voltage correction required for each bus is less than a specified precision index, that is, tolerance.

After the bus voltages $V_2, V_3, V_4, \cdots, V_n$ are thus solved, the bus power at the slack bus can be determined from

$$\frac{P_i - jQ_i}{V_i^*} = Y_{11}V_1 + Y_{12}V_2 + Y_{13}V_3 + \cdots + Y_{1n}V_n \tag{7.19}$$

or

$$P_i - jQ_i = Y_{11}V_1V_1^* + +Y_{12}V_2V_1^* + Y_{13}V_3V_1^* + \cdots + Y_{1n}V_nV_1^* \tag{7.20}$$

As the final step, once all the bus voltages are known, the power flows in all transmission lines can be determined to complete the power flow study. For example, consider the line connecting buses i and j, as shown in Figure 7.1, and assume that the line current I_{ij} can be expressed as

$$I_{ij} = I_{series} + I_{shunt} \tag{7.21}$$

where

$$I_{series} = (V_i - V_{ij}) \tag{7.22}$$

thus,

$$I_{ij} = (V_i - V_{ij})y_{ij} + V_i\frac{y_{ij}'}{2} \tag{7.23}$$

where

y_{ij} = admittance of line ij
$\mathbf{y}'_{ij}$ = total line-charging admittance

The real and reactive power flow from bus i to bus j can be expressed as

$$\mathbf{S}_{ij} = P_{ij} + jQ_{ij} = \mathbf{V}_i\mathbf{I}^*_{ij} \tag{7.24}$$

By substituting Equation 7.23 into Equation 7.24,

$$\mathbf{S}_{ij} = P_{ij} + jQ_{ij} = \mathbf{V}_i(\mathbf{V}^*_i - \mathbf{V}^*_j)\mathbf{y}^*_{ij} + \mathbf{V}_i\mathbf{V}^*_i\left(\frac{\mathbf{y}^*_{ij}}{2}\right)^* \tag{7.25}$$

Alternatively, the real and reactive power flow from bus j to bus i can be expressed as

$$\mathbf{S}_{ji} = P_{ji} + jQ_{ji} = \mathbf{V}_j(\mathbf{V}^*_j - \mathbf{V}^*_i)\mathbf{y}^*_{ij} + \mathbf{V}_j\mathbf{V}^*_j\left(\frac{\mathbf{y}'^*_{ij}}{2}\right)^* \tag{7.26}$$

Note that, the line plus transformers can also be represented by the series and the two shunt admittances.

7.9 APPLICATION OF ACCELERATION FACTORS

Sometimes the number of iterations required to converge can be significantly reduced by application of a so-called *acceleration factor*. The correction in voltage from $\mathbf{V}_i^{(k)}$ to $\mathbf{V}_i^{(k+1)}$ is multiplied by such a factor in order to bring the new voltage closer to its final value. Therefore, the accelerated new voltage can be expressed as

$$\begin{aligned} V_{i(\text{acceleration})^{(k+1)}} &= \mathbf{V}_i^{(k)} + \alpha(\mathbf{V}_i^{(k+1)} - \mathbf{V}_i^{(k)}) \\ &= \mathbf{V}_i^{(k)} + \alpha \cdot \Delta\mathbf{V}_i^{(k)} \end{aligned} \tag{7.27}$$

However, in the selection of a factor, there is nothing to guarantee a fast convergence. However, at the same time, numerous studies have been made to determine the optimum values of acceleration factors [276, 277] [5,7,10,60,73]. For example, the results of one study is shown in Figure 7.2. The actual value of α depends on the method of solution and the nature of the network system. However, an a value of about 1.6 is usually used in the $\mathbf{Y}_{\text{bus}}$ Gauss–Seidel iterations.

7.10 SPECIAL FEATURES

Today, most of the commercial power flow programs are equipped with automatic adjustment features either to permit the use of off-nominal bus quantities or to simulate the real-life operation on the power system. For example, a transformer provided with off-nominal taps can be used to regulate the voltage on a given bus to a specified level whether or not that transformer physically is capable of such automatic operation. In general, the most commonly used automatic adjustment features are

1. Automatic load-tap-changing (LTC) transformers for voltage control
2. Automatic phase-shifting transformers
3. Area power interchange control

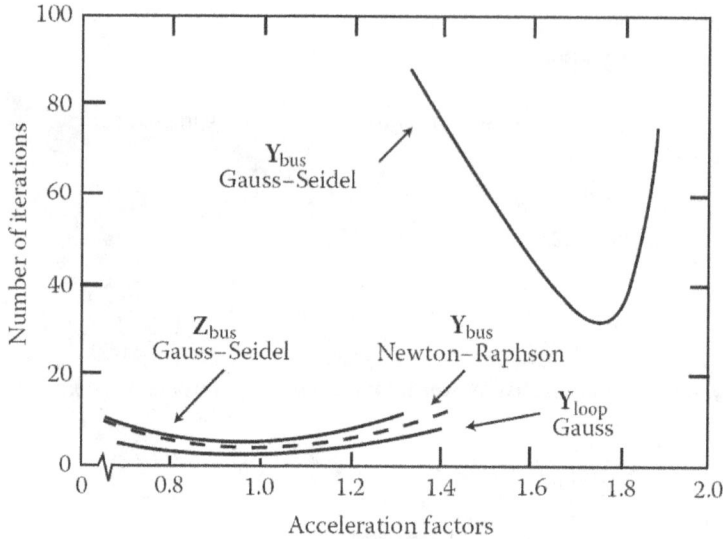

Figure 7.2 Effects of acceleration factors on rate of convergence for power flow solutions. (From Stagg, G. W., and El-Abiad, A. H., *Computer Methods in Power System Analysis*. McGraw-Hill, New York, 1968.)

7.10.1 LTC TRANSFORMERS

The objective of tap changing is to regulate the voltage magnitude at a specific bus, ensuring it remains constant or within certain limits. To achieve this, an automatic LTC (Load Tap Changer) transformer may be connected to the bus. When running the power flow program, it is possible to employ a single tap setting without explicitly specifying the load voltage magnitude.

If the voltage magnitude determined by the power flow program exceeds the given limits, a new tap setting can be selected for the subsequent run. Generally, when representing a manual tap-changing transformer using the automatic tap-changing feature, the power flow program output will indicate the tap setting that achieves the required bus voltage. It is important to note that adjusting the tap setting or ratio will alter the system's admittance/impedance matrix. Therefore, after each tap ratio adjustment, the $\mathbf{Y}_{bus}$ admittance matrix or the $\mathbf{Z}_{bus}$ impedance matrix needs to be updated.

An alternative approach for considering the LTC transformer is to represent it by its impedance or admittance connected in series with an ideal autotransformer, as depicted in Figure 7.3a. This representation can be used to develop an equivalent π circuit, as shown in Figure 7.3b, which can be employed in power flow studies.

7.10.2 PHASE-SHIFTING TRANSFORMERS

Automatic phase-shifting transformers are utilized for power control in a circuit connected in series with the transformer. However, in cases where such equipment is absent in a system, the automatic phase-shifting capability of the power flow program can be employed to determine the required phase shift for a specific power flow in the series circuit. It is important to note that the effect of a phase-shifting transformer can be obtained by replacing a with $\mathbf{a}e^{j\phi}$, where $|\mathbf{a}|$ is unity.

7.10.3 AREA POWER INTERCHANGE CONTROL

When conducting a power flow study for interconnected power systems or areas, the obtained results must fulfill the specified requirements for net power interchange in each area. For instance, power flow studies under normal operating conditions often necessitate a predetermined amount of power

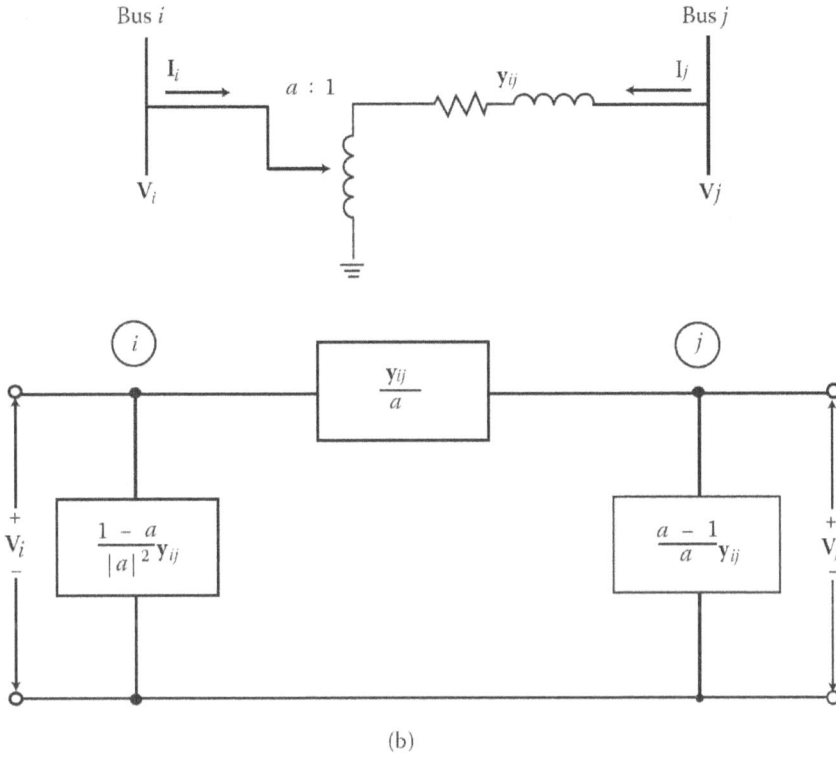

Figure 7.3   LTC transformer representations: (a) One-line circuit; (b) its equivalent circuit.

to be transmitted out of a particular area. Power flow programs equipped with area interchange control incorporate provisions to maintain a constant power outflow from the designated area by regulating the power output of a specific generator, known as the "regulating generator."

After each power flow solution, the output of the regulating generator is adjusted based on the calculated actual net power interchange, which involves algebraically summing the tie-line flows. The adjustment of the regulating generator output is determined by the following equation:

$$P_{(reg)}^{(k+1)} = P_{reg}^{(k)} + \Delta P_T^{(k)} \tag{7.28}$$

where

$$\Delta P_{(T)}^{(k)} = P_{T(sch)} + \Delta P_T^{(k)} \tag{7.29}$$

Thus,

$$P_{(reg)}^{(k+1)} = P_{reg}^{(k)} + \left[P_{T(sch)} - P_T^{(k)} \right] \tag{7.30}$$

where
$P_{(reg)}^{(k+1)}$ = new estimate of power output for regulating generator for the $(k + 1)$th iteration
$P_{reg}^{(k)}$ = power output of regulating generator adjusted for the kth iteration
$P_{T(sch)}$ = scheduled net tie-line flow (or power interchange)
$P_{(T)}^{(k)}$ = actual net power interchange
$\Delta P_{(T)}^{(k)}$ = difference between actual and scheduled power interchanges

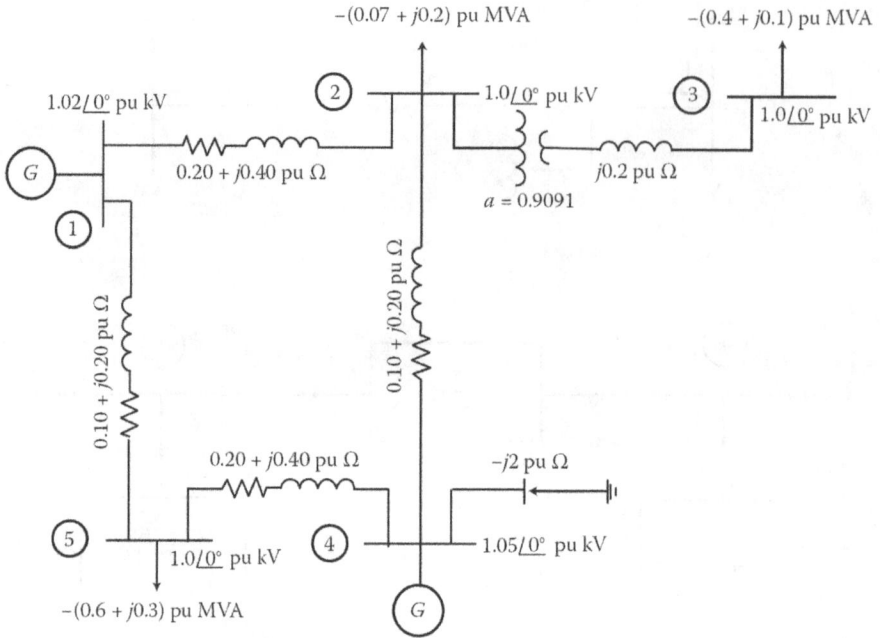

Figure 7.4 Five-bus system for Example 8.1.

The final solution (after several iterations) is obtained when $\Delta P_{(T)}^{(k)}$ becomes less than the specified tolerance limit.

Example 8.1: The one-line diagram of the IEEE five-bus power system, as presented in Figure 7.4, will be analyzed. In this system configuration, the slack bus is designated as generator bus 1, where a constant voltage magnitude of 1.02 pu and an angle of 0° are maintained. Generator bus 4 has a fixed voltage magnitude of 1.05 pu and a specified real power of 1.0 pu, based on a megavolt-ampere (MVA) base of 1.0. The remaining buses in the system are connected to inductive loads, with the load quantities expressed in terms of P and Q and noted as negative to indicate the presence of inductive loads. The initial voltage values for all load buses are set at $1.0\angle 0°$, as shown in Figure 7.4. Additionally, there is a capacitor bank connected to bus 4.

To determine the solution, the Gauss-Seidel method will be employed. By applying this method, the following can be established:

(a) Bus admittance matrix of system
(b) Value of voltage V_2 for the first iteration

Solution:

(a) The primitive branch admittance can be found as

$$y_{12} = \frac{1}{z_{12}}$$

$$= \frac{1}{0.20 + j0.40}$$

$$= 1.0 - j2.0 \text{ pu}$$

$$y_{15} = \frac{1}{z_{15}}$$
$$= \frac{1}{0.10 + j0.20}$$
$$= 2.0 - j4.0 \text{ pu}$$

$$y_{24} = \frac{1}{z_{24}}$$
$$= \frac{1}{0.10 + j0.20}$$
$$= 2.0 - j4.0 \text{ pu}$$

Therefore, the off-diagonal elements of the bus admittance matrix can be expressed as

$$\mathbf{Y}_{12} = \mathbf{Y}_{21}$$
$$= -\mathbf{y}_{12}$$
$$= -1.0 + j2.0 \text{ pu}$$

$$\mathbf{Y}_{15} = \mathbf{Y}_{51}$$
$$= -\mathbf{y}_{15}$$
$$= -2.0 + j4.0 \text{ pu}$$

$$\mathbf{Y}_{23} = \mathbf{Y}_{32}$$
$$= -\mathbf{y}_{23}$$
$$= 0 + j5.5 \text{ pu}$$

$$\mathbf{Y}_{24} = \mathbf{Y}_{42}$$
$$= -\mathbf{y}_{24}$$
$$= -2.0 + j4.0 \text{ pu}$$

$$\mathbf{Y}_{45} = \mathbf{Y}_{54}$$
$$= -\mathbf{y}_{45}$$
$$= -1.0 + j2.0 \text{ pu}$$

The diagonal elements of the bus admittance matrix can be found as

$$\mathbf{Y}_{11} = \mathbf{Y}_{12} + \mathbf{Y}_{15}$$
$$= 3.0 - j6.0$$

$$\mathbf{Y}_{22} = (1.0 - j2.0) + \frac{1}{j0.2} \times \frac{1}{0.9091^2} + (2.0 - j4.0)$$
$$= 3.0 - j12.05$$

$$\mathbf{Y}_{33} = \frac{1}{j0.2}$$
$$= -j5.0$$

$$\mathbf{Y}_{44} = (1.0 - j2.0) + (2.0 - j4.0) + \frac{1}{-j2}$$
$$= 3.0 - j5.5$$

$$\mathbf{Y}_{55} = (2.0 - j4.0) + (1.0 - j2.0)$$
$$= 3.0 - j6.0$$

Therefore, the bus admittance matrix of the system can be expressed as

$$[\mathbf{Y}_{bus}] = \begin{bmatrix} \mathbf{Y}_{11} & \mathbf{Y}_{12} & \mathbf{Y}_{13} & \mathbf{Y}_{14} & \mathbf{Y}_{15} \\ \mathbf{Y}_{21} & \mathbf{Y}_{22} & \mathbf{Y}_{23} & \mathbf{Y}_{24} & \mathbf{Y}_{25} \\ \mathbf{Y}_{31} & \mathbf{Y}_{32} & \mathbf{Y}_{33} & \mathbf{Y}_{34} & \mathbf{Y}_{35} \\ \mathbf{Y}_{41} & \mathbf{Y}_{42} & \mathbf{Y}_{43} & \mathbf{Y}_{44} & \mathbf{Y}_{45} \\ \mathbf{Y}_{51} & \mathbf{Y}_{52} & \mathbf{Y}_{53} & \mathbf{Y}_{54} & \mathbf{Y}_{55} \end{bmatrix}$$

$$= \begin{bmatrix} (3-j6) & (-1+j2) & (0+j0) & (0+j0) & (-2+j4) \\ (-1+j2) & (3-j12.05) & (0+j5.5) & (-2+j4) & (0+j0) \\ (0+j0) & (0+j5.5) & (0-j5) & (0+j0) & (0+j0) \\ (0+j0) & (-2+j4) & (0+j0) & (3-j5.5) & (-1+j2) \\ (-2+j4) & (0+j0) & (0+j0) & (-1+j2) & (3-j6) \end{bmatrix}$$

(b) From Equation 7.15, the value of voltage $\mathbf{V}_2$ for the first iteration can be found as

$$\mathbf{V}_2^{(1)} = \frac{1}{\mathbf{Y}_{22}} \left(\frac{P_2 - jQ_2}{\mathbf{V}_2^{(0)*}} - \sum_{j=1, j\neq 2}^{5} \mathbf{Y}_{2j} \mathbf{V}_j^0 \right)$$

$$= \frac{1}{\mathbf{Y}_{22}} \left(\frac{P_2 - jQ_2}{\mathbf{V}_2^{(0)*}} - \left\{ \mathbf{V}_1^{(0)} \mathbf{Y}_{21} + \mathbf{V}_3^{(0)} \mathbf{Y}_{23} + \mathbf{V}_4^{(0)} \mathbf{Y}_{24} + \mathbf{V}_5^{(0)} \mathbf{Y}_{25} \right\} \right)$$

$$= \frac{1}{3 - j12.05} \left[\frac{-0.7 + j(0.2)}{1.0} - \{ 1.02(-1+j2) + 1.0(0+j5.5) + 1.05(-2+j4) + 1.0(0+j0) \} \right]$$

$$= \frac{1}{3 - j12.05} \left[\frac{-0.7 + j(0.2)}{1.0} - (-3.12 + j11.74) \right]$$

$$= \frac{1}{3 - j12.05} [2.42 - j11.54]$$

$$= 0.9494 \angle -2.1366°$$

$$= 0.94886 - j0.035 \text{ pu}$$

It is better to recompute $\mathbf{V}_2$ from Equation 7.15 by using the corrected value of $\mathbf{V}_2^{(1)*}$. Therefore,

$$\mathbf{V}_2^{(1)*} = 0.9494 \angle -2.1366°$$
$$= 0.94886 - j0.035 \text{ pu}$$

Thus,

$$
\begin{aligned}
\mathbf{V}_2^{(2)} &= \frac{1}{\mathbf{Y}_{22}} \left(\frac{P_2 - jQ_2}{\mathbf{V}_2^{(1)*}} - \left\{ \mathbf{V}_1^{(0)}\mathbf{Y}_{21} + \mathbf{V}_3^{(0)}\mathbf{Y}_{23} + \mathbf{V}_4^{(0)}\mathbf{Y}_{24} + \mathbf{V}_5^{(0)}\mathbf{Y}_{25} \right\} \right) \\
&= \frac{1}{3 - j12.05} \left[\frac{-0.7 + j(0.2)}{0.9495\angle 2.1366°} - \{-3.12 + 11.74\} \right] \\
&= \frac{1}{3 - j12.05} [2.3911 - j11.502] \\
&\cong 0.946\angle - 2.2364° \\
&\cong 0.94533 - j0.0369 \text{ pu}
\end{aligned}
$$

7.11   APPLICATION OF GAUSS-SEIDAL METHOD USING Z_{BUS}

Instead of using the bus admittance matrix, the Gauss-Seidel iterative method can also be employed to solve the power flow problem using the bus impedance matrix [199,201]. In this approach, Equation 7.3 can be directly solved for the bus voltages in relation to the bus currents by utilizing the inverse of the bus admittance matrix. Thus,

$$
\begin{aligned}
\left[\mathbf{V}_{bus}\right] &= \left[\mathbf{V}_{bus}\right]^{-1} \left[\mathbf{I}_{bus}\right] \\
&= \left[\mathbf{Z}_{bus}\right] \left[\mathbf{I}_{bus}\right]
\end{aligned}
\tag{7.31}
$$

where the bus currents can be obtained from Equation 7.5. Therefore, when the bus impedance matrix is known, then for any current vector $[\mathbf{I}^{(k)}]$, it is possible to determine a new voltage vector $[\mathbf{V}^{(k)}]$. After finding the new voltage vector, the associated new current vector can be computed, which in turn provides a new voltage vector.

The iteration process continues until the required voltage correction for each bus is below a predetermined tolerance limit. To enhance convergence characteristics, it is beneficial to introduce a fixed and artificial impedance to ground at each load bus. This approximation helps represent the load based on the estimated voltage at that particular bus. Without such paths to neutral (ground)*, the resulting bus admittance becomes singular, rendering its inverse non-existent.

An alternative approach involves eliminating the equation for the slack bus from Equation 7.31 before inverting the bus admittance matrix. In other words, the swing bus is considered as the reference when determining the bus admittance and impedance matrices. Consequently, the new equation can be formulated as follows:

$$
\left[\mathbf{V}_{bus}\right] = \left[\mathbf{Z}_{bus}\right]^{-1} \left[\mathbf{I}_{bus}\right] + \left[\mathbf{V}_S\right]
\tag{7.32}
$$

where the matrix $\mathbf{V}_s$ is the column vector whose elements are the voltage of the swing bus. Note that, the matrix $\mathbf{Z}$ is different than the one in Equation 7.31 and matrices $\mathbf{V}$ and $\mathbf{I}$ do not have $\mathbf{V}_s$ and $\mathbf{I}_s$, respectively. The iteration process involved in the Gauss–Seidel $\mathbf{Z}_{bus}$ method is similar to the one in the Gauss–Seidel $\mathbf{Y}_{bus}$ method. However, the fundamental Equations 7.5, 7.15, and 7.18 are, respectively, modified as

$$
\mathbf{I}_i = \frac{P_i - jQ_i}{V_i^*} - \mathbf{y}\mathbf{V}_i \text{ for } i = 1, 2, \cdots, n, \quad i \neq s
\tag{7.33}
$$

$$
\mathbf{V}_i^{(k+1)} = \sum_{\substack{j=1 \\ j \neq i}}^{n} \mathbf{Z}_{ij} \left(\frac{P_i - jQ_i}{V_i^{(k)*}} - \mathbf{y}\mathbf{V}_i^{(k)} \right) + \mathbf{V}_s \text{ for } i = 2, 3, \cdots, n
\tag{7.34}
$$

$$Q_i = -\text{Im}\left[\frac{\mathbf{v}_i^{(k)*}}{\mathbf{z}_{ii}}\right]\left(\mathbf{V}_i^{(k)} - \mathbf{V}_s - \sum_{j=1_{j\neq i}}^{n} \mathbf{Z}_{ij}\frac{P_i - jQ_i}{V_i^{(k)*}}\right) \qquad (7.35)$$

While the method exhibits reliable convergence characteristics, it lacks the advantage of iterative bus admittance methods in terms of storage and speed, particularly when dealing with larger systems.

7.12 NEWTON-RAPHSON METHOD

Let's consider a given equation that involves a single variable as follows:

$$f(x) = 0 \qquad (7.36)$$

Since any function of x can be expressed as a power series, the given function can be expanded by Taylor's series about a particular point x_0 as

$$f(x) = f(x_0) + \frac{1}{1!}\frac{df(x_0)}{dx}(x - x_0) + \frac{1}{2!}\frac{df^2(x_0)}{dx^2}(x - x_0)^2 + \cdots + \frac{1}{n!}\frac{df^n(x_0)}{dx^n}(x - x_0)^n = 0 \quad (7.37)$$

If the terms beyond the first derivative are dropped (i.e., assuming convergence after the first two terms), a linear approximation results as

$$f(x) = f(x_0) + \frac{df(x_0)}{dx}(x - x_0) = 0 \qquad (7.38)$$

from which

$$x_1 = x_0 - \frac{f(x_0)}{df(x_0)/dx} \qquad (7.39)$$

However, to prevent any confusion with respect to notation, it is customary to reexpress Equation 7.39 as

$$x^{(1)} = x^{(0)} - \frac{f(x^{(0)})}{df(x^{(0)})/dx} \qquad (7.40)$$

where
$x^{(0)} =$ initial approximation (or estimate)
$x^{(1)} =$ first approximation

Therefore, a recursion formula can be developed so that at the end of the $(k + 1)$th iteration,

$$x^{(k+1)} = x^{(k)} - \frac{f(x^{(k)})}{df(x^{(k)})/dx} \qquad (7.41)$$

or, alternatively,

$$x^{(k+1)} = x^{(k)} - \frac{f(x^{(k)})}{f'(x^{(k)})} \qquad (7.42)$$

Thus,

$$\Delta x = -\frac{f(x^{(k)})}{f'(x^{(k)})} \qquad (7.43)$$

where

$$\Delta x = x^{k+1} - x^{(k)} \qquad (7.44)$$

Therefore, Δx (the amount by which $x^{(k)}$ needs to be modified) can be determined by substituting the value of $x^{(k)}$ into $f(x)$ and $f'(x)$, as shown in Equation 7.43. Thus, an iterative solution process is established.

The method can easily be extended to multivariable nonlinear equations in the following manner. Assume that a set of n nonlinear equations with n unknowns is given as

$$f_1(x_1,x_2,x_3,\cdots,x_n) = 0$$
$$f_2(x_1,x_2,x_3,\cdots,x_n) = 0 \qquad (7.45)$$
$$n(x_1,x_2,x_3,\cdots,x_n) = 0$$

or, in matrix notation,

$$F(x) = 0 \qquad (7.46)$$

where

$$[x] = \begin{bmatrix} x_1 \\ x_2 \\ x_3 \\ \cdot \\ \cdot \\ x_n \end{bmatrix} \qquad (7.47)$$

Here, the objective is to solve for the x values expressed in Equation 7.47 so that

$$f(x) = [0] \qquad (7.48)$$

As has been done previously, if the given functions are expanded according to Taylor's series and the terms beyond the first derivatives are dropped,

$$F(x) = F(x^{(x)}) + [j(x^0)][x - x^{(0)}] = 0 \qquad (7.49)$$

where the coefficient matrix in Equation 7.49 is called the *Jacobian matrix* and can be expressed as

$$[J(x)] \triangleq \begin{bmatrix} \dfrac{\partial f_1}{\partial x_1} & \dfrac{\partial f_1}{\partial x_2} & \cdots & \dfrac{\partial f_1}{\partial x_n} \\ \dfrac{\partial f_2}{\partial x_1} & \dfrac{\partial f_2}{\partial x_2} & \cdots & \dfrac{\partial f_2}{\partial x_n} \\ \vdots & \vdots & \vdots & \vdots \\ \dfrac{\partial f_n}{\partial x_1} & \dfrac{\partial f_n}{\partial x_2} & \cdots & \dfrac{\partial f_n}{\partial x_n} \end{bmatrix} \qquad (7.50)$$

Therefore, from Equation 7.49,

$$[x^{(k+1)}] = [x^{(k)}] - [J(x^{(x)})]^{-1}[F(x^k)] \qquad (7.51)$$

or

$$[\Delta x] = [x^{k+1}] - [x^{(k)}]$$
$$= [J(x^{(x)})]^{-1}[F(x^k)] \qquad (7.52)$$

Example 8.2: Assume as given the nonlinear equations

$$f_1(x_1,x_2) = x_1^2 + 3x_1x_2 - 4 = 0$$

and

$$f_2(x_1,x_2) = x_1x_2 - 2x_2^2 + 5 = 0$$

Determine the values of x_1 and x_2 by using the Newton–Raphson method.

Solution:

The Jacobian matrix is

$$[J(x)] = \begin{bmatrix} \dfrac{\partial f_1}{\partial x_1} & \dfrac{\partial f_1}{\partial x_2} \\ \dfrac{\partial f_2}{\partial x_1} & \dfrac{\partial f_2}{\partial x_2} \end{bmatrix}$$

$$= \begin{bmatrix} 2x_1 + 3x_2 & 3x_1 \\ x_2 & x_1 - 3x_2 \end{bmatrix}$$

As the initial approximation, let

$$[x^{(0)}] = \begin{bmatrix} x_1^{(0)} \\ x_2^{(0)} \end{bmatrix}$$

$$= \begin{bmatrix} 1 \\ 2 \end{bmatrix}$$

Therefore, the first iteration can be performed as

$$[x^{(1)}] = [x^{(0)}] - \begin{bmatrix} 8 & 3 \\ 2 & -7 \end{bmatrix}^{-1} [f(x^{(0)})]$$

$$= \begin{bmatrix} 1 \\ 2 \end{bmatrix} - \begin{bmatrix} \frac{7}{62} & \frac{3}{62} \\ \frac{2}{62} & -\frac{8}{62} \end{bmatrix} \begin{bmatrix} 3 \\ -1 \end{bmatrix}$$

$$= \begin{bmatrix} 0.7741 \\ 1.7742 \end{bmatrix}$$

Thus, after substituting the new values for the second iteration,

$$[x^{(2)}] = \begin{bmatrix} 0.7097 \\ 1.7742 \end{bmatrix} - \begin{bmatrix} 6.7419 & 2.1290 \\ 1.7742 & -6.3871 \end{bmatrix}^{-1} \begin{bmatrix} 0.2810 \\ -0.0364 \end{bmatrix}$$

$$= \begin{bmatrix} 0.6730 \\ 1.7583 \end{bmatrix}$$

Similarly, after substituting the new values for the third iteration,

$$[x^{(3)}] = \begin{bmatrix} 0.6730 \\ 1.7583 \end{bmatrix} - \begin{bmatrix} 6.6210 & 2.0191 \\ 1.7383 & -6.3602 \end{bmatrix}^{-1} \begin{bmatrix} 0.0031 \\ 0.0001 \end{bmatrix}$$

$$= \begin{bmatrix} 0.6726 \\ 1.7582 \end{bmatrix}$$

Finally, after substituting for the last iteration,

$$[x^{(4)}] = \begin{bmatrix} 0.6726 \\ 1.7582 \end{bmatrix} - \begin{bmatrix} 6.6198 & 2.0178 \\ 1.7582 & -6.3602 \end{bmatrix} \begin{bmatrix} 0.6726 \\ 1.7582 \end{bmatrix}$$

$$= \begin{bmatrix} 0.6726 \\ 1.7582 \end{bmatrix}$$

Therefore, it is obvious that the iterations have rapidly converged toward the results of $x_1^{(4)} = 0.6726$ and $x_2^{(4)} = 1.7582$.

7.13   APPLICATION OF NEWTON-RAPHSON METHOD

The Newton-Raphson method is known for its high reliability and rapid convergence, particularly when efficient sparsity programming techniques are employed. It is less sensitive to factors that may hinder or prevent convergence in other power flow methods, such as the selection of the slack bus, the presence of series capacitors, or negative resistances.* Additionally, the convergence rate of the Newton-Raphson method remains relatively unaffected by the size of the system. Both rectangular and polar coordinates can be utilized for representing the bus voltages [206]. The Newton-Raphson method utilizes the bus admittance matrix in its calculations.

7.13.1   POWER FLOW FORMULATION IN RECTANGULAR COORDINATES

Similar to the previous approach, the slack bus, which has a known voltage magnitude and phase angle, is excluded from the iteration process. Consequently, the power at bus i in an n-bus system can be represented as follows:

$$S_i = P_i - jQ_i = \mathbf{V}_i^* \mathbf{I}_i \tag{7.53}$$

or

$$S_i = P_i - jQ_i = \mathbf{V}_i^* \sum_{j=1}^{n} \mathbf{Y}_{ij} \mathbf{V}_j \tag{7.54}$$

Let

$$\mathbf{V}_i \triangleq e_i + jf_i \tag{7.55}$$

$$\mathbf{Y}_{ij} \triangleq G_{ij} + jB_{ij} \tag{7.56}$$

$$\mathbf{I}_i = \sum_{j=1}^{n} \mathbf{Y}_{ij} \mathbf{V}_{ij} \triangleq c_i + jd_i \tag{7.57}$$

where $\mathbf{I}_i$ is defined as the current flowing into bus i. Therefore, after the appropriate substitutions, Equation 7.54 can be reexpressed as

$$P_i - jQ_i = (e_i - jf_i) \sum_{j=1}^{n} (G_{ij} - jB_{ij})(e_j + jf_j) \tag{7.58}$$

or

$$P_i - jQ_i = (e_i - jf_i) \sum_{j=1}^{n} (G_{ij}e_j + jB_{ij}f_j) + j(G_{ij}f_j - B_{ij}e_j) \tag{7.59}$$

Therefore, it can be shown that

$$P_i = \sum_{j=1}^{n} [(e_i G_{ij} e_j + jB_{ij}f_j) + f_i(G_{ij}f_j - B_{ij}e_j)] \tag{7.60}$$

and

$$Q_i = \sum_{j=1}^{n} [(f_i G_{ij} e_j + jB_{ij}f_j) + e_i(G_{ij}f_j - B_{ij}e_j)] \tag{7.61}$$

Note that, P_i and Q_i are functions of e_i, e_j, f_i, and f_j. For each PQ load bus, P_i and Q_i can be calculated from Equations 7.60 and 7.61, respectively, for some estimated values of e and f. After each iteration, the calculated values of $P_{i,\text{calc}}$ and $Q_{i,\text{calc}}$ are compared against the known (or specified) values of $P_{j,\text{spec}}$ and $Q_{j,\text{spec}}$. Similarly, for each PV generator bus, the magnitude of the bus voltage can be calculated from the estimated values of e and f as

$$|\mathbf{V}_i|^2 = e_j^2 + f_i^2 \tag{7.62}$$

Then, the calculated voltage magnitude is compared with its specified value. Thus, the corrected values for the k-th iteration can be expressed as

$$\Delta P_i^{(k)} = P_{i,\text{spec}} - P_{(k)i,\text{calc}} \tag{7.63}$$

$$\Delta Q_i^{(k)} = Q_{i,\text{spec}} - Q_{(k)i,\text{calc}} \tag{7.64}$$

$$\Delta |\mathbf{V}_i|^2 = |\mathbf{V}_{i,\text{spec}}|^2 - |\mathbf{V}_{i,\text{calc}}|^2 \tag{7.65}$$

The resultant values of $\Delta P_i^{(k)}$, $\Delta Q_i^{(k)}$, and $\Delta |\mathbf{V}_i|^2$ can be used to determine the changes in the real and imaginary components of the bus voltages from Equation 7.66. Since the changes in P, Q, and $\mathbf{V}^2$ are related to the changes in e and f by Equations 7.62 through 7.62, it is possible to express them in a general form as

$$
\begin{bmatrix}
\Delta P_2^{(k)} \\
\vdots \\
\Delta P_n^{(k)} \\
\Delta Q_n^{(k)} \\
\vdots \\
\Delta Q_n^{(k)} \\
\Delta |\mathbf{V}_n^{(k)}|^2
\end{bmatrix}
=
\begin{bmatrix}
\frac{\partial P_2}{\partial e_2} & \cdots & \frac{\partial P_2}{\partial e_2} & \frac{\partial P_2}{\partial f_2} & \cdots & \frac{\partial P_2}{\partial e_n} \\
\vdots & & \vdots & \vdots & & \vdots \\
\frac{\partial P_n}{\partial e_2} & \cdots & \frac{\partial P_n}{\partial e_n} & \frac{\partial P_2}{\partial f_2} & \cdots & \frac{\partial P_2}{\partial f_n} \\
\frac{\partial Q_2}{\partial e_2} & \cdots & \frac{\partial Q_2}{\partial e_n} & \frac{\partial Q_2}{\partial f_2} & \cdots & \frac{\partial Q_2}{\partial f_n} \\
\vdots & & \vdots & \vdots & & \vdots \\
\frac{\partial Q_{n-1}}{\partial e_2} & \cdots & \frac{\partial Q_{n-1}}{\partial e_n} & \frac{\partial Q_{n-1}}{\partial f_2} & \cdots & \frac{\partial Q_{n-1}}{\partial f_n} \\
\frac{\partial |\mathbf{V}_n|^2}{\partial e_2} & \cdots & \frac{\partial |\mathbf{V}_n|^2}{\partial e_n} & \frac{\partial |\mathbf{V}_n|^2}{\partial f_2} & \cdots & \frac{\partial |\mathbf{V}_n|^2}{\partial f_n}
\end{bmatrix}
\begin{bmatrix}
\Delta e_2^{(k)} \\
\Delta e_n^{(k)} \\
\Delta f_2^{(k)} \\
\vdots \\
\Delta f_{n-1}^{(k)} \\
\Delta f_n^{(k)}
\end{bmatrix}
\tag{7.66}
$$

In Equation 8.66, the coefficient matrix of partial derivatives is called the Jacobian. Thus, in matrix form, Equation 8.66 can be expressed as

$$
\begin{bmatrix}
\Delta P \\
\hline
\Delta Q \\
\hline
\Delta |V|^2
\end{bmatrix}
=
\begin{bmatrix}
J_1 & J_2 \\
\hline
J_3 & J_4 \\
\hline
J_5 & J_6
\end{bmatrix}
\times
\begin{bmatrix}
\Delta e \\
\hline
\Delta f
\end{bmatrix}
\tag{7.67}
$$

Please note that it is assumed that bus 1 serves as the slack bus, while bus n is the PV generator (or voltage-controlled) bus. Equation 7.67 involves the unknowns, which are the changes (or corrections) in the real and imaginary components of the voltages, denoted as Δe's and Δf's. It can be observed from Equation 7.66 that two equations are required for each bus (excluding the slack bus and reference bus) to incorporate both real and imaginary terms. All elements of the Jacobian matrix are functions of Δe's and Δf's. Thus, they can be computed by substituting the initial assumed values for the first iteration or calculated in the final iteration using the partial derivative equations. Consequently, the unknown values of Δe's and Δf's can be determined from Equation 7.66 by inverting the Jacobian matrix. The obtained values of Δe's and Δf's can then be utilized in the following equation to determine the new voltage estimates for the buses in the next iteration:

$$e_i^{(k+1)} = e_i^{(k)} + \Delta e_i^{(k)} \tag{7.68}$$

$$f_i^{(k+1)} = f_i^{(k)} + \Delta f_i^{(k)} \tag{7.69}$$

The process is repeated until the values of $\Delta P, \Delta Q$, and $\Delta |\mathbf{V}|^2$ determined from Equations 7.63 through 7.65 are less than the specified precision indexes.

It is possible to develop a convenient expression for bus current given by Equation 7.57 as

$$\mathbf{I}_i = \sum_{j=1}^{n} \mathbf{Y}_{ij} \mathbf{V}_j = \sum_{j=1}^{n} (G_{ij} - jB_{ij})(e_j + jf_i) \tag{7.70}$$

or

$$\mathbf{I}_i = \sum_{j=1}^{n} (e_j G_{ij} + jf_i B_{ij}) + j \sum_{j=1}^{n} (f_i G_{ij} - e_j B_{ij}) \tag{7.71}$$

or

$$\mathbf{I}_i = c_i + jd_i \tag{7.72}$$

where

$$c_i = \sum_{j=1}^{n} (e_j G_{ij} + f_i B_{ij}) \tag{7.73}$$

or

$$d_i = \sum_{j=1}^{n} (f_j G_{ij} - e_j B_{ij}) \tag{7.74}$$

$$c_i = e_i G_{ii} + f_i B_{ii} + \sum_{j=1_{j \neq i}}^{n} (e_j G_{ij} + f_j B_{ij}) \tag{7.75}$$

$$d_i = f_i G_{ii} + e_i B_{ii} + \sum_{j=1_{j \neq i}}^{n} (f_j G_{ij} + e_j B_{ij}) \tag{7.76}$$

From Equation 7.60, the real power of bus i can be expressed as

$$P_i = e_i \sum_{j=1}^{n} (e_j G_{ij} + f_i B_{ij}) + f_i \sum_{j=1}^{n} (f_j G_{ij} - e_j B_{ij}) \tag{7.77}$$

or substituting Equation 7.73 and 7.74 into Equation 7.77,

$$P_i = e_i c_i + f_i d_i \tag{7.78}$$

Similarly, from Equation 7.61, the reactive power of bus i can be expressed as

$$Q_i = f_i \sum_{j=1}^{n} (e_j G_{ij} + f_i B_{ij}) - e_i \sum_{j=1}^{n} (f_j G_{ij} - e_j B_{ij}) \tag{7.79}$$

or

$$Q = f_i c_i - e_i d_i \tag{7.80}$$

Thus, the elements of submatrices of the Jacobian matrix, given in Equation 7.67, can be evaluated for the values of P, Q, and $\mathbf{V}^2$ at each iteration as follows. For submatrix J_1, from Equation 7.77, the off-diagonal and diagonal elements, respectively, can be found as

$$\frac{\partial P_i}{\partial e_j} = e_i G_{ij} - f_i B_{ij} \quad i \neq j \tag{7.81}$$

and

$$\frac{\partial P_i}{\partial e_j} = (e_i G_{ii} - f_i B_{ii}) + c_i \quad i = j \tag{7.82}$$

Similarly, for the matrix J_2,

$$\frac{\partial P_i}{\partial f_j} = e_i B_{ij} + f_i G_{ij} \quad i \neq j \tag{7.83}$$

and

$$\frac{\partial P_i}{\partial f_j} = (e_i B_{ii} + f_i G_{ii}) + d_i \quad i = j \tag{7.84}$$

For submatrix J_3, from Equation 7.79,

$$\frac{\partial Q_i}{\partial e_j} = e_i B_{ij} + f_i G_{ij} \quad i \neq j \tag{7.85}$$

and

$$\frac{\partial Q_i}{\partial e_i} = (e_i B_{ii} + f_i G_{ii}) + d_i \quad i = j \tag{7.86}$$

Similarly, for submatrix J_4

$$\frac{\partial Q_i}{\partial f_j} = -(e_i G_{ij} - f_i B_{ij}) \quad i \neq j \tag{7.87}$$

and

$$\frac{\partial Q_i}{\partial f_j} = -(e_i G_{ii} + f_i B_{ii}) + c_i \quad i = j \tag{7.88}$$

Note the similarity between the elements of submatrices J_1 and J_4 and also between submatrices J_2 and J_3. It can be convenient to define the following expressions for diagonal and off-diagonal elements as

$$T_{ij} = e_i G_{ij} - f_i B_{ij} \quad \text{for all} \quad i, j \tag{7.89}$$

and

$$U_{ij} = e_i B_{ij} + f_i G_{ij} \quad \text{for all} \quad i, j \tag{7.90}$$

Therefore, the Jacobian matrix can be expressed as

$$\left[\begin{array}{c|c} J_1 & J_2 \\ \hline J_3 & J_4 \end{array} \right] = \left[\begin{array}{c|c} T & U \\ \hline U & -T \end{array} \right] \left[\begin{array}{c|c} c_i & d_i \\ \hline -d_i & c_i \end{array} \right] \tag{7.91}$$

Note that, the off-diagonal elements in the c_i and d_i submatrices are all zero.

The elements of the submatrices for the voltage-controlled bus i can be found from Equation 7.62 similarly. For submatrix J_5, the off-diagonal and diagonal elements, respectively, can be found as

$$\frac{\partial |\mathbf{V}_i|^2}{\partial e_j} = 0 \quad i \neq j \tag{7.92}$$

and

$$\frac{\partial |\mathbf{V}_i|^2}{\partial e_i} = 2e_i \quad i = j \tag{7.93}$$

Similarly, for submatrix J_6,

$$\frac{\partial |\mathbf{V}_i|^2}{\partial f_j} = 0 \quad i \neq j \tag{7.94}$$

and

$$\frac{\partial |\mathbf{V}_i|^2}{\partial f_i} = 2f_i \quad i = j \tag{7.95}$$

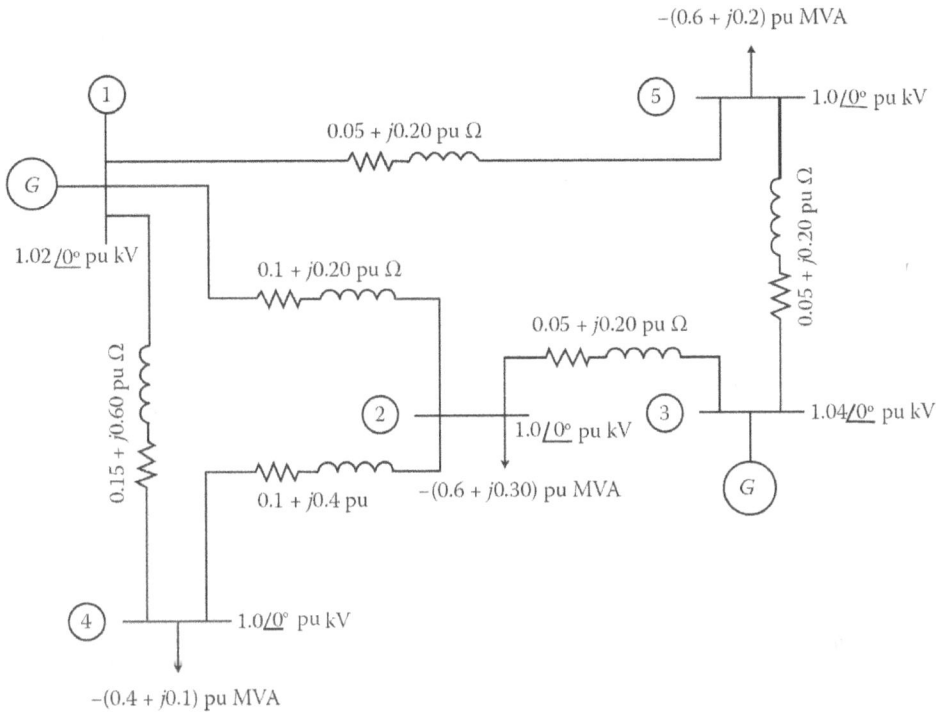

Figure 7.5   Five-bus system for Example 8.3.

Thus, by substituting Equations 7.78 through 7.80 and 7.62 into Equations 7.63 through 7.65, respectively, it is possible to express them as

$$\Delta P_i^{(k)} = P_{i,\text{spec}} - (e_i c_i + f_i d_i) \tag{7.96}$$

$$\Delta Q_i^{(k)} = Q_{i,\text{spec}} - (f_i c_i - e_i d_i) \tag{7.97}$$

$$\Delta |\mathbf{V}_i|^2 = |\mathbf{V}_{p,\text{spec}}|^2 - (e_i^2 + f_i^2) \tag{7.98}$$

Example 8.3: Consider the one-line diagram of a five-bus power system illustrated in Figure 7.5. Generators are connected to buses 1 and 3, with bus 1 designated as the slack bus. Therefore, the voltage magnitude and angle at bus 1, as well as the voltage magnitude at bus 3, will be maintained constant. Buses 2, 4, and 5 are connected to inductive loads, as indicated in the diagram. Note that, since the real and reactive powers are associated with the inductive load buses, they are represented as negative values in the figure. The generator bus (bus 3) has a real power of 1.0 pu based on a megavolt-ampere (MVA) base of 1.0, and its minimum and maximum reactive power limits are 0.0 and 10.0 pu, respectively. We will utilize the Newton-Raphson method in rectangular coordinates to obtain a power flow solution with tolerances of 0.0001 pu for changes in real and reactive bus powers, as well as changes in bus voltages. Please note that the buses shown in Figure 7.5 are not necessary for solving the example, but proper bus ordering can be important for faster convergence in large power systems. Tables 7.2a and 7.3b provide the impedances of the six lines, identified by the connected buses. Table 7.4 contains the real and reactive power values, where positive values indicate power input to the network at each bus, and negative values indicate inductive loads. The voltage values shown in Figure 7.2 represent the initial estimates, with the magnitude and angle of the voltage at bus 3 intended to remain constant. Let the iterative computations begin at bus 2 and determine the value of $\mathbf{V}_2$ for the first iteration.

Table 7.2
Line Impedances

Line (Bus to Bus)	R (Per Unit)	X (Per Unit)
1–2	0.1	0.4
1–4	0.15	0.6
1–5	0.05	0.2
2–3	0.05	0.2
3–5	0.05	0.2

Table 7.3
Power and Voltage Data

Bus	P (Per Unit)	Q (Per Unit)	V (Per Unit)	Remarks
1	–	–	$1.02 \angle 0^0$	Swing bus
2	–0.6	–0.3	$1.00 \angle 0^0$	Load bus (inductive)
3	1	–	$1.00 \angle 0^0$	Voltage bus (inductive)
4	–0.4	–0.1	$1.00 \angle 0^0$	Load bus (inductive)
5	–0.6	–0.2	$1.00 \angle 0^0$	Load bus (inductive)

Table 7.4
Line Admittances and Self- and Mutual Admittances

Line (Bus to Bus)	G (Per Unit)	B (Per Unit)
1–2	0.588235	–2.352941
1–4	0.392157	–1.568627
1–5	1.176471	–4.705882
2–3	1.176471	–4.705882
2–4	0.588235	–2.352941
3–5	1.176471	–4.705882

(a) Determine the ΔP and ΔQ changes for the four buses (2, 3, 4, and 5 after the first iteration).
(b) Determine the Δe and Δf voltage changes for the four buses after the first iteration.
(c) Determine the new voltages for the four buses after the first iteration.
(d) Repeat part (a) for the second iteration.
(e) Repeat part (b) for the second iteration.
(f) Repeat part (c) for the second iteration.
(g) Repeat part (a) for the third iteration.
(h) Repeat part (b) for the third iteration.
(i) Repeat part (c) for the third iteration.
(j) Determine the real and reactive power flows in all transmission lines after the third iteration.
(k) Determine the real and reactive powers summed at each of the five buses after the third iteration.
(l) Determine the total real and reactive power losses in the system after the third iteration.
(m) Determine the bus currents.

Solution:

(a) The computer program output shows that after the first iteration,

$$\Delta \mathbf{S}_2^{(1)} = \Delta P_2^{(1)} + j\Delta Q_2^{(1)} = -0.5411748 + j0.06470662 \text{ pu}$$

$$\Delta \mathbf{S}_3^{(1)} = \Delta P_3^{(1)} + j\Delta Q_3^{(1)} = -0.9021182 + j0.0000000 \text{ pu}$$

$$\Delta \mathbf{S}_4^{(1)} = \Delta P_4^{(1)} + j\Delta Q_4^{(1)} = -0.3921567 - j0.0686270 \text{ pu}$$

$$\Delta \mathbf{S}_5^{(1)} = \Delta P_5^{(1)} + j\Delta Q_5^{(1)} = -0.5294129 - j0.8235341 \text{ pu}$$

(b) The computer program output shows that after the first iteration,

$$\Delta \mathbf{V}_2^{(1)} = \Delta e_2^{(1)} + j\Delta f_2^{(1)} = -0.03651552 - j0.06227919 \text{ pu}$$

$$\Delta \mathbf{V}_3^{(1)} = \Delta e_3^{(1)} + j\Delta f_3^{(1)} = 0.1093983 \times 10^{-6} - j0.004050539 \text{ pu}$$

$$\Delta \mathbf{V}_4^{(1)} = \Delta e_4^{(1)} + j\Delta f_4^{(1)} = 0.06240950 - j0.1263565 \text{ pu}$$

$$\Delta \mathbf{V}_5^{(1)} = \Delta e_5^{(1)} + j\Delta j_5^{(1)} = -0.005154848 - j0.03373529 \text{ pu}$$

(c) The computer program shows that the new bus voltages after the first iteration are

$$\mathbf{V}_2^{(1)} = e_2^{(1)} + jf_2^{(1)} = 0.963485 - j0.06227919 \text{ pu}$$

$$\mathbf{V}_3^{(1)} = e_3^{(1)} + jf_3^{(1)} = 1.040000 + j0.04050539 \text{ pu}$$

$$\mathbf{V}_4^{(1)} = e_4^{(1)} + jf_4^{(1)} = 0.9375905 - j0.1263565 \text{ pu}$$

$$\mathbf{V}_5^{(1)} = e_5^{(1)} + jf_5^{(1)} = 0.9948452 - j0.03373529 \text{ pu}$$

(d) After the second iteration,

$$\Delta \mathbf{S}_2^{(2)} = \Delta P_2^{(2)} + j\Delta Q_2^{(2)} = -0.01451135 - j0.03573412 \text{ pu}$$

$$\Delta \mathbf{S}_3^{(2)} = \Delta P_3^{(2)} + j\Delta Q_3^{(2)} = -0.4950496 \times 10^{-3} - j0.164032 \times 10^{-2} \text{ pu}$$

$$\Delta \mathbf{S}_4^{(2)} = \Delta P_4^{(2)} + j\Delta Q_4^{(2)} = -0.02258435 - j0.0002303781 \text{ pu}$$

$$\Delta \mathbf{S}_5^{(2)} = \Delta P_5^{(2)} + j\Delta Q_5^{(2)} = -0.005330801 - j0.01714361 \text{ pu}$$

(e) After the second iteration,

$$\Delta \mathbf{V}_2^{(2)} = \Delta e_2^{(2)} + j\Delta f_2^{(2)} = -0.01077833 - j0.003310017 \text{ pu}$$

$$\Delta \mathbf{V}_3^{(2)} = \Delta e_3^{(2)} + j\Delta f_3^{(2)} = -0.6285994 \times 10^{-3} - j0.4108835 \times 10^{-2} \text{ pu}$$

$$\Delta \mathbf{V}_4^{(2)} = \Delta e_4^{(2)} + j\Delta f_4^{(2)} = -0.02258435 - j0.002303781 \text{ pu}$$

$$\Delta \mathbf{V}_5^{(2)} = \Delta e_5^{(2)} + j\Delta j_5^{(2)} = -0.02374709 - j0.002142731 \text{ pu}$$

(f) After the second iteration,

$$\mathbf{V}_2^{(2)} = e_2^{(2)} + j f_2^{(2)} = 0.9527062 - j0.06558919 \text{ pu}$$

$$\mathbf{V}_3^{(2)} = e_3^{(2)} + j f_3^{(2)} = 1.039371 + j0.03639655 \text{ pu}$$

$$\mathbf{V}_4^{(2)} = e_4^{(2)} + j f_4^{(2)} = 0.9150062 - j0.1286603 \text{ pu}$$

$$\mathbf{V}_5^{(2)} = e_5^{(2)} + j f_5^{(2)} = 0.9924704 - j0.03587802 \text{ pu}$$

(g) After the third iteration,

$$\Delta \mathbf{S}_2^{(3)} = \Delta P_2^{(3)} + j\Delta Q_2^{(3)} = -0.4726648 \times 10^{-4} - j0.5348325 \times 10^{-3} \text{ pu}$$

$$\Delta \mathbf{S}_3^{(3)} = \Delta P_3^{(3)} + j\Delta Q_3^{(3)} = -0.2393723 \times 10^{-3} - j0.1621246 \times 10^{-4} \text{ pu}$$

$$\Delta \mathbf{S}_4^{(3)} = \Delta P_4^{(3)} + j\Delta Q_4^{(3)} = -0.2399683 \times 10^{-3} - j0.1458764 \times 10^{-2} \text{ pu}$$

$$\Delta \mathbf{S}_5^{(3)} = \Delta P_5^{(3)} + j\Delta Q_5^{(3)} = -0.2741814 \times 10^{-4} - j0.6151199 \times 10^{-4} \text{ pu}$$

(h) After the third iteration,

$$\Delta V_2^{(3)} = \Delta e_2^{(3)} + j\Delta f_2^{(3)} = -0.2121947 \times 10^{-3} - j0.3475162 \times 10^{-4} \text{ pu}$$

$$\Delta V_3^{(3)} = \Delta e_3^{(3)} + j\Delta f_3^{(3)} = -0.4772861 \times 10^{-5} - j0.8642096 \times 10^{-4} \text{ pu}$$

$$\Delta V_4^{(3)} = \Delta e_4^{(3)} + j\Delta f_4^{(3)} = -0.5581642 \times 10^{-3} - j0.1678166 \times 10^{-4} \text{ pu}$$

$$\Delta V_5^{(3)} = \Delta e_5^{(3)} + j\Delta j_5^{(3)} = -0.1013366 \times 10^{-4} - j0.377548 \times 10^{-4} \text{ pu}$$

(i) After the third iteration,

$$V_2^{(3)} = e_2^{(3)} + jf_2^{(3)} = 0.952494 - j0.06562394 \text{ pu}$$

$$V_3^{(3)} = e_3^{(3)} + jf_3^{(3)} = 1.039366 - j0.03631013 \text{ pu}$$

$$V_4^{(3)} = e_4^{(3)} + jf_4^{(3)} = 0.914448 - j0.1286435 \text{ pu}$$

$$V_5^{(3)} = e_5^{(3)} + jf_5^{(3)} = 0.9924603 - j0.03591578 \text{ pu}$$

(J) As can be observed from parts (g), (h), and (i), the power flow problem solution is converged. Therefore, the computer program output gives the real and reactive power flows in all transmission lines after the third iteration as

$$S_{12}^{(3)} = 0.19800 + j0.12264 \text{ MVA}$$

$$S_{14}^{(3)} = 0.24805 + j0.11743 \text{ MVA}$$

$$S_{15}^{(3)} = 0.20544 + j0.08910 \text{ MVA}$$

$$S_{21}^{(3)} = -0.19279 - j0.10179 \text{ MVA}$$

$$S_{23}^{(3)} = -0.57321 - j0.23698 \text{ MVA}$$

$$S_{24}^{(3)} = 0.16600 + j0.03876 \text{ MVA}$$

$$S_{32}^{(3)} = 0.59431 + j0.32138 \text{ MVA}$$

$$S_{35}^{(3)} = 0.40569 + j0.15545 \text{ MVA}$$

$$S_{41}^{(3)} = -0.23719 - j0.07399 \text{ MVA}$$

$$S_{42}^{(3)} = -0.16281 - j0.20601 \text{ MVA}$$

$$S_{51}^{(3)} = -0.20303 - j0.07945 \text{ MVA}$$

$$S_{53}^{(3)} = -0.39697 - j0.12054 \text{ MVA}$$

(k) The real and reactive powers summed at each of the five buses are given by the computer output as

$$\mathbf{S}_1^{(3)} = 0.65149 + j0.32917 \text{ MVA}$$

$$\mathbf{S}_2^{(3)} = -0.60000 - j0.30000 \text{ MVA}$$

$$\mathbf{S}_3^{(3)} = 1.00000 + j0.47683 \text{ MVA}$$

$$\mathbf{S}_4^{(3)} = -0.40000 - j0.10000 \text{ MVA}$$

$$\mathbf{S}_5^{(3)} = -0.40000 - j0.10000 \text{ MVA}$$

(l) The total real and reactive power losses in the system are given as

$$\mathbf{S}_{\text{loss}}^{(3)} = 0.05150 + j0.20600 \text{ MVA}$$

(m) The bus currents are given as

$$\mathbf{I}_2^{(3)} = -0.6052267 + j0.355998 \text{ pu}$$

$$\mathbf{I}_3^{(3)} = 0.9771826 - j0.4235277 \text{ pu}$$

$$\mathbf{I}_4^{(3)} = -0.4135732 + j0.1658478 \text{ pu}$$

$$\mathbf{I}_5^{(3)} = -0.5965176 + j0.2230196 \text{ pu}$$

7.13.2 POWER FLOW FORMULATION IN POLAR COORDINATES

Assume that the basic variables are given in polar coordinates, that is, in terms of magnitude and angles. Let

$$V_i \triangleq |\mathbf{V}_i| \angle \delta_i \tag{7.99}$$

$$V_i \triangleq |\mathbf{V}_i| \angle - \theta_{ij} \tag{7.100}$$

Hence,

$$\mathbf{I}_i = \sum_{j=1}^{n} \mathbf{Y}_{ij} \mathbf{V}_j = \sum_{j=1}^{n} |\mathbf{Y}_{ij}||\mathbf{V}_j| \angle - \theta_{ij} + \delta_j \tag{7.101}$$

Thus,

$$P_i - jQ_i = \mathbf{V}_i^* \mathbf{I}_i = \sum_{j=1}^{n} |\mathbf{V}_i||\mathbf{Y}_{ij}||\mathbf{V}_j| \angle (\theta_{ij} + \delta_i - \delta_j) \tag{7.102}$$

where

$$e^{-j(\theta_{ij} + \delta_i - \delta_j)} \triangleq \cos(\theta_{ij} + \delta_i - \delta_j) - j\sin(\theta_{ij} + \delta_i - \delta_j) \tag{7.103}$$

Hence, the real and reactive powers can be expressed as

$$P_i = \sum_{j=1}^{n} |\mathbf{V}_i||\mathbf{Y}_{ij}||\mathbf{V}_j| \cos(\theta_{ij} + \delta_i - \delta_j) \tag{7.104}$$

and

$$Q_i = \sum_{j=1}^{n} |\mathbf{V}_i||\mathbf{Y}_{ij}||\mathbf{V}_j| \sin(\theta_{ij} + \delta_i - \delta_j) \tag{7.105}$$

It can be shown that the changes in power are related to the changes in voltage magnitudes and phase angles.

7.13.2.1   Method 1. Four Sub-matrices of a Jacobian

$$\begin{bmatrix} \triangle P \\ \hline \triangle Q \end{bmatrix} = \begin{bmatrix} J_1 & J_2 \\ \hline J_3 & J_4 \end{bmatrix} \begin{bmatrix} \triangle \delta \\ \hline \triangle |V| \end{bmatrix} \qquad (7.106)$$

The elements of submatrices of this Jacobian matrix can be found as follows [204–206]. For submatrix J_1, from Equation 8.104, the off-diagonal and diagonal elements, respectively, can be determined as

$$\frac{\partial P_i}{\partial \delta_j} = |\mathbf{V}_i||\mathbf{V}_{ij}||\mathbf{V}_j|\sin(\theta_{ij} + \delta_i - \delta_j) \quad i \neq j \qquad (7.107a)$$

$$\frac{\partial P_i}{\partial \delta_j} = \sum_{j=1}^{n} |\mathbf{V}_i||\mathbf{V}_{ij}||\mathbf{V}_j|\sin(\theta_{ij} + \delta_i - \delta_j) \qquad (7.107b)$$

$$= |\mathbf{V}_i|^2 |\mathbf{Y}_{ii}|\sin\theta_{ii} - Q_i \quad i = j \qquad (7.108)$$

Similarly, for submatrix J_2,

$$\frac{\partial P_i}{\partial |\mathbf{V}_j|} = |\mathbf{V}_i||\mathbf{V}_{ij}||\mathbf{V}_j|\sin(\theta_{ij} + \delta_i - \delta_j) \quad i \neq j \qquad (7.109a)$$

$$\frac{\partial P_i}{\partial |\mathbf{V}_j|} = \sum_{j=1}^{n} |\mathbf{V}_j||\mathbf{V}_{ij}|\sin(\theta_{ij} + \delta_i - \delta_j) + |\mathbf{V}_i||\mathbf{Y}_{ii}|\cos\theta_{ii} \qquad (7.109b)$$

$$= \frac{P_i}{|\mathbf{V}_i|} + |\mathbf{V}_j||\mathbf{Y}_{ij}|\cos\theta_{ii} \quad i = j \qquad (7.110)$$

For submatrix J_3, from Equation 7.105

$$\frac{\partial Q_i}{\partial \delta_j} = |\mathbf{V}_i||\mathbf{V}_{ij}||\mathbf{V}_j|\sin(\theta_{ij} + \delta_i - \delta_j) \quad i \neq j \qquad (7.111a)$$

$$\frac{\partial Q_i}{\partial \delta_j} = \sum_{j=1}^{n} |\mathbf{V}_j||\mathbf{V}_{ij}|\cos(\theta_{ij} + \delta_i - \delta_j) + |\mathbf{V}_i||\mathbf{V}_{ii}|\cos\theta_{ii} \qquad (7.111b)$$

$$= |\mathbf{V}_i|^2 |\mathbf{Y}_{ii}|\sin\theta_{ii} + P_i \quad i = j \qquad (7.112)$$

Similarly, for submatrix J_4,

$$\frac{\partial Q_i}{\partial |\mathbf{V}_j|} = |\mathbf{V}_i||\mathbf{Y}_{ij}|\sin(\theta_{ij} + \delta_i - \delta_j) \quad i \neq j \qquad (7.113a)$$

$$\frac{\partial Q_i}{\partial |\mathbf{V}_j|} = |\mathbf{V}_i||\mathbf{V}_{ii}|\cos\theta_{ii} + \sum_{j=1}^{n} |\mathbf{V}_j||\mathbf{V}_{ij}|\cos(\theta_{ij} + \delta_i - \delta_j) \qquad (7.113b)$$

$$= |\mathbf{V}_i||\mathbf{V}_{ii}|\sin\theta_{ii} + \frac{Q_i}{|\mathbf{V}_i|} \quad i = j \qquad (7.114)$$

In the event that a voltage-controlled bus is present, the general Equation 7.106 has to be modified as

$$\left[\begin{array}{c} \triangle P \\ \hline \frac{\triangle Q}{\triangle |\mathbf{V}|} \end{array} \right] = \left[\begin{array}{c|c} J_1 & J_2 \\ \hline J_3 & J_4 \\ \hline J_5 & J_6 \end{array} \right] \left[\begin{array}{c} \triangle \delta \\ \hline \triangle |\mathbf{V}| \end{array} \right] \qquad (7.115)$$

The elements of the submatrices for the voltage-controlled bus i can be found from the equation

$$|\mathbf{V}_i| = |\mathbf{V}_i| \qquad (7.116)$$

Thus, submatrix J_5,

$$\frac{\partial |\mathbf{V}_i|}{\partial \delta_j} = 0 \quad \text{for all} \quad i, j \qquad (7.117)$$

that is, the J_5 row matrix has only zeros as its elements. Similarly, for submatrix J_6,

$$\frac{\partial |\mathbf{V}_i|}{\partial |\mathbf{V}_j|} = 0 \quad i \neq j \qquad (7.118)$$

and

$$\frac{\partial |\mathbf{V}_i|}{\partial |\mathbf{V}_i|} = 1 \quad i = j \qquad (7.119)$$

that is, the J_5 row matrix also has zeros except one element. Note that, the angle $\triangle \delta$ in Equations 7.106 and 7.115 must be in radians.

It is interesting to observe the similarities between the elements of the submatrices of the Jacobian matrix. Therefore, let

$$\Phi_{ij} \triangleq Q_{ij} + \delta_i - \delta_j \qquad (7.120)$$

$$K_{ij} \triangleq |\mathbf{V}_i||\mathbf{V}_{ij}|\cos\Phi_{ij} = |\mathbf{V}_i|G_{ij} \qquad (7.121)$$

$$L_{ij} \triangleq |\mathbf{V}_i||\mathbf{V}_{ij}|\sin\Phi_{ij} = |\mathbf{V}_i|B_{ij} \qquad (7.122)$$

$$K'_{ij} \triangleq |\mathbf{V}_j|K_{ij} \qquad (7.123)$$

$$L'_{ij} \triangleq |\mathbf{V}_j|K_{ij} \qquad (7.124)$$

Therefore, the elements of each Jacobian submatrix can be redefined. For submatrix J_1, from Equations 7.107 and 7.108, the off-diagonal and diagonal elements, respectively, can be expressed as

$$\frac{\partial P_i}{\partial \delta_j} = L'_{ij} \quad i \neq j \qquad (7.125)$$

$$\frac{\partial P_i}{\partial \delta_i} = L'_{ii} \quad i = j \qquad (7.126)$$

Thus, submatrix J_1 can be expressed as,*

$$[J_1] = [L'] - \left[\begin{array}{ccc} & & 0 \\ & Q_i & \\ 0 & & \end{array} \right] \qquad (7.127)$$

For submatrix J_2,

$$\frac{\partial P_i}{\partial |\mathbf{V}_j|} = K_{ij} \quad i \neq j \qquad (7.128)$$

$$\frac{\partial P_i}{\partial |\mathbf{V}_i|} = K_{ii} + \frac{P_i}{|\mathbf{V}_i|} \quad i = j \tag{7.129}$$

Therefore,

$$[J_2] = [K] + \begin{bmatrix} & & 0 \\ & \dfrac{P_i}{|\mathbf{V}_i|} & \\ 0 & & \end{bmatrix} \tag{7.130}$$

For submatrix J_3,

$$\frac{\partial Q_i}{\partial \delta_j} = -K'_{ij} \quad i \neq j \tag{7.131}$$

$$\frac{\partial P_i}{\partial \delta_j} = -K'_{ij} + P_i \quad i = j \tag{7.132}$$

Thus,

$$[J_3] = -[K'] + \begin{bmatrix} & & 0 \\ & P_i & \\ 0 & & \end{bmatrix} \tag{7.133}$$

For submatrix J_4,

$$\frac{\partial Q_i}{\partial |\mathbf{V}_j|} = L_{ij} \quad i \neq j \tag{7.134}$$

$$\frac{\partial Q_i}{\partial |\mathbf{V}_i|} = L_{ij} + \frac{Q_i}{|\mathbf{V}_i|} \quad i = j \tag{7.135}$$

Hence,

$$[J_4] = [L] + \begin{bmatrix} & & 0 \\ & \dfrac{Q_i}{|\mathbf{V}_i|} & \\ 0 & & \end{bmatrix} \tag{7.136}$$

Thus, the Jacobian matrix can be expressed as

$$\left[\begin{array}{c|c} J_1 & J_2 \\ \hline J_3 & J_4 \end{array} \right] = \left[\begin{array}{c|c} L' & K \\ \hline -K' & L \end{array} \right] + \left[\begin{array}{c|c} -Q_i & \dfrac{P_i}{|\mathbf{V}_i|} \\ \hline P_i & \dfrac{Q_i}{|\mathbf{V}_i|} \end{array} \right] \tag{7.137}$$

Note that, the off-diagonal elements in the submatrices of the second matrix are all zero. The method of solution in the polar coordinates is similar to that in the rectangular coordinates. Therefore, it will not be repeated here.

7.13.2.2 Method 2. Six Sub-matrices of a Jacobian

Research has demonstrated that modifying Equation 7.106 as follows [206] leads to improved efficiency:

$$\left[\begin{array}{c} \Delta P \\ \Delta Q \end{array} \right] = \left[\begin{array}{c|c} J_1 & J_2 \\ \hline J_3 & J_4 \end{array} \right] \left[\begin{array}{c} \Delta \delta \\ \frac{\Delta |\mathbf{V}|}{|\mathbf{V}|} \end{array} \right] \tag{7.138}$$

where submatrices J_1 and J_3 have remained unchanged, as given in Equation 7.137. However, submatrices J_2 and J_4 are to be modified. For example, in the matrix multiplication, the ij-th term in submatrix J_2 becomes

$$\left(\frac{\partial P_i}{\partial |\mathbf{V}_j|}\right)\Delta|\mathbf{V}_j| = |\mathbf{V}_i||\mathbf{Y}_{ij}|\cos(\theta_{ij} + \delta_i - \delta_j)\Delta|\mathbf{V}_j| \tag{7.139}$$

By multiplying the right side of Equation 7.138 by $|\mathbf{V}_j|/|\mathbf{V}_j|$,

$$\left(\frac{\partial P_i}{\partial |\mathbf{V}_j|}\right)\Delta|\mathbf{V}_j| = |\mathbf{V}_i||\mathbf{V}_{ii}||\mathbf{V}_j|\cos(\theta_{ij} + \delta_i - \delta_j)\frac{\Delta|\mathbf{V}_j|}{|\mathbf{V}_j|} \tag{7.140}$$

or

$$\left(\frac{\partial P_i}{\partial |\mathbf{V}_j|}\right)\Delta|\mathbf{V}_j| = K'_{ij}\frac{\Delta|\mathbf{V}_j|}{|\mathbf{V}_j|} \tag{7.141}$$

Since

$$K'_{ij} = |\mathbf{V}_i||\mathbf{Y}_{ij}||\mathbf{V}_j|\cos(\theta_{ij} + \delta_i - \delta_j)$$

Therefore, submatrix J_2 becomes

$$[J_2] = [K'] + \begin{bmatrix} & & 0 \\ & P_i & \\ 0 & & \end{bmatrix} \tag{7.142}$$

Similarly, submatrix J_4 takes the form

$$[J_4] = [L'] + \begin{bmatrix} & & 0 \\ & Q_i & \\ 0 & & \end{bmatrix} \tag{7.143}$$

The voltage-controlled buses can be incorporated in a similar manner as before. Submatrix J_5 retains its previous form, with all elements being zero. As for submatrix J_6, all off-diagonal elements are zero, and the diagonal element corresponds to $|\mathbf{V}_i|$. Thus, the modified Jacobian matrix can be represented as follows:

$$\left[\begin{array}{c|c} J_1 & J_2 \\ \hline J_3 & J_4 \\ \hline J_5 & J_6 \end{array}\right] = \left[\begin{array}{c|c} L' & K' \\ \hline -K' & L' \\ \hline 0 & |\mathbf{V}_i| \end{array}\right] + \left[\begin{array}{c|c} -Q_i & P_i \\ \hline P_i & Q_i \\ \hline 0 & 0 \end{array}\right] \tag{7.144}$$

7.14 DECOUPLED POWER FLOW METHOD

Consider the power flow problem defined in polar coordinating Equation 7.138,

$$\left[\begin{array}{c} \Delta P \\ \Delta Q \end{array}\right] = \left[\begin{array}{c|c} J_1 & J_2 \\ \hline J_3 & J_4 \end{array}\right] \left[\begin{array}{c} \Delta \delta \\ \frac{\Delta|V|}{|V|} \end{array}\right]$$

which is known in the literature [204–206] as

$$\begin{bmatrix} \Delta P \\ \Delta Q \end{bmatrix} = \left[\begin{array}{c|c} H & N \\ \hline J & L \end{array} \right] \begin{bmatrix} \Delta \delta \\ \frac{\Delta |V|}{|V|} \end{bmatrix} \tag{7.145}$$

Typically, in a given power system, the real power flow is regarded as being significantly less sensitive to changes in voltage magnitude compared to changes in voltage angle. As an approximation, the elements of submatrix N (or J_2) can be assumed to be zero. Similarly, the reactive power is considerably less sensitive to variations in voltage angle than voltage magnitude. Hence, the elements of submatrix J (or J_3) can be considered to be zero as well. Consequently, Equation 7.145 can be simplified to:

$$\begin{bmatrix} \Delta P \\ \Delta Q \end{bmatrix} = \left[\begin{array}{c|c} H & 0 \\ \hline 0 & L \end{array} \right] \begin{bmatrix} \Delta \delta \\ \frac{\Delta |V|}{|V|} \end{bmatrix} \tag{7.146}$$

from which

$$[\Delta P] = [H][\Delta \delta] \tag{7.147}$$

$$[\Delta Q] = [L] \left[\frac{\Delta |V|}{|V|} \right] \tag{7.148}$$

which are known as the two *decoupled power flow equations* in the literature [217]. They are solved separately, so that

$$[\Delta \delta] = [H]^{-1}[\Delta P] \tag{7.149}$$

$$\left[\frac{\Delta |V|}{|V|} \right] = [L]^{-1}[\Delta Q] \tag{7.150}$$

It is important to note that the solution requires inverting the H and L matrices, which have dimensions that are approximately one-fourth the size of the full Jacobian matrix. This approach noticeably reduces both the computation time required for each iteration and the amount of computer memory needed.

7.15   FAST DECOUPLED POWER FLOW METHOD

Let's examine the decoupled power flow equations as defined in Equations 7.147 and 7.148, where the elements of submatrices H and L can be represented as follows:

$$[\Delta P] = [H][\Delta \delta]$$

$$[\Delta Q] = [L] \left[\frac{\Delta |V|}{|V|} \right]$$

$$H_{ij} = V_i V_j (G_{ij} \sin \delta_{ij} - B_{ij} \cos \delta_{ij}) \quad i \neq j \tag{7.151}$$

$$H_{ii} = -B_{ii} V_i^2 - Q_i \quad i = j \tag{7.152}$$

$$L_{ij} = V_i V_j (G_{ij} \sin \delta_{ij} - B_{ij} \cos \delta_{ij}) = H_{ij} \quad i \neq j \tag{7.153}$$

$$L_{ii} = -B_{ii} V_i^2 - Q_i \quad i = j \tag{7.154}$$

It can be noted that while the decoupled power flow method significantly reduces memory storage requirements, it still demands substantial computational effort. As a result, Stott and Alsac [219] introduced the *fast decoupled power flow method*. This method is based on the following fundamental assumptions:

1. The power systems have high X/R ratios. Hence,

$$G_{ij}\sin\delta_{ij} \ll B_{ij} \tag{7.155}$$

2. The difference between adjacent bus voltage angle is very small. Thus,

$$\sin\delta_{ij} = \sin(\delta_i - \delta_j) \cong \delta_i - \delta_j = \delta_{ij} \tag{7.156}$$

$$\cos\delta_{ij} = \cos(\delta_i - \delta_j) \cong 1.0 \tag{7.157}$$

3. Also,

$$Q_i \ll B_{ii}V_i^2 \tag{7.158}$$

Therefore, Equations 7.147 and 7.148 can be further approximated as

$$[\Delta P] = [V \times B' \times V][\Delta\delta] \tag{7.159}$$

$$[\Delta Q] = [V \times B'' \times V]\left[\frac{\Delta\delta}{V}\right] \tag{7.160}$$

where the elements of the matrices $[B']$ and $[B'']$ are the elements of the matrix $[-B]$. Thus,

$$B'_{ij} = -\frac{1}{X_{ij}} \quad i \neq j \tag{7.161}$$

$$B'_{ii} = \sum_{j=1}^{n} \frac{1}{X_{ij}} \quad i \neq j \tag{7.162}$$

$$B''_{ij} = -B_{ij} \tag{7.163}$$

The decoupling process in the fast decoupled power flow can be concluded [231] after additional modifications based on the simplifying assumptions as

$$\left[\frac{\Delta P}{V}\right] = [B'][\Delta\delta] \tag{7.164}$$

$$\left[\frac{\Delta Q}{V}\right] = [B''][\Delta V] \tag{7.165}$$

It should be noted that both matrices B' and B'' exhibit properties of being real, sparse, and symmetrical. This is a consequence of excluding shunt susceptance, transformer off-nominal taps, and phase shifts. Consequently, both matrices B' and B'' are always guaranteed to be symmetrical. Due to this symmetry, their constant sparse upper triangular factors can be computed and stored only once at the beginning of the solution process. The method demonstrates exceptional speed and reliability.

7.16   THE DC POWER FLOW METHOD

In certain power system studies, such as reliability studies, a significant number of power flow runs may be required. As a result, a very fast method can be utilized for such studies, prioritizing speed over accuracy due to the linear approximation involved. This method, known as the "dc power flow method," involves calculating real flows by initially solving for the bus angles. In contrast, the exact nonlinear solution is referred to as the "ac solution." The dc power flow method is employed to ensure timely solutions, as the convergence of power flow solutions can pose challenges.

Assume that bus i is connected to bus j over an impedance of Z_{ij}, Thus, the can be expressed as

$$P_{ij} = \frac{V_i V_j}{Z_{ij}} \sin(\delta - \delta_j) \qquad (7.166)$$

where

$$\mathbf{v}_i = |v_i| \angle \delta_i \qquad \mathbf{v}_j = |v_j| \angle \delta_j$$

The following simplifying approximations are made:

$$X_{ij} \cong \mathbf{Z}_{ij} \text{ since } X_{ij} \gg R_{ij}$$

$$|v_i| \cong 1.0 \text{ pu}$$

$$|v_j| \cong 1.0 \text{ pu}$$

$$\sin(\delta_i - \delta_j) \cong \delta_i - \delta_j$$

Hence, the Equation 7.166 can be expressed as

$$P_{ij} \cong \frac{\delta_i - \delta_j}{X_{ij}} \cong B_{ij}(\delta_i - \delta_j) \qquad (7.167)$$

Thus, in matrix form,

$$[P] = [B][\delta] \qquad (7.168)$$

from which

$$[\delta] = [B]^{-1}[P] \qquad (7.169)$$

or

$$[\delta] = [X][P]. \qquad (7.170)$$

In an n-bus system, the $[B]$ matrix has dimensions of $(n-1) \times (n-1)$. The diagonal elements and off-diagonal elements of the $[B]$ matrix can be obtained by summing the series susceptances of branches connected to bus i and setting them equal to the negated series susceptance of branch ij, respectively. By utilizing matrix techniques, the linear Equation 7.170 can be solved for δ. The dc power flow method allows for the execution of numerous power flow runs, which is crucial for comprehensive contingency analysis on large-scale systems. It should be noted that this representation primarily focuses on adequacy assessment related to overload-related system issues.

In summary, the selection of a power flow method involves a trade-off between speed and accuracy. The speed of a method, given a specific level of accuracy, depends on factors such as the size, complexity, and configuration of the power system, as well as the chosen numerical approach.

PROBLEMS

1. Use the results of Example 8.1 and the Gauss–Seidel method and determine the following:
   (a) Value of voltage $\mathbf{V}_3$ for the first iteration
   (b) Value of voltage $\mathbf{V}_4$ for the first iteration
2. Consider the following the nonlinear equations

$$f_1(x_1, x_2) = x_1^2 + 2x_2 - 10 = 0$$

$$f_2(x_1, x_2) = x_1, x_2 - 3x_2^2 + 5 = 0$$

Determine the values of x_1 and x_2 by using the Newton–Raphson method. Use 3 and 2 as the initial approximation for x_1 and x_2, respectively.

Table 7.5
Summary of Bus Classification

Line (Bus to Bus)	R (pu)	X (pu)
1–2	0.2	0.4
1–4	0.2	0.6
1–5	0.1	0.2
2–3	0.1	0.2
2–4	0.2	0.4
3–5	0.1	0.2

3. Use the rectangular form for voltages and the bus admittance quantities as follows and develop an expression for the real power loss in line i to j:

$$Y_{ii} = G_{ij} - jB_{ij}$$

$$V_i = e_i + jf_i$$

$$V_j = e_j + jf_i$$

4. Use the resultant real power loss expression from the solution of Problem 3 and determine the real power loss in line 1–2. Assume that $Z_{12} = 0.1 + j0.3$ pu, $V_1 = 0.98 - j0.06$ pu, and $V_2 = 1.04 - j1.04$ pu.

5. Assume that the branch admittance of a line connecting buses 1 and 2 is given as $1 - j4$ pu and that the bus voltages are given as $0.98 + j0.06$ pu and $1.04 + 0.00$ pu for buses 1 and 2, respectively. Neglect the line capacitance and determine the following:

   (a) Real and reactive powers in line at the bus 1 end
   (b) Real and reactive power losses in the line

6. Assume that line 1–2 given in Problem 3 has a significant amount of line capacitance. Half of the total line-charging admittance, $1/2\ y'$ is given as $j0.05$ pu and is placed at each end of the line in the π representation. Determine the following:

   (a) Real and reactive powers in line at the bus 2 end
   (b) Real and reactive power losses in the line

7. Assume that the data in Table 7.5 have been given for Example 8.3 and determine the following:

   (a) All individual branch admittances
   (b) All elements of bus admittance matrix

8. Use the results of Problem 7 in the application of the Newton–Raphson method in rectangular coordinates to Example 8.3 to determine the following after the first iteration:

   (a) Jacobian submatrix J_1
   (b) Jacobian submatrix J_2
   (c) Jacobian submatrices J_3 and J_5
   (d) Jacobian submatrices J_4 and J_6
   (e) Bus currents in terms of $c_i + jd_i$
   (f) The Δe and Δf voltage changes for four buses
   (g) The ΔP, ΔQ, and $\Delta|V|^2$ changes

9. Modify Example 8.3 using the data in Tables 7.6 and 7.7 and apply the Newton–Raphson method in polar coordinates. Using the second type of formulation of the Jacobian matrix, determine the following:

$$
[\mathbf{Y}_{bus}] =
\begin{array}{c}
\\
1 \\
2 \\
3 \\
4
\end{array}
\begin{array}{cccc}
1 & 2 & 3 & 4 \\
\end{array}
\left[
\begin{array}{cccc}
3 - j9 & -2 + j5 & 0 & -1 + 4 \\
-2 + j5 & 4 - j10 & -2 + j5 & 0 \\
0 & -2 + j5 & 3 - j9 & -1 + j4 \\
-1 + j4 & 0 & -1 + j4 & 2 - j8
\end{array}
\right]
$$

$$
\begin{bmatrix}
\boxed{\Delta P_2} \\
\Delta P_3 \\
\Delta P_4 \\
\Delta Q_2 \\
\Delta Q_3 \\
\Delta |\mathbf{V}_3|^2
\end{bmatrix}
\begin{array}{c}
2 \\
3 \\
4 \\
2 \\
3 \\
4
\end{array}
=
\begin{array}{ccc}
2 & 3 & 4 \\
\end{array}
\left[
\begin{array}{ccc}
\square & & \\
 & J_1 & \\
 & & J_2 \\
\square & & \\
 & J_3 & J_4 \\
 & J_5 & J_6
\end{array}
\right]
\times
\begin{bmatrix}
\Delta e_2 \\
\Delta e_3 \\
\Delta e_4 \\
\Delta f_2 \\
\Delta f_3 \\
\Delta f_4
\end{bmatrix}
$$

(a) Jacobian submatrix J_3 for the second iteration
(b) Jacobian submatrix J_5 for the second iteration

10. Consider a four-bus power system in which bus 1 is the slack bus with constant-voltage magnitude and angle. A generator is connected to bus 3 where the voltage magnitude is maintained at 1.02 pu. Buses 2 and 4 are designated as the load buses. Apply the Newton–Raphson method in rectangular coordinates. Using the following data, given in per units, determine the value of the terms indicated in black (i.e., the elements in $\delta P_2, J_{1(2,2)}$ and $J_{3(2,3)}$) for the next iteration (Table 7.8).

11. Consider the four-bus power system given in Problem 10. Apply the modified version of the Newton–Raphson method in polar coordinates. Assume that the specified voltages are given in polar coordinates as $1.034\angle 0°$, $0.9055\angle -6.3402°$, $1.02\angle 0°$, and $1.0012\angle -2.8624°$ for $\mathbf{V}_1$, $\mathbf{V}_2$, $\mathbf{V}_3$, and $\mathbf{V}_4$, respectively. Determine the values of the terms indicated by bold boxes (i.e., the elements in ΔP_3, $J_{1(2,3)}$, $J_{2(3,2)}$, and $J_{4(2,3)}$)

12. Consider a three-bus power system as shown in Figure 7.6. Assume that bus 1 is the slack bus with a voltage of $1.02 + j0.000$ pu. Bus 2 is the regulated generator bus with a voltage magnitude of 1.02 pu. Use $1.02 + j0.000$ pu and 0.5 pu as its initial voltage and real power values, respectively. Bus 3 is the load bus with an initial voltage of $0.0.95 - j0.05$ pu. Its load is made up of 1.0 pu resistive load and 0.4 pu inductive load. Assume that the current components calculated at each bus are d_2, c_3, and d_3 with values of -0.200, -1.375, and 1.000 pu, respectively. Use the Newton–Raphson method in rectangular coordinates and determine the following:

(a) Current component c_3 for bus 2
(b) Real and reactive power changes, ΔP_3 and ΔQ_3, for bus 3
(c) Submatrices J_1, J_3, and J_5

13. Use the appropriate data given in Problem 12 and the Gauss–Seidel method and determine the following:

(a) Value of voltage $\mathbf{V}_3$ for the first iteration
(b) Value of voltage $\mathbf{V}_2$ for the first iteration

Table 7.6

Conductance, Susceptance, and Admittance Data of the lines for Problem 9

i	j	G_{ij}	B_{ij}	Y_{ij}	Φ_{ij}
1	2	−1	−2	2.236	−116.565°
2	4	−1	−2	2.236	−116.565°
1	4	−0.5	−1.5	1.581	−108.435°
1	5	−2	−4	4.472	−116.565°
2	3	−2	−4	4.472	−116.565°
3	5	−2	−4	4.472	−116.565°
2	2	4	8	8.944	63.435
3	3	4	8	8.944	63.435
5	5	4	8	8.944	63.435
1	1	3.5	7.5	8.2765	63.435
4	4	1.5	7.5	3.808	66.801

Table 7.7

Specified Values, Initial Values, and Calculated Values of P, Q, and $|V|$ at Each Bus for Problem 9

Bus	Specified Values			Initial Values		Calculated Values (After First Iteration)	
	P_i	Q_i	$|V_i|$	$|V_i|$	δ_a	P_i	Q_i
1	−	−	1.02	1.02	0°	−	−
2	−0.6	−0.3	−	1	0°	−0.1	−0.02
3	1	−	1.04	1.04	0°	0.1664	−^{a}
4	−0.4	−0.1	−	1	0°	−0.1	−0.03
5	−0.6	−0.2	−	1	0°	−0.12	−0.24

^{a}Calculated $|V_i|$ = 1.04 after the first iteration.

Table 7.8

For Problem 10

Bus	Specified Values			Calculated Values (After kth Iteration)	
	V_i	P_i	Q_i	c_j	d_i
1	1.03 + j 0.0	−	−	−	−
2	0.9 + j 0.1	−1.2	−0.4	−1.50	−0.85
3	1.02 + j 0.0	1	−	0.96	−0.43
4	1.0 − j 0.05	−0.5	−0.1	−0.45	0.1

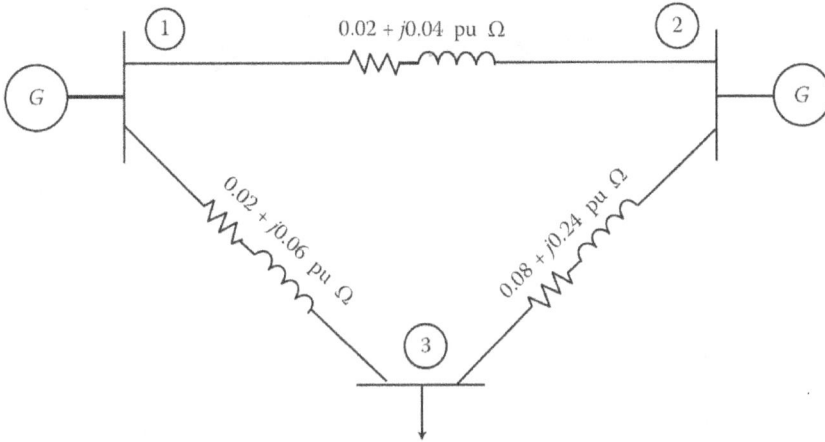

Figure 7.6   Three-bus system for Problem 12.

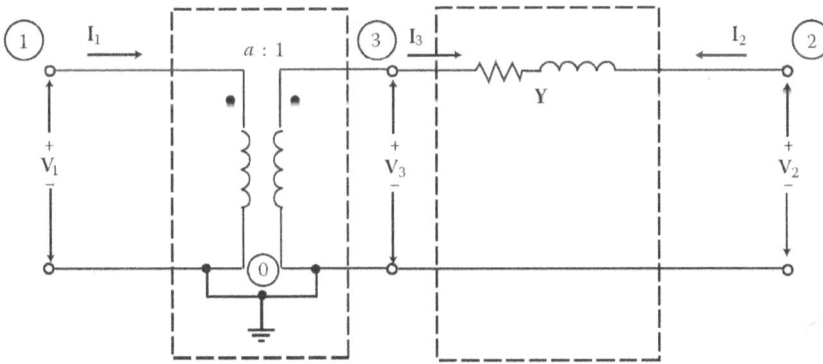

Figure 7.7   LTC transformer connection for Problem 14.

14. Assume that the LTC transformer shown in Figure 7.7 has an off-nominal turns ratio as indicated. (Note that, in the event that the ratio is equal to the system nominal voltage ratio, it is represented as a 1:1 ratio in the per-unit system.) Use the two-port network theory and verify the equation

$$\begin{bmatrix} I_1 \\ I_2 \end{bmatrix} = \begin{bmatrix} \dfrac{Y}{a^2} & -\dfrac{Y}{a} \\ -\dfrac{Y}{a} & Y \end{bmatrix} \begin{bmatrix} V \\ V_2 \end{bmatrix}$$

15. Use the results of Problem 14 and verify the equivalent π circuit representation of the transformer shown in Figure 7.3b.

8 Power System Restoration

It is never too late to mend

Chinese proverbs

8.1 INTRODUCTION

When a blackout occurs, it is crucial for a grid to maintain effective coordination with other control areas to facilitate a systematic restoration within a reasonable timeframe. The knowledge and milestones involved in power restoration can vary across different organizational practices as this field continues to evolve. Power system restoration holds significant importance as one of the critical tasks for electric power grids. After a power outage, dispatchers in the control center collaborate with field crews to reinstate generation and transmission systems, and subsequently resume load and restore service. The efficiency of system restoration plays a pivotal role in ensuring system reliability. It is worth noting that the impact of a blackout intensifies with the duration of the restoration process. The occurrence of blackout events and the aging transmission infrastructure in electrical power systems necessitate greater attention to research and development in system restoration and the corresponding decision support tools.

In the face of a sequence of cascading events, even a robust power system can be susceptible, and the subsequent cascading effects can result in a catastrophic outage. Analyzing past blackouts in power grids, such as the Aug 14, 2003 blackout in the USA and Canada, Aug 28, 2003 blackout in London, U.K., Sep 23, 2003 blackout in Sweden and Denmark, Sep 28, 2003 blackout in Italy, and May 24–25, 2005 blackout in Moscow, Russia [278, 279], provides insights into the generic scenario of a cascading failure. In this scenario, a severe contingency triggers a cascading failure, causing the system's operating state to shift from normal to partial or complete shutdown. During an outage state, the power system loses some or all of its functionality. To restore the operating state back to normal, restorative actions are necessary, such as restarting generating units, re-establishing the transmission network, and efficiently restoring the load. Therefore, the efficiency of system restoration plays a vital role in ensuring system reliability.

The process of system restoration involves a gradual and iterative approach, with utmost consideration given to personnel safety and system stability. When a disturbance occurs, the primary objective of system operation is to preserve system integrity and stability. During an outage state, the focus shifts to restoring services to customers as swiftly as possible while adhering to safety protocols and other constraints. Consequently, the priorities of system operation vary significantly between normal, emergency, and restorative states.

After a major disturbance, remedial actions may be necessary to guide the system towards a secure operating state. To achieve this objective, the system's status, including its topology and the availability of various components, must be monitored and assessed initially. Control actions will then be determined based on the system's current state. In the event that the disturbance results in a power outage, the system restoration procedure will be initiated.

System restoration encompasses the evaluation of system status, optimization of generation capabilities, restoration of transmission and distribution networks, and load pickup. Developing decision support tools for system restoration entails addressing two fundamental challenges. The first crucial consideration is ensuring the generality of these tools, allowing for their portability across systems with minimal customization. Literature reviews and discussions with power system dispatchers from various systems indicate that differences in restoration strategies are closely tied to variations in system characteristics. Consequently, developing a tool that supports power system restoration efforts

DOI: 10.1201/9781003129769-8

Table 8.1
Typical Restoration Stages

Restoration Stages	Purposes	Usual Duration(Hours)	Concern(s) of this Stages
I	Preparation	2-Jan	1. Actions are time critical
			2. Blocks of load are control means to maintain stability
II	System Restoration	3-Jan	Load restoration is only a means to an end
III	Load Restoration	6-Apr	Load restoration is the objective

for a wide range of systems poses a significant challenge. Thus, individual power grids typically establish offline system restoration plans or guidelines.

Another challenge in establishing efficient computational tools lies in the lack of highly efficient optimization methods. The restoration problem necessitates considering numerous practical factors, making it unsuitable for formulation as a single optimization problem. Moreover, these optimization problems often involve combinatorial aspects. For instance, quickly restarting generating units requires solving a dynamic programming-like problem, while maximizing restored loads involves formulating a knapsack problem, which is classified as an NP-hard problem.

8.2 GENERAL RESTORATION PROCEDURE

By reviewing literature and engaging in discussions with utility operators from diverse systems, it has been observed that differences in restoration strategies are closely linked to variations in system characteristics. Developing a comprehensive restoration plan tailored to the specific requirements of each individual power system is essential. However, underlying these diverse restoration strategies are commonalities in what can be referred to as "tactics" incorporated within the overall strategies. Power systems, in general, exhibit similar behaviors and share common characteristics during the restoration process. Consequently, it is feasible to establish a generalized procedure for system restoration [280, 281]. This procedure typically consists of three stages: Preparation, System Restoration, and Load Restoration, as illustrated in Table 8.1.

8.2.1 PREPARATION

During the restoration process, the first stage is preparation, which requires prompt action and time sensitivity. Numerous urgent tasks need to be swiftly addressed in this stage. Typically, the post-disturbance system status is evaluated, the target system for restoration is defined, a strategy for rebuilding the transmission network is chosen, potential subsystems within the system may be identified, initial energy power sources are determined, non-blackstart generating units are activated, and arrangements are made to supply critical loads. In this time-critical phase of restoration, it is crucial to thoroughly plan actions for rapid implementation. Depending on equipment availability and the severity of the disturbance, a post-disturbance or target system is established, which closely resembles the pre-disturbance system.

The availability of an initial power source is a vital aspect of system restoration [280]. Studies have indicated that restoration duration is reduced when there are more available initial power sources [281]. These initial sources refer to local generation options that can be started without relying on external power sources. Typically, the following initial sources may be considered during the restoration process:

- Hydro unit: This category includes run-of-the-river hydro and pumped-storage hydro units, which are designed with blackstart capability and exhibit fast response characteristics. These types of hydro units can typically be started within 5–10 minutes, with a high probability of success.
- Diesel generating unit: Diesel units are generally small and primarily used to meet the startup requirements of larger units. They are not suitable for powering significant loads or energizing transmission lines. However, they have fast response characteristics.
- Gas turbines:
  - a) Aeroderivative gas turbine: This type of gas turbine usually requires local battery power for initial cranking. It can be operated remotely and is capable of quickly picking up loads.
  - b) Larger gas turbine: Restarting a larger gas turbine unit usually requires support from an on-site diesel generating unit. The diesel unit is used to energize an auxiliary bus within the plant. The time required to restart a gas turbine unit depends on the duration it has been offline.
- Tie-line with adjacent systems: If tie-lines are established with adjacent systems, energy from those systems can be utilized as initial sources to crank non-blackstart units, pick up loads, and energize transmission lines. Tie-lines with adjacent systems can be established quickly and reliably due to the system characteristics.

During the restoration process, a specific set of Critical Loads is identified based on the characteristics of the system and operating conditions. These Critical Loads are categorized into three groups. Firstly, high priority loads encompass the loads necessary for cranking drum-type units and pipe-type cable pumping systems. These loads have the highest priority and require immediate attention during the restoration process. Secondly, medium priority loads include loads that are essential for energizing transmission stations and distribution stations. They are considered to be of medium priority in terms of restoration. Lastly, industrial loads are classified as low priority loads. While these loads are important, their restoration can be given lower priority compared to the other load categories.

8.2.2 SYSTEM RESTORATION

To achieve the objective of load restoration in the third stage, the following actions should be executed:

- Energizing skeleton transmission paths
- Re-synchronizing islands
- Restoring of loads due to the stability of generating units and voltage of system
- Preparation for restart of base-load units

In order to fulfill the objectives, certain tasks, such as energizing high voltage lines and synchronizing subsystems, necessitate the utilization of either pre-established reliable guidelines or analytical tools for assessing critical switching actions. During system restoration, it is crucial to monitor important constraints of the power system, including voltages, VAr balance, stability, voltage-frequency response dynamics, unit response capability adequacy, and fault clearing capability.

8.2.3 LOAD RESTORATION

The main goal of the restoration process is to swiftly restore the load and minimize the amount of unserved load. In this final stage, the focus is on fully restoring the load itself. Since the bulk power system has been interconnected in the previous stage, the objective now is to pick up the load. To minimize unserved load, the scheduling of load pickup is based on the total rate of response capability rather than economic dispatch considerations. The system's response rate becomes more effective as more online generators become available that have not reached their capacity limits. Therefore, as the restoration process advances, load restoration can be accomplished in progressively larger increments.

The primary difference between the first stage and the subsequent stages lies in the urgency and time sensitivity of the actions required. In the initial stage, time is of the essence, and quick response is crucial. On the other hand, the key distinction between the initial stages and the third stage is the shift in focus. In the initial stages, the priority is to maintain system stability through controlled load shedding, while in the third stage, the primary objective is load restoration itself [280].

8.3 GENERAL STRATEGY OF RESTORATION WITH DIFFERENT SYSTEMS

In general, the goals of system restoration involve the initial reconstruction of a stable power system, followed by the restoration of all remaining unserved loads. However, the specific restoration strategies can vary based on the unique characteristics, resources, and constraints of each power grid. The classical approach to system restoration is a multistage process that entails distinct objectives and constraints. Throughout this process, dispatchers at the control center collaborate with field crews to gradually reinstate the generation and transmission systems, and subsequently recover load and restore service.

8.3.1 THERMAL SYSTEMS

The system is divided into multiple subsystems, and the process involves the simultaneous reintegration of generation and transmission within each subsystem. The sectionalization is performed based on the following criteria:

- Blackstart capability within each subsystem.
- Steam generators with hot restart capability and the availability of transmission paths.

The restoration procedure is divided into four phases

- Units restart phase: providing station service for the restart of non-blackstart units within each subsystem.
- Reintegration phase: interconnecting of generating stations.

During these phases, only the essential loads are restored based on the generation and transmission needs. As these phases progress, the subsystems become synchronized and interconnected securely.

- Load pick-up phase: loads are picked-up in small increments to avoid excessive under frequency deviations.
- Interconnection phase: high voltage lines are energized and then interconnected.

8.3.2 HYDRO SYSTEMS

Hydro systems typically utilize remote generators connected to loads through long transmission lines [282]. During the system restoration process, several challenges arise, including overvoltage conditions caused by line charging, harmonics, and switching surges. To address these issues, the main grid is divided into multiple subsystems, each energized independently and simultaneously. The primary objective is to rapidly energize the networks within each subsystem. Once the networks are energized, synchronization takes place, and load service is gradually restored in small increments.

8.3.3 HYDRO-THERMAL SYSTEMS

Depending on the system's characteristics, the restoration strategies can vary. In systems equipped with large capacity hydro generators that have ample leading power-factor operation capability, an effective strategy is [282]:

- Energizing the entire bulk power transmission system in one step
- Providing station service to all thermal generating stations.
- Using leading power-factor operation ability of large hydro stations to absorb the charging currents of the complete transmission system.
- High voltages at the receiving end of lines are avoided by manual operation of hydro units' voltage regulators.

The goal is to minimize the time-consuming line sectionalization and switching operations typically involved in system restoration. By utilizing the presence of large hydro units, the interconnection process can be carried out before load pick-up, thus reducing the overall restoration time.

8.3.4 GENERAL GUIDELINE OF RESTORATION

Power system restoration strategies for systems with different characteristics offer various options. Despite the differences, most bulk power supply systems exhibit common characteristics and behave similarly during the restoration process. This enables the establishment of a general procedure and a set of guidelines to enhance system restoration. Multiple restoration strategies share several common guidelines that can be applied [283].

- Identification of the status of the collapsed system, its components, and equipment.
- Restarting and restoring station service to plants, substations, cable pumping plants, compressed air systems, etc.
- Coordinating power plant start-up timings with load pick-ups to bring generators to their stable minimum levels and within the range of major analog controllers.
- Energizing large sections of transmission lines while ensuring acceptable transient and sustained overvoltages.
- Incrementally picking up load without risking a decline in frequency.
- Reintegrating the skeleton of the bulk power supply through time-consuming switching operations.
- Deactivating automatic load shedding and automatic switched capacitors during the initial phases of restoration.
- Maintaining steady-state and transient stability during the restoration process, particularly when dealing with large impedances.
- Minimizing standing angles when closing loops to establish firm transmission paths.
- Ensuring the successful start-up probability of thermal units, especially combustion turbines.
- Maintaining stability for blackstart and peaking combustion turbines, especially when they contribute significantly to generation.

To meet these operational constraints during system restoration, it is necessary to consider the specific characteristics of the power system and its generating units. The restoration planning process takes into account various factors related to the system, such as its topology, capacity, voltage levels, and load demand. Additionally, the characteristics of the generating units, such as their types, capacities, ramping rates, and response capabilities, are also taken into consideration.

By analyzing the system characteristics, including the available generation resources, transmission capacities, and distribution networks, restoration planners can develop an efficient and effective restoration plan. This plan involves determining the optimal sequence for bringing generating units online, energizing transmission lines, and gradually picking up load while ensuring system stability and reliability.

Considering the characteristics of the generating units is crucial for successful restoration. Factors such as their startup time, operating limits, fuel availability, and coordination with other units are taken into account. By understanding these characteristics and incorporating them into the restoration planning process, operators can optimize the utilization of generating units, minimize constraints, and ensure a smooth and reliable restoration of the power system.

In summary, the restoration planning process considers the characteristics of both the power system and its generating units to meet the operational constraints and ensure an efficient and reliable restoration process. By carefully analyzing these characteristics, operators can develop a well-structured restoration plan that takes into account the specific requirements and capabilities of the system and its components.

8.3.5 GENERAL METHODOLOGY OF RESTORATION

Restoring power to a system after a blackout presents a complex challenge that requires careful decision-making and control by both system dispatchers and restoration planners. The main goal in the aftermath of a blackout is to quickly restore the load. Depending on the operational practices, characteristics of generating units' restart capabilities, and specific considerations for power system reintegration, there are several strategy options available for system restoration. These strategies can be broadly classified into five methodologies known as *Build-Upward*, *Build-Downward*, *Build-Inward*, *Build-Outward*, and *Build-Together*. Each methodology is characterized as follows:

- *Build-Upward:* In this strategy, the restoration process involves dividing the system into islands with dedicated blackstart capabilities. Each island is then restored individually, and once restoration is complete for one island, it is resynchronized with the rest of the system. The key actions in this strategy include starting up blackstart units, cranking non-blackstart units, restoring the islands, and synchronizing them together.
- *Build-Downward:* This strategy focuses on re-energizing the network by utilizing blackstart power sources initially. It involves starting up blackstart units and energizing the network. Once the network is energized, the strategy shifts towards providing cranking power to non-blackstart units. The major actions in this strategy include the startup of blackstart units, network energization, and cranking of non-blackstart units.
- *Build-Inward:* This strategy can be employed in systems that have access to tie-line assistance. With the assistance of tie-lines, the strategy involves establishing transmission lines to restart generating units in key generation stations. From this initial system and electric power sources, the restoration process then proceeds. Implementing this strategy requires several actions, including establishing tie-lines, setting up transmission networks, and cranking non-blackstart units.
- *Build-Outward:* To restore the ring network without relying on tie-line assistance, the system restoration process must proceed outward from the ring. The main tasks involved in this strategy are starting up blackstart units, energizing the network (ring network), and cranking non-blackstart units.
- *Build-Together:* This strategy involves establishing the skeleton of a transmission network in stages to provide cranking energy, followed by the restoration of non-blackstart units located near the load.

To address the intricacies of power system structures and the variations in operating conditions, a restoration plan can be developed by combining the five general philosophies of restoration. Regardless of the strategies employed, the ultimate objective of restoration remains the same: to restore the system to its pre-blackout condition as swiftly as possible.

8.4   CHARACTERISTICS OF GENERATING UNITES RELEVANT TO RESTORATION

Restoring generating units is a crucial aspect of the system restoration process. It is vital to maintain a balance between active and reactive power. The timing of restoring generating units usually follows a specific sequence, starting with hydro units, followed by combustion turbines, fossil-fueled units, and finally nuclear units. Within the first 24 hours after a blackout, a substantial portion of hydro, combustion turbine, and fossil generation is typically restored. When it comes to system restoration, the following characteristics of generators are taken into consideration:

- Generator's maximum and minimum output and rate of loading under normal and emergency conditions.
- The dynamic response of prime movers to sudden load pick-up.
- Maximum time intervals for hot restart of drum type of boilers.
- Minimum elapsed time for hot restart of supercritical once-through boilers.
- Blackstart capabilities of combustion turbines.

In addition to these general characteristics, there are special considerations for steam units, combustion turbine units, and nuclear units. These specific types of generating units require additional attention during the restoration process. On the other hand, hydro units are typically regarded as an integral part of the blackstart capability.

8.4.1   STEAM UNIT

Steam units, whether fueled by fossil fuels or nuclear energy, are known for their relatively high efficiency and lower fuel costs. These units are typically utilized for supplying the base load in power systems, as they offer the most economical generation capability. To enhance efficiency and reduce costs, steam units are often designed as larger, high-pressure, and high-temperature units, which exhibit higher cycle efficiencies. However, the maneuverability of these units is limited due to temperature gradient constraints that affect the stress levels of thick high-pressure and temperature metal components. While they may have limitations in load cycling duty, these units are well-suited for operating at the base load with only a few startups and shutdowns per year [284].

Steam units often serve as the largest generation sources within a power system. During power system restoration, the timely restoration of steam units is of utmost importance. However, it is important to note that an unanticipated sudden shutdown differs significantly from a planned shutdown. Additionally, the smoothness of a startup procedure can be influenced by the nature of the preceding shutdown. Apart from the impact of preceding shutdowns, the ability to restart a steam unit quickly can be greatly improved by incorporating design features such as bypass valves and tripping to house load. Furthermore, modern digital systems enable automatic control of auxiliaries through extensive control logic capabilities, further facilitating efficient and swift restarts of steam units.

The duration required for the restart of a steam unit can vary, ranging from a few hours to a day, depending on factors such as the type of start (cold, warm, or hot), fuel type, and steam supply design. When restoring the system promptly after a major outage, it is crucial to prioritize the hot restart of cycling steam units, specifically those of the drum type. These units must be brought back online and supplying loads within a narrow timeframe following the system outage. Failure to do so may result in a delay of several hours before the units can be restarted. This timeframe is critical for system restoration, especially in stations where station service must be provided from another station. Accurate determination of the intervals between unit trip-outs and hot restart is essential. For more detailed information on average startup times after various shutdown durations, please refer to [284].

In summary, when restoring steam units during power system restoration, it is crucial to consider specific critical time intervals that govern the safe restart and operation of these units. These include the maximum time interval within which certain thermal units should be restarted to ensure their safety, as well as the minimum time interval required before a thermal unit can be successfully started. Coordinating these time intervals is essential for the effective restoration of steam units in the power system.

8.4.2   COMBUSTION TURBINE UNIT

A combustion turbine is an engine that utilizes the energy from the combustion gas flow. It consists of an upstream compressor, a downstream turbine, and a combustion chamber in between. There are two categories of combustion turbines: blackstart units and non-blackstart combined cycle units. The starting characteristics of these two turbine types are outlined as follows:

- The blackstart combustion turbine units are capable of starting without the need for external auxiliary power. For typical 15MW units, the cold start process takes approximately 30 minutes, with 15 minutes dedicated to unit warm-up. In emergency situations, the warm-up period can be bypassed, allowing for a 15-minute start-up time. These units can be immediately restarted after shutdown. The starting procedure involves a diesel engine and clutch system, without the use of electric motors. Auxiliary power is provided by a second diesel engine driving a direct current generator, which is normally inactive unless there is a blackout. The turbine unit can operate within a range of 0–15MW, with the operator adjusting the turbine to load conditions using MW setpoint control. The setpoint control must be modified as the load is increased or decreased.
- Non-blackstart combined cycle units, such as the 50MW combustion turbines [285], necessitate an ac source of auxiliary power for their startup process. These units can be brought from a cold start to full load in approximately 15 minutes, which also applies to hot starts. However, before attempting a restart, the turbine must first roll down to a complete stop and then be placed on turning gear. The roll-down procedure typically takes around half an hour. The generator of the unit has the capability to operate from zero to full load. Power adjustments can be made using the MW set point, and the maximum power output is determined by the turbine exhaust temperature, with power increased until the predetermined limit is reached.

8.4.3   NUCLEAR UNIT

When it comes to power system restoration, nuclear units require additional attention and consideration. The Electric System Restoration document by NERC dated December 1, 1992 emphasizes the special treatment needed for nuclear units [286]. Hot restarts are generally not permitted according to Nuclear Regulatory Commission (NRC) startup checklists, and the emergency generators of nuclear units are not allowed to provide auxiliary power to other stations. Restoring off-site power sources to nuclear units requires careful attention, as the restoration of power to the service area load typically needs to be accomplished without the assistance of nuclear units. Therefore, decisions regarding the prerequisites for startup must be made by the personnel responsible for nuclear plants. NRC start-up checklists prohibit hot restarts of nuclear units, and their diesel generators are not authorized to supply auxiliary power to other generating stations. Following a controlled shutdown, nuclear units can usually be restored to service within 24 to 48 hours.

The typical operating modes of a nuclear unit include on-line, hot standby, hot shutdown, and cold shutdown. The preferred mode for restart is hot standby. Generally, the following prerequisites should be taken into consideration when restarting a nuclear unit:

- A minimum of two independent offsite power sources need to be available;
- Adequate actual and unit trip contingency voltages must be observed on the transmission system supplying the nuclear unit;
- Stable system frequency must be present

After the successful restoration of a nuclear unit, system operators need to take into account the distinct characteristics and considerations associated with nuclear units. In terms of system operation and load dispatch, several key factors are typically taken into consideration.

- Given the highest possible priority to restore power to the nuclear units.
- High priority should be assigned to repair transmission lines which connect nuclear units and power system, if they are damaged during faults.
- Once A.C. power becomes available, procedures should specify the circuit breaker operations required to restore A.C. power.

Additionally, nuclear units are usually situated in remote locations and are characterized by high power output. However, they cannot receive power from off-site sources until the high-voltage transmission system is restored. To ensure the safety and stability of the power system, several transmission stability factors need to be taken into account before energizing the Extra High Voltage (EHV) transmission. These requirements address various concerns related to:

- Steady state overvoltage caused by excessive reactive power supply from the capacitive effect of EHV lines;
- Transient overvoltage caused by traveling wave phenomena;
- Dynamic overvoltage caused by transformer magnetizing inrush;
- Reduction in proper relaying protection reliability due to insufficient fault current.

8.4.4   A GENERIC METHOD FOR GENERATING UNIT'S STARTUP

The generic method for generating unit startup follows a step-by-step approach to ensure safe and efficient operation. It begins with a thorough inspection of the unit and its components. Auxiliary systems are then activated to provide necessary support during startup. The startup sequence involves gradually increasing speed and temperature to desired operating levels. Operators monitor parameters closely and make adjustments as needed. Once operating conditions are reached, the unit is synchronized with the grid and load is added gradually. The method emphasizes systematic and controlled procedures, with a focus on safety, monitoring, and coordination with the power system.

8.4.4.1   A Generic Method of Generating Units

The feasible regions of generating units are determined by both system constraints and physical limitations. To develop a comprehensive decision support system, it is important to consider the characteristics of different generating units. Figure 8.1 illustrates a generic model of a generating unit [287], and Table 8.2 provides a description of its parameters.

Considering the physical limitations of generating units, various parameters are required in addition to the common ones such as capacity, ramping rate, and minimal output. Different types of generating units have specific parameters tailored to their characteristics. For instance, fossil units require startup power requirement and critical maximal interval, while super-critical-once-through (SCOT) units necessitate startup power requirement and critical minimum interval. Blackstart units like hydro units have a startup power requirement of zero and require immediate parallel operation

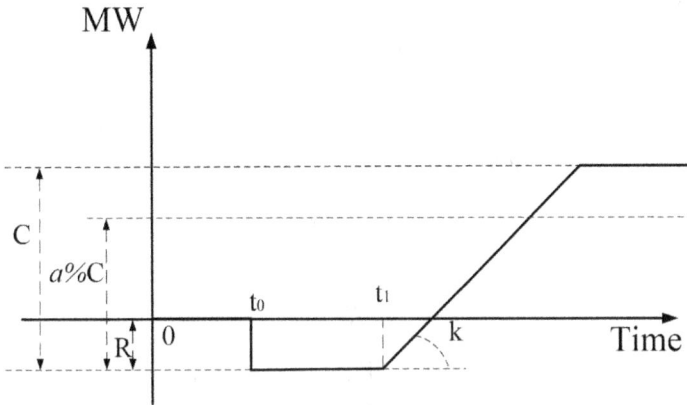

Figure 8.1   Generic model of generating units.

Table 8.2
Parameters of Generating Units

Cap.	Start-up requirement	Ramping rate	Min. output	Cranking to paralleling	Critical max. interval	Critical Min. interval
C	R	R	$\alpha\%$	T_1	T_2	T_3

---

after cranking. To simplify the model, a critical load is represented as a generating unit with a positive startup power requirement and zero ramping rate. Consequently, for a generating unit restarted at time t_0, the following equations are valid.

$$T_3 \leqslant t_0 \leqslant T_2 \tag{8.1}$$

$$T_1 = t_1 - t_0 \tag{8.2}$$

Generally, the generator output at time t may be written as

$$P(t) = \min\{k \times \max[t - (t_0 + T_1), 0], C\} - R \times U(t - t_0) \tag{8.3}$$

where
$U(.)$ is a unit step function, which is defined as

$$U(t) = t \geqslant 0 \tag{8.4}$$

8.4.4.2   Algorithms for Restoring Generating Units

Consider a generating unit or critical load denoted by x_i that is restarted at stage S, and let θ_S represent the set of all restarted generating units and critical loads at stage S. We define $f_S(x_i, \theta_S)$ as the minimum time required to crank all generating units or critical loads after stage S. The recursive computation can be expressed as follows:

$$f_S(x_i, \theta_S) = \min_{x_i \in \theta_S}\{\Delta t_{x_i - x_j} + f_{S+1}(x_j, \theta_{S+1})\} \tag{8.5}$$

where x_j is the last generating unit restarted (so x_j must be in θ_{S+1}). $\Delta t_{x_i + x_j}$ is the time to crank the generating unit (or critical load) x_j from θ_{S+1}. At each stage S, the technical constraints are represented as:

$$PF(\Omega_{E(S)}, P_{G(S)}, Q_{G(S)}, P_{CL(S)}, Q_{CL(S)}, P_{DL(S)}, Q_{DL(S)}) = 0 \tag{8.6}$$

$$P_\pi \in \mathrm{FR_P(II)}, Q_\pi \in \mathrm{FR}_Q(\mathrm{II}), \mathrm{II} = G_{(S)}, CL_{(S)}, \text{ and } DL_{(S)} \tag{8.7}$$

$$\underline{V_B} \leqslant V_B \leqslant \overline{V_B}, \qquad \forall B \in \Omega_{E(S)} \tag{8.8}$$

$$\underline{P_L} \leqslant P_L \leqslant \overline{P_L}, \qquad \forall L \in \Omega_{E(S)} \tag{8.9}$$

where energized block set $\Omega_{E(S)}$ includes all the energized buses and lines at stage S, $P_{G(S)}$, $Q_{G(S)}$, $P_{CL(S)}$, $Q_{CL(S)}$, $P_{DL(S)}$ and $Q_{DL(S)}$ are vectors of real power of generating units, reactive power of generating units, real power of critical loads, reactive power of critical loads, real power of dispatchable loads, and reactive power of dispatchable loads, respectively.

As introduced in previous chapter for power flow analysis. Here in restoration where power flow module is employed to verify the combinations of steps. $PF(\cdot)$ is the power flow equations. $\mathrm{FR}_{P(II)}$ and $\mathrm{FR}_{Q(II)}$ denote feasible regions of real power and reactive power of the set II. G(S), CL(S) and DL(S) are sets of generating units, critical loads and dispatchable loads at stage S, respectively. π represents any one of these three sets. P_{pi} and Q_{pi} are real power and reactive power that belong to set π, respectively. V_B is the voltage at bus B, and $\underline{V_B}, \overline{V_B}$ are the corresponding lower and upper limits. P_L is the real power flow on line L, and $\underline{P_L}, \overline{P_L}$ are the corresponding lower and upper limits.

In this model, (8.6) represents the power flow equations at each stage of restoration; (8.7) shows that the real power and reactive power of each generating unit, critical load and dispatchable load should stay within the feasible regions at each stage; (8.8) and (8.9) indicate that the voltage at each bus and power flow through each line should stay within limits. More constraints may be involved in this algorithm.

To solve this complex multistage optimization problem, a method with two interacting sub-problems is proposed. In these two sub-problems, an energized block of the system, i.e., $\Omega_{E(S)}$, is determined by the primary problem, while the operating point for the block, i.e., $P_{G(S)}$, $Q_{G(S)}$, $P_{CL(S)}$, $Q_{CL(S)}$, $P_{DL(S)}$ and $Q_{DL(S)}$, is specified by the secondary problem. With this framework, practical constraints related to system operation are incorporated.

8.5   CONSTRAINTS DURING RESTORATION

A restoration strategy study assessed the feasibility of restoring a power system under steady state and transient conditions. It examined generation-load balance, voltage control, and steady state over-voltage to ensure proper system operation. The study also evaluated the capability of blackstart units to handle reactive power from transmission system charging currents. Feasibility and compliance with operational limits were verified at each restoration step. The robustness of tested blackstart units in compensating for unavailable critical components was assessed. This study provided valuable insights for a successful and feasible restoration strategy.

The steady state analysis of restoration includes:

- Balance of generation and load
- Voltage control and steady state overvoltage analysis
- Capability of the blackstart units to absorb reactive power produced by charging currents of the transmission system.

- For each step of restoration, it is necessary to ensure feasibility and compliance with required operational limits.
- Verification of the robustness of the tested blackstart units to ensure its ability to compensate for the unavailability of key components to be used in the plan

An important insight from the steady state analysis of restoration is the emphasis on maintaining a balance between generation and load. This ensures that the restored power system can meet the demand for electrical power. Additionally, voltage control and steady state overvoltage analysis are crucial in ensuring that the system operates within acceptable voltage limits. Another key aspect is the evaluation of the blackstart units' capability to absorb reactive power generated by the charging currents of the transmission system. This helps in maintaining system stability during the restoration process. Furthermore, feasibility and compliance with operational limits must be ensured at each step of restoration to prevent any operational issues. Lastly, verifying the robustness of the tested blackstart units is vital to ensure their ability to compensate for the unavailability of critical components required for the restoration plan. These insights highlight the importance of meticulous analysis and planning in achieving a successful and reliable power system restoration. The dynamic analysis of blackstart units includes following issues:

- Load-frequency control
- Voltage control
- Large induction motor starting
- Motor starting sequence assessment
- Self-excitation assessment
- System stability
- Transient overvoltage

In conclusion, the dynamic analysis of blackstart units plays a crucial role in the successful restoration of power systems. By considering factors such as load-frequency control, voltage control, induction motor starting, motor starting sequences, self-excitation, system stability, and transient overvoltage, the reliability and stability of blackstart units can be ensured during the restoration process. Through a comprehensive examination of these dynamics, the feasibility and effectiveness of blackstart units in supporting system restoration can be assessed and optimized. This analysis enables power system operators to make informed decisions and implement strategies that enhance the resilience and efficiency of the restoration process, ultimately leading to the swift recovery of power supply in the event of a blackout.

8.5.1   ACTIVE POWER BALANCE AND FREQUENCY CONTROL

During the restoration process, it is crucial to ensure that the system frequency remains within permissible limits, considering factors such as turbine resonance, system stability, and protection settings. This requires a careful approach to load pickup, taking into account the inertia and response of the restored and synchronized system. Operators must also consider the frequency response of prime movers when there is a sudden increase in load. While small load increments may prolong the restoration duration, large increments carry the risk of frequency decline and another system outage. Manual control is often employed for prime movers at this stage, and a dynamically calculated guideline can be useful in determining the allowable load pickup based on generator capability. It is preferable to maintain constant frequency control for speed governors associated with blackstart units, and as additional units are added, droop control becomes the preferred control mode for speed governors. This ensures a balanced and stable operation during the restoration process.

8.5.2   REACTIVE POWER BALANCE AND OVERVOLTAGE CONTROL

The generator reactive capability (GRC) curves provided by manufacturers for operation planning often have a wider range than what can be practically achieved during actual operation. These curves are typically based solely on the design parameters of synchronous machines and do not account for plant and system operating conditions that may limit their applicability. The concern regarding GRC stems from the importance of reactive power for maintaining voltage support during the transmission of large power blocks.

In the early stages of the restoration process, it is essential to maintain system voltages within allowable limits, which are typically set lower than normal. Several measures are employed to achieve this, including energizing fewer high voltage lines, operating generators at minimum voltage levels, overriding switched static capacitors, connecting shunt reactors, adjusting transformer taps, and picking up loads with lagging power factors. Another consideration during the reenergization of cables is the capacity of the energizing system to handle the charging current of the cable.

It is important to note that cables are operated at loads well below their surge impedance loading, and their charging currents per mile are approximately ten times higher than those of overhead lines of the same voltage class. Additionally, approximately 2 MW of load can be picked up per MVAr of charging. These factors highlight the significance of managing voltages and cable charging currents during the restoration process.

8.5.3   SWITCHING TRANSIENT VOLTAGE

The energization of equipment under blackstart conditions can lead to higher overvoltages compared to normal operation. In the reintegration phase, it is preferable to energize a substantial portion of the high-voltage transmission system, taking into account the limits imposed by switching transient voltages. Energizing smaller sections can cause delays in the restoration process. However, when energizing a large section, there is a potential risk of damaging the insulation of the equipment. Temporary overvoltages can occur due to switching surges, which may arise from various factors:

- Switching circuits that saturate the core of a power transformer, such as when cables and transformers are energized together.
- The harmonic rich transformer inrush currents interact with the harmonic resonances of the power system.
- The resonant frequencies are a function of the series inductance associated with the system's short circuit strength and by the shunt capacitances of the cables and lines.
- Higher inductances (relatively weak systems) and higher capacitances (long cables) yield lower resonant frequencies and a higher chance of temporary overvoltage.

In the process of system restoration, the challenge lies in finding a method to energize a multi-section line without surpassing the equipment's basic impulse level. Some utilities choose to energize the line section by section, focusing on one circuit of the double circuit line at a time, while keeping the generator voltage at the minimum excitation level permissible. This approach helps mitigate both line charging currents and transient voltage magnitudes but may result in a longer restoration period. To streamline the process and provide clear instructions for energizing transmission lines, a straightforward methodology is required to establish practical guidelines.

8.5.4   SELF EXCITATION

In order to regulate the overvoltage caused by charging currents along a line or cable, the charging requirements must be carefully managed. In some cases, the blackstart unit is relied upon to absorb the reactive power, which can be significant. However, there is a risk of self-excitation if the charging current is proportionally high compared to the size of the generating unit. This can lead to an uncontrolled increase in voltage, potentially resulting in equipment failure. Self-excitation can also occur from the load end if the supply is inadvertently lost, such as when a transmission line or cable is opened at the sending end, leaving the line connected to a large motor or group of motors. To prevent generator self-excitation during the restoration process, various strategies can be employed as follows:

- Using the generator with less xd to energized EHV transmission line.
- Using two or more generators to charge one EHV transmission line.
- Changing the impedance of transformer by tap changed.
- Changing the parameters of EHV transmission line by reactive power compensation.

During system restoration, managing the charging currents along a line or cable is crucial to control overvoltage. This involves careful consideration of the charging requirements and the role of the blackstart unit in absorbing reactive power. However, there is a risk of self-excitation if the charging current exceeds the capacity of the generating unit, leading to uncontrolled voltage rise and potential equipment failure. Self-excitation can also occur from the load end if supply is inadvertently lost. To mitigate these risks, strategies such as using generators with lower reactance, employing multiple generators for charging, adjusting transformer impedance through tap changes, and compensating for reactive power in EHV transmission lines can be implemented. These strategies help prevent generator self-excitation and ensure a stable restoration process.

8.5.5 COLD LOAD PICKUP

While blackstart analysis typically focuses on starting large motors, it is also important to consider the pickup of other loads during the restoration process. Understanding the anticipated real and reactive power loads at different stages of restoration is crucial, taking into account factors such as normal peak and light loads, load power factors, power factor corrections, load types, and total connected loads. The challenge lies in utilizing this data to determine the real and reactive power loads as a function of elapsed time.

In particular, when a load has been de-energized for several hours or longer, the inrush current upon re-energization can be significantly higher, reaching up to eight to ten times the normal value. The magnitude and duration of this inrush current depend on the specific load types served by the feeder, including lighting, motors, and thermostatically controlled devices like air conditioners, refrigerators, freezers, furnaces, and electric hot water heaters. Various components of the load contribute to the total inrush current, presenting several factors that need to be considered.

- The component due to the filaments of incandescent lights.
- The starting of motors when the load is picked up.
- Thermostatically controlled loads, which turn on and off automatically to hold temperature to a desired, preset value.

8.5.6 SYSTEM STABILITY

During the restoration process, it is essential to ensure the stability of the system, particularly in terms of voltage and rotor angle. System stability assessments are typically performed to evaluate angular stability when multiple generating units are involved in the restoration stages. However, for plans with a single generating unit, the focus shifts to assessing frequency stability. Additionally, it is important to consider the presence of low-frequency oscillations caused by insufficient oscillation damping, as they can impact system stability.

8.5.7 PROTECTIVE SYSTEMS AND LOCAL CONTROL

Relay settings are typically established based on normal operating conditions. However, during the restoration process, as power system configurations and operating conditions change, there is a possibility of unintended relay operations. In situations where a substantial frequency decline occurs, it becomes necessary to reduce the connected load to mitigate the impact. Under frequency load shedding schemes are often employed for this purpose. When closing network loops, such as reconnecting two subsystems, there can be a significant phase angle difference across the circuit breakers. It is crucial to minimize this difference to safely close the circuit breaker without risking system instability.

8.5.8 SYSTEM SECTIONALIZING AND RECONFIGURATION

After a major outage caused by significant disturbances, the power grid may experience loss of generation, transmission network components, and loads. This leads to a substantial change in the topology of the post-disturbance power grid compared to its pre-disturbance state. The opening of breakers connected to de-energized buses is necessary to ensure safe energization of components without unintended connections. Additionally, to enable parallel restoration or prevent loss of synchronism, the surviving power grid is divided into multiple islands.

To restore the power grid to its normal operation state, the topology of the transmission and distribution networks undergoes modifications throughout the restoration process. Both sectionalizing and reconfiguration schemes are crucial in achieving a smooth, secure, and reliable restoration process. These topological changes require a series of switching actions that adhere to specific security criteria.

This subsection focuses on reviewing research related to system sectionalizing and reconfiguration. The discussion will center on the modeling and solution methods employed in system sectionalizing, network reconfiguration, path selection, and re-dispatch. Each aspect will be elaborated upon in detail, highlighting their significance in mitigating the impact of switching actions.

8.5.8.1 System Sectionalizing

System sectionalizing is utilized to expedite the restoration process for both distribution and transmission systems. However, the impact of sectionalizing differs between these two levels of the power grid. In distribution systems, sectionalizing switches are employed to locate and isolate faults, enabling rapid restoration of the load through alternative paths. In contrast, in transmission systems, following a widespread outage, the power grid is divided into multiple subsystems through sectionalizing. These subsystems can then be restored in parallel and eventually reintegrated or synchronized. The effectiveness of sectionalizing, whether in distribution or transmission systems, relies on the strategic placement and number of sectionalizing points.

8.5.8.2 Network Reconfiguration and Path Selection

The control actions involved in network reconfiguration encompass tasks such as energizing essential transmission paths, restoring critical loads, initiating and stabilizing the operation of generating units, and closing transmission loops. These control actions serve two primary purposes: 1) ensuring reliable energy delivery to non-blackstart units and interrupted loads, and 2) strengthening electrical islands and enhancing reliability by energizing crucial transmission corridors. Consequently, the optimality of network reconfiguration can be assessed based on the energy served or, alternatively, by considering the improvements in reliability and transmission efficiency.

8.5.8.3   Re-dispatch to Mitigate Switching Impact

The switching of transmission lines triggers transient processes in the power grid, leading to potential transient over-voltages, over-currents, or mechanical stress on grid components. To ensure the security of switching actions, dynamic simulation programs are utilized for assessment. If the security criteria are not met, appropriate remedial actions are implemented to mitigate the impact of switching.

8.5.9   LOAD RESTORATION

Customers may experience a loss of electricity supply for two main reasons. Firstly, intentional load shedding by system operators may be implemented to maintain power balance, frequency stability, or transient stability. Secondly, interruptions in load supply may occur due to the loss of generation capacity or transmission corridors during large disturbances or cascading failures. Restoring service to these affected customers, referred to as "load restoration," is a crucial task in the overall restoration process.

During the load restoration phase, the timing depends on restoration strategies and the system's post-disturbance condition. However, certain common features can be identified. Firstly, major generation capacity and the core transmission systems have been restored. Secondly, the objective is to fully and promptly restore the load while ensuring security and reliability. Unlike other restoration phases where load control is emphasized, load restoration focuses on restoring load in larger increments and in parallel to minimize the adverse impact of outages.

Nevertheless, the load restoration process must be carried out with persistent consideration of security and reliability criteria. As load pickup actions can introduce disturbances to the still vulnerable power grid, it is crucial to carefully study steady-state, transient, short-term, and long-term dynamics to prevent further shutdowns or system collapses caused by inappropriate actions. The technical challenges associated with load restoration and relevant research advancements are examined in the following sections.

8.6   RESTORATION STRATEGIES IN INDUSTRY

Restoring power to a system after a blackout presents a complex decision and control challenge for system dispatchers and restoration planners. The main goal after a blackout is to restore the load in the system as swiftly as possible. Depending on the specific conditions of the system, there are different strategy options available for system restoration. This can be observed in the practices of industry leaders such as PJM, AEP, and Hydro-Quebec, where their system restoration strategies are implemented as exemplars.

8.6.1   SYSTEM RESTORATION STRATEGY IN PJM

PJM Interconnection functions as a regional transmission organization (RTO) responsible for coordinating the wholesale electricity movement in various areas, including Delaware, Illinois, Indiana, Kentucky, Maryland, Michigan, New Jersey, North Carolina, Ohio, Pennsylvania, Tennessee, Virginia, West Virginia, and the District of Columbia. As an impartial and independent entity, PJM manages the world's largest competitive wholesale electricity market and ensures the reliability of the largest centrally dispatched grid worldwide. With a membership exceeding 500, PJM comprises power generators, transmission owners, electricity distributors, power marketers, and large consumers.

The approach employed by PJM for system restoration can be described as the "Build-Upward" strategy combined with a focus on critical services. This strategy entails several key tasks, such as initiating blackstart units, starting up non-blackstart units, restoring isolated segments, and synchronizing these segments to restore the system to its pre-disturbance state.

Depending on specific system conditions, PJM also employs strategies referred to as "Build-Inward," "Build-Outward," and "Build-Together." In the case of subsystem stabilization within PJM, requests from neighboring Transmission Owners for cranking power take precedence over restoring additional customer load for the supplying company. The "Build-Inward" strategy is utilized for the power import subsystem, while the "Build-Outward" strategy is implemented for the power flow export subsystem.

8.6.1.1   Restoration Process of PJM

Power system restoration holds significant importance in the operation of an electric power grid. After a power outage occurs, control center dispatchers collaborate with field crews to reinstate the generation and transmission system, as well as to recover load and reinstate service. In the case of the PJM system, if a separation event takes place, PJM dispatchers, along with the Transmission Owners, Generation Owners, and Load Serving Entities, adhere to a comprehensive guideline aimed at efficiently restoring the PJM RTO to its pre-disturbance state [288]. The following are the general steps involved in the restoration process of the PJM grid:

- **Ascertain System Status**. In this step, through communications, the PJM dispatcher determines the transmission and generation loss, equipment damage, and extent of service interruption.
- **Determine Restoration Process**. After the status of the PJM RTO is determined, the restoration strategy is developed.
- **Disseminate Information**. The purpose of this step is to provide updated information of the system status to the appropriate personnel.
- **Implement Restoration Procedure**. When a Transmission Owner/Generation Owner is in a completely isolated or blackout condition, this step is to direct the restart of Generation Owners internal generation and load on-line generation in planned steps while maintaining system load, scheduled frequency, voltage control, and reserves.
- **Member Interconnection**. After Transmission Owners have restarted and desire to interconnect and share reserves or Transmission Owners have coordinated plans to restart while interconnected, this step is to interconnect and control frequency, tie line, voltage schedules, share reserves, and coordinate emergency procedures.
- **PJM Returns to Normal Operation. Re-establish PJM single control center coordination**. This occurs when an ACE can be calculated for the area to be controlled (entire PJM area or portion) and a return to central coordinated operation is desired by PJM and the Transmission Owners. When conditions permit, PJM dispatchers notify all Transmission Owners/Generation Owners that the PJM RTO is returning to normal operation, i.e., free flowing Transmission Owner-to-Transmission Owner ties, Balancing Authority to Balancing Authority ties, generation under AGC control, and return to published regulation and reserve requirements.

The restoration process of the PJM grid involves assessing system status, developing a restoration strategy, sharing information with personnel, restarting generation during isolation or blackout, enabling member interconnection, and returning to normal operation with central coordination. This process includes evaluating transmission and generation loss, developing a strategy, informing relevant personnel, restarting generation while maintaining system parameters, coordinating interconnection and emergency procedures, and returning to normal operation with compliance and notification.

8.6.1.2   Technical Issues of PJM's Restoration

Key considerations arise when creating a restoration plan within the framework of PJM guidelines. These concerns, outlined below, are documented in the Manual 36 – Restoration of PJM [288].

- Allowable frequency bands at different stages of restoration
- Allowable and recommended voltage bands
- Maintaining generation reserves in restoration. This process takes into account the expected response characteristics of the remaining resources when the resource with the greatest output is lost.
- Rebuilding the transmission system
- Restoring the generation fleet to full level
- Running restoration drills

To develop a restoration plan in accordance with the PJM guidelines outlined in Manual 36, several technical aspects must be addressed. These include the implementation of synchronous reserve, generator dynamics (referred to as dynamic reserve), voltage regulation and control, restoration of critical load, and testing of blackstart units [289].

Synchronous Reserve: Synchronous Reserve refers to the reserve capacity required to restore tie-lines or other facilities to their pre-contingency state promptly after an imbalance occurs between load and generation. For the PJM system, Synchronous Reserve can be achieved through on-line generation that can be loaded within 10 minutes or load that can be manually shed within 10 minutes.

Generator Dynamic: To prevent a complete system loss in the event of a generator failure or unexpected large load pickup, PJM has established a conservative guideline. The dynamic reserve is the amount of reserve available to maintain system stability during frequency disturbances. It should be sufficient to withstand the loss of the largest energy contingency and consists of two components: Reserve on Generators, which relies on generator governor action during frequency disturbances, and load with under-frequency relaying, where system load is shed above the frequency at which generators would normally separate during a frequency disturbance.

Voltage Regulation and Control: Smooth implementation of the restoration plan involves permitting wide voltage ranges, typically from 90% to 105% of normal conditions. Static control devices such as shunt reactors and capacitors are utilized to maintain voltages within the lower range and preserve the preferred reactive output range of connected generators. Additionally, during the restoration of the Extra High Voltage (EHV) system, various methods are employed to control line charging current and minimize line overvoltage, such as operating the system in lower voltage ranges, minimizing parallel line energization, and prioritizing restoration of reactive loads.

Restoring Critical Load: According to PJM's definition, only the supply of off-site power to nuclear generating units is considered critical load. These loads experience delays as they are typically connected through EHV lines and served by the main transmission lines. However, in the PJM system, Local Transmission Owners (LTOs) may have their own critical loads to be balanced with connected generation, control voltages, and ensure the restoration process continues.

Blackstart Unit Testing: A blackstart unit is required to independently start and energize a dead bus without external assistance. According to PJM guidelines, blackstart unit testing is included in PJM's Tariff, and there are minimum requirements for blackstart capabilities imposed on LTOs.

In summary, the restoration process within the PJM system involves key considerations such as Synchronous Reserve, Generator Dynamic, Voltage Regulation and Control, Restoring Critical Load, and Blackstart Unit Testing. Synchronous Reserve ensures the capability to restore tie-lines promptly after an imbalance occurs. Generator Dynamic safeguards against system loss during generator failure or unexpected load pickup. Voltage Regulation and Control permit a wide voltage range for smooth restoration, while Restoring Critical Load focuses on nuclear generating units and other critical loads. Blackstart Unit Testing ensures the ability to independently start and energize a dead bus. These guidelines play a crucial role in effective power system restoration within PJM.

8.6.2   SYSTEM RESTORATION STRATEGY IN HYDRO-QUEBEC

The Hydro-Quebec system has a quasi-radial structure, with two remote generation sites located in the northwest and east of Quebec connected to the load in the southwest via long transmission lines. The transmission network consists of 506 substations operating at voltages ranging from 44 kV to 765 kV, with a total of 32,500 km of lines, including 11,422 km at 735 kV. Over 95% of the network's 40,000+ megawatts of generation capacity comes from hydropower. The system also features 17 asynchronous interconnections for power exchanges with neighboring systems. The 735 kV transportation system relies heavily on series compensation, utilizing 34 lines with 25% to 40% compensation, as well as dynamic shunt compensation with 22 static and synchronous compensators of 300 MVA capacity, and 90 switchable shunt reactors of 330 or 165 MVAr.

During system restoration, the Hydro-Quebec system implements a Build-Downward strategy. Challenges in the restoration process include dealing with various overvoltage conditions arising from line charging, harmonics, and switching surges. To address this, the main grid is divided into five networks that are energized independently and simultaneously. The primary focus during this stage is to energize the networks as quickly as possible without prioritizing load restoration. Once the networks are energized, they are synchronized, and service to the load is gradually restored in small increments.

8.6.2.1   Restoration Process of Hydro-Quebec

The restoration strategy for the Hydro-Quebec system differs from that of other systems due to its unique characteristics. In general, most electrical networks prioritize energizing from lower voltage lines to mitigate overvoltage issues during restoration, with high voltage lines being energized later to connect different islands. However, Hydro-Quebec follows a similar strategy but with a distinction. Due to the construction of generating stations situated farther away from load centers and the expansion of the extra-high-voltage transmission system, Hydro-Quebec has chosen to implement a system restoration scenario utilizing the 735-kV transmission system.

To facilitate restoration within the Hydro-Quebec system, the following principles are applied to restore its main grid:

- Partition the system into several islands, energized independently and simultaneously
- Energize the islands as quickly as possible. At this stage, restoring load is unnecessary. To maintain safety of system during restoration, it is necessary to minimize the number of control variable adjustments from stage to stage
- Synchronize the islands after energization of islands
- Restore loads in small increments

The restoration strategy employed by the Hydro-Quebec system reveals insightful approaches to grid recovery. By partitioning the system into independent islands, simultaneous energization becomes feasible, leading to enhanced efficiency in the restoration process. Prioritizing the rapid energization of these islands, with a deliberate focus on speed rather than immediate load restoration, demonstrates a commitment to ensuring system safety. The emphasis on minimizing control variable adjustments between stages further reinforces the dedication to safety measures. The subsequent synchronization of the energized islands fosters coordination and stability within the grid. Lastly, the gradual and incremental restoration of load exemplifies a controlled and stable approach to service recovery. Overall, these principles illustrate the Hydro-Quebec system's commitment to effective and efficient restoration practices, providing valuable insights for power grid management.

8.6.2.2   Criteria for Hydro-Quebec's System Restoration

The restoration strategy employed by Hydro-Quebec involves a staged reenergization of the network. However, this approach presents challenges related to the Ferranti effect during line energization and temporary overvoltages caused by powering transformers. Research indicates that overvoltage issues are more pronounced when the short-circuit level of the power source is low, the transmission lines are long, and the capacity of the transformers being energized is high [290]. These challenges are particularly prominent within the Hydro-Quebec power system. To address voltage constraints, operators have the flexibility to adjust generator voltages, transformer taps, switchable shunt reactors, and the load amount to be restored.

During the restoration process, three types of simulations are typically performed to satisfy constraints. Firstly, steady-state simulations are conducted to identify scenarios that maintain voltages within acceptable limits and minimize operational times. Secondly, harmonic analysis is employed prior to transformer energization to detect any harmful harmonics that could potentially cause equipment damage. Lastly, switching surge studies are carried out to assess potential excessive switching-induced overvoltages.

To address the major concerns of restoration, practical and technical criteria are implemented, ensuring that every operation at each stage remains under strict control to prevent any damage. These criteria are described in detail in the following sections, encompassing both practical and technical aspects of the restoration process [291]. This comprehensive approach ensures a controlled restoration process while mitigating risks and preserving the integrity of the Hydro-Quebec system. The practical and technical criteria for restoration are outlined as follows:

Practical Criteria. The objective of practical criteria is to minimize the duration of restoration process, and the following issues are included.

- *Minimize equipment and switching operations.* To minimize the restoration time and the risks associated with operating equipment, the system should be restored with minimum equipment. Furthermore, at the first stage of restoration, the available energy is limited. Therefore, the number of energy consuming switching operations should be minimized.
- *Minimize travel.* To ensure safety at each stage of restoration, the equipment without remote control should be avoided.
- *Restore progressively.* The objective of each switching operation should proceed progressively toward the final path configuration. Other switching operations should be avoided.

Technical Criteria. Technical criteria concerning generating facilities and transmission facilities are implemented to maintain reliability and efficiency of restoration.

- Generating facilities. Within this category, voltage control, reactive power control, active power control, and auxiliary power supply, are included
- Transmission facilities. Several issues, such as voltage level, reactors on 735kV lines, energizing power transformers, series compensation, static and synchronous compensators, are involved.

The restoration process of the Hydro-Quebec system incorporates practical and technical criteria to ensure a safe and efficient restoration while minimizing duration. Practical criteria focus on minimizing equipment and switching operations, avoiding equipment without remote control, and progressing switching operations gradually. Technical criteria encompass maintaining reliability and efficiency through voltage control, reactive power control, active power control, auxiliary power supply, and considerations for generating facilities, as well as voltage levels, reactors, power transformers, series compensation, and compensators for transmission facilities. Adhering to these criteria effectively manages the restoration process, ensuring a secure and efficient restoration of the Hydro-Quebec system.

8.6.3   RESTORATION PROCESS OF AEP

The restoration process in the context of AEP (American Electric Power) is implemented at the Transmission Dispatch Center level. This process typically consists of five phases aimed at systematically restoring power supply. The phases include assessing the system status, establishing startup and emergency power to power plants, restoring transmission systems, prioritizing the restoration of feeders, and finally restoring the remaining stations and feeders. Specifically, the phases are:

- Assess status of system
- Establish startup and emergency power to power plants
- Restore transmission systems
- Restore priority feeders
- Restore remaining stations and feeders

Following a major fault, immediate initiation of an assessment is crucial in preparation for the restoration process. This assessment encompasses vital information such as the status of generating units, breaker status, station alarms, load/customer outages, and the status of neighboring systems. The second phase is time-sensitive and involves utilizing blackstart and load rejected units as power sources while also establishing paths to crank non-blackstart units. Additionally, cranking power, emergency power, and start-up power are implemented during this phase. In the third phase, maintaining acceptable levels of frequency, generator voltage, station voltage, and spinning reserve becomes a priority to ensure system stability.

To ensure the safety and integrity of the system, steady-state, dynamic, and transient analyses are conducted throughout the restoration process. These analyses are crucial in making informed decisions and maintaining system stability during restoration.

- *Steady State Analysis*. Load flow analyses are conducted to calculate the steady state conditions which may exist during the various steps comprising the restoration plan. By this method, steady state feasibility of sequence steps, system power flows and voltage levels, generators' real and reactive power levels, and switching sequence can be evaluated.
- *Dynamic*. Several issues, such as load/frequency response for blackstart and automatic load rejection (ALR) units, blackstart unit capability, and system stability, should be verified.
- *Transient*. In this issue, network switching performance (overvoltages, etc.) is verified. Abnormal transient conditions, i.e. ferroresonance, harmonic resonance, should be identified. Load/frequency response for blackstart units is also need verified.

By conducting these analyses, the restoration process can be effectively guided and potential risks or issues can be addressed, ensuring a safe and successful restoration of the system.

8.7   EXAMPLES

In this part, two examples will be employed to demonstrate the restoration process.

8.7.1   ILLUSTRATIVE EXAMPLE FOR RESTORING AN INTERCONNECTED SYSTEM

A case based on the New England 39-bus system is employed to demonstrate the concepts of system restoration. This system is shown in Figure 8.2. This system has 10 generating units with a total generation of 7347MW and 2470 MVar. It has a total load of 6855 MW and 2155 MVArs. Suppose there are 3 blackstart units at bus 31, 36, and 37. Based on the criteria of partitioning system into islands that was discussed in section II, this system is divided into 3 islands.

During the optimization process, the duration for restarting all generating units is minimized, and all constraints should be satisfied at the implementation of each restoration actions. The following constraints may be involved:

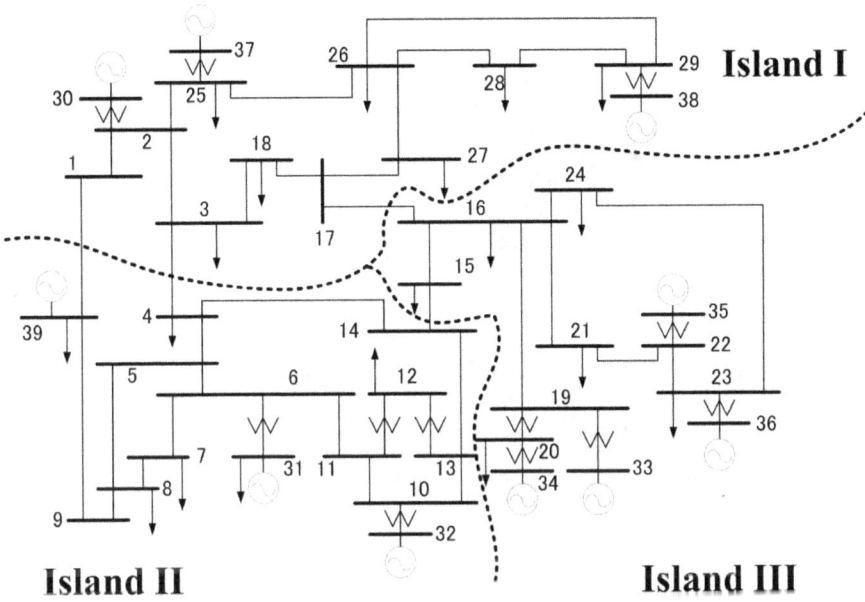

Figure 8.2 New England 39-bus system one-line diagram.

- C1: limit of each generating unit
- C2: steady-state overvoltage
- C3: switching transient overvoltage
- C4: voltage stability
- C5: capacity of each line

The sequence of restoration actions associated with constraints is shown in Figure 8.3. The constraints are described within brackets. By this sequence, all generating units in Islands I are restarted. It should be noted that picking up loads is only a means to maintaining stability of the island. Similarly, the method can be facilitated on Island II and III. The total output of each island is illustrated in Figure 8.4.

In order to restart the system, it requires more actions, including Establish Transmission Grid, Form Electrical Island, Serve Loads in Area, and Synchronize Electrical Islands. The sequence of restoration actions associated with constraints is shown in Figure 8.5 The constraints are described within brackets.

The total output of the entire system before and after synchronization is shown in Figure 8.6

Therefore, the entire system is restarted by a sequence of actions. This methodology can be applied to different power systems with some customization.

8.7.2 ILLUSTRATIVE EXAMPLE FOR RESTORING GENERATING UNITS

This example performs a step-by-step computation. Each step will crank 1 NBS unit or 1 critical load. The major computational task within each step includes:

1. Find the un-energized bus set within a given depth;

Start the black start unit at bus 37
Energize busbar 37(C3)
Energize line 37-25 (C1, C2, C3, C5)
Energize busbar 25 (C3)
Pick up load at 25 (C1, C2, C4, C5)
Energize line 25-2(C1, C2, C3, C5)
Energize busbar 2(C3)
Energize line 2-30(C1, C2, C3, C5)
Energize busbar 30 (C3)
Crank unit at bus 30 (C1, C2, C5)
Energize line 25-26 (C1, C2, C3, C5)
Energize busbar 26 (C3)
Energize line 26-29 (C1, C2, C3, C5)
Energize busbar 29 (C3)
Pick up load at 29 (C1,C2, C4, C5)
Energize line 29-38   (C1, C2, C3, C5)
Energize busbar 38 (C3)
Crank unit at bus 38 (C1, C2, C5)

Figure 8.3 Sequence of restoration actions.

2. Find a list of NBS units and critical loads to be restored;
3. Select one NBS unit or critical load to be restored;
4. Establish a path to the selected component; and
5. Find a feasible operating point to crank the selected component with MW/MVar balance constraints. The following steps present this logic.

- **Step 0:** BS = 73112, Search Depth = 2. Find the un-energized bus set within this range. The bus set is {73108, 73109, 73110, 73111}. See Figure 8.7.
  Within this bus set, three NBS units, (namely, 73108, 73109, and 73110) and one critical load (namely 73111) are found. By checking feasibility, only NBS on bus 73109 (startup requirement 25 MW) and 73110 (startup requirement 12 MW) can be cranked in the next step. By the SRN logic, the NBS with less startup requirement and faster ramping capacity will be cranked with higher priority. The next step will try to crank NBS unit on bus 73110. The time of this step is 60 mins (since the crank to parallel of BS is 60 mins).

- **Step 1:** Crank NBS unit on bus 73110. Find the path to connect bus 73110 by Dijkstra algorithm. The time for building path (73112-73109-73110, 73112-73108) is 5 mins. The energized components are shown in red in Figure 8.8.
  Build the OPF model as follows,

$$\min(P^1_{G.73112} - P^0_{G.73112})$$

  subject to: power flow and operational constraints   (8.10)

  (NBS on bus 73110 is modeled as fixed load demand with startup requirement)

where $P^1_{G.73112}$ is the output of BS on bus 73112 at step 1, $P^0_{G.73112}$ is the output of BS on bus 73112 at step 0. Solve this OPF model. The ramping of BS unit to crank NBS on bus 73110 is 4

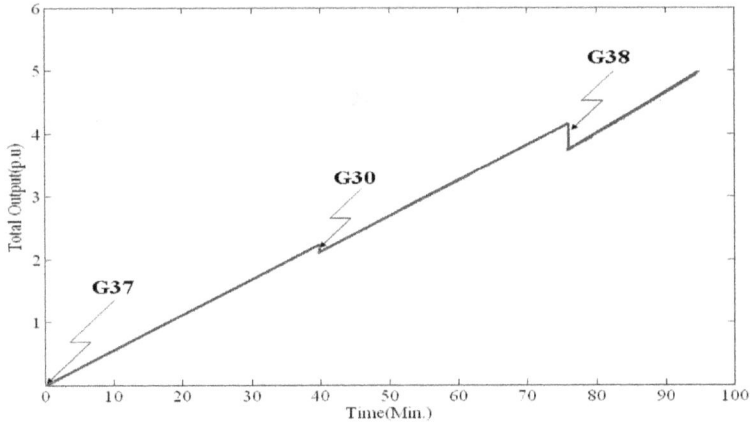

(a) Total output of Island I

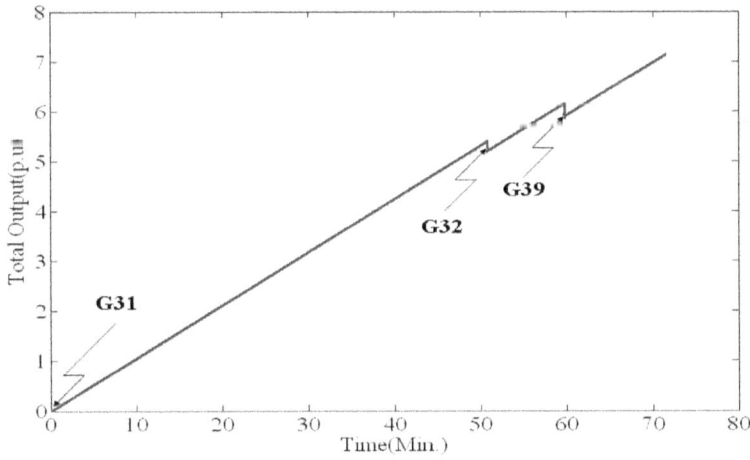

(b) Total output of Island II

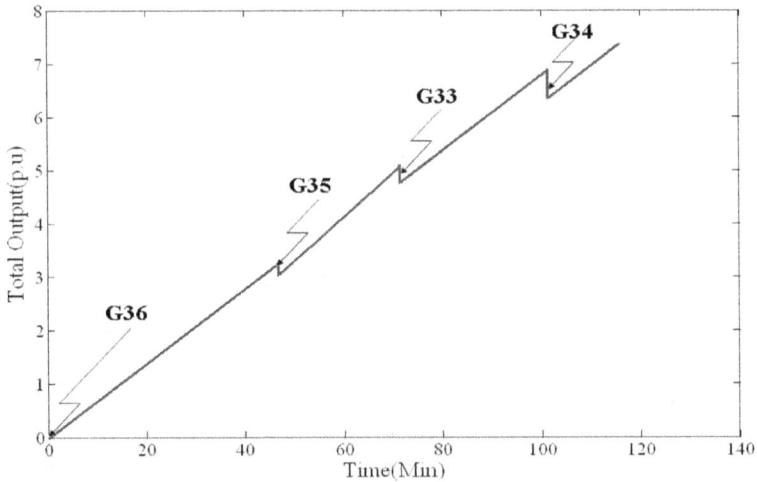

(c) Total output of Island III

Figure 8.4 Total output of each island.

| Energize line 2-1(C1, C2, C3, C5) |
| Energize busbar 1(C3) |
| Connect tie line 1-39 (C1, C2, C3, C5) |
| Synchronize Island I and II (C1, C2, C3, C5) |

Figure 8.5   Sequence of restoration actions.

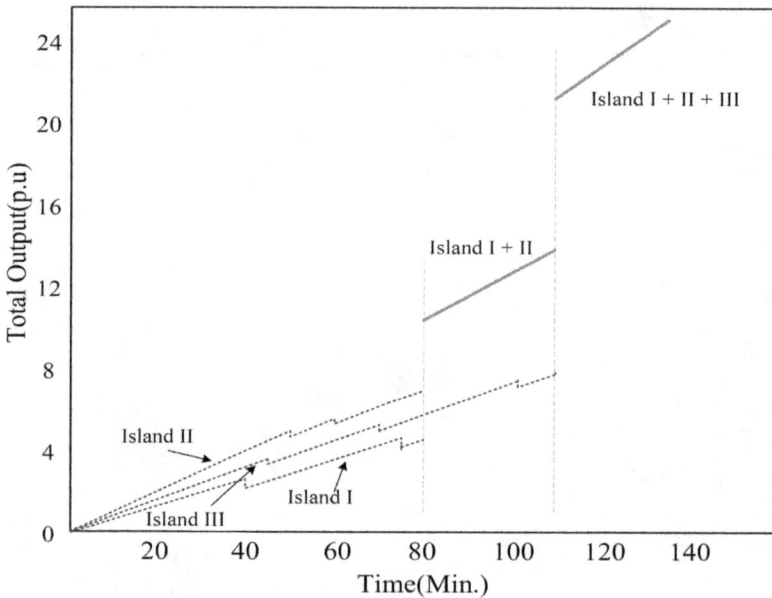

Figure 8.6   Total output of the whole system before and after synchronization.

mins (12MW/3MW/min). The total time duration of this step is 5 mins (ramping time of BS is within the time for building path). At the end of this step, the time is 65 (60 + 5) mins.

- **Step 2:** Search Depth = 2. Within this neighborhood, two NBS units, (namely, 73108 and 73109) and one critical load (namely 73111) are found. The critical load is regarded as ramping rate equals to zero. Based on the SRN logic, the critical load will be cranked after the NBS units are cranked (unless higher user-defined priority than NBSU is set for critical loads). NBS unit on 73108 has a larger ramping rate (3.3MW/min) than that on 73109 (1.8MW/min). Therefore, SRN will crank NBS unit on 73108 at this step. The energized components are shown in red in Figure 8.9.
  Note that SRN treats NBSU and critical load consistently. Namely, critical loads are regarded as zero-ramping-capacity NBSU equivalently. Based on the start up priority rule, critical loads are placed in the tail of the optional NBSU list in this step. On the other hand, dispatchable loads will be picked up, if necessary (i.e., OPF fails to find a feasible power flow solution), in cranking one NBSU or one critical load.

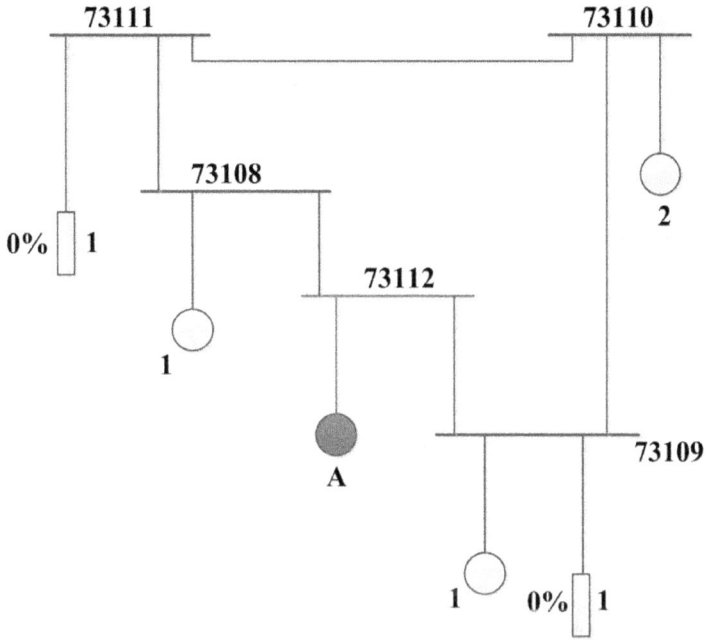

Figure 8.7 Searching for un-energized buses at step 0.

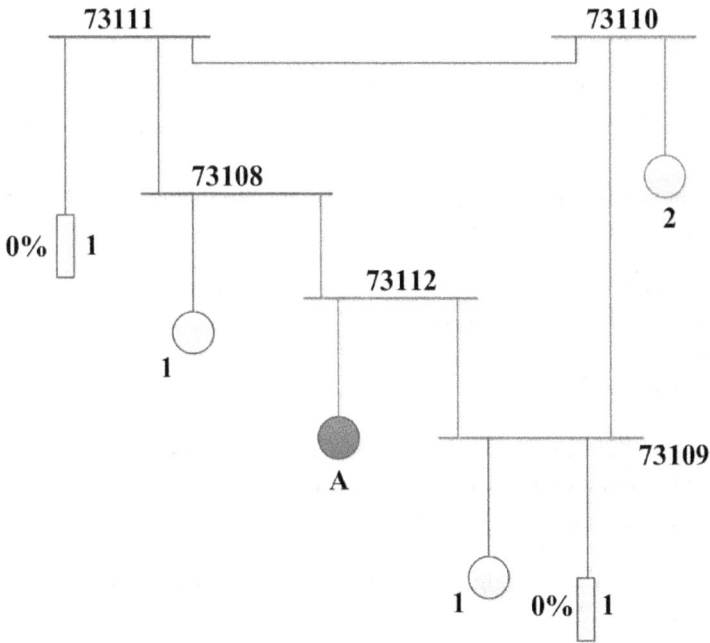

Figure 8.8 System Topology and state of components at step 1.

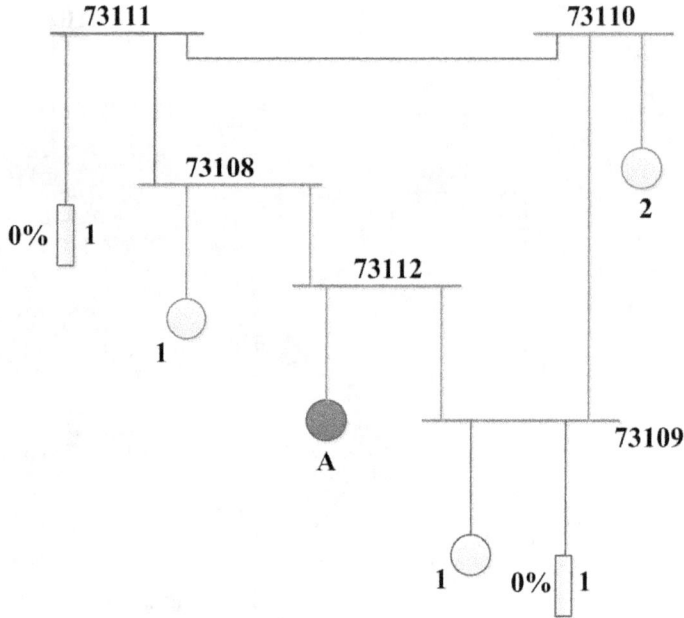

Figure 8.9 System Topology and state of components at step 2.

Build the OPF model as follows,

$$\min(P_{G.73112}^2 - P_{G.73112}^1)^2 + (P_{G.73110}^2 - P_{G.73110}^1)^2$$

subject to: power flow and operational constraints   (8.11)

(NBS on bus 73108 is modeled as fixed load demand with startup requirement)

Solve this OPF model. The ramping time for BS to crank this unit is 4.12 mins. At the end of this step, the time is 69.12 (65 + 4.12) mins.

- **Step 3:** Search Depth = 2. Within this neighborhood, one NBS unit (namely 73109) and one critical load (namely 73111) are found. SRN will crank NBS unit on 73109 at this step. The energized components are shown in red in Figure 8.10.
  Build the OPF model as follows,

$$\min(P_{G.73112}^3 - P_{G.73112}^2)^2 + (P_{G.73110}^3 - P_{G.73110}^2)^2 + (P_{G.73108}^3 - P_{G.73108}^2)^2$$

subject to: power flow and operational constraints   (8.12)

(NBS on bus 73109 is modeled as fixed load demand with startup requirement)

Solve this OPF model. The ramping time for BS to crank all online units is 60 mins. At the end of this step, the time is 129.12 (69.12 + 60) mins.
- **Step 4:** The critical load on 73111 will be cranked at this step. The path 73110-73111 is established. The time for building this path is 5 mins. The energized components are shown in red in Figure 8.11.
  Build the OPF model as follows,

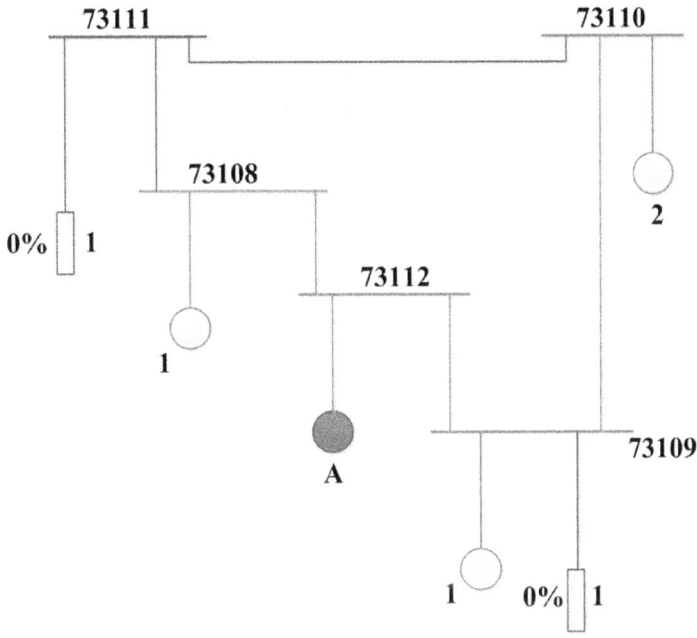

Figure 8.10 System Topology and state of components at step 3.

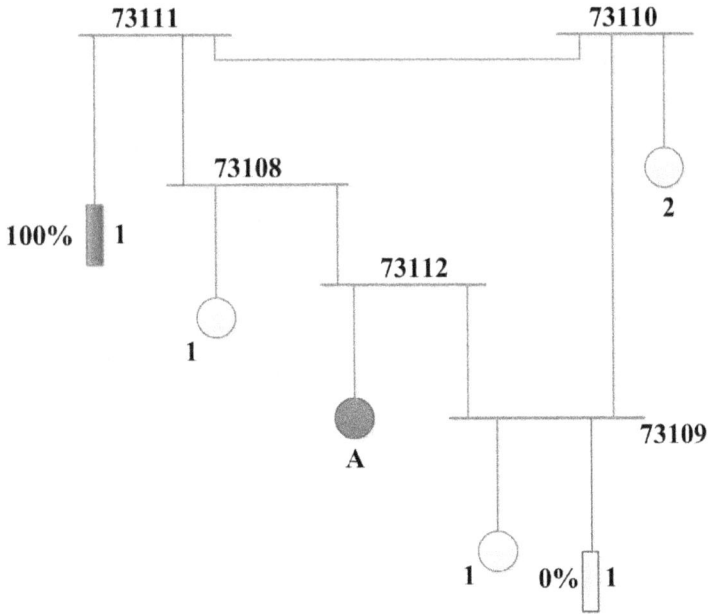

Figure 8.11 System Topology and state of components at step 4.

$$\min(P^4_{G.73112} - P^3_{G.73112})^2 + (P^4_{G.73110} - P^3_{G.73110})^2 + (P^4_{G.73108} - P^3_{G.73108})^2 + (P^4_{G.73109} - P^3_{G.73109})^2$$

subject to: power flow and operational constraints

(NBS on bus 73111 is modeled as fixed load demand with 100% demand)

(8.13)

Solve the OPF model to pick up the critical load (100% demand). At the end of this step, the time is 134.12 (129.12 +5) mins.

References

1. *The National Electric Reliability Study: Technical Study Reports*, DOEIEP-0005.USDOE, U.S. Department of Energy, Office of Emergency Operations, Washington, DC, USA, 1981.
2. F. Li, W. Qiao, H. Sun, H. Wan, J. Wang, Y. Xia, Z. Xu, and P. Zhang, "Smart transmission grid: Vision and framework," *IEEE Transactions on Smart Grid*, vol. 1, no. 2, pp. 168–177, 2010.
3. R. Nuqui and A. Phadke, "Phasor measurement unit placement techniques for complete and incomplete observability," *IEEE Transactions on Power Delivery*, vol. 20, no. 4, pp. 2381–2388, 2005.
4. T. S. Sidhu and Y. Yin, "Modeling and simulation for performance evaluation of IEC61850-based substation communication systems," *IEEE Transactions on Power Delivery*, vol. 22, no. 3, pp. 1482–1489, 2007.
5. D. Fink and H. Beaty, *Standard Handbook for Electrical Engineers*, 11th ed. McGraw-Hill, 1978.
6. I. of Electrical and E. E. C. Report, "The significance of assumptions implied in long-range electric utility planning studies," *IEEE Trans. Power Appar. Syst, Tech. Rep.*, 1980.
7. A. Meliopoulos, R. Webb, R. Bennon, and J. Juves, "Optimal long range transmission planning with ac load flow," *IEEE Transactions on Power Apparatus and Systems*, vol. PAS-101, no. 10, pp. 4156–4163, 1982.
8. A. Keane, L. F. Ochoa, C. L. Borges, G. W. Ault, A. D. Alarcon-Rodriguez, R. A. Currie, F. Pilo, C. Dent, and G. P. Harrison, "State-of-the-art techniques and challenges ahead for distributed generation planning and optimization," *IEEE Transactions on Power Systems*, vol. 28, no. 2, pp. 1493–1502, 2012.
9. S. Heidari, M. Fotuhi-Firuzabad, and S. Kazemi, "Power distribution network expansion planning considering distribution automation," *IEEE Transactions on Power Systems*, vol. 30, no. 3, pp. 1261–1269, 2014.
10. B. Zeng, J. Zhang, X. Yang, J. Wang, J. Dong, and Y. Zhang, "Integrated planning for transition to low-carbon distribution system with renewable energy generation and demand response," *IEEE Transactions on Power Systems*, vol. 29, no. 3, pp. 1153–1165, 2014.
11. J. Li, Z. Li, F. Liu, H. Ye, X. Zhang, S. Mei, and N. Chang, "Robust coordinated transmission and generation expansion planning considering ramping requirements and construction periods," *IEEE Transactions on Power Systems*, vol. 33, no. 1, pp. 268–280, 2018.
12. J. De La Ree, V. Centeno, J. S. Thorp, and A. G. Phadke, "Synchronized phasor measurement applications in power systems," *IEEE Transactions on Smart Grid*, vol. 1, no. 1, pp. 20–27, 2010.
13. S. Sridhar, A. Hahn, and M. Govindarasu, "Cyber–physical system security for the electric power grid," *Proceedings of the IEEE*, vol. 100, no. 1, pp. 210–224, 2011.
14. M. Panteli, C. Pickering, S. Wilkinson, R. Dawson, and P. Mancarella, "Power system resilience to extreme weather: fragility modeling, probabilistic impact assessment, and adaptation measures," *IEEE Transactions on Power Systems*, vol. 32, no. 5, pp. 3747–3757, 2016.
15. M. Panteli, P. Mancarella, D. N. Trakas, E. Kyriakides, and N. D. Hatziargyriou, "Metrics and quantification of operational and infrastructure resilience in power systems," *IEEE Transactions on Power Systems*, vol. 32, no. 6, pp. 4732–4742, 2017.
16. Y. Yang, W. Tang, Y. Liu, Y. Xin, and Q. Wu, "Quantitative resilience assessment for power transmission systems under typhoon weather," *IEEE Access*, vol. 6, pp. 40 747–40 756, 2018.
17. N. E. Wu, M. Sarailoo, and M. Salman, "Transmission fault diagnosis with sensor-localized filter models for complexity reduction," *IEEE Transactions on Smart Grid*, vol. 9, no. 6, pp. 6939–6950, 2017.
18. M. Yan, X. Ai, M. Shahidehpour, Z. Li, J. Wen, S. Bahramira, and A. Paaso, "Enhancing the transmission grid resilience in ice storms by optimal coordination of power system schedule with pre-positioning and routing of mobile dc de-icing devices," *IEEE Transactions on Power Systems*, vol. 34, no. 4, pp. 2663–2674, 2019.
19. J. Yan, B. Hu, K. Xie, J. Tang, and H.-M. Tai, "Data-driven transmission defense planning against extreme weather events," *IEEE Transactions on Smart Grid*, vol. 11, no. 3, pp. 2257–2270, 2019.
20. Y. Zhang, L. Wang, Y. Xiang, and C.-W. Ten, "Power system reliability evaluation with scada cybersecurity considerations," *IEEE Transactions on Smart Grid*, vol. 6, no. 4, pp. 1707–1721, 2015.

21. P. Lau, W. Wei, L. Wang, Z. Liu, and C.-W. Ten, "A cybersecurity insurance model for power system reliability considering optimal defense resource allocation," *IEEE Transactions on Smart Grid*, vol. 11, no. 5, pp. 4403–4414, 2020.

22. P. Lau, L. Wang, Z. Liu, W. Wei, and C.-W. Ten, "A coalitional cyber-insurance design considering power system reliability and cyber vulnerability," *IEEE Transactions on Power Systems*, vol. 36, no. 6, pp. 5512–5524, 2021.

23. H. Ranjbar, S. H. Hosseini, and H. Zareipour, "Resiliency-oriented planning of transmission systems ans distributed energy resources," *IEEE Transactions on Power Systems*, vol. 36, no. 5, pp. 4114–4125, 2021.

24. V. Venkataramanan, P. S. Sarker, K. S. Sajan, A. Srivastava, and A. Hahn, "Real-time federated cyber-transmission-distribution testbed architecture for the resiliency analysis," *IEEE Transactions on Industry Applications*, vol. 56, no. 6, pp. 7121–7131, 2020.

25. F. Bouffard and F. D. Galiana, "Stochastic security for operations planning with significant wind power generation," *IEEE Transactions on Power Systems*, vol. 28, no. 2, pp. 306–316, 2008.

26. S. M. Mazhari, H. Monsef, H. Lesani, and A. Fereidunian, "A multi-objective pmu placement method considering measurement redundancy and observability value under contingencies," *IEEE Transactions on Power Systems*, vol. 28, no. 3, pp. 2136–2146, 2013.

27. A. Bagheri, C. Zhao, F. Qiu, and J. Wang, "Resilient transmission hardening planning in a high renewable penetration era," *IEEE Transactions on Power Systems*, vol. 34, no. 2, pp. 873–882, 2018.

28. A. L. Motto, J. M. Arroyo, and F. D. Galiana, "A mixed-integer lp procedure for the analysis of electric grid security under disruptive threat," *IEEE Transactions on Power Systems*, vol. 20, no. 3, pp. 1357–1365, 2005.

29. W. Yuan, J. Wang, F. Qiu, C. Chen, C. Kang, and B. Zeng, "Robust optimization-based resilient distribution network planning against natural disasters," *IEEE Transactions on Smart Grid*, vol. 7, no. 6, pp. 2817–2826, 2016.

30. M. Vaiman, K. Bell, Y. Chen, B. Chowdhury, I. Dobson, P. Hines, M. Papic, S. Miller, and P. Zhang, "Risk assessment of cascading outages: Methodologies and challenges," *IEEE Transactions on Power Systems*, vol. 27, no. 2, pp. 631–641, 2012.

31. R. A. Jabr, "Robust transmission network expansion planning with uncertain renewable generation and loads," *IEEE Transactions on Power Systems*, vol. 28, no. 4, pp. 4558–4567, 2013.

32. K. Garifi, E. S. Johnson, B. Arguello, and B. J. Pierre, "Transmission grid resiliency investment optimization model with socp recovery planning," *IEEE Transactions on Power Systems*, vol. 37, no. 1, pp. 26–37, 2022.

33. B. Kocuk, S. S. Dey, and X. A. Sun, "New formulation and strong misocp relaxations for ac optimal transmission switching problem," *IEEE Transactions on Power Systems*, vol. 32, no. 6, pp. 4161–4170, 2017.

34. E. B. Fisher, R. P. O'Neill, and M. C. Ferris, "Optimal transmission switching," *IEEE Transactions on Power Systems*, vol. 23, no. 3, pp. 1346–1355, 2008.

35. F. D. Galiana, D. T. McGillis, and M. A. Marin, "Expert systems in transmission planning," *Proceedings of the IEEE*, vol. 80, no. 5, pp. 712–726, 1992.

36. H. Liu, R. A. Davidson, and T. V. Apanasovich, "Statistical forecasting of electric power restoration times in hurricanes and ice storms," *IEEE Transactions on Power Systems*, vol. 22, no. 4, pp. 2270–2279, 2007.

37. Y. Wang, C. Chen, J. Wang, and R. Baldick, "Research on resilience of power systems under natural disasters—a review," *IEEE Transactions on Power Systems*, vol. 31, no. 2, pp. 1604–1613, 2015.

38. G. Huang, J. Wang, C. Chen, J. Qi, and C. Guo, "Integration of preventive and emergency responses for power grid resilience enhancement," *IEEE Transactions on Power Systems*, vol. 32, no. 6, pp. 4451–4463, 2017.

39. N. E. R. Council, "Tenth annual review of overall reliability and adequacy of the north american bulk power systems," in *NERC*, 1980.

40. T. Gonen and H. Bekiroglu, "Some views on inflation and a phillips curve for the u.s. economy," in *Annual Conference of the Decision Sciences Institute Proceedings*, 1977, pp. 328–331.

41. E. P. R. Institute, *Transmission Line Reference Book: 345 kV and Above*. Electric Power Research Institute, 1979.

42. G. E. C. P. EHV. and E. E. Institute, *EHV Transmission Line Reference Book*. Edison Electric Institute, 1968, no. 68.

43. E. P. R. Institute, *Transmission Line Reference Book: 115–138 kV Compact Line Design*. Electric Power Research Institute, 1978.
44. ——, *Transmission Line Reference Book: HVDC to ±600 kV*. Electric Power Research Institute, 1978.
45. T. Gonen, *Electrical Power Distribution System Engineering*. CRC Press, 2008, vol. 2.
46. ——, *Electrical Power Transmission System Engineering: Analysis and Design*. CRC Press, 2009, vol. 2.
47. ——, "Electric power distribution system engineering McGraw-Hill Inc," *New York*, 1986.
48. T. Gonen and P. Anderson, "The impact of advanced technology on the future electric energy supply problem," in *EASCON'78; Electronics and Aerospace Systems Convention*, 1978, pp. 117–121.
49. J. R. Eaton, "Electric power transmission systems," 1972.
50. O. Elgerd, *Electric Energy Systems Theory: An Introduction*. McGraw-Hill Education, 1971.
51. O. I. Elgerd, *Basic Electric Power Engineering*. Addison-Wesley, 1977.
52. E. Clarke, *Circuit Analysis of AC Power Systems*. General Electric Co, 1943, vol. 1.
53. IEEE, "Graphic symbols for electrical and electronics diagrams," *IEEE Std. 315-1971 [or American National Standards Institute (ANSI) Y32.2-1971]*, 1971.
54. AIEE, "Aiee standards committee report, electr. eng. (am. inst. electr. eng.)," AIEE, Tech. Rep., 1946.
55. E. T. B. Gross and E. M. Gulachenski, "Experience of the new england electric company with generator protection by resonant neutral grounding," *IEEE Transactions on Power Apparatus and Systems*, no. 4, pp. 1186–1194, 1973.
56. IEEE, "Graphic symbols for electrical and electronics diagrams," *IEEE Stand. 315-1971 for American National Standards Institute (ANSI) Y32.2-1971)*, 1971.
57. C. F. Wagner and R. D. Evans, *Symmetrical Components*. McGraw-Hill, 1933.
58. I. Travis, "Per unit quantities," *Electrical Engineering*, no. 56, p. 143–151, 1937.
59. H. Skilling, *Electrical Engineering Circuits*, 2nd ed. Wiley, 1966.
60. J. W. Nilsson, *Introduction to Circuits, Instruments, and Electronics*. Harcourt, Brace, and World, 1968.
61. W. Stevenson, *Elements of Power System Analysis*, 4th ed. McGraw-Hill, 1981.
62. J. Neuenswander, *Modern Power System*. International Textbook Co, 1971.
63. P. Anderson, *Analysis of Faulted Power Systems*. Iowa State University Press, 1973.
64. C. A. Gross, *Power System Analysis*. Wiley, 1979.
65. E. Clarke, *Circuit Analysis of AC Power Systems*. General Electric Company, 1950, vol. 1.
66. L. Woodruf, *Electrical Power Transmission*. Wiley, 1952.
67. K. J. Cox and E. Clarke, "Performance charts for three-phase transmission circuits under balanced operations," *Transactions of the American Institute of Electrical Engineers*, vol. 76, pp. 809–817, 1957.
68. J. Zaborsky and J. Rittenhouse, *Electric Power Transmission*. Rensselaer Bookstore, 1969.
69. O. I. Elgerd, "Electric energy systems theory: An introduction," 1971.
70. T. Gonen, J. Nowikowski, and C. Brooks, "Electrostatic unbalances of transmission lines with 'n' overhead ground wires, part i," in *Proceedings of Modeling and Simulation Conference*, vol. 17, no. 2, April 24-25, 1986, pp. 459–464.
71. J. Nowikowski, C. Brooks, and T. Gonen, "Electrostatic unbalances of transmission lines with "n" overhead ground wires—part ii," in *Proceedings of Modeling and Simulation Conference*, vol. 17, no. 2, April 24-25, 1986, pp. 465–470.
72. T. Gonen, S. Yousif, and X. Leng, "Fuzzy logic evaluation of new generation impact on existing transmission system," in *Proceedings of IEEE Budapest Tech'99 Conference*, August 29–September 2, 1999.
73. W. I. Bowman and J. M. McNamee, "Development of equivalent pi and t matrix circuits for long untransposed transmission lines," *IEEE Transactions on Power Apparatus and Systems*, vol. 83, pp. 625–632, 1964.
74. A. E. Kennelly, *The Application of Hyperbolic Functions to Electrical Engineering Problems*, 3rd ed. McGraw-Hill Book Company, Incorporated, 1925.
75. B. Weedy, *Electric Power Systems*, 2nd ed. Wiley, 1972.
76. H. M. Rustebakke and C. Concordia, "Self-excited oscillations in a transmission system using series capacitors," *IEEE Transactions on Power Apparatus and Systems*, no. 7, pp. 1504–1512, 1970.
77. L. Kilgore, L. Elliott, and E. Taylor, "The prediction and control of self-excited oscillations due to series capacitors in power systems," *IEEE Transactions on Power Apparatus and Systems*, no. 3, pp. 1305–1311, 1971.

78. L. Kilgore, E. R. Taylor, D. G. Ramey, R. G. Farmer, and A. L. Schwalb, "Solutions to the problems of subsynchronous resonance in power systems with series capacitors," in *Proceedings of the American Power Conference*, vol. 35, 1973, pp. 1120–1128.

79. C. Concordia, J. Tice, and C. E. J. Bowler, "Sub-synchronous torques on generating units feeding series-capacitor compensated lines," in *Proceedings of the American Power Conference*, vol. 35, 1973, pp. 1129–1136.

80. C. Schifreen and W. Marble, "Changing current limitations in operation of high-voltage cable lines," *Transactions of the American Institute of Electrical Engineers*, pp. 803–817, 1956.

81. R. T. Wiseman, "Discussions to charging current limitations in operation of high-voltage cable lines," *Transactions of the American Institute of Electrical Engineers*, pp. 803–817, 1956.

82. R. Roper, *Kurzchlussstrime in Drehstromnetzen, 5th Ger. ed. (translated as Short-Circuit Currents in Three-Phase Networks)*. Siemens Aktienges, 1972.

83. E. Nasser, "Zum problem des fremdschichtuberschlages an isolatoren," *ETZ-A*, vol. 83, pp. 356–365, 1962.

84. J. Kaminski, "Long-time mechanical and electrical strength of suspension insulators," *IEEE Transactions on Power Apparatus and Systems*, vol. 82, pp. 446–452, 1963.

85. A. Rumeli, "The mechanism of flashover on polluted insulation," Ph.D. dissertation, University of Strathclyde, 1967.

86. E. Nasser, *Fundamentals of Gaseous Ionization and Plasma Electronics*. John Wiley and Sons, 1971.

87. I. W. Group, "A Survey of The Problem of Insulator Contamination in The United States and Canada-Part I," *IEEE Transactions on Power Apparatus and Systems*, vol. PAS-90, no. 6, pp. 2577–2585, Nov. 1971.

88. "A Survey of the Problem of Insulator Contamination in the United States and Canada Part II-Geographical Aspects," *IEEE Transactions on Power Apparatus and Systems*, vol. PAS-91, no. 5, pp. 1948–1954, Sept. 1972.

89. E. Nasser, "An annotated bibliography on the problem of insulator contamination of the electric energy system: Special report," *Technical report*, 1973.

90. IEEE, "IEEE tutorial course: Application of power circuit breakers," *Institute of Electrical and Electronics Engineers*, 1975.

91. J. S. T. Looms and F. H. Proctor, "The development of an epoxy-based insulator for UHV," in *IEEE Power Engineering Society Summer Meeting*. IEEE, 1976, pp. A76 342–6.

92. G. Karady and G. Lamontagne, "Electrical and contamination performance of synthetic insulators for 735 kv transmission lines," in *IEEE Power Engineering Society Summer Meeting*, no. 1976, 1976, pp. 502–505.

93. IEEE Recommended Practice for Grounding of Industrial and Commercial Power Systems (IEEE Green Book), in *ANSI/IEEE Std 142–1982* , pp.1–135, 17 Sept. 1982, doi: 10.1109/IEEESTD.1982.119223.

94. J. Zaborszky, "Efficiency of grounding grids with nonuniform soil," *Transactions of the American Institute of Electrical Engineers*, vol. 74, pp. 1230–1233, 1955.

95. T. Udo, "Minimum phase-to-phase electrical clearances for substations based on switching surges and lightning surges," *IEEE Transactions on Power Apparatus and Systems*, no. 8, pp. 838–845, 1966.

96. J. Sverak, "Optimized grounding grid design using variable spacing technique," *IEEE Transactions on Power Apparatus and Systems*, vol. 95, no. 1, pp. 362–374, 1976.

97. W. D. Stevenson, *Elements of Power System Analysis*, 4th ed. McGraw-Hill, 1982.

98. A. Warrington, *The Protective Relays. Theory and Practice*. Chapman & Hall, 1969, vol. 2.

99. P. Bellaschi, R. Armington, and A. Snowden, "Impulse and sixty cycle characteristics of driven grounds—ii," *Electrical Engineering*, vol. 61, no. 6, pp. 349–363, 1942.

100. J. G. Anderson, "Monte carlo computer calculation of transmission-line lightning performance," *Transactions of the American Institute of Electrical Engineers. Part III: Power Apparatus and Systems*, vol. 80, no. 3, pp. 414–419, 1961.

101. O. W. Andersen, "Laplacian electrostatic field calculations by finite elements with automatic grid generation," *IEEE Transactions on Power Apparatus and Systems*, pp. 682–689, 1973.

102. U. G. Biegelmeier, "Die recleatung der z-schwelle des herzkammerflimmerns fur die festigun von beruhrungsspanungs greuzeu bei den schutzman bradhmer gegen elektrische unfallen," *E and M*, vol. 93, no. 1, pp. 1–8, 1976.

103. M. Abe, N. Otsuzuki, T. Emura, and M. Takeuchi, "Development of a new fault location system for multi-terminal single transmission lines," *IEEE Transactions on Power Delivery*, vol. 10, no. 1, pp. 159–168, 1995.

104. F. Numajiri, "Analysis of transmission line lightning voltages by digital computer," *Electrical Engineering in Japan*, vol. 84, no. 8, pp. 53–63, 1964.

105. H. Lee and A. M. Mousa, "Gps travelling wave fault locator systems: investigation into the anomalous measurements related to lightning strikes," *IEEE Transactions on Power Delivery*, vol. 11, no. 3, pp. 1214–1223, 1996.

106. A. V. C. Warrington, *Protective Relays: Their Theory and Practice.* Chapman and Hall, 1962, vol. 1.

107. T. Funabashi, H. Otoguro, Y. Mizuma, L. Dube, and A. Ametani, "Digital fault location for parallel double-circuit multi-terminal transmission lines," *IEEE Transactions on Power Delivery*, vol. 15, no. 2, pp. 531–537, 2000.

108. F. V. Lopes, P. Lima, J. P. G. Ribeiro, T. R. Honorato, K. M. Silva, E. J. S. Leite, W. L. A. Neves, and G. Rocha, "Practical methodology for two-terminal traveling wave-based fault location eliminating the need for line parameters and time synchronization," *IEEE Transactions on Power Delivery*, vol. 34, no. 6, pp. 2123–2134, 2019.

109. W. Chen, O. Malik, X. Yin, D. Chen, and Z. Zhang, "Study of wavelet-based ultra high speed directional transmission line protection," *IEEE Transactions on Power Delivery*, vol. 18, no. 4, pp. 1134–1139, 2003.

110. F. Martin and J. A. Aguado, "Wavelet-based ann approach for transmission line protection," *IEEE Transactions on Power Delivery*, vol. 18, no. 4, pp. 1572–1574, 2003.

111. D.-J. Zhang, Q. H. Wu, Z. Q. Bo, and B. Caunce, "Transient positional protection of transmission lines using complex wavelets analysis," *IEEE Transactions on Power Delivery*, vol. 18, no. 3, pp. 705–710, 2003.

112. S. M. Brahma and A. A. Girgis, "Fault location on a transmission line using synchronized voltage measurements," *IEEE Transactions on Power Delivery*, vol. 19, no. 4, pp. 1619–1622, 2004.

113. S. M. Brahma, "Fault location scheme for a multi-terminal transmission line using synchronized voltage measurements," *IEEE Transactions on Power Delivery*, vol. 20, no. 2, pp. 1325–1331, 2005.

114. D. Spoor and J. G. Zhu, "Improved single-ended traveling-wave fault-location algorithm based on experience with conventional substation transducers," *IEEE Transactions on Power Delivery*, vol. 21, no. 3, pp. 1714–1720, 2006.

115. O. Naidu and A. K. Pradhan, "Precise traveling wave-based transmission line fault location method using single-ended data," *IEEE Transactions on Industrial Informatics*, vol. 17, no. 8, pp. 5197–5207, 2020.

116. E. Vázquez, J. Castruita, O. L. Chacón, and A. Conde, "A new approach traveling-wave distance protection—part i: Algorithm," *IEEE Transactions on Power Delivery*, vol. 22, no. 2, pp. 795–800, 2007.

117. P. Jafarian and M. Sanaye-Pasand, "A traveling-wave-based protection technique using wavelet/pca analysis," *IEEE Transactions on Power Delivery*, vol. 25, no. 2, pp. 588–599, 2010.

118. H. Livani and C. Y. Evrenosoglu, "A machine learning and wavelet-based fault location method for hybrid transmission lines," *IEEE Transactions on Smart Grid*, vol. 5, no. 1, pp. 51–59, 2014.

119. J. Izykowski, E. Rosolowski, P. Balcerek, M. Fulczyk, and M. M. Saha, "Accurate noniterative fault-location algorithm utilizing two-end unsynchronized measurements," *IEEE Transactions on Power Delivery*, vol. 26, no. 2, pp. 547–555, 2009.

120. C.-S. Yu, "An unsynchronized measurements correction method for two-terminal fault-location problems," *IEEE Transactions on Power Delivery*, vol. 25, no. 3, pp. 1325–1333, 2010.

121. F. Lopes, K. Silva, F. Costa, W. Neves, and D. Fernandes, "Real-time traveling-wave-based fault location using two-terminal unsynchronized data," *IEEE Transactions on Power Delivery*, vol. 30, no. 3, pp. 1067–1076, 2014.

122. O. Naidu and A. K. Pradhan, "A traveling wave-based fault location method using unsynchronized current measurements," *IEEE Transactions on Power Delivery*, vol. 34, no. 2, pp. 505–513, 2018.

123. F. Lopes, D. Fernandes, and W. Neves, "A traveling-wave detection method based on park's transformation for fault locators," *IEEE Transactions on Power Delivery*, vol. 28, no. 3, pp. 1626–1634, 2013.

124. M. Majidi, M. Etezadi-Amoli, and M. S. Fadali, "A sparse-data-driven approach for fault location in transmission networks," *IEEE Transactions on Smart Grid*, vol. 8, no. 2, pp. 548–556, 2015.

125. R. J. Hamidi, H. Livani, and R. Rezaiesarlak, "Traveling-wave detection technique using short-time matrix pencil method," *IEEE Transactions on Power Delivery*, vol. 32, no. 6, pp. 2565–2574, 2017.

126. F. M. Demagalhaesjunior and F. V. Lopes, "Mathematical study on traveling waves phenomena on three phase transmission lines part i: Fault-launched waves," *IEEE Transactions on Power Delivery*, vol. 37, no. 2, pp. 1151–1160, 2022.

127. F. Magalhaes and F. Lopes, "Mathematical study on traveling waves phenomena on three phase transmission lines part ii: Reflection and refraction matrices," *IEEE Transactions on Power Delivery*, vol. 37, no. 2, pp. 1161–1170, 2022.

128. J. Ding, X. Wang, Y. Zheng, and L. Li, "Distributed traveling-wave-based fault-location algorithm embedded in multiterminal transmission lines," *IEEE Transactions on Power Delivery*, vol. 33, no. 6, pp. 3045–3054, 2018.

129. H. Chalangar, T. Ould-Bachir, K. Sheshyekani, S. Li, and J. Mahseredjian, "Evaluation of a constant parameter line-based twfl real-time testbed," *IEEE Transactions on Power Delivery*, vol. 35, no. 2, pp. 1010–1019, 2019.

130. C. Zhang, G. Song, T. Wang, and X. Dong, "An improved non-unit traveling wave protection method with adaptive threshold value and its application in hvdc grids," *IEEE Transactions on Power Delivery*, vol. 35, no. 4, pp. 1800–1811, 2019.

131. R. Liang, N. Peng, L. Zhou, X. Meng, Y. Hu, Y. Shen, and X. Xue, "Fault location method in power network by applying accurate information of arrival time differences of modal traveling waves," *IEEE Transactions on Industrial Informatics*, vol. 16, no. 5, pp. 3124–3132, 2019.

132. H. A. Abd el Ghany, A. M. Azmy, and A. M. Abeid, "A general travelling-wave-based scheme for locating simultaneous faults in transmission lines," *IEEE Transactions on Power Delivery*, vol. 35, no. 1, pp. 130–139, 2019.

133. A. Warrington and C. Van, "Reactance relays negligibly affected by arc impedance," *Electrical World*, vol. 98, no. 12, pp. 502–505, 1931.

134. P. M. Anderson, *Analysis of Faulted Power Systems*. Wiley-Interscience, 1995.

135. H. Kyser, *Elektrikle Energi Nakli (translated into Turkish by M. Dilege)*. Istanbul Technical University Press, 1952.

136. H. Langer, "Messungen von erderspannungen in einem 220 kv umspanwerk," *Electrotechnische zietschrift*, vol. 75, no. 4, pp. 97–105, 1954.

137. J. Neuenswander, *Modern Power Systems*. International Textbook Company, 1971.

138. WEC, *Electrical Transmission and Distribution Reference Book*. Westinghouse Electric Corporation, 1964.

139. ANSI, "Schedules of preferred ratings and related required capabilities for AC high-voltage circuit breakers rated on a symmetrical current basis," *American National Standards Institute*, 1971.

140. T. Gonen, *A Practical Guide for Calculation of Short-Circuit Currents and Selection of High Voltage Circuit Breakers*. Black & Veatch Co, 1977.

141. H. H. Farr, *Transmission Line Design Manual*. US Department of the Interior, Water and Power Resources Service, 1980.

142. F. W. Peek, "Electrical characteristics of the suspension insulator: Part 1," *Transactions of the American Institute of Electrical Engineers*, no. 31, pp. 907–930, 1912.

143. F. W. Peek, "Electrical characteristics of the suspension insulator: Part 2," *Transactions of the American Institute of Electrical Engineers*, no. 39, pp. 1685–1705, 1920.

144. A. Elek, "Hazards of electric shock at stations during fault and method of reduction," *Ontario Hydro Research News*, vol. 10, no. 1, pp. 1–6, 1958.

145. K. Rotter and U. Biegelmeier, "Elektrische wilderstrande und strome in menschlicken korper," *E and M*, vol. 89, p. 104–109, 1971.

146. G. Bodier, "La securite des personnes et la question des mises à la terra dans les postes de distribution," *Bulletin de la Société Francaise des Electriciens*, vol. 74, no. 7, p. 545–562, 1947.

147. R. Rudenberg, "Transient performance of electric power systems," *McGraw-Hill*, 1950.

148. ——, "Electrical shock waves in power systems," *Harvard University*, 1968.

149. L. Ferris, "Effects of electrical shock on the heart," *Transactions of the American Institute of Electrical Engineers*, vol. 55, no. 1236, p. 498–515, 1936.

150. R. J. Heppe, "Step potentials and body currents near grounds in two-layer earth," *IEEE Transactions on Power Apparatus and Systems*, no. 1, pp. 45–59, 1979.

151. C. F. Dalziel, "Electric shock hazard," *IEEE Spectrum*, vol. 9, no. 2, pp. 41–50, 1972.
152. W. R. Lee and C. F. Dalziel, "Lethal electric currents," *IEEE Spectrum*, vol. 6, pp. 44–50, 1969.
153. C. F. Qalziel, "A study of the hazards of impulse currents," *Transactions of the American Institute of Electrical Engineers*, vol. 72, no. 2, pp. 1032–1042, 1953.
154. Lightning and I. Subcommittee, *AIEE Lightning Reference Book*. American Institute of Electrical Engineers, 1937.
155. W. R. Lee, "Death from electric shock," in *Proceedings of the Institution of Electrical Engineers*, vol. 113, no. 1. IET, 1966, pp. 144–148.
156. G. S. Subcommittee, "Electrostatic effects of overhead transmission lines. part ii. methods of calculation," *IEEE Transactions on Power Apparatus and Systems*, no. 2, pp. 426–444, 1972.
157. T. Gönen and H. Bekiroglu, "Electrical safety in industrial plants: Some considerations," *IEEE*, vol. 13, no. 5, pp. 4–7, 1977.
158. IEEE, "IEEE recommended practice for industrial and commercial power system analysis," *IEEE*, 1980.
159. G. S. Subcommittee, "Electrostatic Effects OF Overhead Transmission Lines PART I-Hazards and Effects," *IEEE Transactions on Power Apparatus and Systems*, vol. PAS-91, no. 2, pp. 422–426, March 1972.
160. IEEE, "IEEE standard. 2000, IEEE guide for safety in AC substation grounding," *IEEE*.
161. E. D. Sunde, "Earth conduction effect in transmission system," 1968.
162. H. Dwight, "Calculation of resistances to ground," *Transactions of the American Institute of Electrical Engineers*, vol. 55, pp. 1319–1328, 1936.
163. F. Dawalibi and D. Mukhedkar, "Optimum design of substation grounding in two-layer earth structure. part i. analytical study," *IEEE Transactions on Power Apparatus and Systems*, no. 2, pp. 252–261, 1975.
164. D. Mukhedkar and F. Dawalibi, "Optimum design of substation grounding in two-layer earth structure. part ii. comparison between theoretical and experimental results," *IEEE Transactions on Power Apparatus and Systems*, no. 2, pp. 262–266, 1975.
165. F. Dawalibi and D. Mukhedkar, "Optimum design of substation grounding in two-layer earth structure. part iii. study of grounding grids performance and new electrodes configuration," *IEEE Transactions on Power Apparatus and Systems*, no. 2, pp. 267–272, 1975.
166. T. Gonen, *Electrical Power Transmission System Engineering*, 2nd ed. CRC Press, 2009.
167. IEEE, *IEEE Guide for Safety in AC Substation Grounding*. Institute for Electrical and Electronics Engineers, 1976.
168. J. Sverak, "Safe substation grounding. part i," *IEEE Transactions on Power Apparatus and Systems*, no. 9, pp. 4281–4290, 1981.
169. P. Prache and J. James, *Uzak Mesafe Yeralti Hatlari Dersteri (Translated into Turkish by Daarafakioglu)*. Istanbul Technical University Press, 1958.
170. H. Nunnally, "Computer simulation for determining step and touch potentials resulting from faults or open neutrals in urd cable," *IEEE Transactions on Power Apparatus and Systems*, no. 3, pp. 1130–1136, 1979.
171. W. W. Lewis, "The protection of transmission systems against lighting," 1965.
172. C. L. Fortescue, "Method of symmetrical co-ordinates applied to the solution of polyphase networks," *Transactions of the American Institute of Electrical Engineers*, vol. 37, no. 2, pp. 1027–1140, 1918.
173. J. R. Carson, "Wave propagation in overhead wires with ground return," *The Bell System Technical Journal*, vol. 5, no. 4, pp. 539–554, 1926.
174. E. T. B. Gross and M. H. Hesse, "Electromagnetic unbalance of untransposed transmission lines," *Transactions of the American Institute of Electrical Engineers. Part III: Power Apparatus and Systems*, vol. 72, no. 6, pp. 1323–1336, 1953.
175. G. I. Atabekov, *The Relay Protection of High Voltage Networks*. Pergamon Press, 1960.
176. H. E. Brown, *Solution of Large Networks by Matrix Methods*. Wiley New York, NY, USA, 1975.
177. G. O. Calabrese, *Symmetrical Components Applied to Electric Power Networks*. Ronald Press Company, 1959.
178. E. Clarke, "Simultaneous faults on three-phase systems," *Transactions of the American Institute of Electrical Engineers*, vol. 50, no. 3, pp. 919–941, 1931.
179. ——, *Circuit Analysis of A-C Power Systems*. General Electric Co, 1961, vol. 1.
180. ——, *Circuit Analysis of A-C Power Systems*. General Electric Co, 1961, vol. 2.

181. J. E. Clem, "Reactance of transmission lines with ground return," *Transactions of the American Institute of Electrical Engineers*, vol. 50, no. 3, pp. 901–915, 1931.

182. W. C. Duesterhoeft, M. W. Schulz, and E. Clarke, "Determination of instantaneous currents and voltages by means of alpha, beta, and zero components," *Transactions of the American Institute of Electrical Engineers*, vol. 70, no. 2, pp. 1248–1255, 1951.

183. G. Kron, "Tensor analysis of networks wiley," 1939.

184. P. Anderson, "Analysis of simulatenous faults by two-port network theory," *IEEE Transactions on Power Apparatus and Systems*, no. 5, pp. 2199–2205, 1971.

185. E. Harder, "Sequence network connections for unbalanced load and fault conditions," *Electric Journal*, pp. 481–488, 1977.

186. J. Hobson and D. Whitehead, *Symmetrical Components, in Electrical Transmission and Distribution Reference Book*. WEC, 1964, vol. 2.

187. D. Beeman and D. Beeman, *Industrial Power Systems Handbook*. McGraw-Hill New York, 1955, vol. 195.

188. T. Gonen and M. Haj-Mohamadi, "Electromagnetic unbalances of six-phase transmission lines," *International Journal of Electrical Power Energy Systems*, vol. 11, no. 2, pp. 78–84, 1989.

189. T. Gönen, *Electrical Power Transmission System Engineering*. CRC Press, Roca Baton, Florida, 2009.

190. R. B. Shipley, *Introduction to Matrices and Power Systems*. John Wiley and Sons, 1976.

191. L. Dunstan, "Digital load flow studies," *Transactions of the American Institute of Electrical Engineers*, vol. 73, no. 3A, pp. 825–831, 1954.

192. J. Ward and H. Hale, "Digital computer solution of power-flow problems," *Transactions of the American Institute of Electrical Engineer*, vol. 75, no. 3, pp. 398–404, 1956.

193. A. Glimn and G. Stagg, "Automatic calculation of load flows," *Transactions of the American Institute of Electrical Engineers. Part III: Power Apparatus and Systems*, vol. 76, no. 3, pp. 817–828, 1957.

194. C. Trevino, "Cases of difficult convergence in load-flow problems," in *IEEE Summer Power Meeting*, 1971.

195. D. R. Hayes and F. J. Hubert, "A rapid digital computer solution for power system network load-flow," *IEEE Transactions on Power Apparatus and Systems*, pp. 934–940, 1971.

196. W. W. Maslin, "A power system planning computer program package emphasizing flexibility and compatibility," in *IEEE Summer Power Meeting*, 1970.

197. R. Podmore and J. M. Undrill, "Modified nodal iterative load flow algorithm to handle series capacitive branches," *IEEE Transactions on Power Apparatus and Systems*, pp. 1379–1387, 1973.

198. J. A. Treece, "Bootstrap gauss-seidel load flow," in *Proceedings of the Institution of Electrical Engineers*, vol. 116, no. 5. IET, 1969, pp. 866–870.

199. P. P. Gupta and M. W. H. Davies, "Digital computers in power system analysis," *Proceedings of the IEE-Part A: Power Engineering*, vol. 109, pp. 383–404, 1961.

200. A. Brameller and J. K. Denmead, "Some improved methods for digital network analysis," *Proceedings of the IEE-Part A: Power Engineering*, vol. 109, pp. 109–116, 1962.

201. H. E. Brown, G. K. Carter, H. H. Happ, and C. E. Person, "Power flow solution by impedance matrix iterative method," *IEEE Transactions on Power Apparatus and Systems*, vol. 82, pp. 1–10, 1963.

202. L. L. Freris and A. M. Sasson, "Investigations on the load-flow problem," in *Proceedings of the Institution of Electrical Engineers*, vol. 114. IET, 1967, p. 1960.

203. H. E. Brown, G. K. Carter, H. H. Happ, and C. E. Person, "Z-matrix algorithms in load-flow programs," *IEEE Transactions on Power Apparatus and Systems*, pp. 807–814, 1968.

204. J. E. Van Ness, "Iteration methods for digital load flow studies," *Transactions of the American Institute of Electrical Engineers. Part III: Power Apparatus and Systems*, vol. 78, no. 3, pp. 583–588, 1959.

205. ——, "Convergence of iterative load-flow studies," *Transactions of the American Institute of Electrical Engineers. Part III: Power Apparatus and Systems*, vol. 78, no. 4, pp. 1590–1595, 1959.

206. J. E. Van Ness and J. H. Griffin, "Elimination methods for load-flow studies," *Transactions of the American Institute of Electrical Engineers. Part III: Power Apparatus and Systems*, vol. 80, pp. 299–304, 1961.

207. W. F. Tinney and C. E. Hart, "Power flow solution by newton's method," *IEEE Transactions on Power Apparatus and Systems*, no. 11, pp. 1449–1456, 1967.

208. J. P. Britton, "Improved area interchange control for newton's method load flows," *IEEE Transactions on Power Apparatus and Systems*, pp. 1577–1581, 1969.

209. T. W. Dommel, H. W. Dommel, F. W. Tinney and W. L. Powell, "Further developments in newton's method for power system applications," *IEEE Power Engineering Society Winter Meeting*, 1970.

210. B. Stott, "Effective starting process for newton-raphson load flows," in *Proceedings of the Institution of Electrical Engineers*. IET, 1971, pp. 983–987.

211. J. P. Britton, "Imporved load flow performance through a more general equations form," *IEEE Transactions on Power Apparatus and Systems*, pp. 109–116, 1971.

212. N. M. Peterson and W. S. Meyer, "Automatic adjustment of transformer and phase-shifter taps in the newton power flow," *IEEE Transactions on Power Apparatus and Systems*, pp. 103–108, 1971.

213. N. Sato and W. F. Tinney, "Techniques for exploring the sparsity of the network admittance matrix," *IEEE Transactions on Power Apparatus and Systems*, vol. 82, no. 3, pp. 944–949, 1983.

214. W. P. Tinney and J. W. Walker, "Direct solutions of sparse network equations by optimally ordered triangular factorization," in *Proceedings IEEE*. IEEE, 1967, pp. 1801–1809.

215. G. W. Stagg and A. G. Phadke, "Real time evaluation of power system contingencies detection of steady state overloads," *IEEE Power Engineering Society Summer Meeting*, no. 70, 1970.

216. S. T. Despotovic, B. S. Babic, and V. P. Mastilovic, "A rapid and reliable method for solving load flow problems," *IEEE Transactions on Power Apparatus and Systems*, pp. 123–130, 1971.

217. B. Stott, "Decoupled newton load flow," *IEEE Transactions on Power Apparatus and Systems*, no. 5, pp. 1955–1959, 1972.

218. N. M. Peterson, W. F. Tinney, and D. W. Bree, "Iterative linear AC power flow solution for fast approximate outage studies," *IEEE Transactions on Power Apparatus and Systems*, no. 5, pp. 2048–2056, 1972.

219. B. Stott and O. Alsac, "Fast decoupled load flow," *IEEE Transactions on Power Apparatus and Systems*, no. 3, pp. 859–869, 1974.

220. Y. Wallach, "Gradient methods for load-flow problems," *IEEE Transactions on Power Apparatus and Systems*, no. 5, pp. 1314–1318, 1968.

221. A. M. Sasson, "Nonlinear programming solutions for load-flow, minimum-loss, and economic dispatching problems," *IEEE Transactions on Power Apparatus and Systems*, no. 4, pp. 399–409, 1969.

222. R. H. Galloway, J. Taylor, W. D. Hogg, and M. Scott, "New approach to power-system load-flow analysis in a digital computer," in *Proceedings of the Institution of Electrical Engineers*, vol. 117, no. 1. IET, 1970, pp. 165–169.

223. Y. P. Dusonchet, S. N. Talukdar, H. E. Sinnot, and A. H. El-abiad, "Load flows using a combination of point jacobi and newton's methods," *IEEE Transactions on Power Apparatus and Systems*, no. 3, pp. 941–949, 1971.

224. A. M. Sasson, C. Trevino, and F. Aboytes, "Improved newton's load flow through a minimization technique," *IEEE Transactions on Power Apparatus and Systems*, no. 5, pp. 1974–1981, 1971.

225. H. W. Dommel and W. F. Tinney, "Optimal power flow solutions," *IEEE Transactions on Power Apparatus and Systems*, no. 10, pp. 1866–1876, 1968.

226. H. E. Brown, "Contingencies evaluated by a Z-matrix method," *IEEE Transactions on Power Apparatus and Systems*, no. 4, pp. 409–412, 1969.

227. H. Dommel and N. Sato, "Fast transient stability solutions," *IEEE Transactions on Power Apparatus and Systems*, no. 4, pp. 1643–1650, 1972.

228. B. Fox and A. M. Revington, "Network calculations for online control of a power system," in *IEEE Conference on Computers in Power System Operation & Control*, 1972, pp. 261–275.

229. M. S. Sachdev and S. A. Ibrahim, "A fast approximate technique for outage studies in power system planning and operation," *IEEE Transactions on Power Apparatus and Systems*, no. 4, pp. 1133–1142, 1974.

230. O. Alsac and B. Stott, "Optimal load flow with steady-state security," *IEEE Transactions on Power Apparatus and Systems*, no. 3, pp. 745–751, 1974.

231. B. Stott, "Review of load-flow calculation methods," *Proceedings of the IEEE*, vol. 62, no. 7, pp. 916–929, 1974.

232. A. M. Sasson and H. M. Merrill, "Some applications of optimization techniques to power systems problems," *Proceedings of the IEEE*, vol. 62, no. 7, pp. 959–972, 1974.

233. T. E. DyLiacco, "Real-time computer control of power systems," *Proceedings of the IEEE*, vol. 62, no. 7, pp. 959–972, 1974.

234. J. M. Undrill, F. P. DeMello, T. E. Kostyniak, and R. J. Mills, "Interactive computation in power system analysis," *Proceedings of the IEEE*, vol. 62, no. 7, pp. 1009–1018, 1974.

235. A. J. Korsak, "On the question of uniqueness of stable load-flow solutions," *IEEE Transactions on Power Apparatus and Systems*, no. 3, pp. 1093–1100, 1972.

236. W. E. Bosarge, J. A. Jordan, and W. A. Murray, "Nonlinear block SOR-newton load flow algorithm," *IEEE Transactions on Power Apparatus and Systems*, no. 6. pp. 1810–1811, 1973.

237. L. S. VanSlyck and J. F. Dopazo, "Conventional load flow not suited for real-time power-system monitoring," *IEEE Transactions on Power Apparatus and Systems*, no. 1. pp. 26–28, 1974.

238. B. Borkowska, "Probabilistic load flow," *IEEE Transactions on Power Apparatus and Systems*, no. 3, pp. 752–759, 1974.

239. J. F. Dopazo, O. A. Klitin, and A. M. Sasson, "Stochastic load flows," *IEEE Transactions on Power Apparatus and Systems*, vol. 94, no. 2, pp. 299–309, 1975.

240. H. H. Happ, "Diakoptics—the solution of system problems by tearing," *Proceedings of the IEEE*, vol. 62, no. 7, pp. 930–940, 1974.

241. M. Sachdev and T. Medicherla, "A second order load flow technique," *IEEE Transactions on Power Apparatus and Systems*, vol. 96, no. 1, pp. 189–197, 1977.

242. F. F. Wu, "Theoretical study of the convergence of the fast decoupled load flow," *IEEE Transactions on Power Apparatus and Systems*, vol. 96, no. 1, pp. 268–275, 1977.

243. A. H. Lewis, "Large scale system effectiveness analysis," U.S. Dept. of Energy, Systems Control, Inc., Palo Alto, CA, Tech. Rep., 1978.

244. R. J. Brown and W. F. Tinney, "Digital solutions for large power networks," *Transactions of the American Institute of Electrical Engineers. Part III: Power Apparatus and Systems*, vol. 76, no. 3, pp. 347–351, 1957.

245. R. H. Jordan, "Rapidly converging digital load flow," *Transactions of the American Institute of Electrical Engineers. Part III: Power Apparatus and Systems*, vol. 76, no. 3, pp. 1433–1438, 1957.

246. U. A. Conner, "Representative bibliography on load-flow analysis and related topics," in *1973 IEEE Winter Power Meetings*, 1973, pp. C73–104.

247. J. M. Bennett, "Digital computers and the load-flow problem," *Proceedings of the IEE-Part B: Radio and Electronic Engineering*, vol. 103, no. 1S, pp. 16–25, 1956.

248. A. M. Sasson and F. J. Jaimes, "Digital methods applied to power flow studies," *IEEE Transactions on Power Apparatus and Systems*, no. 7, pp. 860–867, 1967.

249. M. A. Laughton and M. W. H. Davies, "Numerical techniques in solution of power-system load-flow problems," in *Proceedings of the Institution of Electrical Engineers*, vol. 111, no. 9. IET, 1964, pp. 1575–1588.

250. E. C. Ogbuobiri, W. F. Tinney, and J. W. Walker, "Sparsity-directed decomposition for gaussian elimination on matrices," *IEEE Transactions on Power Apparatus and Systems*, no. 1, pp. 141–150, 1970.

251. J. Carpentier, "Application de la methode de newton au calcul des resaux mailles," in *Proceedings of the Power System Computation Conference*, 1963.

252. Y. Tamura, H. Mori, and S. Iwamoto, "Relationship between voltage instability and multiple load flow solutions in electric power systems," *IEEE Transactions on Power Apparatus and Systems*, no. 5, pp. 1115–1125, 1983.

253. G. Kron, *Diakoptics: The Piecewise Solution of Large-Scale Systems*. MacDonald, 1963.

254. H. H. Happ and J. M. Undrill, "Diakoptics and networks," *IEEE Transactions on Systems, Man, and Cybernetics*, no. 4, pp. 405–406, 1971.

255. ——, "Multicomputer configurations and diakoptics: Real power flow in power pools," *IEEE Transactions on Power Apparatus and Systems*, no. 6, pp. 789–796, 1969.

256. H. H. Happ, "Diakoptics and piecewise methods," *IEEE Transactions on Power Apparatus and Systems*, no. 7, pp. 1373–1382, 1970.

257. H. Happ and C. Young, "Tearing algorithms for large-scale network programs," *IEEE Transactions on Power Apparatus and Systems*, no. 6, pp. 2639–2649, 1971.

258. H. H. Happ, *Piecewise Methods and Applications to Power Systems*. Krieger Publishing Company, 1980.

259. G. T. Heydt, "Stochastic power flow calculations," in *IEEE Power Engineering Society Summer Meeting*, no. 1975, 1975, pp. 530–536.

260. R. Allan, B. Borkowska, and C. Grigg, "Probabilistic analysis of power flows," in *Proceedings of the Institution of Electrical Engineers*, vol. 121, no. 12. IET, 1974, pp. 1551–1556.

261. V. Ajjarapu and C. Christy, "The continuation power flow: a tool for steady state voltage stability analysis," *IEEE Transactions on Power Systems*, vol. 7, no. 1, pp. 416–423, 1992.

262. Z. Feng, V. Ajjarapu, and B. Long, "Identification of voltage collapse through direct equilibrium tracing," *IEEE Transactions on Power Systems*, vol. 15, no. 1, pp. 342–349, 2000.

263. B. Lee, H. Song, S.-H. Kwon, G. Jang, J.-H. Kim, and V. Ajjarapu, "A study on determination of interface flow limits in the kepco system using modified continuation power flow (mcpf)," *IEEE Transactions on Power Systems*, vol. 17, no. 3, pp. 557–564, 2002.

264. H. Song, B. Lee, S.-H. Kwon, and V. Ajjarapu, "Reactive reserve-based contingency constrained optimal power flow (rccopf) for enhancement of voltage stability margins," *IEEE Transactions on Power Systems*, vol. 18, no. 4, pp. 1538–1546, 2003.

265. F. Malange, D. Alves, L. Da Silva, C. Castro, and G. Da Costa, "Real power losses reduction and loading margin improvement via continuation method," *IEEE Transactions on Power Systems*, vol. 19, no. 3, pp. 1690–1692, 2004.

266. A. M. Azmy, "Optimal power flow to manage voltage profiles in interconnected networks using expert systems," *IEEE Transactions on Power Systems*, vol. 22, no. 4, pp. 1622–1628, 2007.

267. C.-W. Ten, A. Ginter, and R. Bulbul, "Cyber-based contingency analysis," *IEEE Transactions on Power Systems*, vol. 31, no. 4, pp. 3040–3050, 2015.

268. Z. Yang, C.-W. Ten, and A. Ginter, "Extended enumeration of hypothesized substations outages incorporating overload implication," *IEEE Transactions on Smart Grid*, vol. 9, no. 6, pp. 6929–6938, 2018.

269. Z. Yang, Y. Liu, M. Campbell, C.-W. Ten, Y. Rho, L. Wang, and W. Wei, "Premium calculation for insurance businesses based on cyber risks in IP based power substations," *IEEE Access*, vol. 8, pp. 78 890–78 900, 2020.

270. D. K. Molzahn, "Computing the feasible spaces of optimal power flow problems," *IEEE Transactions on Power Systems*, vol. 32, no. 6, pp. 4752–4763, 2017.

271. F. Feng, Y. Zhou, and P. Zhang, "Quantum power flow," *IEEE Transactions on Power Systems*, vol. 36, no. 4, pp. 3810–3812, 2021.

272. M. Jereminov, D. M. Bromberg, A. Pandey, M. R. Wagner, and L. Pileggi, "Evaluating feasibility within power flow," *IEEE Transactions on Smart Grid*, vol. 11, no. 4, pp. 3522–3534, 2020.

273. M. Jereminov, A. Pandey, and L. Pileggi, "Equivalent circuit formulation for solving AC optimal power flow," *IEEE Transactions on Power Systems*, vol. 34, no. 3, pp. 2354–2365, 2019.

274. A. Pandey, M. Jereminov, M. R. Wagner, D. M. Bromberg, G. Hug, and L. Pileggi, "Robust power flow and three-phase power flow analyses," *IEEE Transactions on Power Systems*, vol. 34, no. 1, pp. 616–626, 2019.

275. A. Agarwal and L. Pileggi, "Large scale multi-period optimal power flow with energy storage systems using differential dynamic programming," *IEEE Transactions on Power Systems*, vol. 37, no. 3, pp. 1750–1759, 2022.

276. G. W. Stagg and A. H. El-Abiad, *Computer Methods in Power System Analysis*. McGraw-Hill, 1968.

277. W. D. Stevenson, *Elements of Power System Analysis*. McGraw-Hill New York, 1982, vol. 4.

278. G. Andersson, P. Donalek, R. Farmer, N. Hatziargyriou, I. Kamwa, P. Kundur, N. Martins, J. Paserba, P. Pourbeik, J. Sanchez-Gasca *et al.*, "Causes of the 2003 major grid blackouts in north america and europe, and recommended means to improve system dynamic performance," *IEEE Transactions on Power Systems*, vol. 20, no. 4, pp. 1922–1928, 2005.

279. Y. V. Makarov, V. I. Reshetov, A. Stroev, and I. Voropai, "Blackout prevention in the united states, europe, and russia," *Proceedings of the IEEE*, vol. 93, no. 11, pp. 1942–1955, 2005.

280. M. Adibi and L. Fink, "Overcoming restoration challenges associated with major power system disturbances-restoration from cascading failures," *IEEE Power and Energy Magazine*, vol. 4, no. 5, pp. 68–77, 2006.

281. M. Adibi and N. Martins, "Power system restoration dynamics issues," in *2008 IEEE Power and Energy Society General Meeting-Conversion and Delivery of Electrical Energy in the 21st Century*. IEEE, 2008, pp. 1–8.

282. M. Adibi, P. Clelland, L. Fink, H. Happ, R. Kafka, J. Raine, D. Scheurer, and F. Trefny, "Power system restoration-a task force report," *IEEE Transactions on Power Systems*, vol. 2, no. 2, pp. 271–277, 1987.

283. M. Adibi, J. Borkoski, and R. Kafka, "Power system restoration - the second task force report," *IIEEE Transactions on Power Systems*, vol. 2, no. 2, pp. 927–932, 1987.

284. F. De Mello and J. Westcott, "Steam plant startup and control in system restoration," *IEEE Transactions on Power Systems*, vol. 9, no. 1, pp. 93–101, 1994.

285. D. Lindenmeyer, H. Dommel, and M. Adibi, "Power system restoration—a bibliographical survey," *International Journal of Electrical Power & Energy Systems*, vol. 23, no. 3, pp. 219–227, 2001.

286. M. Adibi, G. Adsunski, R. Jenkins, and P. Gill, "Nuclear plant requirements during power system restoration," *IEEE Transactions on Power Systems*, vol. 10, no. 3, pp. 1486–1491, 1995.

287. Y. Hou, C.-C. Liu, K. Sun, P. Zhang, S. Liu, and D. Mizumura, "Computation of milestones for decision support during system restoration," vol. 10, no. 3. IEEE, 2011, pp. 1399–1409.

288. P. Manual, "System restoration," 2010.

289. R. Kafka, "Review of pjm restoration practices and nerc restoration standards," in *2008 IEEE Power and Energy Society General Meeting-Conversion and Delivery of Electrical Energy in the 21st Century*. IEEE, 2008, pp. 1–5.

290. R. Naggar, L. Cauchon, S. Henault, and S. T. Phan, "The recré system: restoration planning for hydro-québec's power system," in *2006 IEEE Power Engineering Society General Meeting*. IEEE, 2006, pp. 1–8.

291. F. Levesque, S. T. Phan, A. Dumas, and M. Boisvert, "Restoration plan—the hydro-québec experience," in *2008 IEEE Power and Energy Society General Meeting-Conversion and Delivery of Electrical Energy in the 21st Century*. IEEE, 2008, pp. 1–6.

Index

Note: Page numbers in **bold** and *italics* refer to tables and figures, respectively.

For Product Safety Concerns and Information please contact our EU
representative GPSR@taylorandfrancis.com
Taylor & Francis Verlag GmbH, Kaufingerstraße 24, 80331 München, Germany

www.ingramcontent.com/pod-product-compliance
Lightning Source LLC
Chambersburg PA
CBHW080648220326
41598CB00033B/5134

* 9 7 8 0 3 6 7 6 5 5 0 7 5 *